Mining, Society, and a Sustainable World

Jeremy P. Richards
Editor

Mining, Society, and a Sustainable World

 Springer

Editor
Dr. Jeremy P. Richards
University of Alberta
Dept. Earth & Atmospheric
Sciences
1-26 Earth Sciences Bldg.
Edmonton, AB, T6G 2E3
Canada
jeremy.richards@ualberta.ca

ISBN 978-3-642-01102-3   e-ISBN 978-3-642-01103-0
DOI 10.1007/978-3-642-01103-0
Springer Heidelberg Dordrecht London New York

Library of Congress Control Number: 2009932256

Cover illustration: Cover photographs (clockwise from top):
1. Artisanal miners working in tailings pond of AngloGold Ashanti's Obuasi Mine, Obuasi, Ghana (photo credit: K. Slack, 2005).
2. Open pit copper mining operation, Zaldívar, Chile (photo credit: J.P. Richards, 1995).
3. Artisanal miners panning for diamonds Democratic Republic of the Congo (photo credit: M. Mazalto, 2008).
4. International Space Station photograph of the Bingham Canyon copper mine, Utah (Public domain image from NASA Earth Observatory: 2007).

Cover design: Bauer, Thomas

Printed on acid-free paper

Springer is part of Springer Science+Business Media (www.springer.com)

Foreword

"I sit on a man's back, choking him and making him carry me, and yet assure myself and others that I am very sorry for him and wish to ease his lot by all possible means – except by getting off his back." *Leo Tolstoy - Writings on Civil Disobedience and Non-Violence (1886).*

In today's world where sustainable development has become a critical security concept for the well-being of the environment and society, the man Tolstoy depicts might well be interchangeable for either the planet in terms of its carrying-capacity or its beneficiary, society.

While it is arguable that mining is neither inherently sustainable nor unsustainable (O'Faircheallaigh, this volume), exploration, production, and consumption of non-renewable resources over time makes the industry ultimately unsustainable if it results in negative socio-economic impact (Waye et al., this volume). This invariably leads to definitions of sustainability in terms of the financial benefits that can accrue from transforming natural capital into human capital, theoretically creating intergenerational benefits (ibid.). Such a definition of sustainability is inherently utilitarian, assuming the English political philosopher Jeremy Bentham's suggestion that human nature avoids pain for the pursuit of pleasure, and that legislators should therefore base decisions on the greatest happiness for the greatest number of people (Bentham 1996).

Gro Harlem Brundtland defined sustainable development as "Development that meets the needs of the present without compromising the ability of future generations to meet their own needs," which Waye et al. (this volume) suggest creates sufficient ambiguity to create competing claims as to what is sustainable. On the basis of mining being sustainable if revenues are collected to promote sustainable objectives elsewhere in the community (ibid.), one therefore needs to ask: To what extent is the environment perishable? What beneficial trade-offs would make this acceptable?

Rio Tinto Chairman Paul Skinner notably defined the role of business by standing monetarist economist Milton Friedman on his head. Rather than using Friedman's quote, "The business of business is business," Skinner told the 2004 annual gathering of the Businesses for Social Responsibility conference in New York: "The business of business is sustainable business." (http://www.riotinto.com/media/speeches_2268.asp).

"If our business disappears, so does our contribution to the world's social, environmental and economic needs," said Skinner. There would be no return on shareholder investment and no employment of people, let alone development of their skills and experience. Provision of society's basic essentials and necessities would cease, as would the company's support for economic and infrastructure development, healthcare, and education programmes, Skinner asserted. "Similarly, if we go out of business we will not find new methods and better technologies for combating climate change and preserving or repairing the environment," added Skinner (ibid.).

This point straddles the delicate bridge between corporate social responsibility (CSR) and sustainable development where, quite simply CSR should require companies not to amass their profits by externalizing their costs onto society and the environment. Were these costs instead factored back into a company's balance sheets and monetized, investors would need to re-examine social and environmental performance in ascertaining the real value of the company and its products. In essence this is the reasoning behind the triple bottom-line (e.g., Elkington 1994).

If the analogy of the man carrying Tolstoy refers directly to the planet's carrying-capacity to maintain consumption and production, then "steady-state stewardship" and "ecosystem viability" (Ernst, this volume) are vital concepts – but are they attainable? Mining's ecosystems impact is highly visible, concentrated scars that directly generate deforestation, erosion, soil degradation, and toxic discharge (Mazalto, this volume).

If the world's population is anticipated to increase over 50% by 2050, pushing toward 10 billion people, that 20% of the current population which already consumes 70% of the world's resources (Ernst, this volume) may require some introspection. Indeed, at current extraction rates, it is not unlikely that global copper and zinc resources could be exhausted considerably before 2100, with a similar scenario facing all Australia's mineral resources (Mining Environmental Management July 2008).

Although the argument exists that other deposits may exist elsewhere, notably the ocean floor, achieving any kind of stability in mineral resource development requires immediate research in science and technology to achieve efficient recycling and conservation of natural materials (ibid.). Nevertheless, one should caution with architect and sustainability guru William McDonough's observations that much of society's efforts toward sustainable development are doing "bad but less badly" rather than actually doing good (McDonough and Braungart 2002).

Notably, the mining and metals industry could do well to note McDonough's concept of recycling versus downcycling. Recycling restores the product back to its original integrity, whereas most of what society actually calls recycling is in fact "downcycling", whereby the material becomes of a lesser quality grade, and after a few cycles of downcycling is no longer re-usable (ibid.).

One of the first applications of the term "sustainability" was in Germany in 1713, when von Carlowitz prescribed that the silver mines of Freiberg should not witness the felling of trees faster than re-growth. Such intragenerational and intergenerational resource planning has also been evident in the forestry cycle of Cyprus, which has consumed its forests sixteen times over while sustaining 3,000 years of the island's copper production (Wagner and Wellmer, this volume).

Such sustainability touches on basic human rights, enabling people to achieve economic prosperity and strive toward social justice, basic aims of the 1992 Rio Earth Summit's Agenda 21, and also the Brundtland Report 'Our Common Future' five years earlier that first clearly defined sustainable development (ibid.).

To return to the non-renewable and unsustainable pattern of consumption, the 1,000 year-old Rammelsberg polymetallic and precious metals mine, which was exploited from 968AD to 1988, reveals that a once steady annual average production of 20,000 tonnes/year tripled to 60,000 tonnes/year around 1850, and 100,000 tonnes/year production by the 1930s, before tripling again over the next two decades to 300,000 tonnes/year by the early 1950s (ibid.). In short, the past fifty years has witnessed more consumption than during the previous collective history of mankind, heralding what has now been described as a new geological period in history – the Anthropocene period, starting in the 1800s where mankind began to have a significant global impact on planet Earth (ibid.). Yet these production increases occurred in the industrialized countries and outstripped world population growth regardless of the West's own slow demographic growth (ibid.).

Other common assumptions surround thinking on mining's effects on society and its role in sustainable development. To state that countries like Canada, the USA, and Australia were built on mining, creating the assumption that mineral development is an automatic development driver, could be erroneous. Rather than mineral development per se, the economies of Canada, Australia, and the USA benefited from the "institutional capital" that enabled capital accumulation and its subsequent (distributive) benefits (Slack, this volume).

Indeed, academics and writers elsewhere question the relevance of systemic economic thinking when faced with the world's ecological and social challenges, in particular questioning the viability of traditional concepts of growth (e.g., Victor 2008; Brown and Garver 2009). Rather than measured in terms of benefits, growth is measured "in terms of exchanges of money" (Brown and Garver 2009, p. 9), and, according to this logic, even the cost of cleaning up the Exxon Valdez oil spill off Alaska (US$500 million) added to the region's GDP growth (ibid.). Even for measuring mining's impact on society, GDP growth does not factor distribution, meaning "inequity, poverty and outright starvation often can, and do, rise at the same time that overall economic activity increases." (Ibid p.10.)

However, society's own anthropological journey from Stone Age through Bronze Age to Iron Age and beyond is a trans-historical demonstration of how mineral development is inextricably linked to societal development. From a development perspective, defining economic and societal sustainability through the transformation of natural capital into social or economic capital pivots around the essential and conflicting problem surrounding the "Resource Curse" thesis, which argues that natural resources have a negative effect on economic growth and development (e.g., Auty 1993; Sachs and Warner 1995). However, Davis (this volume) shows that positive growth is likely to be of more benefit to resource-based nations than negative growth; he thus takes issue with the Resource Curse theory, suggesting instead that attention should focus more on the problems of boom–bust cycles in resource-based economies.

An export-dependent economy would likely lead to unequal economic development unless it has a developed internal economy (Crowson, this volume). Good governance in host countries is likely an essential pre-requisite to investment (Hilson and Maconachie, this volume) and, certainly, voluntary principles such as the Extractive Industries Transparency Initiative (EITI) require some existing good governance in order to progress further accountability and transparency measures, including for World Bank and Western government purposes (ibid.).

Nevertheless, one could argue that even the best intentioned voluntary guidelines and principles to promote sustainable societal and economic impacts on a region, even in developed countries, do not make for good governance, nor are they destined to succeed. Closure of the Polaris and Nanisivik mines in northern Canada witnessed a quick fade-out of economic benefits when it came to wealth creation in the Arctic Inuit communities, due to an overemphasis at project conception stage on the benefits of the multiplier effects (Bowes-Lyon et al., this volume).

In the structural global economy where the World Bank has been a major proponent of mining and resource development for developing countries, mining has seldom provided such envisaged benefits as direct employment of the local population, because mining recruits few direct employees and, invariably, local inhabitants do not possess the education or skills required to operate and manage a capital intensive mining project. Instead, an assumed multiplier effect is anticipated to create indirect jobs created by the project's existence (Slack, this volume). However, the ultimate experience of the Arctic communities highlights the need to realistically factor-in the multiplier effect (Bowes-Lyon et al., this volume).

In Peru, the World Bank and International Monetary Fund's Poverty Reduction Strategy Papers assumed that mining would create employment, but did not elaborate any strategy. Because neither institution monitors the impacts of the projects they fund, no stakeholder transparency regarding monitoring, information-sharing, and results can be created for constructing future cost-benefit analyses, either at national or project-level (Slack, this volume). The World Bank's own Extractive Industries Review recommended that it cease funding oil and coal projects due to the lack of benefits created, advice the Bank chose to ignore (Davis, this volume).

Ultimately while there may be no growth that is actually "bad for the poor", positive growth is assumed to be better than negative growth. However, lessons could be learned for implementing pro-growth policies in extractive economies, and a sustained extraction economy requires foreign direct investment to support continued domestic exploration and resource development, rather than to concentrate on political and institutional shortcomings (Davis, this volume).

Limits to growth can vary. Despite Chile's status as the world's largest copper producer, Chile's severely depleted groundwater reserves will lessen the country's ability to resolve its internal wealth inequality, where the poorest 10% owns only 1.2% of Chile's wealth compared to the richest 10% who own half that nation's wealth (Slack, this volume).

Similarly, boom–bust cycles need to be studied (Davis, this volume). Botswana recently found itself at the front end of a boom cycle based on diamond mining, and needs to learn how to encourage its extractive industry to grow in a steady and persistent manner (ibid.).

To avoid the Resource Curse and provide positive socio-economic impacts requires constructive interventions. Clusters in the form of government, company, and voluntary sector partnerships, such as Chile's mining clusters that encouraged growth of small and medium-sized enterprises (SME) are recommended to deepen governance reforms (McPhail, this volume; also see Singh and Evans, this volume). Such reforms can be achieved especially where regional or local development agencies engage to co-ordinate economic diversification and poverty reduction, which could include planning infrastructure development to enable local development (ibid.).

Reinvestment of government revenues is critical, and Ghana is one example of a country that has done little to return investment to its mining communities (McPhail, this volume). This leads to the assertion that if policy-makers adopted the thinking of nineteenth century American economist Henry George, where land is a public good whose use should be taxed such that revenues can be recycled via regulatory and taxation policy instruments for the public good at local and national levels, then benefits can in fact be derived through mining a non-renewable and at first sight unsustainable resource (Waye et al., this volume). Ultimately, mining can be considered sustainable if the derived revenues from mining are collected and used to promote sustainable objectives at community and national levels, particularly through regulation and taxation measures that encourage the investment climate (ibid.).

Government transparency and clear, enforceable regulations would appear to be paramount requirements if developing countries are to benefit from development of their natural resources. An improved investment climate for mining can contribute to better macro-economic policies, and mining was the core of economic recovery in the four countries studied by the International Council on Mining and Metals (ICMM) (Tanzania, Peru, Ghana, and Chile; McPhail, this volume). The ICMM identified six factors that play a critical role in poverty reduction, the most important being adequacy and fairness of tax regimes and revenue allocation systems. In contrast, land use and property rights conflicts, environmental damage, artisanal and large-scale mining conflict, and mine closure issues can cause or extenuate poverty (ibid.).

On an environmental or ecological level of analysis, Wagner and Wellmer (this volume) recommend that the Earth's carrying-capacity would benefit from a four-tier hierarchy of resources for political decision-making, especially where future population growth will require more arable land that in turn will cause more deforestation and soil erosion. Four natural resources, water, soil, clean air, and energy are considered, and the authors identify a principle of vertical substitution: "Lower value resources should wherever possible replace higher value resources." The lowest level 4 represents waste and residues from primary use that can be used in secondary applications; level 3 identifies bulk raw materials from the Earth (such as clay, basalt, and granite), nitrate from the air for fertilizer, and boron, potassium, rock salt, and magnesium from the sea; level 2 includes other less abundant natural resources and pure scrap; while the highest level is energy (Wagner and Wellmer, this volume).

Mineral development has been a force for social transformation, but the "take, make, consume, waste" paradigm has not been sustainable, both degrading the environment and creating a wide gap between rich and poor (Singh and Evans, this volume). While resource development is neither a de facto curse nor panacea, production complexes or "clusters" can help stimulate sustainable economic growth in the regions and areas where they are based. These clusters can take the form of groups of industries whose linkages mutually reinforce and enhance their competitive advantage, and which are composed of diverse groups that incorporate primary, secondary, and supporting industries (ibid.).

Singh and Evans (this volume) define sustainable development as containing three key propositions: economic growth, environmental stewardship, and CSR. For clusters to work in developing countries, policy-makers must recognize how resource-based economies move from a simple extraction mode (in which they export resources while importing equipment) to a balanced economy that includes export of associated goods and services, and which is neither import nor export-dependent (ibid.).

Far from proving to be a resource curse, South Africa's mining experience has enabled the country's economic diversification in which, since 1993, the non-mineral sector has played a slightly larger role in the country's GDP (Dane, this volume). In addition, the 2002 Mineral Resources and Petroleum Development Act has promoted longer term sustainability (ibid.). Under this legislation, social and labour plans are required to be developed, and Dane uses the Landau colliery for his case study, where poor planning historically has not benefited the local population. Recent stakeholder engagement between communities, governments, NGOs, and mining companies, among other private sector players, has helped transform the mining industry, aligning its socio-economic goals with the government. Enhancing black empowerment through SMEs and other roll-out to business hubs should also help sustain the local economy after the mine's closure in 2025 (ibid.).

O'Faircheallaigh (this volume) identifies nine policy process variables as a precursor to on-going comparative public policy studies to illustrate how the policy-making process can guide mining towards a sustainable future. Weak public policy in Papua New Guinea allowed the OK Tedi copper mine to wreak environmental damage on two local rivers, while colonial treatment by the UK, Australia, and New Zealand of Pacific island-state Nauru's phosphate mines did not enhance that country's economy prior to its independence in 1968. The Nauru government's own policies of investment created an unsustainable wealth once the phosphate ran out, because no viable internal economy was promoted in the aftermath of independence (ibid.). Indeed, it could appear that there is an illusory link between the practice of policy-making for the minerals and energy industry, and the theoretical conversion of natural capital from an unsustainable physical resource into social capital that provides for long-term livelihoods, and that can subsequently be deemed sustainable.

Away from the public policy process, greening of the mining supply chain and processes emphasises the reciprocal causation between change of management attitudes and commitment to re-thinking how they do business in order to be more eco-efficient. Suppliers need to understand the role they can play in educating and partnering with a mine's management team, thereby often improving the longevity of a mine's lifecycle (Guerin, this volume). Greater stakeholder engagement is needed in developing new sustainable processes as part of an integrated response from management that will demonstrate a clear commitment to sustainable development. Management needs to work closely with suppliers to identify new processes and initiatives, whether for improving air quality, energy efficiency, and materials efficiency, while minimizing waste and improving water efficiency and waste management (ibid.). In turn, this should not only reduce costs, but will enable companies to contribute more to a region's economic and social fabric (ibid.).

Similarly, mine closure has changed in focus from factors of salvaging and removing equipment to the whole ecosystem, requiring the consideration of social and economic issues where governments would be well-advised to publish guidelines and regulations (Otto, this volume). Public policy can facilitate best practice by encouraging good mine design, such as properly engineered tailings dumps that will prevent costly acid drainage while permitting the future recovery of metals. Cost estimates for closure and reclamation should be included in mine plans (ibid.).

The voluntary Equator Principles adopted by financial institutions for extractive and other projects request that their borrowers abide by environmental and other regulations, and devise decommissioning plans to reduce the negative environmental and public effects mine closure would otherwise cause. The lesson for jurisdictions is the need to seek financial assurance for efficient closure and reclamation, whether through surety bonds, certificates of deposit, trust funds, insurance policy, or simply depositing cash with the jurisdiction's Treasurer, to secure against the mine claiming effective bankruptcy which would prevent the mine from being properly reclaimed (Otto, this volume).

It can start to be seen how public policy can turn mining into a development lever (Mazalto, this volume), at least economically. But government effectiveness is paramount, and intervention primarily rests at the national institutional level, with informal powers or influence residing at transnational or multilateral levels such as the OECD, World Bank, and so forth, who help devise CSR policies (ibid.).

The Democratic Republic of Congo demonstrates this point when the World Bank returned its attention to the country after 2001, the extractive industry having abandoned it in the early 1990s (Mazalto, this volume). In this case, responsibility for resolving environmental problems is not so clear-cut: is it the national government, private sector, or international finance institution that should bear the responsibility? Congo's new democratic government implemented a Mining Law at national level and, to reverse the loss of a country's environmental resources, the

United Nations Environment Programme and UN Development Programme established a Poverty and Environmental Facility to reduce poverty in a bid to enhance implementation of the UN's Millennium Development Goals (ibid.). Arguably, because the state itself is too weak to enforce its own regulations across the country—particularly in the North and South Kivu artisanal mining regions, where commodity price rises for the metal tantalum (coltan; essential to mobile communications and computers) has been and remains the cause for armed conflict and human rights atrocities—the development of norms at transnational level benefit the transnational community (ibid.).

The environmental–development axis is critical because socio-economic development requires an environmental input. Not only is sustainable development a global phenomenon, but it requires some deconstruction in terms of priorities and the public policy-making decisions required to mitigate externalized social and environmental costs. For example, should ecosystems and biodiversity take priority (Waye et al., this volume) over any other issue?

There can be little doubt that as a wealth resource, mineral resources should enable growth. Regardless of whether the natural resources are located in a developed country such as Canada (e.g., having to balance the pros and cons of the environmental costs of mining Alberta's oil sands), or in a Least Developed Country such as Tanzania, any responsible government should carefully weigh the environmental costs against likely social and economic benefits. Growth for and in itself may bring only short-term and artificial benefits, especially if they subsequently exacerbate climate change. One could argue that there are alternatives, for example, to fossil fuels, but there are no alternatives to water, which may yet become a scarce resource essential to life due to over-consumption and depletion due to serious climate change.

In the structural economy, good governance requiring accountability and transparency is required to mediate competing claims for resources historically made by companies, foreign governments, host governments, financial institutions, and the local or regional inhabitants. Where national governments are strong enough to prioritize how to distribute the benefits of their natural resources, transforming these into social capital, they will need to decide what are the essential components of future sustainable development in their region or society. For example, is it combating malaria, HIV/AIDS, and tuberculosis; advancing education; rolling-out economic benefits to foster SME growth and create internal markets; infrastructure development; or adapting to climate change?

In a globalizing society, sustainable development is a global and critical security issue, where to a greater or lesser degree we are all Tolstoy feeling sorry for the man carrying us on his back, and wanting to ease his lot by all possible means.

In order to do this, a serious debate about modern and future security needs to take place in order to harmonize energy and mineral resource security with environmental, human, political, economic, and societal security.

Montréal, Québec, Canada   Felix von Geyer

References

Auty RM (1993) Sustaining Development in Mineral Economics: The Resource Curse Thesis. Routledge, London, UK

Bentham J (1996) An Introduction to the Principles of Morals and Legislation; An Authoritative Edition by J.H. Burns and H.L.A. Hart, Clarendon Press, Oxford p.11

Brown PG, Garver G (2009) Right Relationship: Building a Whole Earth Economy. Berret-Koehler Publishers Inc; San Francisco

Dixon K (2008) Countdown – Are the Earth's mineral resources running out? Mining Environmental Management Magazine. Mining Communications Ltd, London. pp. 25-27 July 2008

Elkington, J (1994) Towards the sustainable corporation: Win-win-win business strategies for sustainable development. California Management Review 36(2) 90–100

McDonough W, Braungart M (2002) Cradle to Cradle: Remaking the Way We Make Things. North Point Press

Sachs JD, Warner AM (1995) Natural Resource Abundance and Economic Growth. NBER Working Paper 5398

Skinner PD (2004) Address to Businesses for Social Responsibility (BSR) Conference, New York, 11 November 2004. http://www.riotinto.com/media/speeches_2268.asp

Tolstoy L (1987), Writings on Civil Disobedience and Non-Violence (originally published in 1886). New Society Publishers, Gabriola Island, British Columbia, Canada

Victor PA (2008) Managing Without Growth – Slower by Design, Not Disaster. Edward Elgar Publishing, Cheltenham, UK

Acknowledgments

Firstly I would like to thank all the authors and co-authors of this book who have given freely of their time, experience, expertise, and vision to this critical global dialogue on sustainability through the fascinating and insightful chapters they have written. Obviously, without their contributions this book would not exist.

An additional vital and essential step in crafting authoritative literature is the peer-review process, and I gratefully acknowledge the following people who served as reviewers for the various chapters: J. Chadwick, C. Coumans, T. Deligiannis, M. DeWit, G. Ernst, R. Gallinger, J. Gammon, T. Guerin, G. Hilson, T. Lane, A. Macdonald, R. Maconachie, R. Moran, G. O'Connor, O. Östensson, G. Plumlee, C. Samdup, M. Scoble, T. Schroeter, B. Skinner, K. Slack, I. Thomson, C. Turner, M. Veiga, F.-W. Wellmer, N. Westoll, and nine anonymous reviewers. I am particularly grateful to Simon Handelsman, who handled the review process for the two chapters by Bowes-Lyon et al. and Waye et al. on which I was a co-author.

I thank Chris Bendall at Springer for proposing to me the idea of compiling a book on sustainable development and mining, and for his unwavering encouragement throughout the long and uncertain process of bringing such an endeavour to fruition.

For moral support and for putting up with long weekends spent editing I thank my wife Lee Ewert, and for operational support I thank the Social Sciences and Humanities Research Council of Canada for a Research Development Initiatives grant.

Edmonton, Alberta, Canada   Jeremy P. Richards
24 February, 2009

Contents

Part III   Mining and the Environment

Part IV   Mining and Societal Issues

Contributors

Léa-Marie Bowes-Lyon Department of Earth and Atmospheric Sciences, University of Alberta, Edmonton, AB, T6G 2E3, Canada, leamarie@justemail.net

Phillip Crowson Centre for Energy, Petroleum and Mineral Law & Policy, University of Dundee, Dundee DD1 4HN, Scotland, p.c.f.crowson@dundee.ac.uk

Anthony Dane Anglo American, Johannesburg, South Africa, imbovane@gmail.com

Graham A. Davis Division of Economics and Business, Colorado School of Mines, Golden, CO 80401, USA, gdavis@mines.edu

Joseph A. Doucet School of Business, University of Alberta, Edmonton, AB, T6G 2R6, Canada, joseph.doucet@ualberta.ca

W.G. Ernst Department of Geological and Environmental Science, Stanford University, Stanford, CA 94305-2115, USA, wernst@stanford.edu

Jim Evans Blue Heron Consultants, Oshawa, ON, L1G 6X8, Canada, jim.evans@bluheron.net

Turlough F. Guerin Telstra Corporation Limited, Melbourne 3000, Australia, turlough.guerin@hotmail.com

Gavin Hilson School of Agriculture, Policy and Development, The University of Reading, Reading RG6 6AR, UK, g.m.hilson@reading.ac.uk

Roy Maconachie Institute for Development Policy and Management (IDPM), School of Environment and Development, The University of Manchester, Manchester, M13 9PL, UK roy.maconachie@manchester.ac.uk

Marie Mazalto Groupe de Recherche sur les Activités minières en Afrique (GRAMA), Université du Québec à Montréal (UQAM), Montréal, QC, H3C 3P8, Canada, marie.mazalto@cirad.fr

Tara M. McGee Department of Earth and Atmospheric Sciences, University of Alberta, Edmonton, AB, T6G 2E3, Canada, tmcgee@ualberta.ca

K. McPhail International Council on Mining and Metals, London W1H 6LR, UK, Kathryn.McPhail@ICMM.com

Ciaran O'Faircheallaigh Griffith University, Nathan, QLD 4111, Australia, Ciaran.Ofaircheallaigh@griffith.edu.au

James M. Otto Mineral Policy, Law and Economics, Boulder, CO 80305, USA, jim.otto@comcast.net

Jeremy P. Richards Department of Earth and Atmospheric Sciences, University of Alberta, Edmonton, AB, T6G 2E3, Canada, Jeremy.Richards@ualberta.ca

Indira Singh Ministry of Northern Development and Mines, Thunder Bay, ON, P7E 6S7, Canada, Indira.singh1@ontario.ca

Keith Slack Oxfam America, Washington, DC 20005, USA, kslack@OxfamAmerica.org

Felix von Geyer 5021 rue Saint-Urbain, Montréal, Québec, H2T 2W4, Canada, felix.vongeyer@gmail.com

Markus Wagner 1, B undesanstalt für Geowissenschassften und Rohstoffe, Hannover D-30655, Germany, Markus.Wagner@bgr.de

Arianna Waye Department of Economics, University of Alberta, Edmonton, AB, T6G 2H4, Canada, aewaye@ualberta.ca

Friedrich-Wilhelm Wellmer Neues Sachlichkeit 32, Hannover D-30655, Germany, fwellmer@t-online.de

Denise Young Department of Economics, University of Alberta, Edmonton, AB, T6G 2H4, Canada, Denise.Young@ualberta.ca

Introduction

Few people would deny that mined materials are an essential, if not foundational part of modern economies and civilizations. But in the same way that few meat eaters like to think of what happens in slaughterhouses, few book readers like to think of clear-cut forests, and few vegetable eaters like to think of the effects of fertilizers, pesticides, weed killers, and monoculture on the environment, few of us, as beneficiaries of the products of mining, like the idea of mining itself. And indeed it is true that historically, and continuing today in some parts of the world, the mining industry has been responsible for egregious human rights abuses and unnecessarily negative environmental impacts. Such malpractices date back at least to the Romans, who used slave labour in most of their mines and caused severe (if localized) environmental degradation.

However, the modern mining industry has learned from almost 40 years of environmental activism that it can no longer pollute with impunity, and it has found itself relatively able to respond to these concerns over environmental degradation by approaching them as engineering problems with scientific solutions. Where the mining industry has been less successful to date is in regard to the social impacts of its activities. Social activism, in its modern form, has appeared relatively recently (within the last 20 or so years), and has focused primarily on the rights of indigenous people, who commonly find themselves in the path of large-scale mining developments, often in the more remote parts of the world. Staffed primarily by geologists, engineers, and accountants, mining companies have struggled to understand and deal with these "soft" issues, increasingly to their detriment (an increasing number of mining projects have been halted by social opposition in recent years: for example, Tambo Grande in Péru, and Esquel in Argentina).

This book aims to explore these broader implications of the mining and minerals industry, from social, economic, and sustainable development perspectives, with a view to identifying the most pressing issues, reviewing current best practices, and proposing ways forward. The book is structured in four parts: (1) the role of mining in developed and developing economies; (2) the role of mining in sustainable development; (3) mining and the environment; and (4) mining and society.

The authors of the sixteen chapters contained herein are all practitioners or researchers in these fields, and have all approached their topics with vision and foresight. My only request to each author prior to writing was to be positivist and solution-oriented in their thinking, although some topics, such as the parlous state of mining in the Democratic Republic of the Congo (DRC), leave one hard pressed for optimism.

The first part of the book looks at mineral resources and economics from a global perspective. The first three chapters by Philip Crowson, Graham Davis, and Kathryn McPhail all tackle the "Resource Curse" paradigm, which holds that natural resource extraction hurts rather than helps build a nation's economy. Crowson examines a number of historical examples of mineral booms, and concludes that, while some did indeed result in impoverishment, others were more positive and provided the foundation for continued post-mineral growth. Crowson concludes that the key to a successful outcome depends mainly on institutional factors in the host country, and especially a government's control of rent-seeking behaviour.

Davis focuses on the impacts of resource extraction on the poor. Like Crowson, he finds examples of both positive and negative impacts, but notes that these impacts commonly relate to national economic trends such as growth spells and recessions. He finds that the poor fare similarly in extractive as in non-extractive economies during positive growth spells, in most cases benefiting but in some cases not. Conversely, the poor tend to fare badly during periods of negative growth. Thus, the poor will be disadvantaged in countries afflicted by the Resource Curse, wherein negative economic growth is caused by poor management of the extractive sector.

Kathryn McPhail describes an initiative undertaken by the International Council on Mining and Metals (ICMM) to explore the factors that control whether or not resource-rich developing countries experience the Resource Curse, and to identify ways in which the mining industry can contribute to poverty reduction at national and local levels. The study concluded that keys to success were mainly institutional (as found by Crowson), including improvements in governance, legislative reform, and fiscal management. However, McPhail notes that industry also has an important leadership role in promoting these outcomes, especially at regional and local levels.

Keith Slack takes a different viewpoint on resource-based development, and calls on development planners to explain more clearly how mining will contribute to national development, and to assess more robustly the industry's long-term costs and benefits. He argues that much of what is called "sustainable development" in relation to mining is a pretense, and while there may be other legitimate reasons for supporting mine development, raising expectations in terms of sustainability may not serve the industry's or a country's best interests.

Markus Wagner and Fred Wellmer consider natural resources in terms of a hierarchy of value and substitutability. Energy resources form the highest tier in this hierarchy, naturally occurring deposits of raw materials the next, materials in unlimited abundance such as sand or components that can be derived from air or seawater form the third level, and waste and residual materials comprise the fourth, lowest level. The authors advocate conservation of top tier resources and substitution where possible with materials from lower levels, in order to maximize resource efficiency. They also argue that resource efficiency should be measured in terms of energy efficiency to reflect the supremacy of energy in the hierarchy of natural resources.

The second part of the book focuses on mining in relation to sustainable development. Gary Ernst continues the theme of the pre-eminence of energy as the key to human development, and discusses limitations on energy availability and use in thermodynamic terms. He identifies only nuclear fusion and solar-based energy (sunlight, but also wind, waves, biofuels) as being effectively unlimited or truly renewable. However, he warns that it may not be energy supply that ultimately limits human development, but rather the impacts that our activities have on the biosphere. Like Wagner and Wellmer, Ernst advocates conservation, resource substitution, and improved efficiency as the keys to long-term sustainability.

Arianna Waye and coauthors start with the tenet that mining can contribute to sustainable development if the revenues generated are appropriately invested. They then examine the effects of various external fiscal and geopolitical parameters on the willingness of companies to invest in mining activities. A key finding, well known but rarely acted upon, is that the actual level of taxation is not as important as the stability of tax regimes in guiding investment decisions. This suggests that more revenue could be generated for sustainable development objectives from the extractive industries when political, regulatory, and fiscal environments are stable.

Again starting from the principle that natural resources can be the springboard for sustainable development, Indira Singh and Jim Evans examine the use of industry "clusters" for maximizing the economic benefits than can be derived from primary resource industries such as mining, agriculture, and forestry. Key to this concept is that focused, cluster-based economic activity can develop beyond the primary sector, and may be sustained even after the original resource has been depleted. This is the ultimate test of a resource industry's contribution to sustainable development—does the benefit survive after mining ceases?

The third part of the book examines the impact of mining on the environment. As noted above, modern mining companies are quite deeply engaged in improving their environmental performance, and the first two chapters in this part examine these trends. The third chapter exposes the problems continuing in other parts of the world where the environmental revolution has yet to take hold.

Turlough Guerin views the environmental performance of the mining industry from the perspective of fuel and lubricant supply and handling. Wastage and spillage of these components are major factors in a mine's environmental performance, and Guerin approaches this problem in terms of a petroleum-based hydrocarbon lifecycle, seeking eco-efficiencies at each stage in the cycle.

Jim Otto examines trends in legislation and practice for mine closure and reclamation, and considers the roles of various stakeholders, such as mining companies, employees, nearby communities, and governments. Standards, certainly in developed countries, have changed dramatically since the walk-away practices of only a few decades ago, and today companies talk about "planning for closure" from the outset, before mining even begins. Some jurisdictions also require bonds to be posted against future closure and reclamation costs, and Otto examines the pros and cons of the various methods used to determine the amount of money required to cover these costs at the end of the mine life.

Marie Mazalto takes us on a visit to the Democratic Republic of the Congo to examine the state of the environment in this resource-rich country's mining regions. The ravages of two decades of civil war and political unrest have gutted the once world class mining operations of the Copperbelt in the southwest of the country, and have fuelled terrible violence in the coltan mining areas in the east. Environmental considerations have been far from the minds of people trying simply to survive under these conditions, and widespread pollution has resulted. Since the return of relative peace to the Congo, The World Bank and other international financial institutions have also returned to the DRC with new tools to improve "good governance", and mining investments have recommenced. Mazalto shows that, based on data from an audit conducted for the Bank, extensive damage has been done to the environment in these mining regions, which may be next to impossible to repair. Close cooperation between donor organizations and the DRC government will be required if improvements are to be seen in this new phase of resource development.

The final part of the book examines the interplay between mining and society, at both local and national scales. Turlough Guerin leads this part with a chapter on stakeholder engagement and eco-efficiency as a means of promoting corporate responsibility in the minerals sector. Guerin examines several examples of positive engagement between mine or processing plant operators and local communities that have resulted in improved efficiencies, for example in water conservation and waste management. However, he notes that the will to do this must be instilled in senior management, who must make the pursuit of such opportunities a priority.

Léa-Marie Bowes-Lyon and co-authors have taken a retrospective view of the socio-economic impacts of mining at two sites in the Canadian High Arctic. The Polaris and Nanisivik lead-zinc mines both operated over similar 20-year time-spans and both closed in 2002. Polaris was located at some distance from the nearest community, Resolute, whereas Nanisivik was located just outside the community of Arctic Bay. Surveys of these mining communities revealed that, while the mines had contributed to increased incomes and other modest benefits for members of these communities while in operation, few of these benefits have persisted since mine closure. As such, this model for mine operations in remote regions has not contributed to *sustainable* development at the local scale. Recommendations for improved performance in the future include better community consultation and participation, and provision of training in business and technical skills that can be transferable to other economic activities after mining ceases.

Anthony Dane has also examined the socio-economic effects of mining, this time in relation to an operating coal mine in South Africa. He reports on a study undertaken by the parent company, Anglo American, using the methodology of ICMM's Resource Endowment Initiative, to measure the contribution of the mine to the Millennium Development Goals and the creation of sustainable communities (post mine closure). While many successes are reported, Dane notes (like Bowes-Lyon et al.) that the main areas of deficiency have been in community engagement, as well as engagement with other stakeholders such as government and non-governmental organizations, which has led to sub-optimal redistribution of the wealth generated by mining.

Ciaran O'Faircheallaigh examines the critical role of public policies and so of public policy processes in shaping sustainable outcomes from the extractive resource industries. Poorly developed policies can result in mine failures or unsustainable consequences, whereas well-crafted policies can support social, economic, cultural, and environmental sustainability. He identifies nine process variables that are critical to successful policy development and implementation, including stakeholder engagement in the process, and provision of adequate funding and time. He argues that radical change is required in existing policy processes if public policies are to enhance the potential of mining to contribute to sustainable development.

Gavin Hilson and Roy Maconachie close out the book with a critical analysis of the effectiveness in sub-Saharan Africa of the Extractive Industries Transparency Initiative (EITI), which was launched in 2002 at the World Summit on Sustainable Development in Johannesburg. The stated aim of the EITI is to improve economic development in resource-rich countries through "good governance", but Hilson and Maconachie doubt that the EITI on its own can produce the required changes, especially in relation to reducing corruption and improving resource management.

Thus, this book explores a wide range of issues surrounding the role of mining and minerals in the context of development, sustainable or otherwise. While several chapters note considerable progress, especially in regard to environmental performance, improved social outcomes, in particular for indigenous peoples or impoverished populations, remain largely conceptual, and fully unrealized in parts of sub-Saharan Africa. The outlook of this book is intended to be positivist, and it is hoped that the problems identified in each of the chapters will serve as a spur to the pursuit of solutions, rather than simply as a justification for opposing mining. Environmental and social activism is a valuable piece of the picture, and is largely responsible for the dramatic improvements in industry performance over the last 40 years, especially in relation to the environment. But the stick must be balanced with a carrot, and encouragement is often lacking for those companies who make real efforts to change. We here offer kudos to those who are making this effort, but warn that "sustainable development" is not a final destination but a path—there will always be room for further improvement.

Jeremy P. Richards

Part I
The Role of Mining in Developed and Developing Economies

The Resource Curse: A Modern Myth?

Phillip Crowson

*Is there any thing whereof it may be said, See, this is new? It
hath been already of old time, which was before us.
(Ecclesiastes: Chapter 1, verse 10)*

Abstract This chapter concentrates on the economic importance and impact of
mineral development rather than its social and environmental effects. It first examines statistics on mineral dependency, demonstrating that many commonly used
measures underestimate the contribution of minerals to economic activity. It then
looks at the ways in which the development of mineral wealth can affect resource-
rich countries for good or ill. Some historical examples of mineral development
show that the issues raised today were no less relevant in times past. Changes in
transport and communication technologies, and in the capital intensity of minerals
extraction and processing over the past century and beyond have, however, reduced
the prospects for strong multiplier effects of mining in many host countries. Far
more than in the past, the main economic benefits to host countries are likely to be
through their fiscal receipts and capture of mineral rents. The extent to which those
benefits are realized consequently depends on the nature of their institutions and
governance.

1 Introduction

In recent years a substantial body of literature has examined the impact of mineral
development in resource rich countries. Rather than provide here a long list of references on this and later comments, readers are directed to Paul Stevens' useful survey of the extensive literature (Stevens 2007a). Commentators are divided between
those who argue with conviction that mineral wealth is a blessing and those who

P. Crowson (✉)
Centre for Energy, Petroleum and Mineral Law & Policy, University of Dundee, Dundee,
DD1 4HN, Scotland
e-mail: p.c.f.crowson@dundee.ac.uk

J.P. Richards (ed.), *Mining, Society, and a Sustainable World*,
DOI 10.1007/978-3-642-01103-0_1, © Springer-Verlag Berlin Heidelberg 2009

maintain with vehemence that its development is an unmitigated curse. These opposing viewpoints, and especially those that maintain the existence of a resource curse, a term first coined by Richard Auty in 1993 (see Stevens 2007a), are often a cloak for more visceral beliefs about the role and desirability of mineral development by multinational companies to serve global markets. Much of the recent commentary is concerned with oil and gas rich states rather than with putative hosts for hard-rock minerals, and the inherent differences between petroleum and non-fuel minerals are brushed aside. Sudden access to any new source of wealth nonetheless creates problems for economic management and for social stability, whatever the size of the unit involved, from an individual through to a nation. In that regard new mineral wealth is no different from any other apparent windfall.

The debate about the economic impact of mining, like so much else today, adopts too short a time perspective. This is not to follow Mao Tse-Tung's belief that it is too early to ascertain the effects of the French Revolution, but to recognize that short and medium term impacts may often differ from those that mature over longer periods. Policy makers and commentators increasingly seek instant answers and solutions. What matters, however, is not the short term, but how any potentially adverse economic and social impacts of development are eventually resolved. In some states, their resolution may be inhibited by underlying political, social, and ethnic strains and stresses. In those countries the curse is not mineral wealth but related to far more deep-seated, and often insoluble issues. Many mineral-rich countries have avoided or minimized any adverse consequences of natural resource development, both historically and more recently. Other countries may not, however, be able or willing to copy them.

2 The Importance of Mining

2.1 The Nature of Mining as an Economic Activity

The main distinguishing features of mining are its dependence on depleting natural resources, and, in some instances, its scale relative to the host economy. Extractive industries also differ from others in the importance of economic rent in their value added. Unless manufacturing and service industries enjoy a unique competitive advantage such as a tariff barrier or patent protection, their economic rents are gradually competed away. The relative costs of production of the mining industry are dictated not so much by the costs of labour and intermediate goods and services used, as by the characteristics of the underlying resources. Those persist over the life of a project, although potential economic rent can be enhanced by skilful management, or whittled away through mismanagement. A marginal project whose production costs, including the opportunity cost of its capital, equal its revenues, earns no rent.

Most attention is focused on the development of major export-oriented mines by multinational companies, but by no means all mining and minerals investments are of that nature. Construction materials and industrial minerals of all types are mined, often by domestic companies for local or regional markets, and there is a large artisanal mining sector in many countries. Much foreign investment is in gold mining, whose economics differ in many respects from those of the major metals, even if its potential impacts do not.

Mining is an important contributor to economic activity in many countries, both developed and developing, although the measures of its importance that are most widely used tend to understate its role. Most econometric analysis of the relationship between minerals extraction and economic development (e.g., Sachs and Warner 1995, 1997a, b; World Bank 2002; Collier and Goderis 2007a, b) has concentrated on export data and on readily available price series. One frequently used source in such econometric analysis is the World Bank's publication, World Development Indicators (World Bank 2007), which shows annual exports of ores and metals as percentages of total merchandise exports. The coverage of the World Bank's series is relatively narrow. It embraces just Section 27 (crude fertilizers, minerals not else-where specified), Section 28 (metalliferous ores, scrap), and Section 68 (non-ferrous metals) of the Standard International Trade Classification. It omits coal, precious stones, and gold (but not gold ores and concentrates), as well as the first-stage products of mining, such as inorganic chemicals (which include some primary products like alumina), manufactured fertilizers, coke, cement, and some construction materials. Paradoxically iron and steel are not included, whereas non-ferrous metals and their semi-fabricated products are. Non-ferrous waste and scrap, which arise mainly in consuming countries, are included. The narrow coverage of the data used naturally influences the analyses, probably introducing a bias against picking up any potential benefits of mining.

2.2 Mineral Exports

Table 1 shows the size and composition of mineral exports, much more broadly defined, from a wide range of countries, developed as well as developing. Everything covered in the World Bank's definition is included as well as the other minerals and their first stage products mentioned in the previous paragraph, except that both ferrous and non-ferrous waste and scrap are excluded. The precise definitions are shown below the table. Iron and steel are not included, although their raw materials are. Developing countries are included where their mineral exports exceeded $20 million in 2004 and data are available. Lack of data from 1978 onwards, notwithstanding the attention paid to its mining sector by the World Bank and other international agencies, explains the omission of the Democratic Republic of the Congo. The broad coverage throws up a few apparent oddities, such as India's large export of precious stones, based on its gem cutting industry, and Trinidad and Tobago's exports of inorganic chemicals.

Table 1 brings out the importance of precious metals and gemstones for many developing countries. They accounted for one fifth of the $304 billion of exports

Table 1 The size and percentage composition of mineral exports in 2004

| Country | Percentages of total mineral exports | | | | | | | | Value |
	Precious metals and stones	Metallic ores and concentrates	Non-ferrous metals	Fertilizers	Crude minerals	Inorganic chemicals	Coal and coke	Mineral manufactures	Total $ million
Algeria	0	0	12	28	1	59	0	0	228
Argentina	8	49	21	5	2	9	1	6	1,854
Armenia	95	0	5	0	0	0	0	0	232
Australia	14	33	18	0	1	1	31	0	31,828
Belarus	0	0	2	73	4	2	1	18	1,186
Bolivia	27	42	25	0	2	2	0	1	466
Botswana (2001)	93	5	0	0	2	1	0	0	2,284
Brazil	5	53	22	1	4	4	0	10	10,814
Bulgaria	0	3	78	6	3	3	1	6	1,225
Burundi	99	1	0	0	0	0	0	0	44
Canada	19	12	41	9	3	6	6	5	26,093
Central African Rep (2005)	23	0	66	0	11	0	0	0	173
Chile	3	37	56	1	0	3	0	0	17,168
China (inc Taiwan)	7	2	29	5	4	13	26	13	29,473
Colombia	24	0	2	1	1	2	64	6	2,886
Cote d'Ivoire (2003)	26	1	1	37	7	9	0	21	57
Dominican Republic (2001)	2	0	9	14	29	8	0	38	27
Finland	3	0	63	4	4	15	0	10	2,928
Gabon	0	97	0	0	0	0	0	2	156
Ghana	77	11	2	0	7	3	0	1	174
Greece	0	8	62	3	8	1	0	17	1,706
Guatemala	0	0	12	16	9	19	0	42	53
Guinea (2002)	28	51	0	0	3	18	0	0	503

Table 1 (continued)

| Country | Percentages of total mineral exports | | | | | | | | Value |
	Precious metals and stones	Metallic ores and concentrates	Non-ferrous metals	Fertilizers	Crude minerals	Inorganic chemicals	Coal and coke	Mineral manufactures	Total $ million
Guyana	87	12	0	0	1	0	0	0	217
India	61	22	7	0	4	2	0	4	17,165
Indonesia	3	32	22	1	1	3	34	4	8,228
Ireland	1	50	10	1	6	6	7	19	984
Jamaica (2002)	0	99	0	0	0	1	0	0	718
Jordan	6	0	3	67	1	17	0	5	865
Kazakhstan	10	20	47	1	3	11	8	0	3,475
Kenya	6	0	11	0	56	4	0	22	162
Krygyzstan	89	0	2	0	1	1	0	6	325
Macedonia	1	1	10	1	23	5	2	57	69
Madagascar	35	25	5	0	32	1	0	1	28
Malaysia	17	1	47	8	3	10	0	14	2,540
Mali (2001)	100	0	0	0	0	0	0	0	432
Mauritius	75	0	1	16	2	4	0	3	64
Mexico	22	15	23	0	7	11	0	21	4,016
Mongolia	39	52	1	0	4	0	3	0	583
Morocco	3	3	5	45	3	39	0	2	1,898
Mozambique	0	1	99	0	0	0	0	0	367
Namibia (2003)	62	18	10	0	6	0	0	3	267
New Caledonia	0	100	0	0	0	0	0	0	302
New Zealand	19	1	72	0	3	1	0	3	885
Nicaragua	91	0	1	0	3	4	0	1	50
Niger (2003)	1	98	0	0	1	0	0	0	115
Norway	3	0	79	0	6	10	0	1	6,312
Oman	2	6	21	0	14	9	0	48	201

Table 1 (continued)

| Country | Percentages of total mineral exports | | | | | | | | Value |
	Precious metals and stones	Metallic ores and concentrates	Non-ferrous metals	Fertilizers	Crude minerals	Inorganic chemicals	Coal and coke	Mineral manufactures	Total $ million
Pakistan	3	19	3	8	12	5	2	48	112
Papua New Guinea (2003)	66	33	0	0	0	0	0	0	1,190
Peru	38	32	28	0	0	1	0	1	7,201
Philippines	9	8	61	7	3	3	0	9	973
Poland	5	1	30	5	2	5	43	10	6,371
Portugal	7	15	23	5	6	5	0	39	1,354
Romania	0	9	47	25	2	10	1	6	1,049
Russian Federation	8	3	51	14	1	5	16	2	20,852
Saudi Arabia	16	2	11	33	3	21	0	14	1,389
Senegal	0	0	1	24	9	56	0	9	321
Serbia and Montenegro	0	2	77	3	1	4	1	12	470
South Africa	48	10	16	1	2	5	17	2	14,513
Spain	2	2	26	4	9	6	2	49	8,357
Sri Lanka	55	2	34	0	2	2	0	6	404
Sudan	98	2	0	0	0	0	0	0	48
Suriname (2001)	0	99	0	0	0	1	0	0	240
Sweden	9	22	48	2	3	6	1	10	3,768
Tajikistan(2000)	6	0	94	0	0	0	0	0	398
Tanzania	97	0	2	0	0	0	0	0	628
Thailand	39	0	20	1	11	5	0	24	2,721
Togo	0	0	0	33	0	0	0	67	145
Trinidad and Tobago	0	0	0	10	0	88	0	2	1,047
Tunisia	1	2	1	50	3	32	0	11	917
Turkey	3	7	23	1	14	3	0	49	2,924

Table 1 (continued)

| Country | Percentages of total mineral exports | | | | | | | | Value |
	Precious metals and stones	Metallic ores and concentrates	Non-ferrous metals	Fertilizers	Crude minerals	Inorganic chemicals	Coal and coke	Mineral manufactures	Total $ million
Uganda	93	0	0	1	2	0	0	3	66
Ukraine	0	23	11	19	6	14	23	4	3,881
United States	38	5	19	8	5	11	7	7	37,890
Venezuela	4	8	58	5	1	7	9	8	1,627
Zambia	4	4	86	4	1	0	0	1	738
Zimbabwe	7	40	32	1	11	1	6	3	521
Totals	20	16	29	5	3	7	12	7	304,005

Definitions:

Precious metals and stones: S.I.T.C Rev 3 (UN Code 30262)—289, Precious metal ores and concentrates; 667, Pearls and precious stones; 681, Silver, platinum and other metals of the platinum group; 971, Gold, non-monetary, excluding ores.

Metallic ores and concentrates: S.I.T.C Rev 3—281, Iron ore and concentrates; 283, Copper ores, concentrates; 284, Nickel ores, concentrates; 285, Aluminium ore, concentrate etc.; 286, Uranium, thorium ores, etc.; 287, Ore, concentrates, base metals.

Non-ferrous metals: S.I.T.C Rev 3—682, Copper; 683, Nickel; 684, Aluminium; 685, Lead: 686, Zinc; 687, Tin; 689, Miscellaneous non-ferrous base metals.

Fertilizers: S.I.T.C Rev 3—272, Fertilizers, crude; 562, Manufactured fertilizers.

Crude minerals: S.I.T.C Rev 3—273, Stone, sand and gravel; 274, Sulphur, unroasted iron pyrites; 277, Natural abrasives not elsewhere specified; 278, Other crude minerals.

Inorganic chemicals: S.I.T.C Rev 3—522, Inorganic chemical elements; 523, Metallic salts, inorganic acids.

Coal and coke: S.I.T.C Rev 3—321, Coal, not agglomerated; 322, Briquettes, Lignite, Peat; 323, Coke, Semi-coke, Retort carbon.

Mineral manufactures: S.I.T.C Rev 3—661, Lime, cement, construction materials; 662, Clay, refractory, construction materials; 663, Mineral manufactures not elsewhere specified.

Source: United Nations, United Nations Common Database, June 2007, ESDS International, (MIMAS) University of Manchester.

covered in the table, with gold and gemstones, including diamonds, making up just over four fifths of their share. Inorganic chemicals, which include alumina, are also important in some African countries, like Senegal and Guinea. The United States, Australia, China, Canada, and the Russian Federation are easily the largest global exporters of mineral products, collectively supplying 48% of the total covered.

2.3 The Shares of Minerals in Total Exports

When considering the impact of mining on an economy the shares of minerals in total exports are more relevant than their absolute scale. Table 2 compares the shares of minerals broadly as defined in Table 1, with the World Bank's more narrowly

Table 2 Comparisons of percentage shares of minerals in total exports in 2000

	Minerals and products (%)[a]	Ores and metals (%)[b]
Africa		
Botswana	92.5	7.0
Central African Rep.	35.5	7.6
Egypt	8.6	3.9
Ghana	61.7	18.8
Madagascar	3.9	1.8
Morocco	19.8	8.8
Namibia	52.0	11.1
Niger	32.4	27.7
Senegal	12.9	4.8
South Africa	22.2	9.4
Tanzania	26.9	0.5
Togo[c]	21.5	25.5
Tunisia	11.4	1.5
Zambia[c]	59.2	62.3
Zimbabwe	13.4	11.0
Americas		
Bolivia	34.9	24.6
Brazil	12.8	9.8
Canada	6.9	4.4
Colombia	9.6	0.7
Guyana	44.1	20.7
Jamaica	58.4	4.0
Nicaragua	3.8	0.4
Peru	49.5	39.3
Suriname	75.3	0.1
Trinidad and Tobago	10.5	0.2
United States	4.4	1.9
Asia		
China	4.7	1.9
India	20.0	2.8
Indonesia	8.3	4.9
Israel	34.7	1.3

Table 2 (continued)

	Minerals and products (%)[a]	Ores and metals (%)[b]
Kazakhstan	22.9	18.2
Lebanon	23.0	7.4
Turkey	4.8	2.6
Australasia		
Australia	33.4	17.5
Papua New Guinea	58.9	51.3
Europe		
Czech Republic	4.7	2.0
Greece	11.1	7.2
Italy	4.4	1.4
Poland	9.7	5.0
Russia	11.9	9.3
Spain	5.0	2.2
Switzerland	8.6	5.7
Ukraine	16.8	12.0
United Kingdom	5.3	2.4

[a] As defined in Table 3.

[b] World Bank data.

[c] The higher shares given by the narrower World Bank data are not entirely clear. A partial reason is that they include non-ferrous waste and scrap whereas those are excluded from the broader definition.

Sources: Table 3 and World Bank, World Development Indicators (WDI), April 2007, ESDS International, (MIMAS), University of Manchester.

defined figures for the shares of ores and metals in total exports in 2000 for those countries where there are more than two percentage points difference. In nearly all instances the shares given by the World Bank are lower, in some cases substantially, because of the wider definition of minerals and products used in Table 1. The major differences between the two columns for countries like, for example, Botswana, the Central African Republic, Ghana, Namibia, and Tanzania reflect the narrower coverage of the World Bank's data, and above all its exclusion of exports of gold and diamonds. These large differences for many African countries suggest that the general use of the World Bank's data for econometric analysis of the relationships between mineral dependency and economic performance may indeed vitiate the validity of the results obtained. It would be necessary to show that exports of gold, diamonds, and other mineral products excluded from the World Bank's estimates have moved in step with, and are subject to exactly the same influences as the products that are included before any satisfactory conclusions could be drawn from analyses based on the World Bank's data alone. The use of readily available figures is seldom an adequate substitute for careful interpretation and understanding of the basic data.

Minerals, broadly defined, accounted for over 50% of total exports for eight of the countries for which export data were available, including some not shown in Table 2 but listed in Table 3. Five of these were in Africa. There were fourteen

Table 3 Percentage shares of mineral exports in gross domestic product and total exports, 1970 and 2000

	GDP (1970)	GDP (2000)	Exports (1970)	Exports (2000)
Africa				
Algeria	0.5	0.3	2.4	− −
Angola	− −	− −	1.8	− −
Botswana	− −	40.1	− −	92.5
Cameroon	2.1	1.1	10.7	6.1
Central African Republic	6.7	6	41.1	35.5
Congo	2.1	− −	18.3	− −
Côte d'Ivoire	0.4	0.4	1.1	1.1
Congo DR	12.7	− −	− −	− −
Egypt	0.1	0.4	1.4	8.6
Gabon	5.1	0.9	11.5	1.7
Ghana	2.5	16.3	11.9	61.7
Guinea	− −	13.6	− −	63.6
Kenya	− −	0.7	− −	4.9
Liberia	44.9	− −	73.2	− −
Madagascar	0.7	0.8	5.1	3.9
Morocco	4.1	4.4	33.6	19.8
Mozambique	− −	1.6	− −	17.1
Namibia	− −	20.1	− −	52
Niger	0	5.1	0.2	32.4
Nigeria	0.4	0	4	0
Senegal	2.1	2.7	11.7	12.9
South Africa	− −	5	− −	22.2
Tanzania	− −	2	− −	26.9
Togo	5.7	5.9	26.3	21.5
Tunisia	3.6	3.4	28.2	11.4
Zambia	54.9	12.2	98.1	59.2
Zimbabwe	− −	3.5	− −	13.4
Americas				
Argentina	0	0.4	0.6	4.5
Bolivia	15.4	5.1	97.7	34.9
Brazil	0.7	1.2	10.5	12.8
Canada	4.2	2.7	21.4	6.9
Chile	12.2	11.7	87.6	45.7
Colombia	0.2	1.5	2.1	9.6
Guatemala	0.1	0.3	0.6	1.9
Guyana	26.5	31	52.5	44.1
Honduras	1.4	1.1	5.5	4.9
Jamaica	− −	9.4	− −	58.4
Mexico	0.7	0.5	17	1.9
Nicaragua	0.9	0.6	3.9	3.8
Peru	7	6.5	48.4	49.5
Suriname	− −	42.6	− −	75.3

Table 3 (continued)

	GDP (1970)	GDP (2000)	Exports (1970)	Exports (2000)
Trinidad and Tobago	3.2	5.5	5.5	10.5
United States	0.3	0.3	8.2	4.4
Venezuela	1.5	1.3	6.2	4.7
Asia				
China	— —	1	— —	4.7
India	0.5	1.8	14.6	20
Indonesia	1.2	3.3	10.8	8.3
Iran		0.1	0.1	0.1
Israel	5.6	9.4	38.6	34.7
Japan	0.3	0.2	2.9	2.3
Jordan	1.2	2.8	22.1	12.6
Kazakhstan	— —	11.1	— —	22.9
Korea S	0.7	0.9	7.4	2.6
Lebanon	— —	1	13.7	23
Malaysia	9	1.8	22.8	1.6
Oman	— —	0.7	— —	1.2
Philippines	3.3	1.1	21.5	2.1
Singapore	1.4	2.2	1.7	1.5
Thailand	1.6	1.7	15.8	3
Turkey	0.3	0.7	9.2	4.8
Australasia				
Australia	3	5.3	28	33.4
New Caledonia	— —	8.1	— —	36
New Zealand	0.1	1.4	0.7	5.4
Papua New Guinea	— —	35.3	— —	58.9
Europe				
Cyprus	— —	0.5	— —	4.4
Czech Republic	— —	2.4	— —	4.7
Finland	0.8	1.6	4.1	4.3
France	0.6	0.6	5.2	2.6
Germany	— —	1.1	9.1	3.7
Greece	0.9	1	17.1	11.1
Iceland	4	4.2	14.2	19.1
Ireland	1.7	1.1	6.4	1.4
Italy	0.5	1	4.2	4.4
Netherlands	1.7	1.5	4.8	2.7
Norway	10.8	2.4	56.2	6.5
Poland	— —	1.8	— —	9.7
Portugal	1.2	0.8	9.9	3.7
Russia	— —	4.8	— —	11.9
Spain	0.4	1	7.4	5

Table 3 (continued)

	GDP (1970)	GDP (2000)	Exports (1970)	Exports (2000)
Sweden	2.1	1.1	10.6	2.9
Switzerland	1	2.6	4.2	8.6
Ukraine	— —	7.8	9.7	16.8
United Kingdom	1.5	1	— —	5.3
Yugoslavia	— —	— —	15.6	— —

Total exports from the IMF series of total merchandise exports f.o.b (UN Code 6190), GDP from the World Bank, Mineral exports are here defined as SITC 2 (UN Code 30264)—271 Fertilizers, crude; 273 Stone, sand and gravel; 274 Sulphur and unroasted iron pyrites, 277 Natural abrasives, not elsewhere specified, 278 Other crude minerals, 281 Iron ore and concentrates, 286 Ores and concentrates of uranium and thorium, 287 Ores and concentrates of base metals, not elsewhere specified, Ores and concentrates of precious metals, waste, scrap, 661 Lime, cement, and fabricated construction materials, 662 Clay and refractory construction materials, 663 Mineral manufactures, not elsewhere specified, 667 Pearl, precious and semi-precious stones, unworked or worked, 322 Coal, lignite and peat, 323 Briquettes; coke and semi-coke; lignite or peat; retort carbon, 522 Inorganic chemical elements, oxides and halogen salts, 523 Other inorganic chemicals; compounds of precious metals, 562 Fertilizers, manufactured, 681 Silver, platinum and other metals of the platinum group, 682 Copper, 683 Nickel, 684 Aluminium, 685 Lead, 686 Zinc, 687 Tin, 688 Uranium depleted in U235, thorium, and alloys, not elsewhere specified; waste and scrap, 689 Miscellaneous non-ferrous base metals, employed in metallurgy, 971 Gold, non-monetary (excluding gold ores and concentrates).

Sources: United Nations, United Nations Common Database, June 2007, ESDS International, (MIMAS), University of Manchester; World Bank, World Development Indicators (WDI), April 2007, ESDS International, (MIMAS), University of Manchester.

countries where minerals accounted for between 20 and 50% of total exports, and twelve whose minerals made up between 10 and 20% of the total. In another eleven, minerals provided between 5 and 10% of total merchandise exports. Australia was the only developed country whose mineral industries provided more than 20% of total exports. Increases in prices and volumes since 2000 have raised the shares of minerals in the total exports of many countries.

2.4 Mineral Exports and Gross Domestic Product

Table 3 shows the shares in 1970 and 2000 for a different range of countries, together with the shares of mineral exports in Gross Domestic Product (GDP). Where countries lack substantial downstream industries supplied by domestic minerals industries, the latter shares are a reasonable approximation to the overall importance of mining in the domestic economy. Elsewhere they understate the mineral sector's importance. Dashes denote gaps in the underlying sources of data.

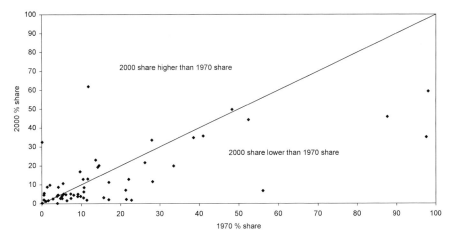

Fig 1 Mineral exports as shares of total exports, 1970 and 2000 (data from Table 3)

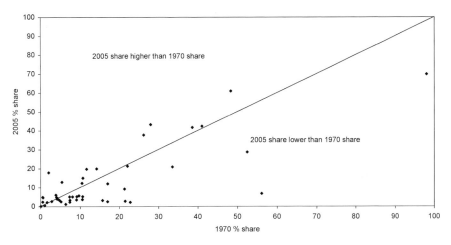

Fig 2 Mineral exports as shares of total exports, 1970 and 2005 (data from Table 3)

2.5 *Changes in Export Shares*

The share of minerals in total exports fell in more countries than it increased between 1970 and 2000, as brought out more clearly in Fig 1. In some instances the fall in export share was dramatic.

The data for exports in 2005, when prices were generally much higher than in 2000, are still incomplete, but enough are available to show that the drop in share was not just a consequence of unduly depressed prices in 2000. Certainly, the balance between countries with falling and rising shares between 1970 and 2005, illustrated in Fig 2, is more even than for the shorter period, but there were still some pronounced declines over the period.

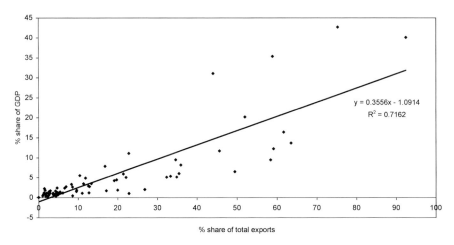

Fig 3 Shares of mineral exports in GDP and total exports in 2000 (data from Table 3)

2.6 Shares of Exports and Gross Domestic Product

As Fig 3 demonstrates for 2000, the share of minerals in exports is only an approximation to the share of mineral exports in GDP. The latter is a more precise measure of dependence on minerals.

In two countries, Botswana and Suriname, mineral exports accounted for over 40% of GDP in 2000, and a further three countries had mineral exports accounting for between 20 and 40% of GDP. They made up between 10 and 20% of GDP in another five, and in twelve they provided between 5 and 10%.

Changes in the share of minerals both in total exports and in Gross Domestic Product partly reflect shifts in the prices of mineral products relative to other prices, but they also mirror changes in volumes. The volatility of mineral exports is illustrated by changes in their share of the gross domestic product of three Latin American countries, as given in Fig 4.

2.7 Mineral Dependency and the Growth of GDP

The relationship between mineral dependency and the growth of GDP lies at the heart of much of the debate about the economic benefits and costs of mining. Cross sectional econometric analysis (e.g., Sachs and Warner 1995, 1997a, b) has demonstrated an inverse correlation between resource dependency, usually measured by the share of resource-based products in exports, and the growth of per capita incomes. Without gainsaying the validity of that analysis for the data used and periods covered, Fig 5 shows no inverse correlation between the share of mineral exports in GDP in 2000 and the annual average rates of growth of GDP between 1990 and 2005 for the countries included in Table 3. These include many, but not all of the

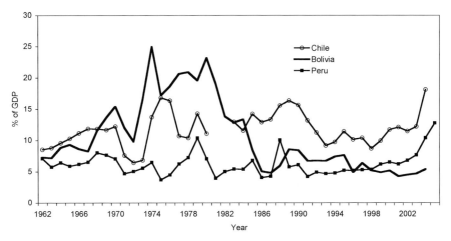

Fig 4 Mineral exports as percentages of GDP 1962–2005 (data from Table 3; note that export data for Chile in 1981 and 1982 are not available in the source)

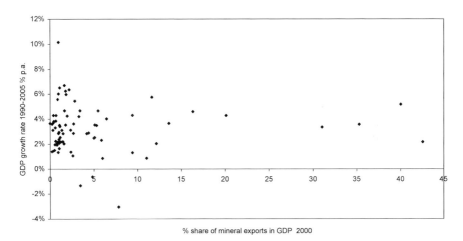

Fig 5 GDP growth and mineral exports (data from Table 3 for export shares). GDP growth rates are based on the World Bank series for GDP at market prices in constant prices, in national currencies from United Nations, United Nations Common Database (UN Code 29916), June 2007, ESDS International, (MIMAS), University of Manchester

fast growing economies with low mineral dependency, but adding countries with few or no mineral exports would probably not materially alter the results for this period. Clearly, much depends on the period used for analysis, and many of the analyses were made in the mid to late 1990s when mineral markets were weak and the prices of many mineral products were apparently on a long-term downward trend.

3 The Impact of Mineral Development

Mining is an economic activity competing for land, labour, and capital with all other forms of economic activity.[1] Its contribution to a country's economic welfare—its value added—is intrinsically no different from that of any other productive industry, including service industries. In many mining ventures, and especially in those developed by multinational companies, that value added may be created largely through exports outside the host country. Dependence on foreign markets may raise some policy issues for host countries, but it does not, in itself, affect the economic contribution of mining. That is partly determined by the disposition of the value added between all the actors involved, whether domestic or foreign, private sector or governmental. The nature and extent of the linkages formed with other sectors of the economy, whether as a user of goods and services or as a supplier to downstream industries, also help determine the economic impact of mining on the economy.

3.1 Economic Impact

Especially in small developing countries, there may be limited scope for purchasing intermediate goods and services within the country. Supplies of all types will be sourced overseas so that multiplier effects from their purchase on local incomes and employment will be minimal. The chances of local purchases will increase the longer established the operation becomes, and they will also rise with the number of projects. Particularly in its early stages, much of a project's value added may also be remitted overseas in the form of interest payments and dividends to shareholders. Many of the employees may be expatriates, and a large part of their wages and salaries will be saved abroad or spent on imports. That means that the main economic benefits will accrue to the host nation through tax receipts. Those benefits will be slow to appear where accelerated depreciation and other forms of tax holiday are allowed, as they are in many countries.

In practice this presents far too narrow a view of the economic impact of very many extractive industry developments. Most become integrated to varying degrees with the local economy. Particularly during the construction stages of projects, but also once they are fully operational, local labour will be employed. The indigenous workers' incomes will be spent primarily on local goods and services, giving rise to wider effects through income and employment multipliers. Large projects will normally require substantial infrastructure, such as power supplies, transport networks, ports, and facilities for their employees and their families that will also benefit the local economy. Most investors will aim to source goods and services locally, as long as they are cost-competitive. Local workers will be trained in necessary skills, particularly in maintenance trades, and some will transfer their newly acquired skills elsewhere in the economy. Labour turnover is often fairly high in consequence.

---

[1]This section is based on part of a monograph, Crowson (2007): Crowson, Phillip, Adding Public Value: The Limits of Corporate Responsibility, Oxford Policy Institute, April 2007 (www.opi.org).

3.2 Local Impacts

Apart from their contributions to national incomes and tax revenues, the main impacts of extractive industry projects are at the local level. Mining is usually an irreversible use of land, particularly where extraction is from open-pit rather than underground mines. Aside from the mine itself, there are waste dumps and tailings ponds. Large volumes of water are normally required for processing. Air and water quality can suffer from waste emissions unless these are properly controlled, and noise may also be a problem for any local community. Fragile ecosystems and existing social structures may be disrupted by extractive industry operations. These potential concerns can be overcome or mitigated by careful and sympathetic management, but that is not always forthcoming. Local communities may welcome the opportunities for employment and increased wealth offered by mineral development, regardless of any environmental costs. Nonetheless, tensions inevitably develop in many cases between the local communities who bear the social and environmental costs, and national governments who secure the economic and fiscal benefits. The investors often find themselves at the centre of these tensions.

3.3 Realisation Versus Potential

That the extractive industries potentially create wealth does not necessarily mean either that the potential will be realised, or that when it is it will be used wisely or well. Nor does it mean that the potential will be harnessed by the host country rather than leak overseas. Where investments are made by foreign companies, some leakages abroad of loan repayments, interest, dividends, and salaries of expatriate staff are inevitable. Those are not inconsistent with the host country maximizing its share of a project's economic rents as opposed to its value added.

3.4 The Construction Phase

Even during their construction phase, large projects raise issues for national economic management. The influx of capital and the associated consumer spending boost demands for goods and services that may not be locally available in sufficient quantities. In consequence, both imports of consumer goods and price inflation will tend to rise, especially where the capital expenditure is large relative to the national economy. Whereas unskilled workers may be recruited from the ranks of the under- or unemployed, skilled and semi-skilled labour will be attracted from existing industries, probably forcing up wage rates in the locality. Established industries will have to raise their own wage rates to compete or lose workers. This will reinforce any inflationary tendencies.

Typically, far more workers are required during the construction of large projects, and especially of major mining projects with their associated infrastructure, than during their subsequent operation. Thus, much of the surge in demand for labour will be temporary, with the excess labour released once the construction is finished. If there is a large increase in unemployment relative to demand for labour elsewhere in the economy, real wages will tend to fall back, with the burden of the fall probably borne by the less skilled. One side effect might be a rise in wage differentials between the skilled workers who are still in demand, and the less skilled who are in over-supply. These possible impacts can lead to difficult re-adjustments for the local economy, even when it has always been known that the needs for labour would be transitory.

3.5 The Operating Phase

When resource projects move into their operating phase, the geographical balance of costs and benefits within the host country will remain uneven. Local interests will naturally press for most of the benefit to accrue locally in the form of increased public expenditure in order to offset any environmental or social costs. Yet governments have to act in the wider public interest, which may require a very different geographical pattern of spending. National priorities may sharply conflict with local interests, but those may have to be appeased in order to preserve the income-earning potential of the resource project.

3.6 Dutch Disease

Once a project is fully operating, there will be a sharp rise in national exports and value added. The immediate impact will fall on the trade account, which will tend to move into surplus. Unless that is offset by appropriate national economic management, the exchange rate will appreciate. This may cause problems for pre-existing industries, which will be made less competitive against imports and in export markets. The rise in the exchange rate will reinforce any reductions in competitiveness caused by rising wages and other domestic costs associated with the project's construction and operation. The co-existence of rampant export earnings from newly established resource projects with declining competitiveness and reduced activity in pre-existing industrial sectors is typified as the "Dutch disease", after the example of the Netherlands coping with its newly established natural gas industry in the early 1960s. It is a sudden, sharp, sustained rise in net overseas earnings (a rise in exports or a fall in imports) that triggers any structural imbalance in an economy, rather than the development of natural resources per se. It is not an inevitable consequence of natural resources investment, but it is always a risk, even in broadly based mature economies, let alone in low-income developing nations. The main symptoms of the disease, an appreciating exchange rate and a decline in traditional industries, can be alleviated or even offset by appropriate national economic management, but that is not always forthcoming.

3.7 Volatility

It is not just the "lumpiness" and large size of natural resource projects that compli-
cate economic policy and management in resource-based economies. The earnings
from most minerals also tend to be volatile and unpredictable. Non-fuel minerals
production is rarely integrated with downstream facilities overseas, so that sales
volume is not guaranteed. The volume of demand for most products, although not
for gold, fluctuates with business activity or personal incomes in the economies of
the consuming nations. Where the products involved are commodities traded on terminal markets, like the major base and precious metals, prices fluctuate even more
widely than demand. The annual average price of copper, for example, rose over
fourfold between 2002 and 2006, and the swing from trough to peak was much
greater. From the host country's viewpoint, the main impact of volatile earnings
will be on tax revenues. In non-fuel mining, a common fiscal system combines a
relatively low (typically up to 3%) royalty on turnover, with some form of income
or profits tax. The latter may sometimes allow for elevated rates of tax on "super"
or "abnormal" profits. In years of weak prices, the tax receipts may be confined to
the royalty element, which will itself be depressed by the level of prices. By contrast fiscal receipts can burgeon in the boom years, subject to the existence of tax
holidays and accelerated depreciation.

3.8 Alternatives?

Many resource-based developing countries may have no realistic alternative to
exploiting their mineral wealth as a means of improving their living standards. They
may have a variety of possible disadvantages, such as small rural populations, little
viable agricultural land, adverse climatic conditions, geographical isolation, or rudimentary transport links to major markets. Eco-tourism, which might in some areas
be regarded as a possible substitute for the extraction of petroleum, gas, or minerals,
is unlikely to yield the same economic potential. It may in any case be inhibited
by some of the same factors that prevent competitive manufacturing, as well as by
institutional issues.

3.9 Governance

Neither the volatility of commodity earnings and tax revenues, nor the possible
emergence of structural imbalances militates against the wealth creating potential
of resources extraction and primary processing. Any correlation between resource
dependence and sluggish economic performance does not necessarily indicate
causality. Many of the factors that inhibit broad-based economic development may
also contribute to weak governance. This is the major reason for the "resource
curse", that is, for many resource-based developing countries failing to benefit
from the development of their mineral resources (Stevens 2007a, b). To argue
that any potentially adverse economic impact of large natural resource projects on

developing host countries could be offset or circumvented by appropriate government policies presupposes the existence of effective institutions and competent governments. Yet these are seriously lacking in many countries, and these are often the same countries that are resource dependent.

The term "weak governance" embraces a wide range of political, social, and institutional failings. In essence the administrative and legal systems are too weak to check and control the activities of powerful interest groups. In some instances the institutions of government are dominated by one or more groups to serve their own objectives. That is more common in autocracies than in democracies, but it is not an automatic consequence of autocratic government. Some democracies may be even worse governed than dictatorships, because their governments may pander unduly to sections of their electorates, and be less able to limit rent-seeking.

3.10 Property Rights

Many of the issues that arise concern the extent and ill-defined nature of property rights in many developing host nations. In the developed mineral-rich countries, such rights are usually clearly defined and enshrined in law. There are well-established legal avenues for pursuing any uncertainties or disagreements, and the decisions of the courts are accepted, subject to the right of appeal. Such avenues are often non-existent or extremely tortuous in some developing countries, where the decisions of the courts may be governmentally guided or they may be over-ridden, and the judiciary is not always an impartial arbiter.

3.11 Rent-Seeking and Patronage

Rent-seeking behaviour, the attempt to cream off any surpluses over the opportunity costs of production, is universal. Witness the activities of powerful trade unions, the large bonuses paid on Wall Street and in the City of London, and the fees charged by private equity firms. In the mature industrial economies, such rent-seeking is usually constrained and brought under control relatively quickly. Legislative or regulatory actions are eventually taken to constrain the more egregious excesses. In many developing countries, however, members of the government and its administrative organs may be capturing rents, not for the good of the nation as a whole, but for sectional interests. The machinery of government is not only unable to control such behaviour but may itself be indulging in it.

Nepotism and patronage, whereby jobs and offices are allocated not by merit but by influence, are widespread. One result is a frequent turnover of staff in ministries and government agencies, depending on which faction is presently in favour. Even in countries like Chile and Turkey, the senior management of state companies are politically appointed and are liable to change with the government.[2] Rapid staff turnover presents problems for the continuity and consistency of policy and its application, even where the individuals concerned are well qualified. All too often the staff are poorly qualified. Many developing countries suffer shortages of qualified manpower with appropriate administrative and managerial skills. Those who have developed such skills, often through study abroad, may be effectively barred from advancement because jobs are not awarded on merit. Commonly, they decide to remain overseas once they have qualified, because the opportunities are too limited in their home countries. When they do go back home, skilled managers and administrators seek work in the private sector because the working conditions, broadly defined, are more congenial and it is financially more rewarding. A corrupt public administration is seldom attractive for honest employees who can gain advancement elsewhere.

Throughout history the power of the purse has been one of the most effective checks on the actions and ambitions of an over-mighty state. Even if they are running a command or slave economy, governments can only act within the taxable capacity of those over whom they rule. That capacity is strictly limited in subsistence economies. If too many feathers are plucked the geese start to hiss too loudly,[3] and seek the redress of grievances before granting additional funds. Over time, governments have been forced to yield and share power in order to raise sufficient income from existing industries and populations. The development of large revenue-earning resource projects can weaken this pressure on governments. Taxes paid by such projects may greatly boost existing government incomes, effectively bypassing already weak democratic constraints. To the extent that this occurs, resource projects will not just be the victims of weak governance, but they may also potentially prolong it.

The national earnings from resource projects may be channeled into conspicuous consumption by the ruling elite, rather than into productive investment or social welfare. The conspicuous consumption may include expensive military hardware disproportionate to the host country's defence requirements, and prestige projects of all types as well as luxury homes and consumer goods. At the extreme they may be diverted into private bank accounts offshore, as in the kleptocracy over which President Mobutu long presided in Zaïre. Natural resource companies have no control over such grand larceny, which can apply to domestic and state-owned companies as much as to foreign owned projects.

In practice such wholesale theft of resource revenues is rare, and it is probably overshadowed by other forms of corruption. Many surveys, such as those of Transparency International (2008), show that resource-rich countries are adjudged

---

[2]This is not just a feature of developing countries. Many government officials and the senior staff of government agencies are politically appointed in many OECD countries, including the United States, and, increasingly, the United Kingdom.

[3]After the statement attributed to Louis XIV's finance minister, Jean Baptiste Colbert: "The art of taxation consists in so plucking the goose as to obtain the largest amount of feathers with the least possible amount of hissing".

to be amongst the most corrupt. Yet venality is by no means confined to resource-rich countries, or indeed to developing countries. Moreover, a positive correlation between a nation's possession of natural resources and its degree of corruption again does not indicate any form of causality. Multi-national natural resource companies are jealous of their reputations and are beholden to shareholders and regulatory agencies in their home countries. The lengthy gestation and pay-back periods of their projects mean that they need to weather possible changes of government during their lifetimes. That is difficult if it can be demonstrated that they have bribed to obtain licences or favours. The temptations to corruption are much greater where companies are seeking to secure sales contracts, in the defence sector for example, than where they are making long-term investments in natural resources.

It is questionable whether there is a strong link between the prevalence of petty corruption, which may be deeply rooted in cultural and social traditions in many developing countries, and poor economic performance. Both China and India, and other Asian countries before them, where petty corruption is common, have exhibited strong economic growth. Eighteenth century England, where the industrial revolution took off, and the nineteenth century United States were not exactly models of governmental probity and incorruptibility. The problems arise where large scale corruption seriously distorts the creation of wealth and the allocation of resources. What might be described as "cultural corruption", or the customary giving of gifts in return for favours, does not automatically lead on to such large-scale corruption, although the path between the two may be primrose-strewn.

3.12 Subsistence Mining

Some of the problems of governance ascribed to minerals extraction are not specifically features of large-scale investment by well-capitalised companies, whether domestic or foreign. Rather they are by-products of, and in some instances are symbiotically linked with, small-scale mining of low-volume, high-value products. These are mainly alluvial gold and diamonds, other gemstones, and ores of tin, niobium, and tantalum. These can be extracted from near surface deposits by manual labour with limited and rudimentary equipment. The products require relatively little, if any, processing to a saleable state, are easily transported, have a high value to weight ratio, and can be readily sold. They are most common in Africa, although they also exist in Asia and parts of Latin America. By no means all small-scale mining is closely linked to crime and lawlessness. Much provides necessary employment and incomes in subsistence economies. The revenues can be subverted, however, to finance and foment civil unrest and military adventurism. African warlords have supported their activities with the proceeds of gold and diamond mining. The sale of "blood diamonds" and their international trade have now been largely controlled, although not completely eliminated, by the certification of sources of origin through the "Kimberley Process" (www.kimberleyprocess.com; Bone 2004).

4 Some Examples from History

Most of the inverse correlations found between resource dependency and economic growth are based on periods between 1960 and the mid-1990s, when resource-based economies were subjected to some major exogenous shocks. These included the oil price spikes of the 1970s, which adversely affected demand for all mineral products and prompted the persistence of over-capacity for many years, the wave of liberalizations and subsequent privatizations during the period, and the break-up of the Soviet Union. It has been widely acknowledged that the inverse relationship has seemingly not operated historically, or for all economies (Sachs and Warner 1995). Yet the historical development of successful mineral-rich countries is often viewed through rose-tinted spectacles. In the initial years, even decades, of their mineral development, they faced many of the same issues that concern developing mineral-rich states today. This is illustrated by the initial experiences of Brazil and Australia, which suffered similar economic distortions and dislocations in their early decades of mining. That their subsequent performance differed reflects mainly institutional factors.

Although mining may persist for generations in individual districts local supplies of ore eventually become depleted, or otherwise uneconomic. That can be seen in all the examples quoted in the remainder of this section. The mining districts have usually suffered economic decline and depopulation unless alternative sources of wealth creation have developed. One key question has always been the extent to which the profits and rents accruing from mining have been retained within the local economy and invested productively. Such investment has seldom taken place in the mining localities, narrowly defined, but within the region or country. A second important issue concerns the extent to which mines have been able to draw on the host region for their inputs of goods and services.

4.1 Eighteenth Century Brazil

Brazil's experiences illustrate both the benefits of mineral development and the problems raised. Gold was discovered in various parts of the Brazilian interior, in what is now Minas Gerais, in the late seventeenth century (Russell-Wood 1987). Brazil was then a Portuguese colony, reliant on slave-based sugar production. The subsequent gold rush opened up the country's interior and pushed its frontiers westwards (Capistrano de Abreu 1907). The discovery of diamonds in 1729 further stimulated mineral development. Mining encouraged migration from overseas and from the coast to the interior. Large sections of the coast were virtually depopulated, with slaves from the sugar plantations being transferred to work the gold and diamond mines of the interior. One side effect of the depopulation of the northeast was the consolidation of the large landowners' control of coastal plantations, and the ossification of its social structure (McCann 1997). Brazil's centre of gravity shifted to the south and west. The population of Minas Gerais rose from a negligible share of

Brazil's total in 1700 to over one-fifth by the early 1780s. Gold exports were largely channeled through Rio de Janeiro, which prospered to the extent that in 1763 it displaced Salvador, in the northern state of Bahia, as the colonial capital (Mansuy-Diniz Silva 1987).

Mining prompted the creation of transport links between the interior and the north and south of the country, and urban development in inland Brazil, based on the mining camps. The mines were, however, mainly located in mountainous areas with difficult access and limited agricultural land. Transport depended primarily on mules. The demand for food and animals for transportation and meat therefore had major repercussions outside the mining districts, and agricultural production increased elsewhere in the country (Mueller and Baer 1997). Although much of the wealth created by mining was remitted overseas, some was used to build urban public works and endow charitable foundations, such as hospitals. It also financed extensive smuggling of foreign manufactures through adjoining countries.

In addition to its impact on Brazilian development, the wealth created by gold mining had major effects on Metropolitan Portugal. The Portuguese government took one-fifth of all gold mined, and forcibly attempted to prevent smuggling. Declining terms of trade of Portugal's traditional colonial products in the face of rising competition during the seventeenth century had encouraged Portugal to develop domestic industries to meet local and colonial needs. The revenues from gold mining then provided ample foreign exchange for purchasing overseas manufactures, especially from Britain, as well as conspicuous consumption. Nascent domestic industries in both Portugal and Brazil were unable to compete, and withered in the face of foreign competition that was boosted by tariff concessions to Britain early in the century. Although Portugal's policies were changed during the 1750s to encourage Brazilian economic development and reduce Britain's dominance of its trade, their beneficial effects were partly offset by the expulsion of the Jesuits in 1759 (Alden 1984). During the last quarter of the century, Portuguese fears of possible Brazilian independence led to a complete reversal of policy and to the wholesale closure of Brazilian industries at a time when gold revenues were declining. Production had peaked during the middle years of the eighteenth century, and was in sharp decline by its end (Schmitz 1979). Portuguese political considerations completely overrode any concerns about the health of the Brazilian economy, and swamped any potentially beneficial impacts of a century of extracting substantial mineral wealth. In summary, Portugal experienced classic symptoms of Dutch disease and of weak governance.

4.2 The Victorian Gold Rush, Australia

Australia's first gold rush initially had adverse effects on the colonial economy, but these were outweighed by subsequent benefits. Gold was first discovered in commercial quantities in early 1851 in the Bathurst district of New South Wales (Blainey 2003). The impact on the nascent state of Victoria was dramatic, with a wholesale shift of men of all classes both from Melbourne and from farms to the gold fields. Both labour shortages and sharp wage rises threatened the colony's economic viability. In desperation, the Victorian government offered a reward for the discovery of gold within 200 miles of Victoria, a prize soon claimed with the discoveries near Ballarat, announced in 1851. The subsequent Victorian gold rushes initially had further devastating effects on the local economy (Scott 2007). A large proportion of the recently established city of Melbourne's able-bodied male population left for the diggings, and Melbourne's property prices collapsed. Meantime, severe labour shortages led to a tripling of wages, and the prices of basic necessities rose sharply. Agricultural activity was severely threatened by labour shortages (Vyas 2007). Fortunately the government soon took effective action to encourage people to return to the land.

The immediate adverse impacts were soon offset by the economic benefits, with the gold fields becoming the driving forces of the Australian mineral economy (Duffy 1994). Their nearness to ports, fertile soil, and temperate climate enabled successful miners to develop businesses or farms in the area. Much of the profit of gold mining was ploughed back into commercial enterprises and services in local towns. As the mines went deeper and more capital was invested in them so their demands for fuel, stores, and equipment increased. Timber was required to line shafts and support drives, and wood to fuel boilers. Foundries and machinery companies were established to manufacture and maintain capital equipment, which became increasingly more complex. The demands of the miners for food, clothing, housing and equipment directly stimulated agricultural and industrial development, with large multiplier effects creating added indirect impacts. Rising domestic demand for meat and hides, and for horses, both for transport and to work machinery, created increased and more stable farm incomes. A wide variety of goods were manufactured and supplied locally. Their great distance from other centres of population, and the costs and delays of international transport protected local farmers and manufacturers supplying the Victorian gold fields such as Ballarat from overseas and interstate competition.

The construction of railways from the ports to the goldfields in the early 1860s opened up the interior and further stimulated farming. The lure of gold encouraged immigration, not just from Britain but from a wide variety of countries. After the initial rush most middle class immigrants returned to the cities, and some moved into farming. Migrants tended to be adaptable, vigorous, and well educated. They were prominent in creating and maintaining the religious, educational, and social infrastructure, and their broad social mix hastened the growth of democracy. Not only did immigration directly increase the population and greatly expand local markets, but migrants contributed to high marriage and birth rates in the 1850s and 1860s. Australia's population almost quadrupled between 1851 and 1861, and Victoria contained over half Australia's population in the late 1850s.

The development of gold mining transformed Victoria from a pastoral economy subject to the vagaries of fluctuating export markets for wool into a much more diversified and balanced economy. Melbourne rapidly developed into a commercial and financial centre that rivaled and then outstripped Sydney. Although the mining localities themselves may have declined when their ore ran out, the re-investment of profits within the state laid the foundations for its sustainable growth.

4.3 Gold in South Island, New Zealand

Like Victoria, New Zealand's government offered rewards for gold discoveries. It was found in small quantities on North Island in 1856–1857, but major gold exploration and development did not begin until 1861 when gold was discovered in Otago on South Island (Mein Smith 2005). Exploitation of South Island's West Coast fields soon followed from 1864, and New Zealand's gold output averaged some 17 t/annum in the decade beginning in 1862 (Schmitz 1979). The development of gold mining in South Island occurred whilst North Island was immersed in the Maori Wars, and without gold, New Zealand's early economy would not have developed as quickly as it did. As in Victoria, gold attracted a wave of migrants, which led to investment in transport infrastructure that opened up the interior, and created demands for food, clothing, and other supplies from local farms and industries. Once mining moved into its mechanized phase, local foundries and engineering firms were established to supply and service the operations. Even more than in Victoria, distance largely insulated these firms from overseas competition. The profits of mining were largely invested locally in farming and commerce, ensuring that economic activity could continue to flourish when the easily won deposits were depleted. Although the tented encampments and some boom towns soon died, many of the settlements survived to service agriculture. Dunedin rose to preeminence amongst New Zealand's cities in the late nineteenth century, in the same fashion as Melbourne in Australia, on a foundation of earnings from gold mining. (The Encyclopedia of New Zealand 2007).

In summary, the local retention of profits and the gold industry's dependence on domestic suppliers ensured that the economic benefits of mining were largely kept within New Zealand. The pastoral industries that were being developed by settlers were struggling to survive and prosper before the gold discoveries boosted the population, fostered local needs and prompted greatly improved shipping links with other countries.

4.4 Cornish Copper Mining

Cornwall's rise and decline during the eighteenth and nineteenth centuries demonstrates how the wealth created by mining does not necessarily benefit the mining region, rather than the host country, over the longer term. Cornish copper mine production took off in the mid eighteenth century, but was hampered by inadequate means of draining deep mines. This prompted considerable and far-reaching technological innovations in pumping, mining, and processing equipment, and improved working methods. Many of the technical developments pioneered in Cornwall were successfully transferred to deep underground mines elsewhere in the world. Early in the nineteenth century new ports and transport links were developed to serve the mines, and engineering and ancillary industries were set up locally. Production rose strongly, and for much of the period between 1830 and 1850 the Cornwall and West Devon district was the world's largest producer of copper, and remained an important supplier until the late 1860s. Its output of contained copper averaged rather over 11,500 t/annum between 1830 and 1865 (Barton 1978). The opening of new mines and rising output increased the opportunities for employment, and led to the expansion of existing towns and the creation of new settlements. Educational, scientific, and technical facilities were set up and social infrastructure of all types created. Local agriculture was correspondingly stimulated, with its output partly coming from miners' smallholdings. Mine owners and landlords used their profits to build or re-model their houses, estates, and gardens (Cornish Mining World Heritage Site 2007). Between 1801 and 1841 the population of Cornwall grew slightly faster than that of England as a whole (University of Portsmouth, Department of Geography). From the perspective of the mid-nineteenth century it would have seemed that Cornwall's mineral developments had given the county a sustainable broad-based future.

Yet this would have been a short-lived perspective. Copper mining created great wealth but the district increasingly faced competition from richer and lower cost mines overseas, and most mines closed down when copper prices collapsed in the late 1860s. Cornish mines required expensive pumping to remain dry, and their underground workings became increasingly complex as they followed the narrow veins of payable material. Commonly, their equipment was outmoded and their working conditions poor. Some mines successfully converted to tin production, which had continued throughout the copper boom, but many of those were eventually depleted. Although the production of mine equipment and ancillary services continued, Cornwall entered a prolonged economic depression. Much of the infrastructure, including some of the towns, was specific to mining. The county was too remote from major population centres and markets, and from sources of other industrial raw materials to support a sustainable manufacturing base. Even during copper's heyday it was more economic to ship Cornish ore to South Wales for smelting than to import Welsh coal. The decline can be traced in Cornwall's population, which grew more slowly than in the rest of England between 1841 and 1861, and then fell absolutely in each decade between 1861 and 1901 (Office for National Statistics 2001). Most of the sustained benefits from Cornish copper mining were outside the county. The collapse of employment opportunities in Cornwall's mines fostered large scale emigration of Cornish miners throughout the world. Cornwall's loss was the global mining industry's gain. The Cornish experience remains relevant for today's mineral producing regions.

4.5 Chilean Nitrates

As in Africa in recent decades, mineral wealth has historically occasioned or financed wars. The War of the Pacific between Chile, Bolivia, and Peru in 1879–1883 was one instance. It was triggered by the Bolivian imposition of export taxes on a Chilean nitrate producer in defiance of earlier treaties. Fighting broke out in 1879 when Chilean entrepreneurs and mine owners in Peru and Bolivia resisted new taxes, the formation of monopoly companies, and other impositions. Chile was overwhelmingly victorious, gaining territory from both Bolivia and Peru (Martín 1974; Drake 1994). A few years later, in 1891, disagreements between the Chilean President and Congress led to a civil war. Whilst broader political issues were the key, disagreements over policies towards the nitrate industry played an important part. British mine-operators actively supported the victorious congressional rebels (Monteon 1982; Drake 1994).

Chile's seizure of the nitrate fields of Tarapacá from Peru and Antofagasta from Bolivia unleashed a period of strong growth, both in nitrate production and in the domestic Chilean economy. Prior to the Pacific War, Chilean companies had been heavily involved in nitrate production, mainly using primitive labour-intensive methods, but they effectively made way for foreign investors in subsequent years. Foreign capital, management, and improved technology boosted output and productivity. European (mainly British, but some German, and later United States) companies soon controlled a substantial share of the industry, with more capital intensive and productive operations than those of domestic Chilean companies (O'Brien 1982). This dependence on foreign companies inhibited the development of local industry, because the companies imported their equipment and engineering spares from overseas. Although Chile mined coal, the nitrate producers imported coal from the UK and Australia, arguing that Chilean coal was unsuitable for nitrate processing. That was because their plants were designed to use hard coal, which could be imported cheaply as ballast in the ships that exported nitrates (Monteon 1982). The main benefits of nitrate production to the Chilean economy were through taxation and its provision of a growing domestic market for agricultural output. All food-stuffs and supplies for workers, their dependants, and the numerous animals (mules and horses) used in the northern desert provinces had to come from further south. In the 1870s, traditional Chilean exports of wheat had suffered from foreign competition and weak prices, so that rising demands from the north gave agriculture a new lease of life. It was, however, highly labour intensive, and the development of new domestic markets averted the need to invest in more productive agricultural machinery in order to survive. In that sense, the rise of nitrate production entrenched the power and wealth of large landowners (O'Brien 1982).

In the early twentieth century, Chile was the predominant global producer of nitrates, which increasingly replaced guano as a source of nitrogenous fertilisers and for the manufacture of explosives. Its exports almost tripled between the civil war and the First World War (Monteon 1982), and nitrate operations accounted for an average 25% of Chile's gross domestic product over that period (Edwards and

Edwards 1994). Government revenues took the form of an export duty that captured between one quarter and one third of the value of exports (Monteon 1982), and sales of additional lands for nitrate production. The objective of these was to increase exports and, at times, reduce the power of existing producers to combine to fix prices. The export duties were the main channel for transferring wealth from nitrate production into the Chilean economy. They accounted for roughly half of the government's ordinary budget revenues over the 40 years from 1880 to 1920 (Edwards and Edwards 1994). This dependence meant that government income moved cyclically with exports that fluctuated according to world market conditions, and this contributed to financial instability. Whilst part of the government's revenue was devoted to education and public works, a substantial proportion was spent on the military and the bureaucracy. The Santiago district took a disproportionate share of government spending, with the expansion of transport facilities largely aimed at improving the shipment of agricultural products to the capital. Education was heavily biased towards the humanities rather than science, engineering, and technology, and did not touch most of the population. Whilst Chile's gross domestic product expanded substantially during its nitrate age, most of the population gained little benefit. The nitrate revenues relieved the upper class of tax burdens and fuelled a nepotistic system of patronage and corruption.

In summary, the Chilean economy benefited greatly from nitrate production, but the country's largely inflexible social and institutional structure inhibited the maximisation of benefits. Chile suffered from rent-seeking behaviour by the ruling oligarchy that restricted the bulk of the gains to a small proportion of the population. Although the remoteness of the nitrate deposits in an inhospitable desert region distant from the main centres of population and from cultivable land inevitably limited the scope for integrating the nitrate sector with the rest of the economy, its reliance on foreign capital, know-how, and management further restricted its multiplier effects.

The initial boom in nitrate production fortuitously coincided with a run-down of Chile's domestically owned copper industry and with a collapse of its wheat exports. Equally fortuitously, the demise of nitrate mining was accompanied by a resurgence of the copper industry based on US capital and new technology. That demise highlights the dangers of undue dependence by any country on a single product. The development of synthetic nitrate production during the First World War by German and British scientists rapidly undermined its international markets. Although ample resources remained, they soon became uneconomic, Chilean production withered, and most of the production facilities and the accompanying infrastructure, including railroads and towns, were abandoned to the desert. Chile's production more than kept pace with global demand during the boom years, but its experience provides an object lesson in the necessity of keeping abreast of technical change. It also emphasises that demand rather than resource availability is the main driving force of mineral industry developments.

5 Discussion

The five historical examples described above are typical of many others. Mining was labour intensive, relying on muscle power rather than machinery. The need for labour, together with the lure of gold in gold mining districts, prompted migration to the mining areas and to the host countries. The large work forces required regular supplies of food and basic necessities, including shelter and clothing, let alone more sophisticated products. The development of mines created or expanded local markets that, in their initial stages, could only be supplied at great cost and with logistical ingenuity. High prices for supplies fostered local production of wide ranges of goods and services, so that the multiplier effects on employment and incomes tended to be relatively large. Once local industry had developed, and especially agriculture and commerce, it commonly became self-perpetuating through its own multiplier effects. The initial participants in the nineteenth century gold rushes needed a minimum of equipment, and the miners themselves often built it. The rough and ready nature of much mining equipment and machinery of the period meant that skilled mechanics or blacksmiths could fix most problems without recourse to overseas expertise or suppliers. Although capital intensity soon rose with the introduction of machinery, that remained relatively easy to maintain by local skilled workers. Even the Cornish copper industry mainly depended on local engineering companies. The Chilean example, however, shows that the multiplier effects were more limited where the mining industry relied on imported capital intensive equipment and know-how.

The average scale of new mines has risen markedly since the nineteenth century, with an especially pronounced rise since 1990 (Crowson 2003). Today's new projects are typically much larger and more capital intensive, employing far fewer workers. Unless several mines are developed simultaneously, the long term labour force is usually insufficient to support much development of local agricultural production or commercial services. The locations in which mines are being developed today are generally far less amenable to permanent human settlement than many of the mining districts of the past. Meantime, the capital equipment and processes used in minerals extraction and processing have become technically complex. Parts are often not repaired but replaced and replacement spares can commonly be supplied only by the original manufacturers, who enjoy economies of scale, and fitted by specialists. Computer-based systems are serviced by the original suppliers. That means that there is much less need for local skilled workers or maintenance workshops of the same size or nature as in former times. This changed nature of the industry's capital equipment has further reduced the scope for backward linkages between mines and the local economy. Domestic engineering and metal fabrication are now much less likely to develop to serve the mining industry. Because most new mines need to be cost-competitive in order to serve international markets, they can seldom, if ever, afford to subsidize local engineering and maintenance facilities.

The effects of changes in mine size and complexity have been reinforced by dramatic improvements in transport and communications. The real costs of transport and of telecommunications have fallen markedly since the Second World War, let alone since the nineteenth century (Crowson 2006). Costs are, however, much less important than transit and communications times in explaining backward linkages between minerals extraction and host economies. In the early 1880s, the typical voyage time between North West Europe and Chile or Australia was some three months. Most countries were then already linked by telegraph cables, with the first effective transatlantic cable operating from 1866, but their capacity was restricted and transmission rates were slow. Even where countries had telegraphic linkages, communication with the remote areas where most mines were located was slow and often hazardous. In consequence, mining companies that relied on overseas sources of equipment and supplies would have had to wait several expensive months. They inevitably sought local suppliers, and that led to the development of ancillary industries, particularly for repairs and maintenance. Those, in turn, had multiplier effects on incomes and employment in the local economy.

The contrast with today is marked. Communications are now virtually instantaneous throughout the world. In consequence the needs and problems of a remote mine can be immediately communicated to equipment suppliers and relevant experts. Provided that suppliers can produce any required parts and equipment within reasonable lead times (and manufacturing times have also generally fallen), the delays between equipment failure and replacement have been greatly reduced since the late nineteenth century. If a manufacturer has a spare part readily available it could be at the most remote mine site within a matter of days, or even hours.

In effect the time required to transmit details of requirements for equipment to overseas manufacturers, and then for those requirements to be met, acted as insurmountable barriers to trade, regardless of transport costs or local tariffs. Local suppliers could not just operate behind that time barrier, but were also essential to the viability of the mines. That was as true in nineteenth century Australia and New Zealand as in eighteenth century Brazil. Today, the time barrier is substantially less important, and what remains has been offset, at least partially, by the increased technical complexity and scale of operations of minerals extraction and processing. Reductions in transport costs have further reduced the potential competitiveness of local suppliers.

In summary, the forces that have contributed to globalization, and that have encouraged the development of remote mineral deposits, may have simultaneously weakened the likelihood of backward linkages developing between mining and supplying industries in many otherwise susceptible economies. Where mines are serving export markets, either directly or through downstream operations, the cheapest and quickest means of sustaining output will usually be followed. Resource companies therefore tend to rely far less on local suppliers and service trades than they once did. Reduced transport costs and transit times also mean that the demands of workers and their dependants can be more cheaply and effectively satisfied from imports than from local companies. Significant multiplier linkages for economies, rather than for localities, are likely to be much more limited than in the past. This in turn means that the main economic benefits of mining to many countries will be through their potential contributions to government revenues, rather than multiplier linkages. This is a distinct change from many historical mineral developments, suggesting that many of the success stories of the past are largely irrelevant to most host countries today. Chile's experience with its nitrates industry is more relevant in that foreign capital, equipment, and management were instrumental in its success. The ruling elite was as avidly rent-seeking as those of many mineral-rich countries today. The main economic benefits of the nitrate industry to Chile were transmitted via government tax revenues, which were largely captured by the upper classes rather than the population at large.

No matter how effective historic mine development was in creating wealth and fostering sustainable economic growth, population and economic activity drifted away from mining localities when their ore was depleted. That was especially true of Chile's nitrates industry once it withered in the face of competition from synthetics. In most instances, attempts to create competitive successor industries in mining districts are likely to fail unless those industries can tap into some inherent comparative advantages of their own. The subsidization of such industries in order to maintain redundant mining settlements is not sustainable. It effectively dissipates the past fruits of mining in supporting uneconomic activities. Furthermore, it also brings into question attempts to develop downstream industries based on domestic mining in order to "add value", unless they have substantial local markets or can be based on large ore deposits with potentially long lives. All too often such attempts may be storing up problems for the future, and the future usually arrives sooner than expected. Costs tend to rise even before ore deposits are exhausted, thereby reducing the competitiveness of dependent downstream industries. Once local ore is depleted, the downstream industries will become reliant on imported raw materials if they are to remain in production. Economic development would be more sustainable if it were based on investing mineral rents in successor rather than complementary industries to mining.

Given that the main economic benefits of mining will accrue to host countries through tax receipts of all types, the extent to which mineral projects support economic development will depend mainly on institutional factors, just as they did in Chile during the nitrates era. There are few, if any, effective international sanctions on sovereign governments' domestic actions, so that there is never any guarantee that mineral revenues will be used wisely even in established democracies. No matter how much multilateral institutions of all types may advise and attempt to influence host governments, weak institutions can rarely be changed overnight, even where change is accepted as desirable. Rent-seeking behaviour will always flourish where governments are weak or dominated by sectional interests. Whether it is correct to call such behaviour a curse, rather than a frequent consequence of access to wealth that has no or only limited accompanying obligations, is debatable. Even an imperfect mobilization of mineral wealth is usually preferable to the sterilization of mineral deposits by leaving them undeveloped. Again the rapid collapse of Chile's nitrates industry in the face of competition from synthetics provides a strong cautionary tale.

References

Alden D (1987) Late Colonial Brazil, 1750–1808. In Bethell L (ed) Colonial Brazil, Chapter 7 Cambridge University Press, Cambridge.

Barton DB (1978) A History of Copper Mining in Cornwall and Devon. Bradford Barton Ltd, Truro

Blainey G (2003) The Rush that Never Ended, 5th edn. Melbourne University Press, Melbourne

Bone A (2004) Conflict diamonds: The De Beers group and the Kimberley process, Chapter 11. In: Bailes AJK, Frommelt I (eds) Business and Security, Public-Private Sector Relationships in a New Security Environment. Oxford University Press, Oxford

Capistrano de Abreu J (1907, English edn 1997) Chapters of Brazil's Colonial History 1500–1800. Oxford University Press, Oxford

Collier P, Goderis B (2007a) Commodity Prices and Growth: Reconciling a Conundrum. Department of Economic, University of Oxford, Oxford

Collier P, Goderis B (2007b) Prospects for commodity exporters: Hunky Dory or Humpty Dumpty. World Economics 8(2), April–June 2007

Cornish Mining World Heritage Site (2007) Cornwall & Scilly Historic Environment Service. www.cornish-mining.org.uk consulted January 12th 2008

Crowson P (2003) Mine size and the structure of costs. Resources Policy 29: 15–36

Crowson P (2006) Comment. Natural resource-based economic development in history. World Economics 7(1), January–March 2006, Henley-on-Thames

Crowson P (2007) Adding Public Value: The Limits of Corporate Responsibility. Oxford Policy Institute, Oxford, April 2007. (http://www.opi.org)

Drake PW (1994) Chapter 1. The Historical Setting, Chile, Federal Research Division, Library of Congress, Washington, DC. (Library of Congress Call Number, F3058. C5223 1994)

Duffy M (1994) The Contribution of Mining to Australia's Development. International Council on Metals and the Environment, Ottawa

Edwards S, Edwards AC (1994) Chapter 3. The Economy, Chile, Federal Research Division, Library of Congress, Washington, DC. (Library of Congress Call Number, F3058. C5223 1994)

Encyclopedia of New Zealand (2007), Ministry of Culture and Heritage. www.teara.govt.nz/ EarthSeaAndSky/MineralResources/GoldAndGoldMining consulted August 9th, 2007

Kimberley Process: www.kimberleyprocess.com

Mansuy-Diniz Silva A (1987) Imperial re-organisation, 1750–1808. In Bethell L (ed) Colonial Brazil, Chapter 6. Cambridge University Press, Cambridge

Martín L (1974) The Kingdom of the Sun, A Short History of Peru. Charles Scribner's Sons, New York

McCann FD (1997) Gold Mining Replaces Cane Farming, in Chapter 1, Historical Setting, Brazil, Federal Research Division, Library of Congress, Washington, DC (Library of Congress Call Number F2508. B846 1998).

Mein Smith P (2005) A Concise History of New Zealand. Cambridge University Press, Cambridge

Monteon M (1982) Chile in the Nitrate Era, The Evolution of Economic Dependence, 1880–1930. University of Wisconsin Press, Madison, Wisc.

Mueller CC, Baer W (1997) Chapter 3, The Economy, Brazil, Federal Research Division, Library of Congress, Washington, DC. (Library of Congress Call Number F2508. B846 1998).

O'Brien TF (1982) The Nitrate Industry and Chile's Crucial Transition: 1870–1891. New York University Press, New York

Office for National Statistics (2001). 200 Years of the Census in Cornwall and the Isles of Scilly. ONS, Hampshire, UK

Russell-Wood AJR (1987) The gold cycle c. 1690–1750. In Bethell L (ed) Colonial Brazil, Chapter 5. Cambridge University Press, Cambridge

Sachs JD, Warner AM (1995) Natural Resource Abundance and Economic Growth. Harvard Institute for International Development, Development Discussion Paper No 517, October 1995, and NBER working paper No. 5398, December 1995

Sachs JD, Warner AM (1997a) Fundamental Sources of Long-Run Growth. American Economic Review, Papers and Proceedings, May 1997

Sachs JD, Warner AM (1997b) Sources of slow growth in African economies. Journal of African Economies 6: 335–376

Scott T (2007) The impact of gold on Australia. Victorian Cultural Collaboration. www.sbs.com.au/ gold/story.html consulted August 9th, 2007

Schmitz CJ (1979) World Non-Ferrous Metal Production and Prices 1700–1976. Frank Cass & Co Ltd, London

Stevens P (2007a) Resource impact: Curse or blessing? A literature survey. CEPMLP Internet Journal 13, Article 14. (www.dundee.ac.uk/cepmlp/journal/html/Vol13/Vol 13-14.html)

Stevens P (2007b) Resource impact: Curse or blessing? CEPMLP Internet Journal 14, Article 1. (www.dundee.ac.uk/cepmlp/journal/html/Vol13/Vol 14_1.html)

Transparency International (2008) Corruption Perceptions Index, http://www.transparency.org/ surveys/index.html/cpi

United Nations (2007) United Nations Common Database, June 2007, ESDS International, (MIMAS) University of Manchester

University of Portsmouth, Department of Geography, Great Britain Historical GIS Project. www.visionofbritain.org.uk/data, consulted January 12th 2008

Vyas K (2007) Gold in Victoria. Victorian Cultural Collaboration www.sbs.com.au/gold/story.html consulted August 9th, 2007

World Bank (2002) Treasure or Trouble? Mining in Developing Countries, World Bank, Washington, DC.

World Bank (2007) World Development Indicators (WDI), April 2007, ESDS International, (MIMAS), University of Manchester, Manchester

Extractive Economies, Growth, and the Poor

Graham A. Davis

Abstract There is mixed evidence regarding the relationship between the extractive intensity of economic activity and the level of human development. Some studies find that mineral- and energy-intensive economies have higher levels of development than economies without a substantial extractive sector, whereas others find that they have lower development levels. Those that find the negative relationship commonly infer that there are ongoing, dynamic effects at work, such as an erosive effect of resource wealth on institutional quality, or a structural shift that reduces manufacturing employment and, with this, manufacturing's special development-enhancing impacts. These studies also infer that the lower development levels equate to negative impacts on the poor. Even in growing extractive economies, the poor are thought to be made worse off as a result of the extractive activities. None of these studies, however, specifically examines how growth spells in extractive economies affect the poor in those economies. This chapter examines this relationship via a simple comparison of a series of growth spells in extractive and non-extractive economies. It shows that the poor in growing extractive economies are as likely or more likely to benefit from that growth than are the poor in growing non-extractive economies. Thus, there is no evidence that positive growth in extractive economies is any worse for the poor than positive growth in non-extractive economies. What hurts the poor in any economy is negative growth, and resource extraction is bad for the poor only if it increases, for whatever reason, the frequency of negative growth in an economy.

1 The Resource Curse Conundrum

There is mixed evidence regarding the relationship between the intensity of non-renewable resource extraction in an economy and the level of human development

G.A. Davis (✉)

Division of Economics and Business, Colorado School of Mines, Golden, CO 80401, USA
e-mail: gdavis@mines.edu

J.P. Richards (ed.), *Mining, Society, and a Sustainable World*,
DOI 10.1007/978-3-642-01103-0_2, © Springer-Verlag Berlin Heidelberg 2009

in that economy.[1] Davis (1995) found that extractive economies as a whole have higher levels of development than economies without a substantial extractive sector, whereas others have found that extractive economies as a whole have lower levels of development (e.g., Bulte et al. 2005). None of these studies, however, directly examines how the boom–bust cycle of extractive activity affects the poor. Addressing the claims of the resource pessimists, this chapter compares the quality of a series of historical growth spells in extractive and non-extractive economies, with the goal of determining whether extractive economies experience a type of growth that is particularly bad for the poor.

A brief history of the debate over the development performance of the extractive economies provides context for this research. There are two general methods by which extractive economies have been evaluated: their *level* of human development, poverty, or per capita gross domestic product (GDP) relative to the level of that same indicator in a group of peer economies; and their *growth* in human development, poverty, or per capita GDP relative to growth of that same indicator in a group of peer economies. Via a series of case studies, geographer Richard Auty and economist Alan Gelb were among the first to raise modern concerns that there is a "Resource Curse",[2] whereby selected extractive economies appeared to suffer from inferior GDP growth as a result of their mineral and energy endowments (Gelb and Associates 1988; Auty 1993). The fear was that this inferior economic growth was leading to inferior levels of development, even to the extent that these economies were being made worse off as a result of their extractive activities. However, Davis (1995) investigated the level of development of the extractive developing economies as a group, and found them to have higher income levels and better human development indicators than non-extractive developing economies. The extractive economies also appeared to be pulling away from the other developing economies in terms of development performance over the 1970–1991 sample period.

Subsequent empirical studies of resource extraction and development performance have been equivocal. In a report for OXFAM America, UCLA political scientist Michael Ross found that extractive economics underperform in terms of broad human development indicators, and in several institutional and political metrics (Ross 2001). He also found that these countries had higher rates of poverty than their non-extractive peers. Though his data and statistical analysis have been challenged as being unreliable (Davis and Tilton 2002), a follow-up study in 2003 reiterated these findings (Ross 2003). Ross concluded in these studies that extractive activity

---

[1]Economies with substantial non-renewable mineral and energy resource extraction activity will henceforth, in the interest of brevity, be referred to as extractive economies. Others use the term extractive economies to denote economies devoted to primary production inclusive of agricultural activities. That is not the intended use here.

[2]The "Resource Curse" is a term employed to represent below-par economic performance in extractive economies. The main implications of a Resource Curse are lower than expected real per capita economic growth and increased levels of poverty in these economies. See Davis (1995) and Davis and Tilton (2005) for an extended discussion of these issues.

is bad for the poor. A contemporaneous study of the 49 Least Developed Countries (LDCs) by the United Nations Conference on Trade and Development, asserted that dependence on mineral production was responsible for the large and rising levels of extreme poverty in mineral-exporting LDCs such as the Central African Republic, the Democratic Republic of the Congo (formerly Zaïre), Guinea, Liberia, Niger, Sierra Leone, and Zambia (UNCTAD 2002). The mechanism here was thought to be these countries' slower rates of export growth, and hence slower overall economic growth, due to armed conflict and rent seeking that distracts the government from undertaking appropriate macroeconomic planning. Bulte et al. (2005) also found that extractive economies tended to suffer from lower levels of human development relative to a set of non-extractive peer economies. The World Bank (McMahon and Remy 2001) and the International Council on Mining and Metals (ICMM 2006) have countered these findings by publishing a series of case studies outlining successful development progress in mining economies.

While all of this was going on, an influential and comprehensive set of econometric studies by Jeffrey Sachs and Andrew Warner determined that higher levels of resource extraction in 1970 were indeed associated with lower than expected subsequent rates of economic growth, confirming Auty and Gelb's earlier case-study work (Sachs and Warner 1995, 1997, 2001).[3] Stijns (2005), Ding and Field (2005), and Brunnschweiler and Bulte (2008a, b), on the other hand, found that under different measures of natural resource abundance, resource endowments have either no effect or a positive effect on economic growth. Exemplifying the resource measurement problem, Sala-i-Martin (1997a, b) found that the fraction of primary products in total exports, a measure of overall extractive and agricultural intensiveness, is negatively correlated with economic growth, whereas the fraction of GDP in mining and energy extraction, a measure of purely extractive activity, is positively correlated with economic growth. In a later paper Sala-i-Martin et al. (2004) again found that the fraction of GDP in mining and energy extraction is positively correlated with growth, but that the fraction of primary products in total exports is not correlated with growth. In this later paper Sala-i-Martin et al. (2004) used a slightly different statistical technique. To confuse matters even more, in an IMF interview (IMF 2004), Sala-i-Martin summarized the empirical evidence as showing that minerals and energy extraction *depress* growth, ignoring his repeated empirical findings that minerals and energy extraction are positively correlated with growth. Sala-i-Martin is not the only academic to ignore his own research findings; in a radio interview in 2003 (NPR 2003) Jeffrey Sachs suggested that Iraq needed to increase its oil output, failing to note the Sachs and Warner (1995, 1997, 2001) evidence that this will slow Iraq's long-term growth. Jeffrey Sachs' latest position is that while the evidence points to extractive economies having a slower rate of growth than they should be realizing given their vast resources, the oil-rich countries, at least, tend to

---

[3] Actually, Sachs and Warner examined growth in labor productivity (growth in GPD per worker), which is closely related to what is commonly referred to as economic growth (growth in GDP per capita).

have higher life expectancies, lower child mortality rates, higher electricity use per capita, and more paved roads than oil-poor countries. "In many categories of well-being... oil producers are better off than their oil-poor counterparts" (Sachs 2007, p. 173).

Despite the state of confusion as to what the empirical evidence does or does not show, both with regard to the level of development and the rate of economic growth associated with extractive activity, the idea of a broad Resource Curse has become fixed in most people's minds. Extraction has been cast as a "loser" sector by political economists (see the discussion in Davis 1998), and oil is being blamed for poverty and unrest in the Middle East and North Africa (Surowiecki 2001). Ross (2001, p. 16) concluded that "Oil and mineral dependence produce a type of economic growth that offers few direct benefits to the poor; moreover, oil and mineral dependence make pro-poor forms of growth more difficult, due to the Dutch Disease."[4] The evidence is apparently so compelling that radical policy adjustments are suggested: "We believe the best course of action for poor states would be to avoid export-oriented extractive industries altogether, and instead work to sustainably develop their agricultural and manufacturing sector—sectors that tend to produce direct benefits for the poor, and more balanced forms of growth" (Ross 2001, p. 17). Others have joined the fray: (Christian Aid 2003, p. 4) asserted that "Oil has not only failed to bring benefits to the world's poor communities—it has been decisive in making them poorer." These and other statements had enough traction that the World Bank began questioning its promotion of mineral and energy projects as a path to poverty reduction, and initiated an external Extractive Industries Review in 2001 (World Bank 2003).[5] Along with a series of recommendations as to how the World Bank could better manage its extractive projects to reduce poverty, the Review suggested that it stop funding oil and coal mining projects. Michael Ross served on the Review Advisory Committee. The Bank subsequently decided not to follow this recommendation.

The latest assault on extractive activity, and in particular the multinational companies involved in extraction, is being made by Nobel Prize winner Joseph Stiglitz (2007). Stiglitz interprets the research as providing unequivocal evidence that extractive economies have higher rates of poverty due to their extractive activity.

If we are to truly understand whether extractive activity increases or decreases the level of poverty, the topic of interest to this chapter, two methodological adjustments need to be made. First, when investigating levels, there must be a distinction between the level of human development and the level of poverty. Ross (2001, 2003) and Bulte et al. (2005) used the United Nations' Human Development Index (HDI)

---

[4]The Dutch Disease refers to the macroeconomic adjustments that occur in an open economy during a resource boom. These adjustments include an expanding extractive sector, shrinking manufacturing and agricultural sectors, and an appreciating exchange rate.

[5]The report can be found (as of July 23, 2008) at http://web.worldbank.org/WBSITE/EXTERNAL/TOPICS/EXTOGMC/0,,contentMDK:20306686~menuPK336936~pagePK:148956~piPK:216618~theSitePK:336930,00.html

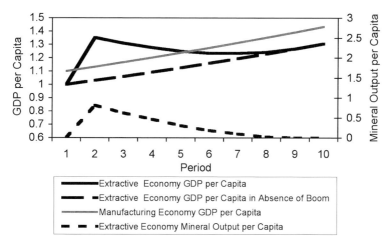

Fig. 1 Real GDP per capita (currency units per capita) and mineral output per capita (weight or volume units per capita) for a poor extractive economy, and real GDP per capita (currency units per capita) for a wealthy manufacturing economy, over 10 periods. The resource boom begins in period 2 and ends in period 9. The mineral output profile is a result of a dynamic optimization of a fixed stock of resources, based on a model by Boyce and Emery (2007)

and other human development indicators in their assessment of extractive economy performance. The HDI includes measures of longevity, educational attainment, and standard of living. Ross contended that the HDI and other development indicators provide an assessment of the condition of the poor. Human development indicators, however, are not a measure of poverty (UNDP 1997, pp. 22–23). This is why the United Nations created the Human Poverty Index (HPI) in 1997 for a group of 78 developing countries. In 1998, it created a parallel Human Poverty Index (HPI-2) for selected industrial countries, noting that there is "no pattern between the HDI and human poverty" (UNDP 1998, p. 29). Thus, whatever the relationships between mineral or oil dependence and development performance, it tells us little about the relationship between extractive activity and the welfare of the poor.

Second, when investigating trends rather than levels, which is what most of the research does, it is useful to separate out impacts of extractive activity on the rate of economic growth from its impacts on the poverty-reducing quality of that growth. Extractive economies typically go through a boom–bust cycle over the life of the extractive activity, wherein there is initially rapid economic growth as the resource begins to be exploited, followed by a drop off and even negative growth as the resource is exhausted (Sachs and Warner 1995; Rodriguez and Sachs 1999; Bravo-Ortega and de Gregorio 2007). To illustrate this, Fig. 1 presents a comparison of the growth paths of a stylized developed manufacturing economy, and a slightly less productive but otherwise identical developing economy that discovers resources and extracts them until the are exhausted. The graphic represents a macroeconomic model of an optimally managed extractive economy, with no rent seeking or other ill effects of the resource boom (Boyce and Emery 2007). Because of the mechanics

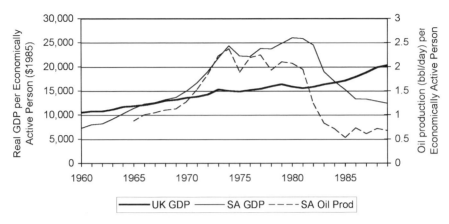

Fig. 2 Annual real GDP (1985 PPP dollars) per economically active person for Saudi Arabia (SA GDP) and the United Kingdom (UK GDP) from 1960 to 1989; daily oil production per economically active person for Saudi Arabia (SA Oil Prod) is shown from 1965 to 1989

of national income accounting, the boom–bust profile of GDP per capita follows the boom–bust profile of extractive output per capita. The extractive economy's real rate of growth during the resource boom is initially higher than that of the richer manufacturing economy, and then becomes negative as extractive output slows. Note that during part of the resource boom the extractive economy's level of income per capita is higher than that of the manufacturing economy.

Figure 2 depicts the actual oil boom–bust cycle in Saudi Arabia, a developing economy, from 1960 to 1989. Saudi Arabia's economic growth was at first positive as per capita oil output grew, and then turned negative after 1981 as per capita oil output fell precipitously. By comparison, the UK, a developed manufacturing economy, grew at a relatively constant rate. Interestingly, previous analyses of Saudi Arabia's faltering economic growth since the 1970s have looked to profligate government spending as the cause (Karl 1997; Stiglitz 2007), with no mention of the natural boom–bust cycle of oil production. That is, researchers have been viewing Fig. 2 without the benefit of seeing the oil output profile superimposed on the graphic.

It is widely accepted that positive growth is good for the poor, whereas negative growth is not (Dollar and Kraay 2002; Kraay 2006). Poverty should therefore decrease during the initial phase of the extractive boom, and then increase as extraction wanes and economic growth turns negative. UNCTAD (2002) showed that LDCs that were mineral economics saw increasing levels of poverty during the bust cycle from 1990–1999, while non-extractive economies saw decreasing levels of poverty as their economies grew. The decline in extractive output per capita depicted in Fig. 2 hit more than just Saudi Arabia in the 1980s (and 1990s)—the per capita extractive output of the other mineral- and energy-intensive economies also fell (Manzano and Rigobón 2007). Given the tight relationship between extractive output and GDP (Fig. 1), this would seem to indicate that the "problem" for the poor in the extractive economies is not deleterious political and social outcomes of resource extraction, but rather the failure of extractive output to keep up with population growth.

Ross (2001, 2003), on the other hand, argued that the growth that occurs in extractive economies is not good for the poor, no matter whether that growth is positive or negative. This distinction between the pro-poor quality of growth and rate of growth in extractive economies is thus an important consideration when trying to understand the Resource Curse. Ross himself has named as a research priority studies that track the distribution of income over time in countries undergoing a mineral boom (Ross 2007, p. 251). Is it the case that poverty persists or worsens even in a growing extractive economy? Botswana, the only country ever to graduate from LDC status, is commonly presented as an example of where extraction-led growth has given rise to poverty reduction (e.g., Acemoglu et al. 2003), but this may be the exceptional case. In the next section, I begin an introductory foray into this type of analysis by comparing the quality of growth spells across a large set of developing and developed economies, both extractive and non-extractive, over a 43-year time period. My emphasis is not on human development, but on the poor, and on how growth in extractive economies may or may not be as beneficial to the poor as growth in non-extractive economies.

2 Measuring the Quality of Extractive-Economy Growth

This section analyzes the poverty-reducing quality of extractive-economy growth using World Bank data. These data track income growth by income quintile across 67 non-extractive economies and 21 extractive economies from 1956 to 1999.[6] The World Bank data are based on periodic individual or household surveys of income or consumption, collected at the national or subnational level. From these surveys economy-wide mean income level by income quintile is computed. Repeated sampling allows the change in mean income level to be tracked over time for each income quintile. Prior to computing these changes, the income data were converted to real international dollars using 1985 PPP (purchasing power parity) exchange rates.

Each intertemporal income comparison within an economy is called a growth spell. There are 240 growth spells in the sample. Some economies in the sample have multiple recorded growth spells, whereas others only have one. Because of the

---

[6]The 21 extractive economies in the sample, based on the ratio of mineral exports to GDP in 1995 (Ross 2001), are Sierra Leone (1968–1989), Zambia (1959–1996), Mauritania (1988–1993), Niger (1960–1992), Chile (1968–1992), Jordan (1980–1997), Bolivia (1968–1990), Peru (1961–1994), Ghana (1987–1997), Bulgaria (1984–1989), Norway (1962–1989), Australia (1967–1986), Yemen (1992–1998), Nigeria (1959–1997), Gabon (1960–1995), Algeria (1988–1995), Venezuela (1962–1993), Ecuador (1968–1994), Malaysia (1970–1995), Indonesia (1976–1999), and Cote d'Ivoire (1985–1993). The period over which each country's growth spells were measured is given in parentheses.

irregular spacing of the income surveys for each country, the growth spells range from 5 to 37 years, and average 7 years. The earliest growth spell starts in 1956 and the latest ends in 1999, which is the period over which most of the extraction-based economies rapidly expanded their extractive activity, providing a relevant set of data for the analysis at hand.

Economic growth can have two effects on the poor: an income level effect, and an income distribution effect. In measuring income level effects, it is often necessary to first define the poverty level (e.g., $0.50/day, $1.00/day, $2.00/day, etc.), and then to define how the quantity of poverty is measured with respect to that level (Watt's index, headcount, poverty gap, squared poverty gap, etc.). Different poverty levels and poverty measures lead to different results, commonly creating ambiguity as to whether poverty has risen or fallen as a result of a growth spell. To make this analysis as robust as possible, I have adopted the requirement that a growth spell be deemed good for the poor only if poverty is measured to drop regardless of the poverty measure used (Watt's index, headcount, poverty gap, squared poverty gap, etc.), and regardless of where the poverty line is drawn ($0.50/day, $1.00/day, $2.00/day, etc.); the improvement in poverty must be unambiguous.

Poverty is also reflected in non-monetary ways, such as health, which may depend on relative rather than absolute income levels (Deaton 2003). This is the income distribution effect. The "Kakwani" definition of pro-poor growth takes this second facet of poverty measurement into account, requiring that a *pro-poor* growth event reduce income poverty (an *anti-poverty* effect) and income inequality (an *anti-inequality* effect) at the same time (Kraay 2006). Along the same lines, I define "unambiguous pro-poor growth" as growth, be it positive or negative, that reduces income poverty no matter how income poverty is measured (i.e., the growth is unambiguously anti-poverty), while at the same time reducing income inequality across the entire population (i.e., the growth is also unambiguously anti-inequality). Requiring that the growth event be unambiguously pro-poor avoids the criticism that the results are conditional on the poverty measure or cut-off used, or that they ignore income distribution effects.

Not all growth is unambiguously pro-poor: it can worsen income inequality across the entire population, which is *pro-inequality* growth; it can unambiguously increase poverty, which is *pro-poverty* growth; or it can have an ambiguous effect on income and inequality, where the categorization would depend on the poverty measure and poverty level used. The labels for the types of growth spells are given in Table 1. *Pro-poor positive growth* and *pro-poor negative growth* are the unambiguously desirable types of growth (from the perspective of the poor), whereas *immiserizing positive growth* (positive growth that results in the poor being made unambiguously worse off) and *not pro-poor negative growth* (negative growth that results in the poor being made unambiguously worse off) are the undesirable types of growth. The other categorizations (not pro-poor positive growth, inconclusive positive growth, or inconclusive negative growth) can have both positive and negative effects on the poor, and are therefore not of interest here because they do not help us to resolve unambiguously the quality of an extractive economy growth spell.

Table 1 Qualities of economic growth under positive and negative growth spells

Positive growth spell quality	
Pro-poor	Income of the poor unambiguously increases, inequality unambiguously decreases
Not pro-poor	Income of the poor unambiguously increases, inequality unambiguously increases
Immiserizing	Income of the poor unambiguously decreases, inequality unambiguously increases
Inconclusive	Categorization depends on chosen poverty level
Negative growth spell quality	
Pro-poor	Income of the poor unambiguously increases, inequality unambiguously decreases
Not pro-poor[a]	Income of the poor unambiguously decreases, inequality unambiguously increases
Inconclusive	Income of the poor unambiguously decreases and inequality unambiguously decreases, or categorization depends on chosen poverty level

[a]Equivalent to immiserizing in the positive growth spell case.

Ross's (2001, 2003) charge, perhaps the most pointed of those who decry extraction-based development, is that extractive economies suffer from a type of growth that is bad for the poor. Certainly, if extractive economies enjoy fewer positive growth spells as a result of the Resource Curse, there will likely be fewer pro-poor growth events. But the question at hand is rather: are the growth spells that extractive economies experience, be they positive or negative, any worse for the poor than those experienced by non-extractive economies? Or, using my growth spell terminology, are there relatively fewer cases of unambiguously pro-poor positive growth, and relatively more cases of unambiguously immiserizing positive growth, in the extractive economies? There are certainly reasons to think that this may be the case. Mineral and energy booms have been thought to crowd out jobs for unskilled or semi-skilled workers (Ross 2003, p. 7), to reduce manufacturing jobs that favor women or older workers (Ross 2007) and result in less income inequality (Leamer et al. 1999), and to reduce agricultural sector jobs, a sector that is suggested to have special importance in reducing income poverty in a growing economy (Ross 2003, p. 8, 2007; World Bank 2008).

On the other hand, extractive activity decreases unemployment and provides economic stimuli in rural areas (Wallace 1999; McMahon and Remy 2001). Evidence from Cote d'Ivoire shows that rural incomes increase with decreased distance to paved roads and public markets (Klugman 2002, p. 45). This is infrastructure that is associated with extractive activity. Extractive sector booms may also lead to new government jobs through mineral and energy revenues. There is some evidence that countries with a larger government labor force have lower income inequality (Ross 2007). Income diversification within traditionally agricultural households has also been found to be important for avoiding the poverty trap (Ellis and Allison 2004), and the rural nature of mining would facilitate this, either directly or through remittances. If there is enough extractive activity within an economy, these local effects are likely to compound to such an extent that they can be measured as pro-poor growth at the national level.

2.1 Empirical Results

Table 2 tabulates the quality of the 240 growth spells in the data sample according to type of economy—extractive or non-extractive—as defined by (Ross 2001). Although I do not necessarily agree with Ross's typology, I use it because it was through this typology that he determined that extractive-economy growth was bad for the poor. The tabulations are based on an analysis of each growth spell using Son's "poverty growth curve" method (Son 2004). To implement this method, the World Bank income data are converted into a poverty growth curve, which represents rate of growth of mean income by ascending cumulative income quintile, $g(p)$. For example, $g(20)$ represents the rate of growth of the mean income of the poorest 20% of the population in an economy, $g(40)$ represents the rate of growth of the mean income of the poorest 40% of the population, and so on. Economy-wide income growth, which is what national income accounts measure and which is often referred to as growth in GDP per capita, is represented by $g(100)$. The patterns in the growth in income across the various income quintiles allow each growth spell to be tabulated according to the categories listed in Table 1.[7]

The extractive economy cohort contains 48 growth spells, 35 with positive growth and 13 with negative growth, whereas 192 growth spells were analyzed for non-extractive economies, 162 with positive growth and 30 with negative growth. These spells therefore represent both booms and busts in both types of economy.

Focusing on the positive growth spells, 46% (16 of 35) of these events in the extractive economies were unambiguously pro-poor, whereas only 37% (60 of 162) were unambiguously pro-poor in the non-extractive economies. Positive growth spells in the extractive economies therefore appear to be more likely to be unambiguously pro-poor than the positive growth spells in the non-extractive economies. The chance that a positive growth spell is unambiguously immiserizing is the same in both types of economy, at roughly 5%. The other growth spells cannot be conclusively categorized as good or bad for the poor.

Based on these data there is no evidence that positive growth in an extractive economy, when it happens, tends to be bad for the poor. As with findings in previous research, growth is generally good for the poor, and there is no evidence that it is less frequently good for the poor in extractive economies than in non-extractive economies.

---

[7]For more on this growth spell categorization method, see (Son 2004) and (Davis 2007). Davis provides graphical depictions of three sample poverty growth curves. I would like to thank Hyun Son for providing me with her poverty growth curve data.

Table 2 The pro-poorness of economic growth in extractive and non-extractive economies

	Positive growth spells		Negative growth spells		Total
	Count (%)		Count (%)		
	Extractive economies	Non-extractive economies	Extractive economies	Non-extractive economies	
Pro-poor	16 (46)	60 (37)	0 (0)	3 (10)	79
Not pro-poor	5 (14)	33 (20)	n/a	n/a	38
Immiserizing[a]	2 (6)	6 (4)	7 (54)	16 (53)	31
Inconclusive	12 (33)	63 (39)	6 (46)	11 (37)	92
Total	35 (100)	162 (100)	13 (100)	30 (100)	240

[a]Defined as "not pro-poor" in the case of negative growth, because "immiserizing" growth is a term traditionally restricted to positive growth spells.

Table 3 Summary of growth and poverty in extractive and non-extractive economies

	Positive growth spells		Negative growth spells		Total
	Count (%)		Count (%)		
	Extractive economies	Non-extractive economies	Extractive economies	Non-extractive economies	
Anti-poverty	27 (77)	126 (78)	0 (0)	3 (10)	156
Pro-poverty	2 (6)	6 (4)	12 (92)	25 (83)	45
Inconclusive	6 (17)	30 (19)	1 (8)	2 (7)	39
Total	35 (100)	162 (100)	13 (100)	30 (100)	240

What of the negative growth spells? Are these worse for the poor in extractive economies than for the poor in non-extractive economies? Though the sample set is small, the negative growth spells suffered by the extractive economies are unambiguously not pro-poor 54% of the time, about the same frequency as for the non-extractive economies (53%; Table 2). Thus, as expected intuitively, negative growth is rarely pro-poor in either type of economy.

This finding of a higher frequency of positive growth spells that are pro-poor in extractive economies than in non-extractive economies has focused on a measure that combines poverty and income inequality effects. However, it is also useful to look separately at poverty and income inequality (Tables 3 and 4). Kraay (2006), who included a number of extractive economies in his growth spell sample, found that there were no cross-country differences in the impact of growth on the level of income poverty. In this study, the conclusion is similar when comparing the frequency of unambiguous poverty improvements in extractive and non-extractive economies. Positive growth spells were unambiguously anti-poverty in 77% of the episodes in extractive economies, and in 78% of the episodes in non-extractive economies (Table 3). Some 6% of the positive growth spells were unambiguously pro-poverty in the extractive economies, compared with 4% in the non-extractive

Table 4 Summary of growth and inequality in extractive and non-extractive economies

	Positive growth spells		Negative growth spells		Total
	Count (%)		Count (%)		
	Extractive economies	Non-extractive economies	Extractive economies	Non-extractive economies	
Anti-inequality	16 (46)	60 (37)	3 (23)	8 (27)	87
Pro-inequality	9 (26)	59 (36)	7 (54)	16 (53)	91
Inconclusive	10 (29)	43 (27)	3 (23)	6 (20)	62
Total	35 (100)	162 (100)	13 (100)	30 (100)	240

economies. In a sample that is probably too small to have much reliability, negative growth spells were unambiguously pro-poverty in 92% of the cases for extractive economies, and in 83% of the cases for non-extractive economies. The other results are inconclusive, meaning that whether or not the growth spell was good for the poor would depend on the poverty measure used and the income level by which poverty is defined.

The other aspect of growth spells is the concomitant change in relative income, or income inequality, within a country. Kraay (2006) found that different countries have different relative income effects associated with growth spells, and that there is no evident explanation as to why growth spells in different countries have these different relative income effects. According to my data, positive growth spells in the extractive economies have an increased tendency to be unambiguously anti-inequality (46% of cases) than in non-extractive economies (37% of cases; Table 4).[8] They are unambiguously pro-inequality in 26% of the cases, whereas non-extractive economy positive growth spells are unambiguously pro-inequality in 36% of the cases. There is parity between the extractive economies and non-extractive economies in the negative growth spells, though the sample is too small to carry much weight. The other results are inconclusive, meaning that income inequality is not reduced or increased across the entire population profile, a requirement for the growth spell to be classified as unambiguously anti-inequality or pro-inequality.

The conclusion from this empirical analysis is that when extractive economies experience positive growth, there is no prima facie evidence that the growth is any worse in its impact on the poor than growth in non-extractive economies. If anything, positive growth in an extractive economy appears to have a greater chance of being pro-poor than positive growth in a non-extractive economy, due to the former's favorable impacts on income distribution. This finding is especially signifi-

---

[8]All anti-inequality positive growth spells were also anti-poverty growth spells, and hence the number of anti-inequality positive growth spells in Table 4 matches the number of pro-poor growth spells in Table 2.

Table 5 Simple correlation between rate of growth in a growth spell and the quality of that growth spell

Growth spell quality	All growth spells	Positive growth spells	Negative growth spells
Pro-poor	0.22 ($n = 79$)	0.00 ($n = 76$)	0.27 ($n = 3$)
Immiserizing[a]	−0.55 ($n = 31$)	−0.21 ($n = 8$)	−0.31 ($n = 23$)

[a] "Not pro-poor" in the case of negative growth.

cant given that the growth spells in the dataset are measured over the period when many of the 21 extractive economies in the sample were intensifying their extractive activity, and given the frequent claims that extraction-based growth leads to worsening income distribution (e.g. Stiglitz 2007).

In the light of the evidence that extractive economies may suffer from a decreased or even negative rate of growth (the Resource Curse), I now look at whether the level of growth within a positive or negative growth spell makes a difference. Is slower positive growth worse for the poor than faster positive growth? Is more negative growth worse for the poor than less negative growth? Table 5 reports the results of a calculation of the simple correlations between the rate of growth in a growth spell, measured on a continuous absolute scale, and growth spell quality across the 240 growth spells in the sample. In performing this calculation, growth spell quality is identified as the binary 1,0 (pro-poor, other) when categorizing pro-poor growth spells, and the binary 1,0 (immiserizing, other) when categorizing immiserizing growth spells. In the first column we see that, amongst all growth spells, the higher the rate of growth the higher the probability of that growth spell being unambiguously pro-poor and the lower the probability that the spell is unambiguously immiserizing. However, when the sample is divided into separate positive and negative growth spell cohorts, the correlation coefficient values indicate that a higher rate of positive growth amongst the positive growth cohort reduces the probability of that growth spell being unambiguously immiserizing, but it has no effect on the chance that the growth spell will be unambiguously pro-poor. The latter effect comes about because of inequality effects, where very high rates of positive growth do not necessarily reduce income inequality, which is half of the pro-poor criterion. Higher growth rates amongst the negative growth cohort (negative growth rates that are closer to zero) are better for the poor in that they increase the probability of the growth event being unambiguously pro-poor and decrease the chance that it is unambiguously not pro-poor.

Summarizing these findings, faster positive growth is beneficial in its reduced chance that the growth is unambiguously bad for the poor, and a lower absolute rate of negative growth is beneficial in its improved chance that it will be unambiguously good for the poor and its reduced chance that it will be unambiguously bad for the poor.

As shown in Tables 2, 3, and 4, a negative growth spell has a lower probability of being unambiguously anti-poverty, anti-inequality, or, combining the two, pro-poor, than a positive growth spell. It also has a higher probability of being unambiguously

pro-poverty, pro-inequality, or, combining the two, not pro-poor. Given the low frequency of immiserizing positive growth, the main implication from this analysis is
that positive growth is good for the poor, even in extractive economies. The most
important way to avoid bad growth outcomes for the poor in extractive economies
is to avoid negative growth, or at least have a low absolute rate of negative growth.

2.2 Development in Indonesia, Nigeria, and Zambia

This section uses the growth spell categorization methods of this chapter to assess
the often contrasted development experiences of Indonesia, Nigeria, and Zambia,
three extractive economies. Table 6 lists the measured growth spells and growth
spell qualities for these countries. Indonesia is an economy that has been heralded as
escaping the Resource Curse through effective government policy aimed at pro-poor
initiatives (Bevan et al.1999; Ross 2007). In this analysis all of Indonesia's four measured growth spells between 1976 and 1999 were unambiguously anti-poverty, and
two were also unambiguously anti-inequality (making them unambiguously pro-
poor overall). Perhaps these favorable outcomes are a result of exceptional development programming. Yet in the light of the preceding analysis it cannot be ignored
that all four of these growth spells were positive, some with very high rates of economic growth.

On the other hand, Nigeria, which is often mentioned as epitomizing the
Resource Curse (Ross 2007), had three measured growth spells from 1959 to 1997.
The first growth spell, measured over a 26 year period, was positive, and was unambiguously anti-poverty: real incomes increased across the entire income distribution. The next two growth spells, each lasting 6 years, were negative, and both were
unambiguously pro-poverty. The first of these spells was also unambiguously pro-
inequality, making it unambiguously not pro-poor overall.

Zambia, similarly, had a 17 year positive growth spell, with no unambiguously
negative impacts on the poor, followed by a 15 year negative growth spell, then
another 5 year negative growth spell. The first negative growth spell was unambiguously pro-poverty but unambiguously anti-inequality (poverty worsened but income
inequality improved), whereas the second was unambiguously not pro-poor (poverty
worsened and income inequality worsened). Clearly, the comparative fate of the
poor in these three economies is strongly conditioned on whether the economy was
experiencing positive or negative growth. Positive growth is beneficial to the poor
and negative growth is harmful to the poor, no matter what the structure of the economy.

Finally, it is of interest to dig a little deeper into what might have caused the
varying growth performance of these three economies. Traditional Resource Curse
proponents argue that extractive economies have slower or negative growth due to
poor macroeconomic management. As Figs. 1 and 2 show, however, the boom–bust
nature of extraction also plays a large part in a country's growth profile. Figure 3
maps the growth spells presented in Table 6 onto a graph of the annual mineral

Table 6 Quality of growth spells for Indonesia, Nigeria, and Zambia

Country	Income or consumption survey year	Growth spell index	Annual growth in mean income level, 1985 PPP dollars, by cumulative income percentile					Length of growth spell in years; positive (+) or negative (−) growth	Quality of growth spell
			g(20)	g(40)	g(60)	g(80)	g(100)		
Indonesia	1976								
	1981	A	7.57	9.17	8.56	8.42	8.29	5(+)	Anti-poverty
	1987	B	2.91	2.69	2.55	2.47	2.34	6(+)	Pro-poor
	1993	C	5.77	4.46	4.72	4.67	4.39	6(+)	Pro-poor
	1999	D	1.12	1.01	0.81	0.55	0.59	6(+)	Anti-poverty
Nigeria	1959								
	1985	E	1.64	2.84	3.49	3.53	2.22	26(+)	Anti-poverty
	1991	F	−7.16	−4.38	−2.97	−1.7	−0.35	6(−)	Not pro-poor
	1997	G	−0.05	−1.99	−2.94	−3.73	−1.49	6(−)	Pro-poverty
Zambia	1959								
	1976	H	−1.77	−0.70	0.52	1.43	1.36	17(+)	Inconclusive
	1991	I	−0.73	−1.44	−1.96	−2.47	−3.46	15(−)	Pro-poverty, anti-inequality
	1996	J	−7.35	−5.71	−4.73	−3.79	−1.7	5(−)	Not pro-poor

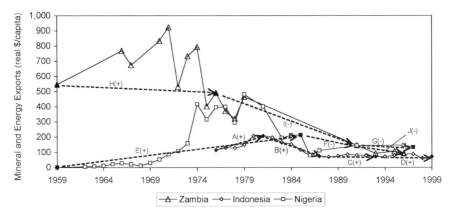

Fig. 3 Annual mineral and energy exports for Indonesia, Nigeria, and Zambia, 1959–1999, in real $US/capita (2000). Data are not available for some years and are interpolated in these years. The corresponding growth spells from Table 6 are outlined in heavy dashed lines and indexed by letters, with the sign in the brackets following the letter indicating whether the growth spell was positive or negative

and energy exports per capita, measured in real US dollars, in each of these three countries. Mineral and energy exports are used as a proxy for domestic production. This period of analysis shows Nigeria's oil boom–bust and Zambia's copper bust. Indonesia's per capita mineral and energy exports were relatively small and stable by comparison.

Indonesia enjoyed its first growth spell (A) on the back of a mild minerals and energy boom. Indonesia's other three spells (B, C, and D) saw mineral and energy exports decline and economic growth slow, yet in each spell economic growth was positive and either anti-poverty or pro-poor. Nigeria initially embarked on a similar positive growth path on the back of increasing oil exports (E), with resultant unambiguous reductions in poverty. But negative growth ensued in the next two growth spells (F and G) as per capita oil exports declined. As noted above, that negative growth was unambiguously unfavorable to the poor. Zambia had a massive per capita copper production collapse after great promise in the early 1960s (Baldwin 1966), with a 20 year period of negative economic growth (growth spells I and J) that was for the first 15 years pro-poverty, and for the last 5 years both pro-poverty and pro-inequality (not pro-poor).

In sum, the differences in development experience across these three extractive economies, as measured by the quality of their growth spells (Table 6), may reflect differing domestic policies or institutional quality aimed at improving the welfare of the poor. It is more likely that the different experiences of the poor across these three countries arise from Indonesia's sustained positive economic growth versus Nigeria's and Zambia's negative economic growth. In the case of Zambia and Nigeria it can be argued that the negative growth spells were a result of the declining per capita resource production and exports. What would Zambia's state of development be if it had not had a 30 year mineral bust? Nigeria's oil boom is credited

for derailing a Nigerian economy that was previously growing well, the oil revenues helping to fund massively inefficient public investment projects and allowing
unsustainable borrowing against future oil revenues (Bevan et al. 1999). Perhaps
the seeds of Nigeria's collapse were indeed sown long before 1979. Even so, this
research leaves little doubt that Nigeria's poor would now be better off if per capita
oil exports had increased enough from 1979 onwards to create sustained positive
economic growth.

3 Explaining Development Optimism and Pessimism for the Extractive Economies

With the proposition that extractive economies follow a boom–bust extraction cycle,
and that economic growth will also follow that cycle (Figs. 1 and 2), previous contra-
dictory analyses of the development performance of extractive economies can now
be seen in a new light. Davis (1995), for example, measured extractive activity as
being relatively good for development, whereas Ross (2003) and Bulte et al. (2005)
measured it as being relatively bad for development. All came to their conclusions
by looking at essentially the same data over the same time period, and all interpreted
resource intensiveness in essentially the same way. In fact, the results are not inconsistent. The differences in development experience come from the fact that Ross
(2003) and Bulte et al. (2005) assessed an economy's development performance
over time conditional on the initial level of income per capita, and in comparison
to the performance expected of economies at that level of income. By doing this
they created a peer group for comparison according to initial income per capita.
Because extractive economies are initially booming economies, with high levels of
GDP per capita as the extractive activities ramp up, the peer groups for the developing extractive economies in these analyses included many developed economies.[9]
Davis's (1995) peer group for developing extractive economies, on the other hand,
was other developing economies.

To understand the impact of this conditioning on the analysis, consider three
stereotypical economies in a sample set, AG (an agricultural economy), MAN (a
manufacturing economy), and MIN (a mining economy). AG is a poor developing
economy that has a sustained low level of income per capita, poor initial development indicators, and is trundling along as an LDC over the period of analysis. In
an analysis conditioned on initial income per capita, AG would be found to be performing at the same level as other LDCs (its income per capita peer group), and no
red flags would be raised. MAN is a developed economy with a high initial level
of income, good initial development indicators, and a stable level of real growth.
In an analysis conditioned on initial income per capita, MAN is also performing at

---

[9]Bulte et al. (2005) also controlled for the relative price of investment goods in an economy, and
so, strictly speaking, they are forming a peer group in both measures. Income per capita turns out
to be the more important variable to control for, so I emphasize it here.

the same level as its peer group (other manufacturing economies), and again no red flags are raised. MIN is a booming extractive economy at the start of the analysis period, with newly found oil or mineral wealth. It was in the AG peer group prior to its resource boom, but is now at an income level such that in an analysis conditioned on initial income it is being compared against MAN economies, even though it is still classified as a developing economy. Its subsequent growth is likely to be slower than MAN due to the unsustainability of its resource boom, and its development indicators are inferior to those in its peer group (developed MAN economies). A statistical analysis, however, shows that MIN differs from its peers only in terms of extractive activity. Hence, MIN is deemed to be underperforming, and the reason for that underperformance is extractive activity.

Figure 2 illustrates how Saudi Arabia would be found to be underperforming via this method of analysis. Note Saudi Arabia's ascent by 1970 into the developed economy peer group, matching the GDP per capita of the UK, on the back of increasing oil production. In Bulte et al.'s (2005) analysis, the UK is one of Saudi-Arabia's peer countries based on their similar PPP GDP per capita levels in 1970 (1970 being the date of peer group formation in these studies).[10] The development performance measurement is taken 21 years later. Saudi Arabia, with its 2001 HDI value of 0.769, was measured by Bulte et al. (2005) to be under-performing relative to the UK, which had a 2001 HDI value of 0.930. They ascribed the relatively poor development performance of extractive economies such as Saudi Arabia to weaker institutions compared to their peer group of developed economies, and blamed the weaker institutions on oil. This is no doubt correct, but unfair. A booming extractive economy's income per capita level is suddenly and artificially inflated by its extractive activity (Figs. 1 and 2). An exercise that then compares the booming economy's institutional capabilities and development levels during the boom (or even two decades later) with those of a nation like the UK, by design sets up the extractive economy to fail.

Other examples include: (1) Peru, which was found to underperform relative to Portugal, Peru's income peer in 1970 due to its booming economy. In this case, mining was blamed (Portugal had little extractive activity in 1970); (2) Iran, which under-performs relative to Japan; (3) Chile, which under-performs relative to Singapore; and so on. Botswana, on the other hand, were it included in the sample, would out-perform its 1970 income peer, Zambia, and would be deemed to do so because of its absence of extractive activity in 1970 compared with Zambia![11]

In another study that uses peer group formation conditioned on initial income, Ross (2001) compared development performance in 1998 across countries with similar incomes per capita in 1998. In this analysis, Kuwait's development performance was inferior to Canada's, Canada being Kuwait's peer because the two had the same

---

[10]Ross (2003) does not provide the list of countries in his data sample, and so I confine my comments in this and the next paragraph to Bulte et al.'s (2005) paper, which does list the countries in the sample.

[11]In 1970, Botswana had a primary production level roughly equivalent to that of Egypt.

1998 PPP GDP/capita; the United Arab Emirates under-performs compared to New Zealand; Botswana, now classified as a mineral economy because of its booming extractive sector, under-performs compared to Costa Rica; and so on.[12] Once again, extractive activity is identified as the common characteristic of the under-performing group. This method of analysis and the conclusions derived from it are clearly problematic.[13]

Davis (1995), on the other hand, selected developing countries such as AG as the peer group for MIN when assessing MIN's development performance. Because of the extractive boom, MIN outperforms AG both in terms of income per capita and development indicators. Saudi Arabia would be compared to developing countries as a whole, with its 2001 HDI value of 0.769 being superior to the developing country average of 0.655 (UNDP 2003).

So the development performance of the extractive economies depends on what one wishes to measure: development performance conditioned on a booming and unsustainable income level, or development performance conditioned on initial development status. The former naturally disposes the extractive economies to negative evaluations, because their booming levels of initial income per capita cause them to be compared to developed nations.

By comparison, the method used in this research does not condition growth quality by income level *or* by development status. It simply looks at the growth quality of 21 extractive economies as a group versus 67 non-extractive economies as a group.

4 Implications for Sustainability

This analysis of 240 growth spells across 88 countries from 1956 to 1999 finds little evidence to support the assertion that extractive economies experience a type of growth that is bad for the poor. According to the growth spell assessment method used here, positive growth tends to be better for the poor than negative growth, and there is nothing to indicate that this tendency is muted in extractive economies. By extension, if natural resource development is the only means a country has to create growth, then it is better doing that than not doing it. One of the strongest messages from the analysis presented in this chapter is that *negative growth is bad for the poor, and any economy, extractive or non-extractive, is likely to have an unsatisfactory trend in poverty levels if it has an extended period of negative*

---

[12]While Ross does not list the countries used in his 2001 study, he was kind enough to send me his raw data file.

[13]Ross (2001) performed a few unconditional analyses of extractive activity and development performance, and found that once conditioning on income is removed, oil dependence is not related to lower human development performance. Mineral dependence is still related to lower human development performance. A troubling feature of this and most other analyses is that they do not control for other initial differences between economies, such as differences in institutional capabilities. Those that do control for initial institutional capability find that there is no negative development effect of being an extractive economy (e.g., Acemoglu et al. 2001).

growth. Zambia and Nigeria are examples, as is Saudi Arabia after 1981 (see Fig. 2). Indonesia and Botswana are counter examples, with extended periods of positive growth and consequently pro-poor outcomes.

To the extent that sustained or increasing per capita resource extraction increases the chances of positive growth, and because positive growth is good for the poor, pro-growth policy in extractive economies should not only be directed at rectifying political and institutional shortcomings, but also directed at enabling a sustained extraction profile. This includes incentives to ensure continued domestic exploration and resource development, and the attraction of foreign direct investment (UNCTAD 2007), as well as investments in knowledge and relevant technology (Wright and Czelusta 2007). This solution is diametrically opposed to that of Ross (2001) and other Resource Curse proponents, who advocate a move *away* from investments in resource extraction. To the extent that the Resource Curse is due to declining extractive output, the cure is not to avoid the initial extractive boom, but to avoid the subsequent decline in extractive output. For example, Botswana was one of the fastest growing economies in the world for three decades, and increased mining output is thought to have been instrumental to that growth (UNCTAD 2007, pp. 144–145; Wright and Czelusta 2007). However, others argue that Botswana's performance was only different from, for example, Somalia (a resource-poor nation) due to Botswana's superior governmental institutions and economic planning (e.g., Acemoglu et al. 2003). Certainly, extractive output and government capability are linked, and the two views are not necessarily inconsistent. My feeling, though, is that there is not enough emphasis on the fact that extractive economies commonly experience the boom–bust cycle depicted in Fig. 1, and that Botswana is on the front end of that cycle. Wright and Czelusta (2007) go even further, arguing that through prudent actions it can stay on the front end of that cycle for at least another 25–30 years.

Wright and Czelusta (2007) and De Ferranti et al. (2002) have emphatically argued that resource-based growth can be sustained over hundreds of years in well-endowed economies, and that where boom–bust outcomes like Fig. 1 have occured, it is because of the lack of will to support exploration, technological progress, and investments in appropriate knowledge to extend and develop the resource base. In other words, while we observe boom–bust cycles, they are not inevitable. Were extractive economies to manage their affairs such that extractive output was sustained, they would likely have the same pro-poor development outcomes as non-extractive economies.

5 Caveats and Areas for Further Research

There are several limitations to the analysis presented here. I have restricted the approach to simple statistical comparisons at the national level, ignoring regional impacts of extraction on the poor. There may well be complicated statistical regularities relating extraction to unfavorable outcomes on the poor that this analysis has been unable to uncover. The small number of extractive economy growth events in the sample has also required that I ignore the possibility that the extraction of different commodities may have different impacts on the poor. The World Bank data consisting of income growth by quintile is also subject to methodological issues, and must be treated with a healthy degree of caution. I do not have direct measurement of the domestic output of the resource sector in each economy, and instead use trade data as a proxy. It would be of interest to determine the extent to which growth spells in extractive economies are correlated with extraction rates, and the extent to which increasing mineral and energy extraction leads to sustained positive growth and development. Finally, it may also be that the extractive economies that have experienced minerals and energy booms, and corresponding rapid economic growth, are now irreversibly "damaged", with no prospect of the poor benefiting from subsequent extraction-led positive growth spells. Indeed, given all of these caveats, a more sophisticated empirical intertemporal analysis of the relationship among growth spells, changes in extractive activity by type of activity (lootable versus non-lootable resources, energy versus minerals), and the pro-poorness of growth is warranted. I leave this for future research, and present these results only as an introductory investigation of whether or not positive growth in extractive economies, where it has been realized, has been any worse for the poor than in non-extractive economies.

One final point should be made: this analysis has only examined domestic outcomes, and a broader geographic perspective is probably warranted. For instance, the lower incidence of poverty in the Middle East and North Africa (MENA) countries (Adams and Page 2003) may be indicative of an additional benefit of extractive activities in surrounding countries. If resource extraction stimulates worker remittances, and thereby increases consumption by the poorest peoples in neighboring countries, then this would further motivate the role for extraction in sustainable development. An example might be South Africa's remittances to Lesotho and Swaziland, countries which, despite the negative development prospects associated with being small, landlocked, and African (Collier 2007), have medium levels of human development (UNDP 2003). Moreover, and a point that is often overlooked, increased availability of mineral and energy resources lowers the real prices of resource-intensive goods such as grains. In this way, increased extraction activities have an additional, indirect effect on the welfare of the poor by increasing the purchasing power of their income.

Overall, these outstanding research issues indicate that we still have much to learn about the positive and negative effects of extractive activities on the wellbeing of the poor. In particular, they caution against drawing definitive conclusions, as some have done, that extractive activity hurts the poor. Indeed, what hurts the poor in countries with mineral and energy wealth, in the end, appears to be the failure of the extractive sectors to grow in a steady, persistent manner. If we are interested in the welfare of the poor in extractive economies, less research is needed into ways to diversify away from extractive activity, and more is needed into how extractive activity can be sustained.

Acknowledgments I would like to thank Arturo Vasquez Cordano for able research assistance during the preparation of this chapter. I also thank David Humphreys, Daniel Jarrett, Jeremy Richards, Marcello Veiga, and especially John Tilton for comments on an earlier draft.

References

Acemoglu D, Johnson S, Robinson JA (2001) The colonial origins of comparative development: an empirical investigation. American Economic Review 91:1369–1401

Acemoglu D, Johnson S, Robinson JA (2003) An African success story. In: Rodrik J (ed) In search of prosperity: analytic narratives on economic growth. Princeton University Press, Princeton, pp. 80–119

Adams RH Jr, Page J (2003) Poverty, inequality and growth in selected middle East and North Africa countries, 1980–2000. World Development 31(12):2027–2048

Auty RM (1993) Sustaining development in mineral economies: the resource curse thesis. Routledge, London

Baldwin RE (1966) Economic development and export growth a study of Northern Rhodesia, 1920–1960. University of California Press, Berkeley, CA

Bevan DL, Collier P, Gunning JW (1999) The political economy of poverty, equity, and growth: Nigeria and Indonesia. Oxford University Press, New York

Boyce JR, Emery JCH (2007) What can exhaustible resource theory can tell us about per capita income growth and levels in resource abundant countries? Unpublished manuscript, University of Calgary, Alberta

Bravo-Ortega C, de Gregorio J (2007) The relative richness of the poor? Natural resources, human capital, and economic growth. In: Lederman D, Maloney WF (eds) Natural resources, neither curse nor destiny. World Bank Group/Stanford University Press, Washington, DC/Palo Alto, CA, pp. 71–99

Brunnschweiler CN, Bulte EH (2008a) Linking natural resources to slow growth and more conflict. Science 320:616–617

Brunnschweiler CN, Bulte EH (2008b) The resource curse revisited and revised: a tale of paradoxes and red herrings. Journal of Environmental Economics and Management 55(3): 248–264

Bulte EH, Damania R, Deacon RT (2005) Resource intensity, institutions, and development. World Development 33(7):1029–1044

Christian Aid (2003) Fueling poverty: oil, war, and corruption. Christian Aid, London

Collier P (2007) The bottom billion. Oxford University Press, New York

Davis GA (1995) Learning to love the Dutch disease: evidence from the mineral economies. World Development 23(10):1765–1779

Davis GA (1998) The minerals sector, sectoral analysis, and economic development. Resources Policy 24(4):217–228

Davis GA (2007) Measuring unambiguously pro-poor growth. Journal of Economic and Social Measurement 32(4):253–261

Davis GA, Tilton JE (2002) Should developing countries renounce mining? A perspective on the debate. Unpublished manuscript, Colorado School of Mines. Available at http://inside.mines.edu/~gdavis/Papers/Davis_and_Tilton_2002.pdf

Davis GA, Tilton JE (2005) The resource curse. Natural Resources Forum 29(3):233–242

De Ferranti D, Perry GE, Lederman D, Maloney WF (2002) From natural resources to the knowledge economy: trade and job quality. World Bank Group, Washington, DC

Deaton A (2003) Health, inequality, and economic development. Journal of Economic Literature XLI(1), March:113–158

Ding N, Field BC (2005) Natural resource abundance and economic growth. Land Economics 81(4):496–502

Dollar D, Kraay A (2002) Growth is good for the poor. Journal of Economic Growth 7:195–225

Ellis F, Allison E (2004) Livelihood diversification and natural resource access. Food and Agricultural Organization of the United Nations. Available at ftp://www.ftp.fao.org/docrep/fao/006/ad689e/ad689e00.pdf

Gelb AH and Associates (1988) Oil windfalls: blessing or curse? Oxford University Press, New York

International Council on Mining and Metals (ICMM) (2006) The challenge of mineral wealth: using resource endowments to foster sustainable development. Available at http://www.icmm.com/document/188

International Monetary Fund (IMF) (2004) Tackling the natural resource curse: an illustration from Nigeria. IMF Survey, March 15:78–80

Karl TL (1997) The paradox of plenty: oil booms and petro-states. University of California Press, Berkeley, CA

Klugman J (ed) (2002) A sourcebook for poverty reduction strategies, Vol 1. World Bank Group, Washington, DC

Kraay A (2006) When is growth pro-poor? Evidence from a panel of countries. Journal of Development Economics 80:198–227

Leamer EE, Maul H, Rodriquez S, Schott PK (1999) Does natural resource abundance increase Latin American income inequality? Journal of Development Economics 59:3–42

Manzano O, Rigobón R (2007) Resource curse or debt overhang? In: Lederman D, Maloney WF (eds) Naturual resources, neither curse nor destiny. World Bank Group/Stanford University Press, Washington, DC/Palo Alto, CA, pp. 41–70

McMahon G, Remy F (eds) (2001) Large mines and the community: socioeconomic and environmental effects in Latin America, Canada, and Spain. International Development Research Centre/World Bank Group, Ottawa/Washington, DC

National Public Radio (NPR) (2003) Interview: Professor Jeffrey Sachs discusses the US plan to privatize the Iraqi economy. September 23

Rodriguez F, Sachs JD (1999) Why do resource-abundant countries grow more slowly? Journal of Economic Growth 4:277–303

Ross ML (2001) Extractive sectors and the poor. OXFAM America, Boston, MA

Ross ML (2003) How does mineral wealth affect the poor? Unpublished manuscript, UCLA. Available at http://www.sscnet.ucla.edu/polisci/faculty/ross/minpoor.pdf

Ross ML (2007) How mineral-rich states can reduce inequality. In: Humphreys M, Sachs JD, Stiglitz JE (eds) Escaping the resource curse. Columbia University Press, New York, pp. 237–255

Sachs JD (2007) How to handle the macroeconomics of oil wealth. In: Humphreys M, Sachs JD, Stiglitz JE (eds) Escaping the resource curse. Columbia University Press, New York, pp. 173–193

Sachs JD, Warner AM (1995) Natural resource abundance and economic growth. NBER Working Paper No. 5398. National Bureau of Economic Research, Cambridge, MA

Sachs JD, Warner AM (1997) Natural resources abundance and economic growth. Unpublished manuscript, November, Harvard University, Cambridge, MA. Available at http://www.cid.harvard.edu/ciddata/warner_files/natresf5.pdf

Sachs JD, Warner AM (2001) Natural resources and economic development: the curse of natural resources. European Economic Review 45:827–838

Sala-i-Martin X (1997a) I just ran four million regressions. NBER Working Paper No. 6252. National Bureau of Economic Research, Cambridge, MA

Sala-i-Martin X (1997b) I just ran two million regressions. American Economic Review Papers and Proceedings 87(2):178–183

Sala-i-Martin X, Doppelhofer G, Miller RI (2004) Determinants of long-term growth: a Bayesian averaging of classical estimates (BACE) approach. American Economic Review 94(4):813–835

Son HH (2004) A note on pro-poor growth. Economics Letters 82:307–314

Stiglitz JE (2007) Making globalization work. WW Norton, New York

Stijns JP (2005) Natural resource abundance and economic growth revisited. Resources Policy 30:107–130

Surowiecki J (2001) The real price of oil. New Yorker, December 3:41

United Nations Conference on Trade and Development (UNCTAD) (2002) The least developed countries report 2002. United Nations, New York

United Nations Conference on Trade and Development (UNCTAD) (2007) World investment report 2007. United Nations, New York

United Nations Development Programme (UNDP) (1997) Human development report 1997. Oxford University Press, New York

United Nations Development Programme (UNDP) (1998) Human development report 1998. Oxford University Press, New York

United Nations Development Programme (UNDP) (2003) Human development report 2003. Oxford University Press, New York

Wallace K (1999) The Yanacocha project: breaking the cycle of need in the rural Peruvian Andes. Engineering and Mining Journal 200(8), August:NA-16CC–NA-16HH

World Bank Group (2003) Striking a better balance: the final report of the extractive industries review. World Bank Group, Washington, DC

World Bank Group (2008) World development report 2008: agriculture for development. World Bank Group, Washington, DC

Wright G, Czelusta J (2007) Resource-based growth past and present. In: Lederman D, Maloney WF (eds) Natural resources, neither curse nor destiny. World Bank Group/Stanford University Press, Washington, DC/Palo Alto, CA, pp. 183–211

The Challenge of Mineral Wealth: Using Resource Endowments to Foster Sustainable Development

K. McPhail

Abstract The purpose of this research was to identify the factors that have allowed some countries to avoid the so-called "resource curse", and to determine practical, collaborative steps that can be taken by companies, governments, local communities, and aid agencies to enhance mining's contribution to poverty reduction. Research was conducted collaboratively with the UNCTAD and the World Bank Group, overseen by an independent advisory group, and tested through two multi-stakeholder workshops. Industry involvement in the initiative took place through an International Council on Mining and Metals (ICMM) working group comprising around 20 companies and chambers of mines. Success depended on three factors: reformed mineral legislation, improved macroeconomic management, and some improvements in governance. Although the robust nature of the processes employed can be demonstrated, the explicit aim of identifying factors that allow certain "successful" countries to avoid the resource curse has inherent limitations.

Companies, governments, donors, and other actors need to work together to help strengthen capacity in mining countries and regions, particularly at the sub-national level. Mining and metals companies have been at the forefront of some of the most innovative multi-stakeholder processes of any industry sector. This paper outlines why some companies collaborated to take a leadership role, and explores how the industry has sought to become more accountable to its stakeholders by demonstrably improving its sustainable development performance.

1 Introduction

In 2004, the International Council on Mining and Metals (ICMM) commissioned independent research into ways in which it could enhance mining's contribution to

K. McPhail (✉)
International Council on Mining and Metals, London, W1H 6LR, UK
e-mail: Kathryn.McPhail@ICMM.com

Parts of this material are taken from an award-winning essay submitted by Kathryn McPhail to an IFC-FT essay competition; the views expressed in the essay may not represent IFC's or the FT's views on any matter.

poverty reduction[1]. This marked the beginning of a long-term project with multi-stakeholder input that is now into its third phase. At the outset, the research team was asked to analyze how the mining sector contributes to national development, to identify strategies that are effective in managing mineral revenues for economic growth and poverty reduction, and to investigate how mining investments can contribute to economic and social development at the national, regional, and local level.

The topic is controversial. Some international non-governmental organizations (NGOs) raise fundamental questions about the poverty reduction and development benefits of mining investments (Campbell et al. 2007; IUCN 2007). The Bank Information Center and Oxfam International in a September 2006 report cited the experience at IFC-supported gold mines in Peru, Ghana, Guatemala, and Kyrgyzstan, as well as the World Bank's own research in mining-dependent countries, which illustrate that the costs to local communities affected by mining often exceed the benefits they receive (Sarin et al. 2006)

Moreover, although economic growth is a key ingredient to poverty reduction, academics such as Sachs and Warner (1995)[2] have proposed the existence of a "resource curse", which can be briefly summarized as follows:

- Large earnings from mineral resources can lead to the "Dutch Disease" phenomenon involving exchange rate overvaluation leading to a decline in the competitiveness of other, non-mineral, economic sectors.[3]
- Dependence on such earnings are problematic if the prices of the minerals in question are volatile in the short-term or subject to sustained decline in the long-term.
- The presence of mineral wealth can encourage governments to adopt misguided industrial policies that offer protectionist barriers to support uncompetitive new activities.
- An economy blessed with abundant but finite natural resources may over-consume. One reason is that incomes in the short term may fail to account properly for the depletion (depreciation) of the nation's natural capital, thereby

---

[1]The research was conducted by independent consultants overseen by an independent advisory group. Research partners were the United Nations Conference on Trade and Development (UNCTAD) and the World Bank Group, both of which added to the development expertise of the research team. Industry involvement in the project mostly took place through an ICMM working group which comprised around 20 representatives from member companies and associations. An International Advisory Group was established and comprised: Georg Kell, Executive Head, United Nations Global Compact Office; Pedro Pablo Kuczynski, Minister of Economy and Finance, Peru, (until mid 2005); Mamadou Lamine Loum, independent consultant and former Prime Minister of Senegal; Warwick J. McKibbin, Member, Board of the Reserve Bank of Australia; Hon. Felix Mutati, Deputy Minister of Finance and National Planning, Republic of Zambia; Jane Nelson, Director, Corporate Social Responsibility Initiative, Kennedy School of Government, Harvard University. Eighteen reports and publications are available on www.icmm.com

[2]For more information, see the literature review at www.icmm.com/document/185

[3]A reviewer pointed out that this leads to a resource curse in economic models only if there is no technological progress associated with extracting minerals, and if there is endogenous technical progress in the non-mineral sectors.

resulting in consumption levels that are unsustainable—the correction, when it comes, is inherently damaging to livelihoods.

• Some countries blessed with natural resources may be more prone to poor governance, and in some cases will experience a "predatory" state characterized by corruption, political conflict, and inequalities largely created by state actions.

The objective of ICMM's research was to identify the critical factors that have allowed some countries to benefit from their substantial resource endowments, and to avoid the so-called resource curse. From the start of the project, the hope was to identify the practical steps that might be taken by the industry and others, such as governments, local communities, and development agencies, to enhance the positive impacts of mineral resource investments.

To ensure a balanced analysis of both successes and failures from mining, and the importance of ensuring that different viewpoints were adequately captured, the research methodology and the subsequent findings from the field work were critiqued by non-governmental organizations, Equator Principles banks[4], academics, labour organizations, and government representatives in two multi-stakeholder workshops. In the first workshop, held at the Lansdowne Club in London, 29–30 November 2004,[5] participants critiqued the research methodology, and the draft toolkit was revised to incorporate their comments. In the second workshop, participants discussed the draft findings from the four country case studies, and their comments were used to strengthen the conclusions and recommendations of the Synthesis Report (ICMM 2006a).

2 Assessing the Performance of Mineral-Dependent Economies and Developing a Toolkit to Document Systematically Positive and Negative Impacts

Phase 1 of the research comprised a comprehensive literature review, a quantitative analysis of the performance of 33 mineral dependent countries across a range of socio-economic indicators (published as the Analytical Framework: ICMM 2006a), and the development of a toolkit to document the impacts (both good and bad) of individual mining projects at local, regional, and national levels.

The intent of the literature review was to capture how differences in institutions and governance may influence socio-economic outcomes, and how these outcomes relate to mining activities. This was achieved by focussing on governance, defined as the capacity of a country's formal and informal institutions to design, implement, and enforce policies that benefit the wider public and improve private sector effectiveness.

---

[4]www.equator-principles.com

[5]Minutes of the proceedings are available at www.icmm.com

The 33 mineral-dependent countries[6] were defined as those where mining constituted 20% or more of exports from 1965–2003. Performance was measured with a range of socio-economic indicators over a 20-year period using six variables: two for economic growth, and four for poverty alleviation (changes in infant mortality, the UN Human Development Index, and two Millennium Development Goals: the number of people below a minimum level of dietary consumption, and the number of rural households with access to improved drinking water). Growth performance was also compared to the preceding decades using data over a 50-year period to see how typical or otherwise the recent past has been in terms of growth. These six variables were also compared against the countries' relevant country peer groups: both by region (e.g., sub-Saharan Africa, Latin America) and by income group (e.g., low, middle, high middle income).

The findings showed that the resource curse is not preordained in mineral rich economies. Almost half[7] of the 33 mining countries outperformed their regional and income comparator countries, whether endowed or not with minerals. In relatively few of the 33 country cases was the economic and poverty story all good or all bad. Six were "better performers" (Chile, Botswana, Malaysia, Tunisia, Ghana, and Mexico), while nine countries (Bolivia, Central African Republic, Congo DR, Liberia, Niger, Papua New Guinea, the Philippines, Sierra Leone, and Zambia) were poor on all counts.

Five of the six better-performing countries achieved a reasonable total governance score (Ghana being the exception with low scores on governance), although there was a great divergence of scores across those five countries. Several of the worst performing countries (e.g., Bolivia, the Philippines) had governance scores that were not too different from countries such as Ghana, even though Ghana was seen to be a better performer. It also should be noted that several of the "in-between" countries scored better on governance than some of the countries classified as better performing (e.g., South Africa, Namibia, Morocco, and Suriname). Significantly, most of the worst-performing countries had very poor governance scores (CAR, Sierra Leone, Liberia, and Congo DR).

This suggests that mining per se does not systematically lead to poor governance performance, and also indicates that the relationship between good governance and good economic and social performance is far from clear cut. Furthermore, the ex-post indicators gave no insight into the underlying causes of the quality of governance.

So, while the performance of the 33 mineral-dependent countries showed that some countries had suffered from the resource curse and related economic ailments,

---

[6]Bolivia, Botswana, Central African Republic, Chile, Colombia, Congo DR, Gabon, Ghana, Guinea, Guyana, Jamaica, Jordan, Liberia, Malaysia, Mali, Mauritania, Mexico, Morocco, Mozambique, Namibia, Niger, Papua New Guinea, Peru, Philippines, Senegal, Sierra Leone, South Africa, Suriname, Tanzania, Togo, Tunisia, Zambia, and Zimbabwe.

[7]Chile, Botswana, Malaysia, Tunisia, Ghana, Mexico, Colombia, Guinea, Jamaica, Mali, Morocco, Mozambique, Namibia, and Senegal.

it also showed that in other countries, natural resource extraction had contributed to long term sustainable and broadly based socio-economic development.

The Resource Endowment Toolkit (ICMM 2006a) was developed to provide a systematic and consistent approach to documenting the impacts (both good and bad) of individual mining projects at a local, regional, and national level in such countries. Key stages in the Toolkit are: defining the country context; profiling the mining activity to be assessed; documenting the country's economic and social performance and outcomes; assessing the project/micro-scale impacts of the mining activities; and assessing the broader macro-scale and governance causes of performance. This, then, was the starting point for further analysis based on country field work.

3 Testing the Toolkit

Phase 2 tested the methodology outlined in the toolkit in a multi-stakeholder workshop, and was refined further by application in four country field-based case studies in Peru, Chile, Ghana, and Tanzania.[8] These studies analyzed, in an objective way, the contribution of the overall mining sector to economic growth and poverty reduction at national and regional/local levels, and also traced the contribution of individual mines at local, regional, and national scales, including the interconnections between these three levels. National-level analysis included quality of governance, macroeconomic policy, and mining-related legislative reform. The findings from the four country case studies were synthesized into a single report along with a series of recommendations for industry, host governments, donor organizations, and NGOs. The emphasis was on the practical steps that all of these organizations can take alone and in partnership (ICMM 2006b). Three of the four countries were selected by the multi-stakeholder workshops in London as examples of positive outcomes.

A key finding in all four cases was that mining had contributed to improved economic policies. Although all four countries had experienced severe macro economic mismanagement in the past fifty years, the implementation of an economic reform package, usually supported by the World Bank and/or the IMF, helped some countries to avoid the resource curse. All four countries experienced higher and more stable GDP growth: Ghana has had an unbroken period of positive per capita growth for over 20 years (see Fig. 1); Tanzania for over 10 years; and Chile for over 20 years. Peru has seen significant growth during a few years since the beginning of the 1990s, albeit with a still high level of instability.

In all four cases, the long-term (50-year) comparisons suggest that the resurgence of mining activity in recent years has been accompanied by smaller problems in most aspects of the resource curse thesis than in the years when mining was stagnant or in decline. For example, in Peru and Ghana, Dutch Disease problems seem to

---

[8]Antamina (Compania Minera Antamina) in Peru, Escondida (BHP Billiton) in Chile, Obuasi (AngloGold Ashanti) in Ghana, and North Mara (Barrick Gold) in Tanzania.

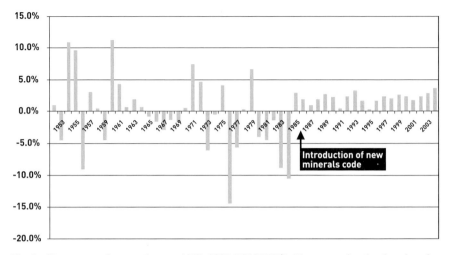

Fig. 1 Ghana per capita growth rates: 1950–2003 (1990 PPP$). The *arrow* signals when the minerals code was reformed
Source: Groningen Growth and Development Centre.

have been avoided, and exchange rate policy movements have largely compensated for inflationary changes. The growth of non-mineral tradeable GDP was found to be positive in real terms in all four countries, and higher than that of their regional comparators in all but one case (Peru). There was also some instability in revenue flows that has persisted in some countries, but the revenue/GDP proportion has been higher, especially in Peru and Ghana.

From this evidence, it can be postulated that: (i) even during periods of economic malaise in all four countries, a better investment climate for mining contributed to improved macroeconomic policies; and (ii) when mining investment enjoyed a revival, after the economic reform programs, mining was one important dimension of the recovery.

A second element of the in-country research was to assess poverty reduction. In two of the four countries, there had been significant gains in reducing poverty at national and local levels. In Chile, poverty had fallen at both the national level and regional levels. Mining is concentrated in Region II, where poverty fell faster and further than in any of Chile's eleven other Regions. This was accomplished without the government redistributing revenues down to the regional or local/community levels. Instead the government had strong policies for employment generation, supported by long-term investment in human capital, such as training and skills development. Driven by rapid mining expansion, Region II's economic growth has been faster than across Chile as a whole: GDP per head reached US$11,996 in 2003, double the national average. In 2000, Region II had the lowest poverty rate as a proportion of the population in the country, while literacy and education levels in the region are also the highest in the nation.

In Ghana, using the Ghanaian Living Standards Surveys for 1991–1999, income poverty has fallen from 52 to 40% in 1999. In the four mining districts, poverty levels were lower than in the remaining 112 districts, outside of Accra. In contrast, the research found that in Peru the economic recovery process has not changed poverty levels significantly. More than half the population continues to live in poverty with nearly a quarter living in extreme poverty of less than one dollar per day. More than 7 million (of Peru's 17 million population), have no access to safe drinking water, 5 million have no access to sewerage, and around 7 million have no access to electricity. In Tanzania, because significant mining investments only began in the late 1990s, the poverty data do not yet show any change.

Notwithstanding the indicators of positive impacts, all four case studies found local problems. The companies were often facing issues of community distrust and escalating expectations. This coincidence of dissatisfaction with economic boom conditions has been faced by other mining countries. It fits with the general policy research finding that increasing horizontal inequality induces social tension and conflict, even as general incomes rise.

The six most common problems found across the four case studies were: (i) the adequacy/fairness of the tax regime for mining in the host country; (ii) the revenue allocation system, particularly when it constrains the efficient and effective (in terms of the public/private share) use of public resources, including those generated by mining taxes and royalties; (iii) conflicts over land use and property rights; (iv) environmental damage and concerns; (v) conflicts between large-scale and artisanal mining; and (vi) the problems associated with mine closure.

In Ghana, there was a redistributive mechanism that provided for 20% of mining taxes to flow to the local level through the Minerals Development Fund. However, local communities reported that the lack of transparency meant that that they had no idea how much money was flowing to the local level or who was benefiting. Moreover, the local government's ability to mediate conflicts over land use, monitor environmental damage, or to create alternative employment opportunities was criticized. Therefore, it seems likely that poverty gains could have been greater if government and donors had given equal attention to strengthening the fiscal management, public administration, and local level political decision-making processes.

Likewise in Peru, there were significant revenues flowing to the regional level, but parallel efforts to build local institutional capacity remained weak and incomplete. The *canon minero* (requiring 50% of corporate income tax to go to mining areas), which was introduced in 2000, raised local expectations. These expectations were compounded by the addition of a royalties tax in 2005 (comprising 1–3% of the value of production) and, in 2007, the Voluntary Support Funds.

4 Lessons from the Research

While some of these issues might have been remedied in part by improved environmental management or community engagement on the part of companies, the case studies, which were signed off by each respective government, indicated that

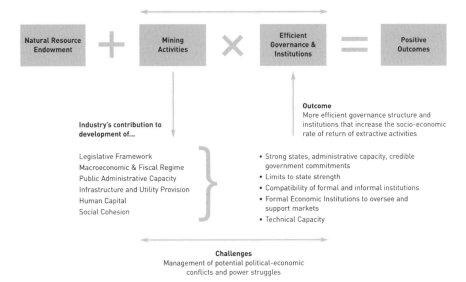

Fig. 2 The composition of effective governance
Source: ICMM (2006a).

good corporate sustainable development practices are often not enough. Governance weaknesses, particularly at the sub-national level, are at the root of many of these problems. In the typical sequence of events, the national government concerned would institute macro-level governance reforms, thus attracting mining investments and fuelling headline economic growth. But it would then fail to follow through with other reforms that might empower local and regional governments, and build their technical capacity and legitimacy. Figure 2 shows how the common features of good governance can be positively impacted by the presence of large-scale mining.

The observation that different country circumstances lead to different outcomes points to the finding that distinctive paths must be found for the achievement of positive outcomes. The core of the approach proposed here is to connect policy reform and institutional change. For example, how has the mining industry been affected by the macroeconomic and fiscal regime at the national level, by government's spatial policies at the regional level, and by local government capacity at the mine site level? Has this permitted the industry to contribute to the development of formal and informal economic institutions that can enhance the positive socio-economic impacts of mining? What are the challenges that have been encountered in this process? And what are the diverse policy implications for different tiers of governance? Mining's impacts at national, regional, and local levels are distinct but interdependent. The specific policy linkages among the three levels are essential determinants of project impact.

4.1 Legal and Regulatory Framework

Starting with the legal and regulatory framework, all four countries have reformed their minerals code to provide a stable regime for foreign direct investment (FDI), with secure property rights. This was a key factor in triggering very large inflows of FDI in all four countries. For example, Chile received US$15.5 billion between 1991 and 2003, and a further US$7 billion in 2004 alone; and Peru received US$9.8 billion between 1992 and 2004. In Ghana, the commercial mining sector suffered a major decline from the early 1940s until the mid-1980s when the minerals code (together with an economic reform package) came into effect. This resulted in more than US$5 billion being invested by foreign companies between 1986 and 2001. Tanzania has experienced the development of one new mine each year from 2000 to 2003, for a total of US$1.3 billion—easily the largest source of FDI in the country.

4.2 Formal Economic Institutions

There is a hierarchy of effects from mining's economic contribution, the greatest benefits accruing at national levels through FDI as noted above, and also exports and taxes. With the economic recovery, all four countries have been able to put in place formal economic institutions and aggregate fiscal regimes that guard the essential requirements of a stable economy. This includes broadly sound macroeconomic policies, reasonably low inflation, stable exchange rates, and aggregate fiscal sustainability. Technical capacity has been improved in central banks, ministries of finance, and often in mining ministries. Although technical capacity has improved at the national level, this has commonly not been achieved to the same extent at the sub-national level. Most noticeably, this applies to fiscal management, general public administration, and local level political decision-making processes. Ideally, strengthening of local institutions should begin at an early stage of mining development, in order to avoid the two-track syndrome identified in the first multi-stakeholder workshop in London.

4.3 Public Administrative Capacity

It can be seen that the impact of mining depends to a great extent on how government revenues are used. In turn this puts emphasis on developing public administrative capacity. All countries have seen strong central-level governance reforms, including effective, efficient, and transparent public expenditure management. However these have yet to filter down sufficiently to local and regional levels.

In Ghana, for instance, relatively little of the tax revenue generated by mining appears to find its way back to mining communities. There is also evidence that the mining companies' social investments (such as those of Obuasi, the Ghanaian mine at the focus of this case study) could be targeted more effectively at poverty alleviation. But even with such changes, a significant gap would have remained, with local institutions having limited capacity, as well as resources, to deliver basic services.

In Peru, a very similar process has occurred: companies have sought to appease local community complaints with additional social investments. However, with local government often failing to operate as a trusted mediating authority between the two sides, such corporate tactics have often proved to be no more than "quick fixes", with tensions re-emerging before long. As in Ghana, central-level governance reforms, including effective, efficient, and transparent public expenditure management, were also shown not to have filtered down to local and regional levels. Furthermore, what may be holding up the development of small-scale, non-mining enterprises in Peru is another governance issue: the need to extend recognition of formal property rights to entrepreneurs in the informal sector rather than just to large firms.

In summary, in Ghana and Peru, the lack of technical capacity at the sub-national level is creating a challenge in terms of ensuring that the funds are used for development purposes. This in turn heightens community anxiety about the impact of mining operations on their livelihoods and employment opportunities. This underscores the importance of partnerships between donor agencies, companies, and governments.

In Chile, the government has adopted a different approach. It has focussed on increased employment opportunities rather than redistributive measures. This has been done by innovative governmental policies at the regional level, providing incentives to companies. For example, Region II has established an "executive committee of the Mining Cluster". This consists of one national government representative, two local government representatives, and the CEOs of two major mining companies. They meet regularly to discuss partnership plans and procurement opportunities. The Government allocated US$15 million over the period 2002 to 2006, and this sum has been matched by the mining companies. To date, the most important of these has been a drive to assist local companies linked to the mining industry to obtain ISO 9000 and 14000 certification. The number of certified companies has grown from 8 in 2002, to 122 at the end of 2004. The cost of certification has been shared between the government, mining companies, and the participating enterprises. It is believed that the certification will make the companies more competitive, particularly in export markets. While it is too early to assess the results in terms of overall increased export success, anecdotal evidence suggests that it has worked for at least some companies.

A final point is that governance weaknesses also underlie many of the most heated local criticisms against (or issues faced by) the companies: for example, regarding indigenous land rights, alleged environmental damage, and conflicts with artisanal miners. Companies must maintain high standards of corporate responsibility on all these issues, and there is scope for improvement in firms' behaviour in some respects. More broadly, without well-functioning local institutions, which are democratically elected and trusted by local people, there is no body that can mediate disagreements effectively and fairly. This governance vacuum has created space both for expectations on companies that may be unfair, and for corporate actions that may lack legitimacy.

5 Recommendations for Enhancing the Socio-Economic Impacts of Mineral Resource Investments

A key finding of the Phase 2 research[9] was that parallel and coherent actions are needed by all concerned groups and organizations: governments (particularly at the sub-national level), local communities, mining companies, development assistance agencies, and the voluntary sector. Pending the establishment of effective local capacity, the study recommended that partnerships between all concerned stake-holders are needed to enhance the positive impacts from mining, and to tackle the negative impacts that have been identified.

Partnerships can help in at least three ways. In terms of practical steps to enhance mining's contribution to poverty reduction, national governments need to take the lead to ensure that there are shared responsibilities for outcomes by integrating the mining sector into national poverty reduction strategies. This in turn will require government departments (for example, treasury and mining) to collaborate more closely, and require chambers of mines and companies to participate in national development dialogues and to link their social investment budgets to one or more of the Millennium Development Goals.[10] In addition, social funds and donor agencies should connect funds better to the special needs of communities affected by mining.

Another issue, flagged by workshop participants during Phase 1 of the initiative, is the need to focus on transparency and tackling corruption. The Extractive Industries Transparency Initiative (EITI)[11] is a multi-stakeholder initiative which has a shared goal of promoting transparency of royalties and taxes paid by companies and received by governments. It promotes the verification and full publication of company payments and government revenues. NGOs and foundations are working in countries such as Peru reviewing public expenditures to see whether and how the investments are benefiting the poor. If the EITI is extended to transparency of revenues and to the sub-national level, this could go a long way towards tackling the anxiety expressed by local communities in both the Ghana and Peru case studies. The IFC is taking a leadership role in Peru, working with government through the National Canon Program to improve municipalities' capacity to invest, and civil society's capacity to monitor the use of, resources.

Another practical step that governments and donors can take addresses an issue identified by a member of ICMM's Advisory Group[12] reflecting on the experience

---

[9] www.icmm.com/resourceendowment

[10] www.un.org/millenniumgoals

[11] www.eitransparency.org

[12] See Footnote 1.

of the Chad-Cameroon pipeline. The experience on this project showed that state and community capacity building often takes longer than expected, while infrastructure construction is often completed ahead of schedule.

Thus, not only does capacity building need to start early, but innovative approaches need to be found to fund such activities because national (and thus local) governments receive few taxes or royalties at the very beginning of the construction period (or at the end). There is an opportunity for governments to monetize assets and bring cash flows forward by, for example, using long-term loans and guarantees offered by the World Bank. These loans could be used to transform future project revenue flows into current capital funding for social and infrastructure investments in local communities, while reserving sufficient funds to deal with subsequent mine closure.

The second opportunity for partnerships to have an effect is to deepen governance reforms, especially public administration reforms. There is a need for more reliable and sustained arrangements for properly sequenced decentralization of fiscal authority to local and regional authorities. Evidence suggests that greater decentralization, accompanied by necessary improvements in local government capacity, would enhance the impacts of mining projects.

The setting up of regional or local development agencies may help co-ordinate efforts towards economic diversification and poverty reduction. Again, governments will need to take the lead in supporting these bodies, perhaps aided by international donor organizations. In Peru, Grupo Propuesta Ciudadana, an NGO, is working with 15 regional governments to support the development of regional development plans and participatory budgeting processes. Companies can undertake governance risk assessments to obtain a better understanding of the institutional landscape in which they operate at national, regional, and local levels, and to provide training in project management. Companies can also plan their infrastructure development (roads, ports, water supplies, power plants) to integrate into these regional or local development plans. If expanded public use of this infrastructure is built into the design stage, the incremental capital cost could be minimal. By allowing privately funded infrastructure to be sized so as to meet the full needs of the regional economy over time, the economic and social spill-over effects can extend far beyond the mining sector, as was demonstrated in Chile.

A third recommendation is to intensify efforts for poverty reduction and dispute resolution mechanisms at local levels. Partnerships can support the development of mining clusters and thereby contribute to economic diversification in a region. In Chile, the "mining clusters" have encouraged the growth of small enterprises more generally. The Escondida mine supported this policy with an innovative program to help local suppliers with quality control that has enabled them to supply to others in the region and to obtain ISO 9000 and 14000 certification. The case studies showed that local procurement in Chile reached almost 80% of local goods and services, whereas in Ghana, where the Obuasi mine has been operating for more than 100 years, local procurement only reached 46%. Such an approach is one of the reasons why Chile has been so successful in capturing the economic benefits from mining. However, more can be done by companies. Workshop participants suggested that business can improve its capacity on the ground if companies can find ways to work together to implement the recommendations.

Locally based non-governmental organizations can work with local communities and indigenous peoples' organizations to help build consensus for how the benefits are equitably shared. They could also help to ensure that the voice of the marginalized is better represented. Workshop participants highlighted the importance of ensuring that the benefits of mining are communicated and delivered to local communities. Participants also confirmed the importance of putting in place dispute resolution mechanisms early in project development. Companies can take the lead by establishing a process to set grievance procedures with the input of all parties, within a wider decision-making framework involving local communities and other actors.

6 Conclusions

ICMM has, through a global process of engagement, tested a methodology with international NGOs, labour organizations, governments, donor agencies, and companies. This validation process has resulted in a Resource Endowment Toolkit (ICMM 2006a) to allow comparability across countries for public policy analysis, and to provide a user-friendly template for companies to document the costs and benefits of mining's socio-economic contribution at local, regional, and national levels.

In order to confirm the validity of the review and explore ways and means of implementing the recommendations, the third phase of the project has now begun. Pilots are planned in three of the case study countries, to see whether and how in-country partnerships between governments, companies, and the voluntary sector can be developed to encourage the uptake of the recommendations emerging from the initiative to date.

Acknowledgments We would like to thank K. Slack, J. Richards, and an anonymous reviewer for their helpful comments which have enriched the content of this chapter. ICMM's Resource Endowment initiative has greatly benefitted from contributions of its partners UNCTAD and the World Bank Group, the ICMM Working Group chaired by J. Groom, as well as the active support and participation of many individuals and companies in each of the four countries Chile, Ghana, Peru, and Tanzania.

References

Campbell B, Belem G, Coulibaly VN (2007) Poverty Reduction in Africa: On Whose Development Agenda? Lessons from Cotton and Gold Production in Mali and Burkina Faso. Research Paper. Oxfam America
ICMM (2006a) The Analytical Framework—Main Report; available at www.icmm.com
ICMM (2006b) Ways Forward—Spotlight 03; available at www.icmm.com
IUCN (2007) Mining in the Philippines—Concerns and Conflicts, the Society of St Columban, in Liaison with PIPLinks, CEESP, and Irish Centre for Human Rights

Sachs J, Warner A (1995) Natural Resource Abundance and Economic Growth. NBER Working Paper No. W5398

Sarin R, Reisch N, Kalafut J, Slack K (2006) Tarnished Gold: mining and the unmet promise of development. Bank Information Center and Oxfam International

The Role of Mining in the Economies of Developing Countries: Time for a New Approach

Keith Slack

Abstract In recent years, the benefits of mining to developing countries have been widely promoted by the industry and institutions such as the World Bank. However, the structural limitations of the global economy that may impede mining-based development have received significantly less attention. Similarly, mining's overall costs and benefits at the local and national levels are rarely given full consideration. This chapter examines these issues, and calls for a new approach to mining in developing countries that would establish a more realistic picture of what mining can and cannot deliver in terms of development benefits.

1 Introduction

Mining's contribution to sustainable economic development in developing countries has been the subject of much debate in recent years. This debate comes amidst dramatically increasing levels of investment in new and existing "mining countries", which has been driven by investment liberalization and strong demand for minerals in the booming economies of China and India. In many countries, this new wave of investment has been accompanied by conflict with local communities and increased environmental damage. Similarly, as minerals prices have reached record highs in recent years, national governments have begun to agitate for a fairer share of the windfall profits reaped by the major mining multinationals.

As the role of mining in economic development has come under sharper scrutiny, the industry and its key promoter, the World Bank Group, have responded by carrying out a number of initiatives and studies designed to respond to concerns expressed by local communities, non-governmental groups, and some governments themselves. These studies include, most recently, a multi-year research project published

K. Slack (✉)
Oxfam America, Washington, DC 20005, USA
e-mail:kslack@OxfamAmerica.org

Views expressed are not necessarily those of Oxfam America or Oxfam International.

in 2006 by the International Council on Mining and Metals (ICMM 2006), which sought to analyze mining's contribution to development in Chile, Peru, Ghana, and Tanzania. This was preceded in 2002 by the Mining, Minerals and Sustainable Development project (MMSD 2002), an industry-led initiative ostensibly designed to define mining's contribution to sustainable development in the run-up to the Earth Summit held in Johannesburg that year. For its part, the World Bank Group commissioned the Extractive Industries Review in 2001–2004, which sought to assess an appropriate role for the institution in supporting mining and oil development in poor countries.

While each of these initiatives has yielded some valuable information on the connections between mining and development, each has missed some larger fundamental questions that must be answered if we are to assess fairly the prospects for mining to contribute to meaningful poverty reduction and sustainable development in poor countries. In this chapter, I will identify some of these key "unasked questions" in this debate, and put forward a series of recommendations that may help development of a more realistic and intellectually honest approach to this overall issue. Answering these questions is of some urgency given the dominant (and in some cases, growing) role that mining plays in the economies of many poor countries, the rapid expansion of the mining "frontier" into new developing countries, and the stated intention of the World Bank Group to increase its support for mining in new counties (Business Report 2007; Dow Jones 2007).

2 Mining and Development: Theoretical Rationales

The essence of the "development" argument for mining in developing countries is that the industry can contribute significantly to economic growth and generate revenues for host country governments. Economic growth is, as nearly all mainstream economists recognize, essential for poverty reduction. If an economy is not expanding, it is not generating the resources and opportunities that are necessary to lift people out of poverty. By the same token, governments must have income in order to support the kinds of essential services that are also critical for poverty reduction, such as education and health care.

Mining proponents usually concede that the industry itself does not generate significant levels of employment (particularly if one were to compare jobs created per dollar invested.) Modern mining is highly capital-intensive, relying on high technology equipment and relatively little labor. Of the labor that it does need, much of it is skilled and thus usually not accessible to unskilled members of local communities who live in the remote areas that are commonly where large-scale mining operations in developing countries are located.

Though acknowledging relatively scant direct employment benefits, mining advocates usually make reference to a "multiplier effect" by which one job in a mine may be linked to several others in related industries, such as companies that supply services to mines. They may also point to "downstream" linkages such as refining or processing industries that may be linked to mining activity (Weber-Fahr et al. 2002).

Two specific cases are cited repeatedly as examples of how mining can contribute to economic development (or at least growth): Chile and Botswana. According to these arguments, Chile has successfully converted its copper resources into robust economic growth and poverty reduction in recent years. This has come about as a result of solid fiscal policies that have enabled the country to invest its copper revenues wisely, and save a significant percentage of them in a stabilization fund that helps smooth out the cyclical rise and fall of copper prices. Similarly, thanks to enlightened policies, Botswana has used its diamond reserves to achieve strong economic growth (particularly in comparison to other African countries), and to invest its diamond income in needed social services.

The limitations of these apparently successful models receive less attention from mining advocates. For example, little acknowledgement is made of Chile's severe income inequality, which has made it one of the most unequal countries in the world: the richest 10% of the population control nearly 50% of the country's wealth, and the poorest 10% control only 1.2% (Scott and Leight 2007). The country's Gini coefficient, which gauges inequality, has remained unchanged for the past fifteen years, a period of high economic growth. The country's inequality has given rise to protests and become an important political issue (Schweimler 2007). Chilean president Michelle Bachelet has stated that inequality is the "principal obstacle" to making Chile a developed country (Scott and Leight 2007, p. 14). Chile's copper-based development has also contributed to significant environmental impacts, which include severely depleted groundwater reserves in the country's main copper-producing region. The Chilean government recently announced that the country's mining region is reaching its limit for sustainable extraction from underground aquifers, a situation that surely puts into question the long-term sustainability of the industry in the country (Cereceda 2007).

The overall replicability of the Chilean model is also not seriously questioned by mining advocates, which is problematic given certain conditions that may be unique to Chile's situation, including the country's dominance of global copper production, and the grouping of its mines in a particular area, which may enable the development of greater employment opportunities in service industries. By the same token, Botswana, though achieving strong recent growth, is almost entirely dependent on a single natural resource (diamonds) which will become more costly to produce and eventually become depleted, a situation that may begin as soon as 2020 (Maganu 2008). It is not clear at this stage what will replace the dominant role of diamonds in Botswana's economy, which faces high transportation, water, and power costs due to its land-locked and arid conditions. Such factors may significantly impede efforts to diversify. Similarly, the country's diamond revenues appear to have done little to arrest the country's rate of HIV/AIDS infection, which is the world's highest and has reduced life expectancy to 35 years (Reuters 2008). Such facts are not widely cited in discussions of the country's successful management of its natural resources.

From a longer historical perspective, the US, Canada, and Australia are often cited as examples of how major economic powers built their economies on mining.

One can see clear examples of this line of argument for mining in documents from the World Bank, which stated in a 2002 report that "it is impossible to argue that Australia, Canada ... and the United States did not base their development on their natural resources." (de Ferranti et al. 2002, p. 6)

The mining industry also makes this case forcefully, declaring in 2002 that:

> the experiences of the US, Canada and Australia in becoming the richest countries in the world while continuing to rely on mining clearly proves that mining is an economic foundation that can reliably help nations escape poverty. (Butler et al. 2002)

Although it may seem logical to extend this historical analogy to current mining-dependent developing countries, there are significant reasons to question its applicability in today's global economy. This will be discussed further below.

In reviewing the theoretical justifications for promoting mining as a development driver, and the alleged empirical examples that demonstrate their successful implementation in practice, it is important to recognize that economic growth and generation of government revenue are undoubtedly important components of the overall sustainable development equation. What is much less clear, however, is precisely *how* these components are intended to translate into long-term *sustainable* development and meaningful poverty reduction. It seems to be simply taken on faith that growth and revenues will produce these benefits. This overlooks critical considerations at both the micro and macro levels. At the micro level, little attention is given to whether mining's benefits will outweigh the inevitable impacts on local communities (which can include displacement and loss of livelihoods).

At the macro level, very little analysis is usually provided as to how mining will, in specific terms, fit into a broader development strategy for a given country. How, exactly, will the revenues generated by mining be used to support industries that will create the jobs that are needed to raise incomes, which in turn is critical to reducing poverty? How will it contribute to economic diversification? Are there aspects of the global economy, such as trade barriers, competition from other countries, and transportation costs that make it less likely that resource-based economies can transition to higher value-added modes of production that would create greater economic opportunity for the poor? The lack of serious attention to these questions—what one might call a "faith-based" approach to mining and development—will also be discussed in more detail below.

3 The Unanswered and Unasked Questions

3.1 The Importance of Raising Incomes and Broader Development Strategy

To provide some additional context to this discussion, it is perhaps helpful to briefly review the importance to poverty reduction of raising incomes and employment levels. There is broad agreement among development economists that income levels must increase in order for poverty levels to decrease. Empirically, the greatest development success stories of the post-war era were the "Asian Tigers" of Taiwan, South Korea, Singapore, and Hong Kong, who built their economies on highly labor-intensive economic activities that employed large numbers of people.

Thus according to prevailing theory and recent empirical examples, in order to have meaningful poverty reduction, jobs must be created and incomes must rise. In a globalized world, there is no other viable option for developing countries. Given the degree to which employment creation is seen as a critical component of broad-based poverty reduction by most mainstream economists, it is all the more jarring to find such little clear articulation of how the growth and revenues generated by mining link to a comprehensive strategy to achieve higher incomes and employment. Such information is not contained, for example, in the World Bank's Country Assistance Strategy (CAS) documents, which set out the Bank's overall approach to promoting development in a given country. In these documents one does see some acknowledgement of the limitations of mining-based development. For example, Peru's CAS for 2003–2006 notes in some detail the small amount of employment that the country's mining sector has generated, and concedes that "affected communities generally do not receive, or perceive, any benefits" from mining (World Bank Group 2004a, p. 25). But it provides few specific suggestions on how this situation can be remedied beyond general calls for increasing transparency and strengthening governmental institutions.

Similarly, the Poverty Reduction Strategy Papers (PRSPs), which are prepared at the behest of the World Bank and International Monetary Fund, and which set out a country's plans for reducing poverty over a three-year period, do not describe in any detail how mining revenues will feed into a discernable strategy for creating employment, raising incomes, and economic diversification. For example, the 2006 PRSP for the Democratic Republic of Congo, a country with one of the world's richest mineral endowments, states the following:

> The strategy advocated by the Government in [the mining] sector also involves streamlining the exploitation of mining resources, providing sustainable support for economic growth and effectively contributing to improving the social conditions of workers in the sector and the Congolese people as a whole. (International Monetary Fund 2007, p. 75)

While improving the social conditions of the Congolese people is certainly a laudable goal, one would expect to find at least a little more detail on how exactly the government and its international financial institution (IFI) backers would intend to use the mining sector to bring this about.

In its own research, ICMM notes a similar lack of attention to mining and poverty issues in Ghana's PRSP, stating that there "seem to be no mechanisms in the Ghana Poverty Reduction Strategy (GPRS) to connect these (debt relief) funds in any reliable manner to the special needs of communities affected by mining." (ICMM 2006, p. 48). More generally, it notes that while donor-supported "anti-poverty strategies invariably refer to 'private sector and SME (small and medium enterprise) development' the actions involved seem to have little connection to the specific needs and potential of mining areas." (ICMM 2006, p. 59).

Also notable for their lack of detail in this area are the documents produced for specific mining projects financed by the International Finance Corporation (IFC), the private sector arm of the World Bank that plays a critical role in supporting mining development in low-income countries. Generally these documents, such as the IFC's Summary of Project Information (SPI), refer simply to revenues to be generated by a project as evidence of its overall development impact. One example is the Marlin Mine project in Guatemala, for which the IFC approved financing in 2004. In describing the project's "development impact", the document makes only brief and general references to community "capacity-building programs", improvement of "community facilities and services", and the company's goal of hiring most of its workforce locally (IFC 2004a). It is worth noting that this weak development rationale was sharply questioned by some members of the IFC's board of directors, who, among other concerns, highlighted the fact that the $261 million project would create only 160 long-term jobs (IFC 2004b).

An additional example of the lack of consideration given by proponents of mining's benefits to economic development can be found in ICMM's 2006 research. The research was undertaken to refute growing questioning by non-governmental organizations (NGOs) such as Oxfam America and others about mining's development value. The research concludes essentially that, under the right conditions, mining can contribute to poverty reduction. Yet the volume does not include specific discussion about how mining should fit into broader strategies to create jobs and increase incomes. Ironically, the study attributes the success of its most successful case study, Chile, to the creation of direct mining-related employment in that country. This despite wide acknowledgement even within the industry and its financial backers that large-scale, open-pit industrial mining does not in general create significant amounts of employment. Thus, the industry's one (apparently) unambiguously successful case of mining-led development seems to prove the general rule that poverty reduction is inherently linked to creating income-generating opportunities—which in the vast majority of cases mining does not produce (Chile, for a variety of reasons, being a possible exception).

3.2 Mining and the Global Economy

The global economy of the early twenty-first century is fundamentally different than that of the early twentieth. This may seem obvious, yet the proponents of mining-based development seem largely to ignore this reality. For them, the role that minerals development played in the development of the US, Canada, and Australia one hundred years ago can be replicated now. The problems with this comparison have been well-documented (Power 2002). The argument is faulty for a variety of reasons, but two in particular are worth re-emphasizing. First, lower transportation costs make it much cheaper to ship unprocessed ore away from mining areas to processing areas overseas. Thus the value-added benefits that could be gained from processing the ore are lost. Additionally, global trade rules now make it illegal for poor countries to protect down-stream industries that could be developed to process mined minerals. These factors help make economic diversification (which even the industry agrees is necessary for mining to contribute to broad-based development) extremely difficult or impossible.

A second critical flaw in this "historical analogy" approach is that it ignores the degree to which these countries had functioning legal, democratic, and financial institutional structures in place coincident with the rise of their mineral sectors. Each country had significant "institutional capital" that allowed financial capital to be accumulated and used productively, and ensured a more egalitarian distribution of income. Such institutions "assured that the potentially large economic rents associated with mineral development did not lead primarily to conflict, corruption and waste" (Power 2002, p. 24).

The economic powerhouse of China is also a critical factor in today's global economy that has direct impact on the prospects for diversification in mining-dependent economies. In some countries, non-mining sectors, such as textiles, have been wiped out by competition from cheap Chinese imports. Zambia is a stark example of this, in which anti-Chinese sentiment has given rise to violent demonstrations (Polgreen and French 2007). Intensive competition from Chinese products makes diversification into labor-intensive manufacturing (again, the path followed by the most successful developing countries in the last fifty years) effectively impossible. Given these factors, it is very difficult to discern how mineral wealth can help move countries up the ladder of development. In fact, some observers note that current high commodity prices and an inability to diversify may be intensifying the dependence of developing countries on natural resource extraction—creating a phenomenon of "de-industrialization" that may ultimately leave developing countries even worse off than previously (Zafar 2007).

3.3 A Broader Understanding of Costs and Benefits

Mining, like all economic development activity, involves costs and benefits—"tradeoffs" that must be made between competing interests and priorities. There will inevitably be winners and losers from mining development, just as there would be from construction of a factory or other industrial operation. What is different about mining, however (and this is particularly evident in the developing world), is the scale and severity of the costs that the industry generates. The sheer magnitude of the disruption that mining can bring to a local community and its surrounding environment places it on a different scale than other kinds of economic development activity. A large industrial mining project can generate hundreds of millions of tons of waste material, displace thousands of people and create permanent damage to land and water resources, rendering them permanently unfit for any other use, including agriculture. Costs for cleaning up mine sites can run into the billions of dollars, which can be a huge burden on cash-strapped developing country governments if they are left to cover these costs. In many developing countries (and even in some developed ones), governmental capacity to regulate these impacts effectively simply does not exist, nor is anything approaching adequate funding for mine closure available. These factors can exacerbate these problems, and virtually guarantee that the long-term costs of mining outweigh the benefits.

Ghana is a particularly compelling case in point on this question. The country, formerly known as the Gold Coast, has been the site of large-scale industrial mining for more than one hundred years. In the 1980s, the World Bank supported opening of the country's mining sector to foreign investment. The IFC invested directly in several projects that were seen as part of the new wave of mining investment in the country. The benefits of this investment have, however, failed to trickle down to the poor communities that inhabit Ghana's mining areas. In 2003, the World Bank's internal review unit analyzed the sector and made the following rather startling assessment, which is worth quoting in some length:

> It is unclear what its [gold mining's] true benefits are to Ghana. Large-scale mining by foreign companies has a high import content and produces only modest amounts of net foreign exchange for Ghana after accounting for all its outflows. Similarly, its corporate tax payments are low, due to various fiscal incentives necessary to attract and retain foreign investors. Employment creation is also modest, given the highly capital intensive nature of modern surfaces mining techniques. Local communities affected by large-scale mining have seen little benefit to date in the form of improved infrastructure or services provision, because much of the rents from mining are used to finance recurrent, not capital expenditure. *A broader cost-benefit analysis of large-scale mining that factors in social and environmental costs and includes consultations with the affected communities, needs to be undertaken before granting future production licences.* (World Bank, 2003, p. 25; emphasis added).

This passage succinctly encapsulates the challenges that many mining-dependent developing countries currently face. It is precisely the kind of "cost-benefit analysis" that the Bank recommends here, one that encompasses the full range of economic, social, and environmental impacts, that all mining-dependent countries should undertake, but which none have done in any meaningful or objective way. Such analyses may become even more acutely needed as competition for natural resources is exacerbated by climate change. Countries will be forced to make choices in order to husband precious resources, like water and arable land. To do this effectively, they will need to understand the full implications of development paths they may choose to pursue, particularly ones that are intensely resource-intensive such as mining.

It is worth noting here that this kind of reflection on the true cost and benefits of mining is not typical of the World Bank, which, to restate, is perhaps the leading global promoter of mining in developing countries. The IFC does not, for example, actually track the net impacts on development and poverty reduction of the projects it funds, including in mining (Bank Information Center and Oxfam International 2006). The corporation employs a mechanism called the Development Outcome Tracking System (DOTS), which collects numerical data on some indicators (such as a company's contribution to health and education services) and compliance with IFC social and environmental policies, but does not report on whether development objectives have actually been achieved: i.e., whether poverty and living standards have improved (or worsened) in a project-affected area (IFC, 2008).

In a recent review of IFC's development impact monitoring, its Compliance Advisor/Ombudman's office pointed out the limitations of DOTS in measuring actual, on-the-ground development benefits of IFC projects, stating that "...the question of whether *local livelihoods* or living standards have been positively or negatively affected by an investment is not directly addressed by IFC's analysis of development results in DOTS." (Office of the Compliance Advisor/Ombudsman, 2008).

There is thus currently no way to know if the benefits to poverty reduction (which the IFC is mandated to promote) actually outweigh the negative social and environmental impacts that have occurred at its mining projects in countries like Guatemala, Peru, Ghana, and Kyrgyzstan.

4 Mining and Development: The Need for a New Approach

The global mining industry now finds itself at an important crossroads. Financially, the industry is in the midst of a huge commodities boom driven by demand for minerals from China and India. Fiscal irresponsibility in the United States and uncertainty about global geo-politics have helped drive gold prices to record highs. In recent years, companies have made record profits and scrambled to find new projects in which to invest. In 2008, profit margins declined considerably for some companies due to high costs of energy and other inputs. Most analysts, however, predict the current trend of high minerals prices will continue for several more years. By the same token, however, the industry and its backers recognize rising levels of discontent amongst mining-affected communities, who believe that they are suffering disproportionately from the impacts of mining and are not receiving enough of the benefits. Indeed, almost every major new or existing project in countries such as Peru, Guatemala, Ghana, Indonesia, and the Philippines is subject at some point to some kind of community protest or expression of concern.

The mining industry has begun to acknowledge these concerns by noting that communities have begun to play the role of "gatekeepers" to new ore bodies. Paul Mitchell, former Secretary General of the International Council on Mining and Metals, articulated these views in a June 2006 presentation to the World Bank:

> [O]ne of the two parts of industry's case for access to land and countries' non-renewable resources, namely that these resources are essential for society and their exploitation provides economic development in the host country. *It follows that if either of these arguments is weak or untrue, then our case for access to land is weakened significantly.* (Mitchell, 2006; emphasis added)

Governments, too, have begun to question whether or not they are getting their fair share of the benefits, as evidenced by recent moves by governments in Peru and Chile to institute royalties, and decision by the governments of Indonesia and the Democratic Republic of the Congo to review and possibly renegotiate contracts with multinational mining firms.

Consumers and investors have also become increasingly concerned about the mining industry's performance on social and environmental sustainability issues. In 2006, for example, Citigroup issued its sustainable mining index, in which it recommended investing in companies with the strongest policies and practices on social and environmental sustainability issues (Jansen et al. 2006). Similarly, more than twenty leading jewellery companies have called on the global mining industry to respect basic human rights and environmental standards in the production of gold (No Dirty Gold Campaign: www.nodirtygold.org).

This scenario suggests that the mining industry is under growing pressure to demonstrate that it does contribute meaningfully to sustainable development. Indeed, the recent ICMM-sponsored study on resource endowments as well as the MMSD project were undertaken with this objective in mind. What is less clear, however, is whether, given the realities of the global economy in the early twenty-first century, such pressure and the responses to it can actually be translated into demonstrable sustainable development on the ground in developing countries.

So why does any of this matter? There are two primary areas of reasons. First, international financial institutions (meaning primarily the World Bank Group) continue to promote mining for its alleged contribution to sustainable development. By doing so, they employ limited resources that might be better used to support other kinds of economic activity that could bring greater benefits to developing countries, and that could create more opportunities for employment and raising incomes. These could include agriculture, manufacturing, or tourism. Second, and more importantly, by choosing to pursue mining-based development, developing countries put at risk other development activities, most notably agriculture, that might bring more benefits to a broader section of their populations. Water and land resources destroyed by mining are unavailable for agricultural production, which is still the primary source of livelihood for most of the poorest populations in developing countries. For example, in a country like Ghana, where the majority of the population still derives its livelihood from small-scale agricultural production, it is far from clear that mining has generated sufficient benefits to compensate for the damage it has done to land and water resources needed for farming, and to the communities that mining has displaced. For this reason the World Bank, and most recently a Ghanaian government human rights commission, have called for a comprehensive cost-benefit analysis of the impact of mining on development in the country (Commission on Human Rights and Administrative Justice 2008).

Mining advocates often argue that in some areas there is simply no viable economic alternative to mining. And, indeed, in a high altitude plateau 4,000 m above sea level or in a remote desert area, it can be difficult to imagine other forms of economic activity that can generate the degree of revenues that mining can provide. But experience shows that mining's track record as a development driver in such places, particularly in developing countries, is highly problematic. Moreover, mining, no matter where it is done, will bring costs and impacts. Therefore it is not adequate simply to ask "What's the alternative?" without also asking "What are the costs and benefits, now and in the future"?

The point here is that, from the standpoint of sustainable development, mining-based economic development implies trade-offs and choices, as do all kinds of economic activity. The scale and the permanence of mining's impacts on a community, particularly in a developing country context, are such that once those choices are made, they may be difficult or impossible to change.

There is thus an urgent need for more comprehensive and transparent consideration of the full costs and benefits that mining is likely to bring to a country. It is not adequate to posit an ideal situation—"under the right conditions"—as economists are wont to do, without acknowledging the likelihood (or lack thereof) that the ideal situation can actually be realized. Thus, it is hypocritical for an institution like the World Bank to acknowledge, for example, that "good governance" is necessary to ensure that mining will contribute to sustainable development in developing countries, while at the same time promoting mining projects in countries like Guatemala, which any reasonably informed observer can see clearly suffers from serious governance weaknesses (e.g., World Bank 2004b).

If mining is to continue to be promoted for its sustainable development value, a clearer articulation of precisely *how* mining will serve this purpose needs to be set out at the start of a potential project, or indeed at the start of industry development as a whole in a given country. Those assumptions and predictions, in turn, should be monitored closely, and evaluated to ensure that they are actually being met. Mining companies, international financial institutions, and civil society all have a role to play in this articulation and monitoring, although the predominant responsibility must lie with governments who are ultimately accountable for delivering on a country's development objectives.

If the sustainable development rationale for mining cannot be clearly and consistently articulated, then the "sustainability pretense" for mining should be dropped. This would not necessarily mean the end of mining, or that mining should not be pursued in developing countries. There may be legitimate, non-developmental reasons for doing so. But an upfront admission that mining is not about sustainable development in a given context, could help generate realistic expectations, and suggest that scarce "development" resources be employed elsewhere. Such upfront intellectual honesty might also help serve the mining industry's interests by establishing more realistic expectations about what mining can and cannot deliver.

5 Policy Recommendations

There are a number of specific policy venues in which this new approach to mining and development can be taken forward. Progress in these areas would provide a good starting point for shifting thinking and practice, and perhaps provide better outcomes for all stakeholders. These specific venues are set out below.

5.1 *Conduct Independent National and Project-Level Cost-Benefit Analyses*

Before a developing country embarks on significant new minerals development, the government should support a broad analysis of the likely potential costs and benefits of the sector to the country as a whole. Such analyses should be independently produced by multi-disciplinary teams, including economists and specialists in poverty reduction, human rights, and environmental issues. These teams should be independent of the mining sector and have no vested interest in the outcome of the analysis. They should have access to reliable baseline data on a range of factors, notably water quality and quantity. Similar analyses should be conducted for every new individual project that is proposed. Both levels of analysis should seek to produce, to the extent possible, detailed and quantitative analysis of the potential benefits of mining set against the likely costs, which should include cumulative loss of local livelihoods, mine clean-up costs, and environmental damage. Additionally, realistic assessments of government capacity to regulate mining's social and environmental impacts and redistribute mining revenues to impacted areas should also be set out clearly. By the same token, these reviews should include frank discussion of local governments' capacities to use and invest mining revenues.

It is important to recognize the technical and political difficulties of carrying out this kind of objective analysis. And, indeed, there are very few precedents on which an analysis of this type could be based. Much of the relevant expertise needed lies within the mining industry itself or within the international financial institutions, which, as noted, in many cases would have a vested interest in the outcome. Politically, governments would no doubt find it very difficult to allow a truly objective analysis that could realistically contemplate a net negative outcome. Nevertheless, experts exist with adequate skills and who are sufficiently independent of the industry to conduct unbiased reviews. There are very few of them, however, so investment would be needed to train additional experts. Funding for training could come from a special independent fund supported by contributions from governments, industry, financial institutions, and NGOs, but administered independently and transparently. Although involvement of industry and NGOs could affect the perceived independence of such a fund, their participation from a financial and political standpoint would be essential. Mechanisms could be established to protect against manipulation of the fund by any particular interest group.

Political opposition to these reviews could perhaps be overcome by creating incentives for governments to allow them to go forward. In Ecuador, a recent effort by NGOs, private foundations, and the government to support protection of environmentally sensitive areas in lieu of oil development could provide one potential model for "incentivizing" transparent analysis and debate about the potential costs and benefits of mining (Amazon Watch and World Resources Institute 2007).

5.2 Produce Sustainable Development Strategies that Articulate the Role of Mining

Closely related to the issue of cost-benefit analyses is the clear articulation of the role of mining in broader development strategies. As noted above, such strategies (as embodied by the World Bank's CAS and PRSPs) provide little to no analysis of how mining development is expected to support broader sustainable development goals, beyond a general contribution to economic growth. This is highly problematic because it ignores critical issues like economic diversification, and creates a situation in which the linkages to the rest of a country's economy are simply assumed rather than clearly articulated. This may help explain why mining in general has had such a poor track record in contributing to *sustainable* development: the assumptions about how mining would support development in a particular country are faulty or not thought through.

Thus, a way to begin to remedy this situation is by starting with development strategy documents, including Country Assistance Strategies (CAS) and Poverty Reduction Strategy Papers (PRSP). These documents should clearly spell out how revenues from mining will be used to support broader development objectives. Given that mining itself will not create significant long-term employment or income-generating opportunities (as noted above), development strategies should explain how mining revenues will be used to help produce durable employment or other opportunities to raise income levels. Relatedly, these strategies should directly address how issues in the larger global economy will affect the assumptions about the contribution of mining to development in a given context. For example, if mining revenue in, say, Mali is to be invested in increasing the productivity of the country's cotton sector, how will trade barriers (subsidies) in developed countries impede that process?

There are some precedents for "ear-marking" resource-based revenues for specific development purposes. For example, in Chad, a petroleum revenue management law, put in place at the behest of the World Bank, directed oil revenues to health, education, and other priority sectors for reducing poverty. However, these arrangements said nothing about development of non-oil-based industries or other economic activity, and the World Bank failed to assist the government to develop an economic diversification strategy. The project itself ultimately failed over the Chadian government's refusal to comply with the conditions set by the Bank. In September 2008, the Bank cancelled the project and Chad paid back its loan. Despite this failure, the project does demonstrate that specifying development usages for resource-generated revenues is possible. Actually enforcing such commitments is, of course, another matter.

5.3 Establish and Monitor Development Indicators for Individual Projects

Recent rhetoric about foreign aid and development in US foreign aid policy and at institutions such as the World Bank has emphasized the need to make such efforts more effective and accountable for delivering on development objectives. The US government's creation in 2002 of the Millennium Challenge Account program, in which countries must meet certain criteria in order to receive US aid, is reflective of this point of view. Increasing the effectiveness of development efforts has also been a key concern of World Bank president Robert Zoellick (Zoellick 2008). Given this rhetoric, it is difficult to understand why there has been almost no serious effort to actually track the net impacts on development of individual mining projects. As noted above, the World Bank's IFC, a key promoter of mining in developing countries, does not measure in a meaningful way the impact of the mining projects it finances on poverty levels or other development-related indicators. The institution seems content with stating that such projects will contribute to economic growth and produce a relatively small number of jobs.

As noted above, if mining is to continue to be promoted for its sustainable development value, the IFC and mining companies must establish development indicators against which the development impact of the project can be assessed. These indicators should be monitored regularly and independently throughout the life of a project, and the results of this monitoring should be made public and debated openly with all stakeholders. Among the issues that should be measured are overall levels of poverty, education, and health in a project area. Indicators should also be developed for environmental and governance issues.

Skeptics of this approach argue that it is difficult to isolate the contributions of an individual mining project to development and poverty reduction. Improvements in development come over time and are attributable to a complex set of interrelated factors. There is undoubtedly some truth to this argument. But large scale mining is increasingly being promoted by the mining industry and by the IFIs specifically for its *development* value, so surely, then, there must be ways to demonstrate whether or not that value is actually being achieved. Furthermore, while the benefits accruing from an individual project may be difficult to quantify, the local level *negative* impacts can be seen very clearly. These include displacement of communities, disruption of local livelihoods, and environmental contamination. Thus, if we are to accept the sustainable development justification for mining, it is imperative to demonstrate that the costs are outweighed by the benefits generated by an individual project.

6 Conclusions

In a time of high minerals prices, ever closer global economic integration, and a global environmental crisis, it is more important than ever that development rationales for minerals-based development be clearly articulated and monitored. Indeed, the current period may be the last chance for some developing countries to convert their mineral wealth into broad-based and *sustainable* development (UN Integrated Regional Information Networks 2007). Thus, if these resources are to be used wisely, and without foreclosing other development options, more attention must be given to these issues. If this does not happen, the current pattern of unsustainable, non-developmental resource exploitation will continue, to the detriment of mining companies, governments, and local communities alike.

References

Amazon Watch and World Resources Institute (September 27, 2007) Ecuador's Pioneering Climate Change Plan Announced at Clinton Global Initiative. Press Release. Accessed at http://www.amazonwatch.org/amazon/EC/yasuni/view_news.php?id=1461

Bank Information Center and Oxfam International (2006) Tarnished Gold: Mining and the Unmet Promise of Development. Bank Information Center, Washington, DC, p. 3. Accessed at http://www.bicusa.org/proxy/Document.9518.aspx

Business Report (March 14, 2007) IFC to Put More into African Mines.

Butler T, Mead G, Turner M (April 3, 2002) Response to "Extractive Sectors and the Poor" by Michael Ross, an Oxfam America report (October 2001); posting to Mineweb (mineweb.com). On file with author.

Cereceda E (2007) Water and Mining: A Thirsty Business. Business News Americas, Santiago, p. 2.

Commission on Human Rights and Administrative Justice (2008) The State of Human Rights in Mining Communities in Ghana. Commission on Human Rights and Administrative Justice, Accra, p. 193.

de Ferranti D, Perry G, Lederman D, Maloney W (2002) From Natural Resources to the Knowledge Economy. The World Bank, Washington, DC

Dow Jones Newswire, IFC to Invest in Asia Mining Projects (April 5, 2007)

ICMM (International Council on Mining and Metals) (2006) Synthesis of Four Country Case Studies. The Challenge of Mineral Wealth: Using Resource Endowments to Foster Sustainable Development. International Council on Mining and Metals, London

IFC (International Finance Corporation) (2004a) Summary of Project Information: Marlin Mining Project. International Finance Corporation, Washington, DC. Accessed at http://www.ifc.org/ifcext/spiwebsite1.nsf/2bc34f011b50ff6e85256a550073ff1c/9e42e13df0ff8b3485256e-61006d226a?opendocument

IFC (International Finance Corporation) (2004b) Summary of Discussion at the Meeting of the Board of Directors of IFC June 23, 2004 (on file with author)

IFC (International Finance Corporation) (2008) IFC's Evaluation Framework. Accessed at http://www.ifc.org/ifcext/devresultsinvestments.nsf/Content/Evaluation_Framework

International Monetary Fund (2007) Democratic Republic of the Congo. Poverty Reduction Strategy Paper. Accessed at http://www.imf.org/external/pubs/ft/scr/2007/cr07330.pdf

Jansen H, Tyrrell M, Heap A (2006) Towards Sustainable Mining: Riding with the Cowboys or Hanging with the Sheriff? Citigroup, London.

Maganu P (2008) Botswana: Soverign Wealth Funds a Safety Net After Mines Run Out. The Reporter (Gaborone), September 11, 2008

MMSD (Mining, Minerals, and Sustainable Development Project) (2002) Breaking New Ground. Earthscan Publications Ltd., London

Mitchell P (June 19, 2006) Mining and the Development of Local and National Economies. Presentation to the World Bank, Washington, DC. PowerPoint presentation on file with author.

Office of the Compliance Advisor Ombudsman (2008) Improving IFC's and MIGA's Local Development Impact at the Project Level. International Finance Corporation, Washington, DC, p. 10.

Polgreen L, French H (August 21, 2007) China's trade in Africa carries a price tag. New York Times, p. 1.

Power T (2002) Digging to Development? A Historical Look at Mining and Development. Oxfam America, Washington, DC.

Reuters (June 12, 2008) Botswana Diversification Hampered by High Costs. Accessed at http://www.mmegi.bw/index.php?sid=4&aid=8&dir=2008/June/Thursday12

Schweimler B (September 23, 2007) Inequality remains in prosperous Chile. BBC News. Accessed http://news.bbc.co.uk/2/hi/business/7006120.stm

Scott B, Leight J (2007) Chile: The Conundrum of Inequality. Harvard Business School, Cambridge, MA

Weber-Fahr M, Strongman JE, Kunanayagam R, McMahon G, Sheldon C (2002) Mining. In: World Bank A Sourcebook for Poverty Reduction Strategy Papers. World Bank, Washington, DC, p. 446. Accessed at http://siteresources.worldbank.org/INTPRS1/Resources/383606-1205334112622/4251_chap25.pdf

UN Integrated Regional Information Networks (2007) Zambia: Mineral tax increase holds no benefit for citizens. Accessed at: http://www.irinnews.org/report.aspx?ReportId=70279

World Bank (2003) Project Performance Assessment Report: Ghana Mining Sector Rehabilitation Project (Credit 1921-GH). Report No.: 26197. 2003. World Bank, Operations Evaluation Department, Washington, DC, p. 25.

World Bank (2004a) Peru: Country Assistance Strategy Progress Report. World Bank, Washington, DC

World Bank (2004b) Striking a Better Balance: The World Bank Group and Extractive Industries: The Final Report of the Extractive Industries Review: World Bank Group Management Response, p. iii. Accessed at: http://siteresources.worldbank.org/INTOGM-C/Resources/finaleirmanagementresponseexecsum.pdf

Zafar A (2007) The Growing Relationship Between China and Sub-Saharan Africa: Macroeconomic, Trade, Investment, and Aid Links. World Bank Research Observer, World Bank, Washington, DC, p. 126.

Zoellick R (April 12, 2008) Statement by Robert Zoellick, President of the World Bank, to International Monetary and Finance Committee. Washington, DC. Accessed at: http://www.imf.org/External/spring/2008/imfc/statement/eng/wb.pdf

A Hierarchy of Natural Resources with Respect to Sustainable Development—A Basis for a Natural Resources Efficiency Indicator

Markus Wagner and Friedrich-Wilhelm Wellmer

Abstract To resolve complex issues and establish guidelines for industry, politicians need data that can be transformed into indicators for policy decisions. Using a four-level hierarchy of natural resources as a base, a meaningful resource efficiency indicator can be developed as a tool for such policy decisions.

According to this concept, sustainable development implies substituting materials at a higher level of the hierarchy, either by material from a lower level, or by resources from the technosphere that replace resources from the same level in the geosphere. Energy resources occupy the highest level of the four-level hierarchy. Most problems concerning natural resources can be solved with enough affordable energy: water can be recycled after use; saline water can be desalinated; soil erosion through deforestation can be reduced by lessening the need for biofuel; cut-off grades in metal deposits can be lowered to increase available reserves; and lower-quality scrap metal can be recycled. The next hierarchy level is represented by raw materials derived from occurrences that developed over geological time and were formed by natural enrichment (e.g., all metal deposits and some non-metallic deposits such as barite or phosphate). This level also includes deposits of the technosphere that can be recycled. The third level comprises materials available in almost unlimited amounts on Earth, such as granite, sandstone, and clay, but also those raw materials that can be produced from air (e.g., nitrogen fertilizer), or from sea water (e.g., boron, potassium, or magnesium). Wood used for construction purposes is included in this third level because it is a renewable resource. The lowest level represents waste and residue materials from the technosphere that are potential raw materials for secondary use. Because energy resources occupy the top of this hierarchy, it makes sense to conserve energy by using more raw materials of lower ranking, rather than materials from the top levels. It then follows that in order to measure resource efficiency it is not appropriate to use a pure indicator, such as "total tonnage of natural resources produced or consumed in relation to the gross national product." Instead, in establishing guidelines for political decisions designed

M. Wagner (✉)
Bundesanstalt für Geowissenschaßften und Rohstoffe, Hannover D-30655, Germany
e-mail:Markus.Wagner@bgr.de

J.P. Richards (ed.), *Mining, Society, and a Sustainable World,*
DOI 10.1007/978-3-642-01103-0_5, © Springer-Verlag Berlin Heidelberg 2009

to improve resource efficiency in a national economy, resource efficiency should mainly be measured in terms of an energy efficiency indicator.

1 Introduction

Measures to increase energy efficiency have assumed an ever higher priority on the agendas of politicians, industrialists, consumers, and researchers since the first oil crisis in 1973. The dependence of most European countries on energy importation, as well as increasing awareness of climate change (symbolized by the Kyoto Protocol of 1997), have accelerated political action, both to save energy and to move into non-carbon energy generation. "Total resource efficiency"—including non-energy materials—was a keynote of the *2002 World Summit on Sustainable Development* in Johannesburg, South Africa.

To create viable policies for complex issues such as climate change and resource efficiency, politicians need data that can be transformed into indicators to establish guidelines for industry. In the following article we have developed a meaningful resource efficiency indicator as a tool for such policy decisions, based on a four-level hierarchy of natural resources. We begin by discussing the basic concept and roots of sustainable development as applied to natural resources and the minerals industry.

2 Basic Considerations

Rapid population growth accelerates the depletion of the Earth's resources of space, soil, water, mineral resources, and energy, leading to irreversible changes in ecosystems, deterioration of living conditions, loss of biodiversity, and envisaged exhaustion of non-renewable, as well as renewable, resources.

Although the Earth's system has always been changing, rapid population growth has dramatically intensified these changes. To avoid collapse of critical functions, a responsible management of the Earth's system is required. In addition to "soft" policies, such as international agreements on emissions and pollutants, interventionary national strategies are needed, such as controlling birth rates, and introducing scientific measures to improve our understanding of geohazards, global climate change, and increasing demand for natural resources. These measures will be used to guide the development of technologies for coping with these challenges.

Humankind has always depended on the use of energy and mineral resources for its technological and cultural evolution. As far back as the Bronze Age metal was used for tool-making, but the large-scale exploitation of non-renewable energy resources only emerged much later with the invention of the steam engine at the beginning of the 18th century. The energy and raw materials then required for heating, smelting, and ship-building called for replacements for wood, a renew-

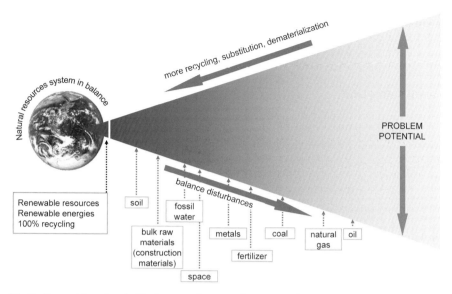

Fig. 1   Concept for using the Earth concerning natural resources: the "wedge of problem potential"

able biomass, as a raw material. Maintaining today's standard of living in both industrialized and developing nations depends entirely on the use of non-renewable energy and mineral resources. This dependency inevitably results in a massive interference in the ability of our planet to adapt to natural and human-induced changes, as well as to sustaining a "natural resources system in balance." By this term we mean a natural resources system that fulfils the intergenerational fairness requirements of the Brundtland (1987) report, as described below.

The scale of these interferences can best be illustrated by a "wedge of problem potential". In Fig. 1, various natural resources are ranked along the wedge according to the magnitude of their potential to disturb the resource equilibrium of the Earth. The width of the wedge is proportional to the disturbances caused: the larger the gap, the greater the problem. Because fossil fuels such as oil, natural gas, and coal are irreversibly consumed, they present the largest problem potential. Oil has been the fossil fuel of first choice because it is readily available and convenient to use. However, oil production is expected to peak soon (e.g., Gerling 2007). Consequently, oil is regarded as having the largest problem potential, and is therefore placed at the widest end of the wedge. Metals, by contrast, occupy a position halfway along the wedge because they can be recycled to a considerable extent. Raw materials that reach the environment in a dispersed state occupy an intermediate position between metals and fossil fuels.

The wedge of problem potential has been widening with the steady growth in consumption of non-renewable energy and mineral resources. The production of raw materials remained more or less constant for centuries, and even after the beginning of the Industrial Revolution in the second half of the eighteenth century, it took about another hundred years before raw material production and consumption commenced its relentless increase, gaining momentum over the years to reach the levels that we are accustomed to today.

Rammelsberg, the famous polymetallic base and precious metal mine in the Harz Mountains of Germany, is a good example to illustrate historical production trends. Figure 2 shows tonnage of ore produced since the mine opened in 968 AD until final closure in 1988. For several centuries, production fluctuated around a long-term average of 20,000 t/year, rising and falling due to historical events such as the Thirty Years' War (1618–1648). Production increased to 60,000 t/year from ~1,850, to 100,000 t/year in the 1930s, and to 300,000 t/year after 1950. Even a cursory examination of the 3,000 year-long production of the "old" metals such as gold, tin, copper, and iron, illustrates that man has consumed more in the last 50 years than in his entire previous history (Wellmer and Becker-Platen 2002).

The epochs into which the Tertiary and Quaternary geological periods are subdivided have the ending -cene, derived from the Greek word kainós which stands for "recent" (e.g., Holocene). Because humankind has become a major geological agent in shaping the future of our planet, the term Anthropocene has been coined to describe the most recent period in Earth's history. The Anthropocene started in the 18th century, when man's activities first began to have a significant global impact on the Earth's climate and ecosystems (Crutzen and Stoermer 2000). Considering our technological advances in the design of earth-moving equipment used in mining and construction man has already reached the same order of excavating capacity—about 35 billion m^3 per year—as natural processes such as erosion (Neumann-Mahlkau 1997).

It is tempting to correlate the rapid increase in production, consumption, and utilisation of mineral and energy resources with the increase in world population, which in October 1999 exceeded 6 billion people (in 1900, about 1.6 billion; in 1950, about 2.4 billion; and in 1965, about 3.2 billion). Up to the end of the last millennium, however, the bulk of mineral and energy resources consumption took place in industrialized countries that had only a very modest population growth. We were living in an "upside-down world": about 25% of the world's population lived in industrialized nations, and consumed 70–80% of the world's energy and mineral resources (with coal as one of the few exceptions). Since the turn of the millennium, however, we have witnessed the start of a new growth cycle, shown in Fig. 3, that is driven by a surge in demand in developing nations with large populations such as China and India. China has become the leading consumer in the developing world, as shown in Table 1. In 2006, it had the highest consumption in every major commodity with the exception of crude oil. Its shares of worldwide copper, aluminium, and steel consumption alone were 20.8, 25.4, and 32%, respectively. In 2006, India joined the five largest consumer nations of steel, crude oil, and coal, displacing Germany (which is the largest consumer in the European Union).

Population growth also brings soil, potable water, and space into the wedge of problem potential of Fig. 1. To satisfy the need for more arable land demanded by

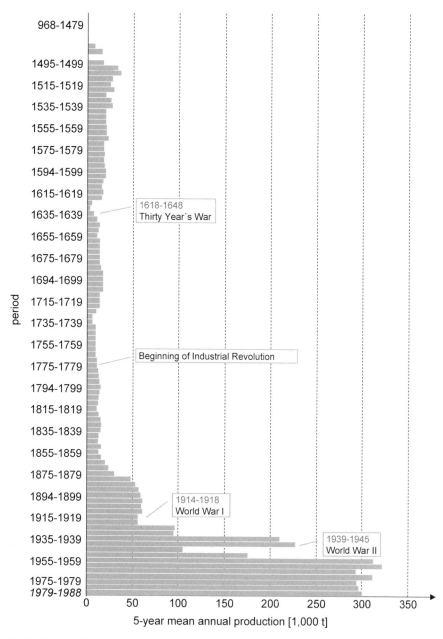

Fig. 2 History of ore production from the Rammelsberg mine (Harz Mountains, Germany) from 968 to 1988 (redrawn and modified after Dennert 1986)

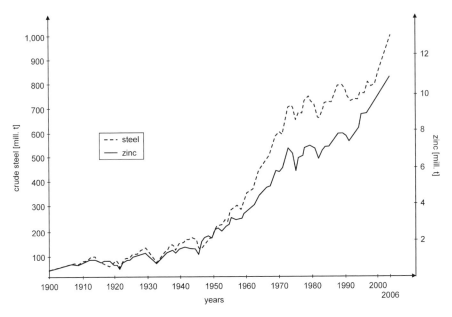

Fig. 3 (a) Worldwide production of steel and consumption of zinc 1900 to 2005

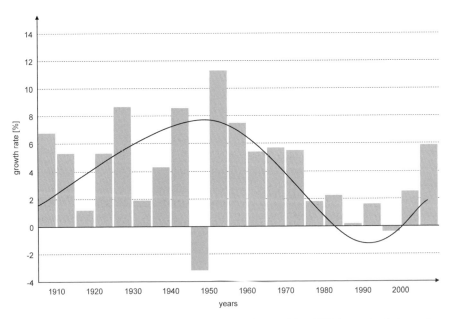

Fig. 3b (b) Five years average growth rate in percent from 1900 to 2005 for steel, relative to the preceding year. Each column represents the average growth rate over a 5 year period

Table 1 Consumption of copper, aluminium, and steel by the populated developing nations Brazil, India, and China, as percentage of global consumption

	1980	1990	2000	2006
Copper				
Brazil (%)	2.6	1.6	2.1	1.9
India (%)	0.8	1.2	1.5	2.6
China (%)	4.2	5.5	12.6	20.8
Aluminium				
Brazil (%)	1.9	1.6	1.9	1.8
India (%)	1.5	2.2	2.3	3.1
China (%)	3.5	4.5	13.6	25.4
Steel				
Brazil (%)	2.0	1.4	2.0	1.6
India (%)	1.6	2.6	3.4	3.8
China (%)	5.8	8.2	16.4	32.0

growing populations, and complicated by the conversion of arable land into land for dwellings, infrastructure, and industrial sites, forests are cut, and irreversible soil erosion increases. Moreover, the exploitation of fossil water resources in arid and semiarid lands intensifies: water is drawn down in excess of the capacity of natural recharge to replenish aquifers.

The purpose of our paper is to propose a responsible action plan for the future consumption of natural resources in a sustainable economy. We believe that the wedge of problem potential can be narrowed successfully by bringing consumption and use of non-renewable resources closer to a level compatible with a "natural resources system in balance." Because more efficient utilization of natural resources is the overall goal, possible routes towards narrowing the wedge are shown in Fig. 1, including developing new technologies for increasing use of renewable resources, and enhancing existing technologies for recycling, greater substitution, and dematerialization. Substitution includes replacing primary resources by both alternative primary resources and by secondary recycled ones.

As a guideline for these effective substitution strategies, we describe a new concept of a four-level hierarchy of natural resources within a framework of sustainable development, beginning with the concept of sustainable development in the minerals and energy industry.

3 Sustainable Development

3.1 Historical Considerations

Sustainable development evolved historically with the use and foreseeable depletion of renewable resources. Centuries ago, uncontrolled deforestation to clear land for farming or to obtain timber for construction, ship-building, and fuel wrought enormous devastation, changing the landscape of whole regions of Europe. In medieval times, wood as a raw material had the equivalent position of fossil fuels in our own age in the wedge of problem potential (Fig. 1).

The earliest practical concept for the utilization of sustainable resources in Germany was inspired by concern about deforestation. The head of mining administration in the important silver-mining district of Freiberg in Saxony, Oberberghauptmann Johann-Karl von Carlowitz, urged in his book *Sylvicultura oeconomica* published in 1713 that tree felling should not exceed re-growth. Being responsible not only for the mining and smelting operations in his district, but also for obtaining timber to construct underground mining structures and for procuring vast amounts of wood to make charcoal used in smelting silver ores, von Carlowitz is generally credited as being the "inventor" of the concept of sustainable development in Germany (Grober 1999); at least, he was the first person to use the German word for "sustainable"—*nachhaltig*—in this context. Although the idea of offsetting timber harvesting by re-planting probably developed simultaneously in several places (such as in the forestry administration of the Hapsburg empire in the southern Black Forest), it was von Carlowitz who published the concept first.

Other early examples for developing sustainable forest economy concepts are (Hausrath 1983):

–In the Siegerland, a former iron ore mining and smelting area in Germany, the so-called "Hauberg" regulations, already introduced in 1562–1565, stipulated harvesting cycles between 16 and 20 years long;
–Further afield, wood cutting for mining and smelting in balance with timber growth was practiced much earlier in the classical copper mining district of Cyprus. Constantinou (1981) calculated that in order to produce the 4 million tons of copper slag in over 3,000 years of copper smelting, the island's forests must have been statistically clear cut at least 16 times.

Another concept of sustainable exploitation dates back to silver mining in the Harz mountains in the 16th century. Duke Heinrich the Younger of Brunswick-Lüneburg revived silver production after the plague had devastated the whole region, by introducing the principle of "posteriority" in order to safeguard the interest of future generations. He and his successors pre-financed from their own pockets the driving of 20,000 m of adits for dewatering the mine workings during the first 80 years after restarting the mining activities, expecting repayment from future production. (The adits were hand-driven with hammer and chisel, this being the only method available in the 16th century before the use of explosives in mining was invented; Dennert 1972.)

3.2 Modern Concepts of Sustainable Development

Sustainable development is an abstract, normative principle, which was defined in the United Nations (UN) Report *Our Common Future* (the "Brundtland Report") as "development that meets the needs of the present without compromising the ability of future generations to meet their own needs" (Brundtland 1987, p. 8). This has become internationally the most widely accepted definition. It has subsequently been expanded to incorporate the notions of maintenance, rational use, and enhancement of the natural resources base with the general aims of facilitating ecological resilience and economic growth, while simultaneously holding out the hope of progress towards international equity (United Nations Environment Programme 1989).

The next step was the Rio Declaration at the UN Conference on Environment and Development (Rio de Janeiro 1992), set out in Agenda 21. The Brundtland Report dealt with "*inter*generational fairness". However, what was needed in addition was measures to guarantee "*intra*generational fairness". With this in mind, the Rio Declaration added three humanitarian objectives to the earlier aims of sustainable development of the Brundtland Report:

(1) to conserve the basic needs of life;
(2) to enable all people to achieve economic prosperity; and
(3) to strive towards social justice.

All three objectives initially were assigned the same priority.

While the three-pillar concept of sustainable development in Agenda 21 was based on clearly-stated ecological, economic, and social justice goals, it left open the question of how to achieve these goals. To address this shortcoming, further amendments have been proposed that would apply to both developed and developing nations:

- A four-cornerstone concept, in which the additional cornerstone is represented by "good governance", was suggested in The Mining, Minerals and Sustainable Development (MMSD) Project presented to the Global Mining Initiative Conference (Toronto 2002). This fourth cornerstone would ensure that both development and use of mineral resources would optimally benefit a nation, especially developing nations (IISD 2002).
- A fourth pillar of "research and development of new technologies" in order to facilitate sustainable global development was proposed by Wellmer and Becker-Platen (2002, 2007). As shown in Fig. 1, the narrowing of the wedge of problem potential—with concomitant minimization of balance disturbances in the natural resources system—requires development of new technologies for the utilization of more renewable resources, and for increased recycling, substitution, and dematerialization.

3.3 Guidelines for Sustainable Development

The definitions and requirements set out in the Brundtland Report and Agenda 21 are rather abstract proposals for the general understanding and acceptance of the tenets of sustainable development. Practical guidelines were needed. In 1993 the Enquete Commission on Protection of Man and the Environment, set up by the German Federal Parliament, undertook to formulate four general rules for the sustainable development of natural resources, which could be applied worldwide (Enquete-Kommission Schutz des Menschen und der Umwelt 1993). Rules 1 and 2 apply to resources, and Rules 3 and 4 deal with the resilience of the environment. Rule 1 is in essence inspired by the ideas of von Carlowitz (1713). Later, in 1994, a fifth rule was added by the German Expert Commission for Environmental Questions (Sachverständigenrat für Umweltfragen, SRU) concerning hazards and unacceptable risks to human health. A literal translation from the publication of the Enquete Commission in German language is given below:

- *Rule 1. Use of renewable resources:* The rate of consumption of renewable resources should not exceed the rate at which they can be regenerated.
- *Rule 2. Use of non-renewable resources:* The consumption of non-renewable resources should not exceed the amount that can be substituted for by functionally equivalent renewable resources, or by attaining a higher efficiency in the use of renewable and non-renewable resources.
- *Rule 3. Material and energy inputs:* Material and energy inputs to the environment should not exceed the capacity of the environment to absorb them with minimal detrimental effects.
- *Rule 4. Rate of anthropogenic input and environmental interference:* The rate of anthropogenic input and environmental interference should be measured against the time required for natural processes to react to and cope with environmental damage.
- *Rule 5. Hazards and unacceptable risks:* Hazards and unacceptable risks to human health caused by human activities are to be avoided.

3.4 Interpretation of Rule 2

The argument for simplifying Rule 2 to emphasize concentration on functionally-equivalent resources and new technologies is as follows: If we analyze *why* we need natural resources, we have to conclude that, with a few notable exceptions (such as nitrogen, potassium, and phosphate used as fertilizers in agriculture), it is not the metal or raw material as such that is important, but a function that is intrinsic to the material properties of the commodity. Thus, in the example of copper used for wiring, its electrical conductivity is its key function. However, other commodities can perform these functions just as well, commonly in conjunction with a fundamentally different technology. Thus, copper telephone wires were until recently extensively used for transmitting information, but have now been largely replaced by glass fiber cables made of silica, the supply of which is virtually inexhaustible. Similarly, wireless transmission of information using directional antennae or satellites obviates the need for copper cables. Each solution requires different materials.

Other examples are in photography and printing. Not so long ago, silver was needed for the capture of pictures, but today, digital cameras with completely different raw material requirements have largely replaced the use of film. In the field of printing, lead was formerly used as type metal, but it has been replaced today by offset or computer printing.

We are in fact in a position to choose alternative solutions to functions using three resource domains:

- All the resources from the geosphere (i.e., primary resources).
- All the resources from the technosphere[1] (i.e., secondary resources created by recycling).
- And last, but not least, the most important resource: human ingenuity and creativity.

Finding the solution for various functions essentially requires two things: optimization of substitution processes as discussed below, or the development of new technologies.

The stimulus to find new solutions comes from two main sources. On the one hand, human curiosity for the unknown and his ingenuity urges man to develop new methods, instruments, technologies, and processes. On the other hand, the price incentive motivates innovation. When there is a shortage of a commodity in a market economy, prices rise, triggering a feed-back control system of raw materials supply (Dalheimer 1999). The expectation of high returns will encourage inventiveness and creativity to find new solutions. On the supply side, this will include the discovery of and production from new deposits, and improving recycling rates. This is supported on the demand side by the initiation of new and more efficient processes, development of substitution technologies, material savings, and the invention of entirely new technologies that fulfil the same functions without the need of using scarce materials, thereby improving natural resources efficiency. The effectiveness of the price feedback control cycle has been illustrated by Wellmer and Becker-Platen (2002) and Wellmer (2003, 2009), exemplified by the molybdenum and cobalt price peaks of 1978, and the tantalum peak of 1980 (ETH-NSSI 2007). In many cases, the feed-back control system works so well that prices, after the price peak activated the control mechanism, may fall back below the former level. This was the case with the molybdenum peak at the end of the 1970s.

As stated above, the concept of finding different solutions does not work for the essential agricultural fertilizers nitrogen, potassium, and phosphate. They are as essential as clean air or water, and there is no replacement for them. For nitrogen and potassium, this is not a problem because the atmosphere is an effectively inexhaustible source of nitrogen and oceans are full of potassium. These two elements, therefore, take up a low position in the natural resources hierarchy, and will

---

[1] Technosphere is defined as the world as created by man, such as surface and subsurface constructions, machines, or waste dumps.

be discussed later. Phosphate, however, is different because there is no unlimited reservoir. Phosphate has one major advantage that can be exploited to achieve more efficient use. Unlike potassium, the solubility and therefore mobility of phosphates is very low. Therefore, with improved fertilizing technology (by adding just the amount that the plants need through computerized precision farming), by improving the phosphate uptake of animals (for example, via the enzyme phytase), and by improved recycling of wastes and manure, a concept for sustainability can also be developed for phosphates. The utilization of phosphate could develop into a complete and nearly closed-cycle system, which would move it into the sustainable, thin end of the wedge of problem potential (Fig. 1; Wellmer and Kosinowski 2003).

As a consequence, it is concluded that the market economy is perfectly capable of finding solutions for functions to fulfil Rule 2 (Sect. 3.3), or to find solutions that lead to sustainable, nearly closed-cycle resource systems. The key to this ability is human creativity and ingenuity.

4 Substitution Processes

4.1 Substitution on the Same Level

Substitution is essential if we want to improve natural resources efficiency, and to bring natural resources systems back into a balanced state (Fig. 1). To highlight the importance of improving natural resource efficiency, von Weizsäcker et al. (1995) coined the axiom "Factor 4", which stands for: "Doubling of wealth while at the same time curbing consumption of natural resources by half". We are of the opinion, however, that it is more important to increase energy efficiency than to increase efficiency in the use of other natural resources, as will be outlined below. The price incentive, coming from the tripling of oil prices in the last four years, sequentially followed by the prices of other energy commodities, was a major stimulus for improvement in energy efficiency. In addition, and working towards the same purpose, is the political ambition of the G8 nations, as the major energy consumers, to accelerate energy conservation and efficiency even faster in order to reduce CO_2 emissions. Statistics about long-term energy savings in industry appear in the press regularly, and are not discussed further here. Instead, we concentrate on substitution processes that can lead to improved energy efficiency.

In the following, we distinguish two kinds of substitution: substitution on the same hierarchical level, which includes the replacement of primary materials from the geosphere with recycled secondary material from the technosphere; and substitution in keeping with a four-level hierarchy of natural resources under the aspect of sustainable development.

As outlined in Sect. 3.4 above, there are no substitutes for the agricultural nutrients nitrogen, potassium, and phosphate. At the top, with the highest potential of substitutability, we place energy resources. Under energy resources we include all fossil fuels (oil, natural gas, and coal), as well as uranium (as fuel for nuclear power

Table 2 Geothermal energy potential (electricity and electricity/heat coupling) in Germany (after Paschen et al. 2003) in comparison to annual actual consumption of 2 EJ electricity and 5 EJ heat

Host/source	Electricity (EJ)	Heat without electricity/heat coupling (EJ)	Heat with electricity/heat coupling (EJ)
Crystalline rocks	1,100	1,600	2,600
Faults	45	65	120
Aquifers	9	23	50

stations), and all renewable energy resources. All sources of energy can directly or indirectly produce light, heat, and power by generating electricity, and can, therefore, in principle be interchanged. Even in Germany, where there is no active volcanism or natural steam generation, and is thus a geological environment with low enthalpy, the renewable energy potential of geothermal energy alone is 500 times higher than the current demand for heat and electricity (Table 2). A recent study for the USA concluded that, at depths between 3 and 10 km in crystalline basement rock formations, the energy potential that can theoretically be exploited by "enhanced geothermal systems" is 13,000 times greater than the annual energy consumption of primary energy for the entire country in 2005 (Massachusetts Institute of Technology 2006). A vast energy potential remains, even if we accept that, due to the second law of thermodynamics, only a fraction of the in situ energy can be won by deep drilling and hydrofracturing of large volumes of rock. In addition, solar power is available in every corner of the world. Solar insolation is 3–4 orders of magnitude larger than terrestrial heat flow. The potential, therefore, exists everywhere in the world to progress on the path for renewable energies, and make the use of fossil fuels and nuclear power merely a bridging technology.

Metals have a relatively high substitution potential. Steel alloy metals such as chromium, vanadium, manganese, nickel, molybdenum, etc. were in the past considered to be strategic because of their importance to a nation's economy and their limited substitution potential. To a high degree, the constraint on substitution was due to standardization, which of course can be changed in times of metal scarcities (Wellmer 1998). Steel alloy metals can replace each other in various applications. The recent drastic price increases stimulated many such substitution processes. The feed-back control system for raw materials supply (Sect. 3.4)—with the commodity price acting as a regulator—can also be observed at work in the current boom cycle. As an example, in one German steel plant, a certain quality of steel used for high quality line pipes required the microalloy elements niobium and vanadium. By adapting and changing the rolling mill procedure, 0.02% titanium could substitute for 0.04% vanadium, leading to a significant cost saving (Bannenberg 2007; Dillinger Hütte, personal communication). Although this is only a small step towards increasing materials efficiency, the accumulation of such small optimisation increments over time can lead to significant material savings. The Eiffel Tower in Paris could be built today with just 2,000 t of steel instead of the 8,000 t used from 1885 to 1889 (Lurgi 1992).

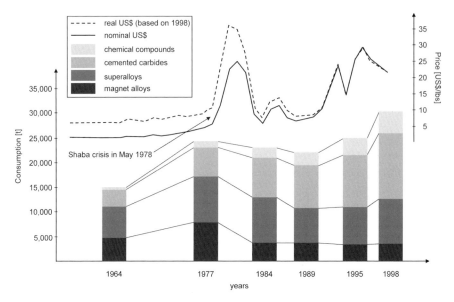

Fig. 4 Cobalt: mean annual prices in US$/lb and consumption in the Western World by main use (Wellmer and Dalheimer 1999)

The feed-back control system of raw material supply can lead to unexpected substitution processes if scarcities of raw materials arise and price spikes create incentives for completely new solutions. This is well illustrated with the example of cobalt during the 1978 cobalt crisis, which was caused by politically motivated interruptions of production in the principal supplier, Zaïre (now the Democratic Republic of the Congo). Due to the crisis, the price of cobalt rose significantly, and new materials were suddenly invented for areas of application where formerly cobalt was considered strategic (because it had been thought to be irreplaceable). In the case of permanent magnets, ferrites were applied, which displaced a significant market share of cobalt (Fig. 4). Whereas before the crisis, 30% of the cobalt supply was being used to manufacture permanent magnets, after the crisis and the ensuing invention of ferrites, the share of cobalt needed for this application fell to only 10% (Wellmer and Dalheimer 1999).

When dealing with substitution at the same level, the replacement of primary materials by secondary recycled materials also has to be considered. This is well illustrated by the recycling of water in industrial processes, or the recycling of metals that are used but not consumed, as exemplified by copper which can be recycled endlessly without loss of quality. Other metals that are less noble (i.e., with a lower redox potential) have certain limitations to the possibilities of recycling. Aluminium and nickel, for instance, are less recyclable due to the many alloys in which these metals are used. The secondary metals will therefore either have to be blended with primary material to maintain quality, or a decrease in quality will have to be accepted

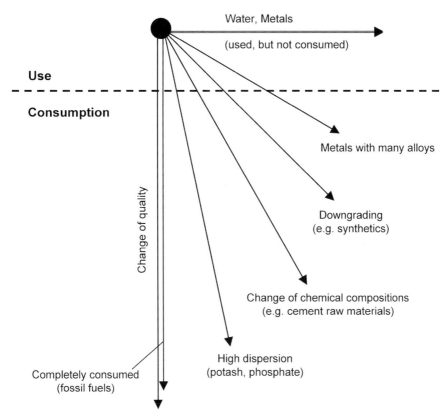

Fig. 5 Changes of quality during recycling of materials from the geosphere and technosphere (from Wellmer and Kosinowski 2005)

(unless it is acceptable for production costs and energy consumption for further refining steps to rise significantly, as discussed in more detail below).

Quality decrease, so-called downgrading, is generally also the fate for recycled synthetic materials. The changes of quality during recycling are shown schematically in Fig. 5. Energy, normally at the top of substitution potential, is in this case at the bottom of the spectrum for recyclability, because it is totally consumed in the process of power and heat generation, increasing entropy, and can therefore never be recycled. Although, according to the first law of thermodynamics, energy cannot be lost, the usable portion required for human activities, the so-called "exergy", is irrevocably consumed und thus irretrievable. Materials that are used in a state of high dispersion, such as fertilizers, can also not be recycled. Examples of highly dispersive use of metals include zinc used in skin creams, titanium in paint, platinum group metals in galvanic processes (Hagelüken et al. 2005), and, so far, indium in flat television screens (Hagelüken 2007, personal communication).

Because the absolute amount of resources in the technosphere increases with consumption (as discussed in Sect. 2), this offers the possibility for increased

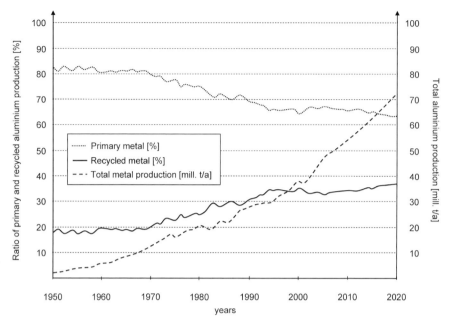

Fig. 6 Total production, and proportion of primary and secondary production of aluminium (source Gerber 2007)

replacement of primary resources from the geosphere through recycling. Aluminium is a good example of this process: Fig. 6 shows how the increased use of secondary, recycled aluminium replaces primary material. In Fig. 1, this corresponds to shifting to the left of the wedge of problem potential, towards the goal of a natural resources system in balance. Therefore, recycling not only makes sense because it saves resources from the geosphere, but also because it encourages the greatest possible use of resources from the technosphere, actually offering the possibility of saving energy. If, as we maintain, energy is our most valuable resource, then finding the optimum recycling rate for metals in order to minimize energy input and curb emissions (like CO_2) into the environment is definitely worth striving for.

Pure secondary metals require less energy for recycling than does the smelting of primary ores and concentrates. Recycling aluminium requires only 12% of the energy needed to produce refined aluminium from primary ores (Krone et al. 1990); for copper the corresponding figure is 15%, and for lead, 35% (Table 3). However, the higher the degree of dispersion of a metal in a secondary raw material, the more energy is necessary to purify it. Figure 7 shows the optimum recycling rates for aluminium used in light-weight packaging material (LPM; Rombach 2006). The total energy usage in processing and re-melting reaches a minimum at ~90% recycling rate, and thereafter rises exponentially as the last atom of Al is chased down.

If the aluminium-bearing fraction from a waste-sorting plant contains less than about 40% Al, with the remainder consisting predominantly of undesirable organic

Table 3 Energy savings by recycling pure secondary material

Metal	Energy savings (%)
Steel	74
Aluminium	88
Copper	85
Lead	65
Paper	64
Plastics	80

Sources: Krone et al. (1990) and Gerber (2007).

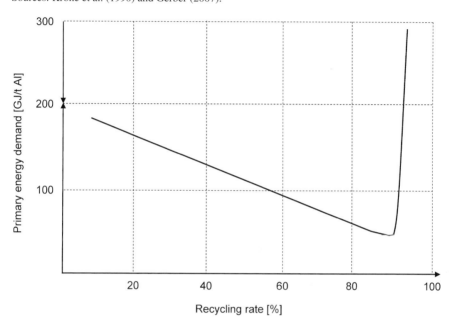

Fig. 7 The optimum recycling rate for aluminium used in light-weight packaging material is 90%. Any higher or lower rate results in higher energy demand (redrawn and modified after Rombach 2006)

waste, this aluminium-bearing fraction cannot be re-melted directly. However, with a combination of mechanical and thermal processing, it is possible to obtain a high-quality fraction with about 99% Al, which can be re-melted to produce a pure aluminium metal product. At the end of this process, ~90% of the aluminium in the original LPM waste has been recovered (i.e., 90% recycling rate; Fig. 7). This recycling rate currently represents an upper efficiency limit due to the increasing energy requirements of additional processing steps necessary for sorting and processing the last remaining wastes. Moreover, if one includes the energy used in the collection of LPM, then the optimum recycling rate (for minimum energy input) is further reduced as the increased effort required to collect remaining wastes produces diminishing returns (Quinkertz et al. 2001).

The energy input for the production of a metal commodity can be minimized by using a mixture of primary and secondary materials as shown by Wellmer and Becker-Platen (2001). There is another aspect for finding an optimum for mixing primary and secondary materials, namely solving the problem of impurities in secondary materials and thereby maximizing the reuse of metal scrap. Verhoef et al. (2004) point out that production systems and consumption patterns yield a mixture of waste materials containing an assortment of metals. The physical combinations and chemical compositions introduced in products and materials cause carry-over of impurities during recycling, which results in off-specification secondary metals and alloys. As a consequence, a fraction of the recovered material is useless, unless one wanted to invest more energy into further refining. The current practice of mixing low-quality secondary metals with primary metals prevents considerable loss of stocks, and optimizes recycling rates. Markets require both high-quality grades and the absence of specific impurities. For example, copper induces brittleness in steel and adversely influences rolling properties. Therefore, in the raw materials market for steel plate, only very low amounts of copper (less than 0.03%) are tolerated in pig iron and steel scrap. This, at present, is a big problem in the European steel scrap market (Garside 2007). Another example would be the recycling of many aluminium alloys, which usually contain various amounts of iron, zinc, and copper. To bring such mixtures of alloys back to required specifications they are either blended or diluted with wrought aluminium alloys with very low impurities. Metallurgists use the term "solution by dilution" (Boin 2008, personal communication).

4.2 Substitution According to a Hierarchy of Natural Resources

4.2.1 Introduction of the Hierarchy of Natural Resources

We propose a four-level hierarchy of natural resources for substitution between levels, with the following general rule applying to all substitution processes: lower value resources should replace higher value resources, whenever possible.

This proposed hierarchy (Fig. 9) was first introduced by Wellmer and Stein (1998), then presented in a poster presentation by the second author of this article at the International Geological Congress 2000 in Rio de Janeiro (Wellmer 2000). The concept was later elaborated by Wellmer and Kosinowski (2005), and reproduced in modified form in the German Federal Government publication "Environmental Data for Germany" (BGR, D_STATIS, UBA 2007).

One can argue about which natural resource is the most important for mankind: water, soil, clean air, or energy. There is no doubt, however, that most problems pertaining to natural resources can be solved provided we have enough affordable energy at our disposal: any sewage can be treated and reused, saline water desalinated, soil erosion by deforestation reduced if we stop cutting down forests for biofuel, air pollution can be drastically reduced, and the CO_2 input into air can be reduced by carbon capture and storage. Depending on energy prices, metal recycling can be increased and lower grade deposits can be brought into production. Unlimited

Fig. 8 Four-level hierarchy of natural resources with respect to sustainable development

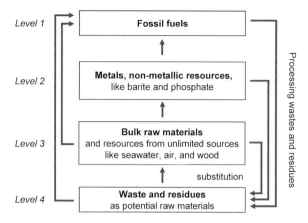

Fig. 9 Recycling proportion of glass in receptacles Germany (Wellmer and Kosinowski 2005). Development of recycling proportion over time defined as the amount of old glass from receptacles recycled in relation to total amount of old glass from receptacles

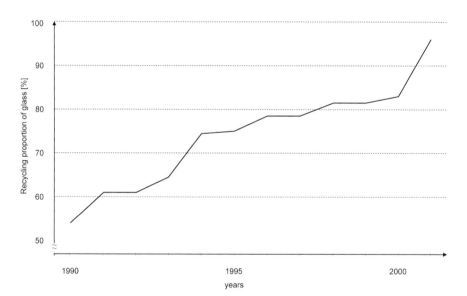

resources like nitrogen in the air or magnesium in sea water can be made available to mankind if there is affordable energy to produce nitrogen fertilizer or magnesium metal from these dispersed sources. Formerly, fertilizers were produced from nitrate ores, a naturally occurring but limited resource. Today, nitrates are mostly synthetically produced from atmospheric nitrogen using the Haber-Bosch process, preferably at sites where cheap natural gas is available. Similarly, magnesium can be extracted out of sea water where cheap hydro-electricity is available.

Table 4 Recycling rates for waste material in Germany in 2002

	Amount (Mt)	Recycled (Mt)	Recycling rate (%)
Demolition rubble (2002)	52.1	35.7	68.5
Road construction material (2002)	16.6	14.2	85.5
Flue-gas desulfurization: gypsum from coal fired power plants (2004)	7.7	6.9	89.6
Ash from hard coal power plants (2004)	5.5	5.4	98

Sources: Schulz (2006) and Ifeu and IAÖ (2006).

Level 1: Energy. Due to its overriding importance and its position at the wide end of the wedge of problem potential in Fig. 1, we place energy at the top of the four-level hierarchy of natural resources outlined in Fig. 8.

Level 2: Natural Ores and Pure Scrap. The next hierarchy level comprises raw materials from deposits that were formed by geological processes of natural enrichment. Included are all metal deposits, and some non-metallic deposits such as phosphates or barite. This level also includes the resources of the technosphere, such as scrap metal or other materials that can be recycled at the same or nearly the same level (as shown in Fig. 5).

Level 3: Bulk Raw Materials. The third level comprises anything that is available in almost unlimited amounts on Earth, such as bulk materials used in the construction industry (e.g., granite, basalt, sand, clay), and raw materials that can be produced from air (e.g., nitrate fertilizer) or from sea water (e.g., boron, potassium, rock salt, or magnesium). For construction purposes, it also includes the renewable resource wood.

Level 4: Wastes and Residues. The lowest level of the hierarchy includes waste and residues, and material from the technosphere as potential raw materials for secondary use (e.g., waste materials derived from beneficiation plants or from burning, smelting, or refining higher value resources). Examples for reuse are: the coarse fractions of rejects from the beneficiation of limestone-hosted metal ores, which can be used in road fill; gypsum from flue gas scrubbing, which can replace primary gypsum; or ash from coal-fired power plants, which can be used for making cement, replacing primary raw materials (Table 4). In accordance with Rule 2 of the guidelines described in Sect. 3.3, to find functional replacements for consumed primary materials, waste should be recovered as much as possible as a replacement of primary resources.

Other concepts have also been used to rank natural resources with respect to depletion and sustainable development. Gordon et al. (2006) predicted that we will see a renewed engineering emphasis on using metals like copper and zinc more efficiently, and accompanied by increased use of abundant alternative materials, principally iron and its alloys, aluminium, and magnesium. In the context of our hierarchy of natural resources in Fig. 8, the prediction of Gordon et al. corresponds to internal substitution on Level 2.

Returning to the general rule that lower value resources should, whenever possible, replace higher value resources, it must be kept in mind that in a market economy this does not of course happen automatically. Only if the hierarchy proposed by us is reflected in prices can we expect the principle to work. Take the example discussed below in Sect. 4.2.2 (Example 3) of installation of improved heat insulation in a building. In the absence of price pressures in the era of cheap energy in the 1960s, insulation materials were used sparingly in spite of the fact that everyone knew of the energy saving potential of heat insulation.

In practice, the implementation of this hierarchy requires long-term planning, even if we make allowance for the economic gains that can be achieved. We again take heat insulation as an example: an investment today can save energy and, therefore, money tomorrow. Companies generally have guidelines about a required payback period for an investment, or a required return on investment. Private consumers, however, normally do not think in terms of payback periods and return on investment, but rather in terms of the time value of money. The time value of money means that money in the hand today is worth more than the same amount in the future (e.g., Wellmer et al. 2007). To overcome this short term thinking, governments in Europe are introducing laws to force private consumers to invest more in energy saving measures with longer payback periods.

4.2.2 Application of the Rule "Lower Value Resources Should Whenever Possible Replace Higher Value Resources"

The principle of vertical substitution resulting from the hierarchy of natural resources, and examples of horizontal substitution increasing energy efficiency, are illustrated with the following examples:

Example 1: Closed-Loop Systems for Non-Energy Materials: Reuse of Residues

At the 2002 World Summit on Sustainable Development (in Johannesburg, South Africa), a Plan of Implementation was agreed upon to promote sustainable patterns of consumption and production. The Plan of Implementation called for a 10-year promotion and development schedule, now referred to as the Marrakech Process. One of its proposals was to promote new models of economic development, such as closed-loop systems for non-energy materials in which losses are minimized. All industrialized nations have programmes to optimize waste streams with the aim of achieving such closed-loop economies. Japan's 3R initiative (reduction, reuse, recycling; Bleischwitz and Bringezu 2007) is an example, whereby the Japanese coal industry has reported a reutilization rate of 97% of coal ash in 2006 (JCOAL 2007).

Other initiatives include the Scientific Panel on Resource-Efficiency and the Environment, planned by the United Nations Environment Programme (UNEP), the European Commission's Thematic Strategy on the Sustainable Use of Natural Resources, and Germany's Waste Avoidance, Recovery, and Disposal Act (BMU 1994). The latter waste management act maximizes the rate of recycling within a framework of high energy prices and strict control of dangerous residues. Residues for disposal are penalized, thus providing an incentive to find alternatives, or to avoid or reuse the residues (see Table 4 for various waste recycling rates in Germany). Figure 9 illustrates the increase in the rate of recycling of glass receptacles since 1990. The waste management act leads to improved reuse, which saves other primary materials higher in the hierarchy, such as quartz sand, soda, and dolomite, but also energy. Because glass is an amorphous substance with a lower melting point than crystalline primary quartz materials, there can be energy savings. CO_2-emissions are reduced, not only by using less energy, but also because it is not necessary to add additional carbonates which also release CO_2 in the process. Bosse (1999) quantified the mineral, energy, and CO_2-savings as compared to glass production from primary natural resources. A contribution of 10% of recycled glass receptacles results in a reduction of 3% of energy usage. In 1997 in Germany, about 1.5 Mt of CO_2 emissions was saved by glass recycling. In the context of the hierarchy of Fig. 8, raw materials from Levels 1, 2, and 3 are replaced by waste from Level 4.

Example 2: Energy Optimisation by Using Waste Material and Horizontal Substitution

The reduction of iron oxide to iron requires energy to smelt the iron ore, and coke is required as a reductant and as a supporting grid in the blast furnace. However, pulverized coal, oil, and synthetic and plastic waste can be used as an energy source as well as coke. Figure 10 shows how the efficiency in German steel plants improved from 1950 to 2006 in terms of decreased consumption of reductants and energy sources such as coke, coal, and hydrocarbons (VDEh 2007). In the context of the hierarchy of natural resources (Fig. 8), there is replacement of energy in Level 1 by waste of Level 4, and horizontal substitution in Level 1.

Example 3: Energy Savings by Better Insulation

Heating and cooling accounts for more than 50% of primary energy consumption in the industrialized world. Improved insulation of houses is a very cost effective method to increase energy efficiency. The insulation material consists of raw materials, such as glass for triple glazing instead of double glazing (as has been required for several years for new buildings in Sweden), or rock wool for improved insulation of walls. The production of these materials requires energy. Consequently, an initial investment in energy and non-energy materials improves long-term energy efficiency in accordance with the hierarchy of natural resources (Fig. 8). This is shown schematically in Fig. 11, where two cases for insulating buildings are compared. Case A is the standard case. Case B benefits from a greater initial investment in insulation material, such as additional window glass or rock wool insulation. The diagram shows the cumulative energy balance over time. There are two elements of the energy balance for both cases: the energy input into manufacturing the insulation material, and the energy consumption for heating. Of course, the energy input into building and insulation material occurs only at the fabrication and construction stage. After construction of the building, energy is constantly used for heating, but

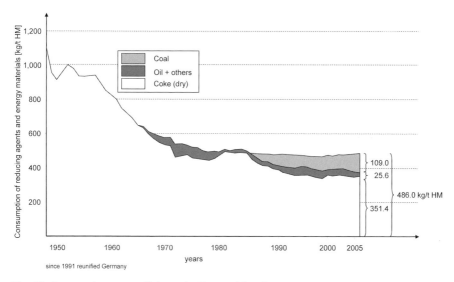

Fig. 10 Increase in energy efficiency in German blast furnaces as measured by the decreasing amount of energy materials and reductant used (coke, coal, and hydrocarbons; VDEh 2007). HM on y-axis = hot metal

at a lower rate for the better insulated case B. The two lines for cases A and B will intersect at some time in the future, at which point the net energy usage of building B will be less than building A, and savings will be made. This concept for saving energy is used below to suggest reasonable indicators for measuring raw material efficiency. In the context of the hierarchy of natural resources, energy from Level 1 is replaced by effectively unlimited resources of Level 3. Using Level 3 resources requires energy, but this leads to a reduction of overall energy consumption.

Example 4: Energy Savings in the Transport Sector

The construction of road bridges and tunnels is normally only considered from the point of view of time efficiency and safety. However, it can also be regarded as a contribution to energy efficiency by shortening driving distances. There is an obvious initial investment in construction materials and steel to build bridges and tunnels, which are produced with the input of raw materials and energy: cement is produced in kilns or rotary furnaces, and steel is produced from iron ore and scrap in blast furnaces and steel converters. However, energy savings are made in the longer term due to the shorter driving distance (Fig. 12a). In this case, the break-even point of Fig. 11 is reached when the energy invested in steel and construction materials plus the energy used to build the tunnel equals the amount of energy saved by vehicles driving a shorter distance. From that point on in time, the energy balance is positive, and, moreover, CO_2-emissions are reduced. So we are replacing energy of Level 1 with natural resources of Level 2 and 3 in the hierarchy of Fig. 8.

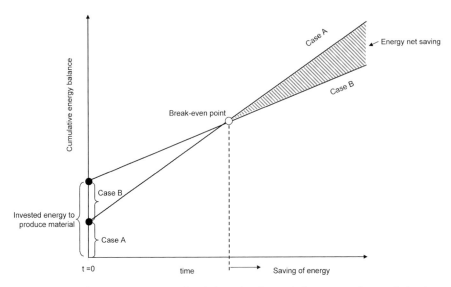

Fig. 11 Cumulative energy usage over time in home heating: early investment of energy in insulating materials during construction may result in a net saving of energy in the future (after Wellmer and Kosinowski 2005)

The same holds true for investments in the construction of better road systems to avoid traffic-jams and stop-and-go-traffic, which are a significant waste of energy, particularly in and around larger cities. However, a counter view holds that more efficient traffic management only encourages more traffic, and thus negates all the energy and CO_2-emissions savings, although this is likely an inevitable result of population and economic growth.

Similar arguments for efficiency can be made for public transportation systems, especially railways. Plans for high speed railway systems in Europe like the TGV in France, the AVE in Spain, or the ICE in Germany are normally only seen from the perspective of time efficiency, but can again also be considered in terms of energy efficiency (Wellmer and Kosinowski 2005). In the early days of railway building, routes were designed to minimize gradients, and so were quite circuitous. By contrast, modern railway lines are designed to connect major cities by the most direct, linear routes, requiring major investments in bridges and tunnels. For example, nearly half of the 324 km high-speed ICE line between Hannover–Göttingen–Kassel–Würzburg in Germany consists of bridges and tunnels (Fig. 12b; Geissler 1994). Once built, these modern rail links provide a convenient and energy-efficient transport system, and encourage the switch from air and private car travel to public transport. For example, the AVE line in Spain from Madrid to Barcelona (635 km) reduced air travel on this route by 18% (Feth 2008). Thus, after the initial investments in materials and energy are accounted for, modern railway systems provide savings in time, energy, and CO_2-emissions.

Building bridges using concrete reinforced with steel rebar instead of solely from steel is another good example of the application of the natural resources hierarchy.

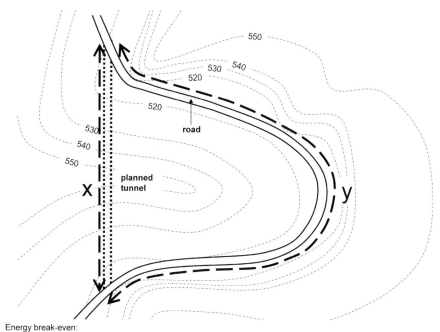

Energy break-even:

Energy for tunnel construction = Saved energy of all cars not driving y - x

Fig. 12 (**a**) Energy savings from building a tunnel to shorten a road route. (**b**) Railway grid in Germany around the city of Hannover with the new high speed ICE line

Table 5 Comparison of steel consumption in various types of bridges (length 60 m) per 1 m^2 traffic area

Type of bridge	Steel consumption (t/m^2)
Pre-stressed concrete bridge	0.25
Concrete pillars with steel arches	0.40
Steel truss construction	0.55

Source: Schröder (2004).

Steel is a commodity on hierarchy Level 2, but the materials for making cement for the concrete are commodities of Level 3, or, if the ash from coal-fired power plants is used in cement production, Level 4. Hence, we are replacing material of Level 2 with materials of Levels 3 or 4. Table 5 shows the reduction of steel consumption by changing from pure steel construction to reinforced concrete.

The use of wood instead of steel used for the construction of pedestrian bridges, airplane hangars, or other constructions requiring wide spans is a similar example. Here, so-called gluelam (glued laminated timber) that bears loads over considerable spans can replace steel. Wood as a renewable resource of Level 3 replaces steel of Level 2 at significant energy and CO_2-emissions savings. Using gluelam as a

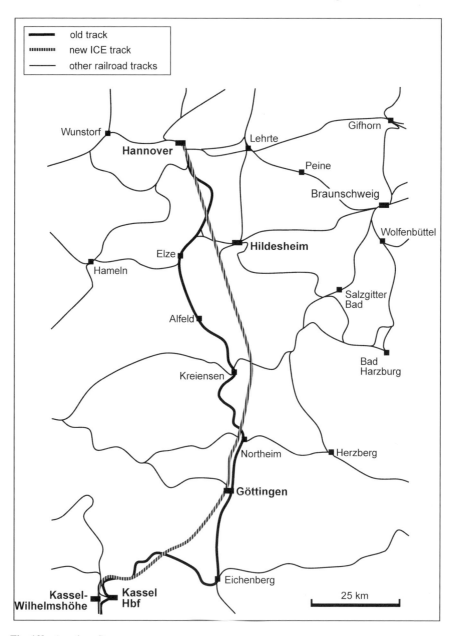

Fig. 12b (continued)

construction material also has a positive CO_2-balance due to the consumption of CO_2 during tree growth by photosynthesis. It has been calculated that 1 m^3 of wood stores 0.9 t CO_2, and substitutes for 1.1 t CO_2 emitted by steel production, for a total saving of 2.0 t CO_2 (Frühwald et al. 2003).

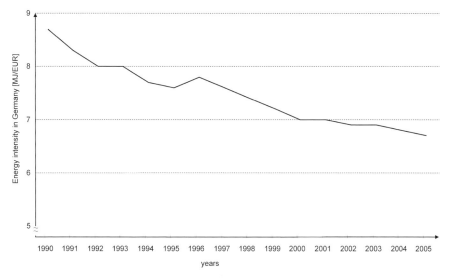

Fig. 13 Indicator: Energy intensity in Germany (BGR, D_STAT, UBA 2007). The ratio of energy consumed in MJ and gross national product in EUR over time

Another example where substitution on the same hierarchy level of natural resources in Fig. 8 leads to energy savings is described by Glimm (2001). He published a study showing how a car constructed from aluminium instead of steel could lead to energy savings. Aluminium mining and smelting requires more energy than steel per tonne, but aluminium and its alloys are lighter than steel, and, thus, fuel savings can be achieved. The principle shown in Fig. 11 can be applied here by replacing time on the x-axis with kilometres driven. Glimm (2001) estimated that the break-even point for a lighter automobile would be at 55,000 km. The initial capital outlay to buy a more expensive car with a lighter aluminium body is compensated for in the long run by fuel savings.

5 Resource Efficiency Indicators

The four examples in Sect. 4.2.2 demonstrate that by applying the concept of a hierarchy of natural resources, significant energy savings can be achieved (Level 1) by investing in non-energy materials and/or using energy more efficiently. One can recoup the cost of energy invested at the beginning (i.e., during construction or manufacture) through reduced energy expenditures during the lifetime of an operation. The investment in non-energy materials, however, cannot normally be recovered unless recycling on the same level is an option. So, figuratively speaking, one can make a saving on energy starting from the break-even point, but the non-energy materials are considered to be mostly sunk capital.

As we have shown, energy savings by substitution are quantifiable, and as such can be expressed as indicators to be used in policy making. Indicators used in the past relied on measuring resource efficiency solely in terms of production or total consumption of raw materials per unit of gross national product. Such measures had the obvious shortcoming of leaving the prospect of energy savings out of the equation. Due to the fact that energy investments made at an early project stage can be recouped (or even exceeded) through energy savings during the life of a product or operation, a total consumption-based indicator can be misleading, and its application in policy making can lead to unintended consequences. For example, taxing primary resource consumption regardless of the position in the natural resources hierarchy could result in some short-term savings but at the greater expense of higher energy consumption in the long term. Total consumption-based indicators are at best only useful in industrialised countries as secondary indicators.

In our opinion, therefore, an energy-based primary resource efficiency indicator (REI) that measures efficiency by consumption of energy in relation to a chosen unit of economic activity would be a far more effective tool for policy-making. There are two options for an REI: the amount of energy consumed per unit of gross national product (GNP; the intensity of use factor of Malenbaum 1978); or the inverse relationship (i.e., the amount of GNP per unit of energy consumed). The first option is the indicator used by the German government (BGR, D_STATIS, UBA 2007), and is referred to as the energy intensity of the whole economy (Fig. 13). The political goal in Germany is to double national energy productivity by 2020 (from 1990) by reducing energy intensity by 50% over the same period (BMU 1998).

6 Final Remarks

The concept of a four-level hierarchy of natural resources should lead to greater resource efficiency and responsible use of non-renewable resources, and thus contribute to the ultimate goal of a sustainable world economy. It could help mitigate potential disturbances of the natural resources system shown in Fig. 1, and also help reduce environmental and social disturbances. These we dealt with under the concept of four cornerstone described in Sect. 3.2. Responsible companies and individuals have learned from past mistakes that a balance needs to be maintained between the conflicting requirements of economy, ecology, and social justice. We are still quite a long way from a sustainable and just economy, but are beginning to appreciate and tackle the challenges that remain to everybody who is producing and consuming energy and exploiting natural mineral resources.

Acknowledgments The authors thank W.G. Ernst and J.P. Richards for their critical review, B. Bognar, D. Large, and J.C. von Maltzahn who critically read the manuscript and made numerous suggestions for improvements, and E. Westphale for preparing the drawings. Shortcomings, of course, are only the fault of the authors.

References

BGR, D_STATIS, UBA, Federal Institute for Geosciences and Natural Resources, Federal Statistical Office, Federal Environment Agency (2007) Environmental Data for Germany. Dessau, Umweltbundesamt (UBA)

Bleischwitz R, Bringezu S (2007) Global resource management. Policy Paper 27, Development and Peace Foundation, Bonn: 12 pp

BMU (Bundesministerium für Umwelt, Naturschutz und Reaktorsicherheit) (1994) Gesetz zur Vermeidung, Verwertung und Beseitigung von Abfällen. Bundesgesetzblatt I 1994: 2705–2728

BMU (Bundesministerium für Umwelt, Naturschutz und Reaktorsicherheit) (1998) Nachhaltige Entwicklung in Deutschland. Entwurf eines umweltpolitischen Schwerpunktprogrammes, Bonn, 147 pp

Bosse HR (1999) Nachhaltiges Wirtschaften durch Hohlglasrecycling. Fakten, Analysen, wirtschaftliche Hintergrundinformationen 4, Federal Institute Geosciences and Natural Resources BGR, Hannover, 2 pp

Brundtland GH (1987) Our Common Future—Report of the World Commission on Environment and Development. Oxford University Press, Oxford, UK

Constantinou G (1981) Geological features and ancient exploitation of the cupriferous sulphide orebodies of Cyprus. Acta of the International Archaeological Symposium, Larnaca, Cyprus 1.-8. June 1981, Perides Foundation (Larnaca): 13–23

Crutzen PJ, Stoermer EF (2000) The "Anthropocene". Global Change Newsletter 41: 12–13

Dalheimer M (1999) Regelkreis zur Rohstoffversorgung. In: Wellmer F-W, Becker-Platen JD (eds) Mit der Erde leben. Springer Berlin Heidelberg, New York

Dennert H (1972) Der Westliche Oberharz als erstes geschlossenes Industriegebiet im Lande Niedersachsen. Erzmetall 25(12): 640-644

Dennert H (1986) Bergbau und Hüttenwesen im Harz vom 16. bis zum 19. Jahrhundert, dargestellt in Lebensbildern früherer Persönlichkeiten. Pieper, Clausthal-Zellerfeld, Germany

Enquete-Kommission Schutz des Menschen und der Umwelt (1993) Verantwortung für die Zukunft—Wege zum nachhaltigen Umgang mit Stoff- und Materialströmen. Economica-Verlag, Bonn, Germany

ETH-NSSI (Natural and Social Science Interface) (2007) Workshop Scarce Raw Materials. Davos, Switzerland, September 1/2, 2007

Feth GG (2008) Wie im Flug vergeht die Zeit. Frankfurter Allgemeine Zeitung, August 1, 2008

Frühwald A, Welling J, Scharai-Rad M (2003) Comparison of wood products and major substitutes with respect to environmental and energy balances. Paper ECE/FAO Seminar: Strategies for the Sound Use of Wood, Poiana Brasov, Romania, March 24–27, 2003

Garside B (2007) ArcelorMittal squeezes scrap suppliers to cut out copper. Metal Bulletin Monthly September 3: 18

Geissler H (1994) Die Tunnel im Nordabschnitt der Schnellbahnstrecke Hannover-Würzburg. Beitr Ber Naturhist Ges Hannover 11: 1–73

Gerber J (2007) Strategy towards the red list from a business perspective. Paper ETH workshop Scarce Raw Materials, Davos, Switzerland, 1/2 September 2007

Gerling P (2007) Commentary on crude oil and natural gas liquids. In: World Energy Council (ed) Survey of Energy Resources, 21st ed., World Energy Council, London: 41–92

Glimm S (2001) Aluminium, ein nachhaltiger Werkstoff. Metall 12: 738–742

Gordon RB, Bertram M, Graedel TE (2006) Metal stocks and sustainability. PNAS 103(5): 1209–1214

Grober U (1999) Der Erfinder der Nachhaltigkeit. Die Zeit, 48, November 25, 1999: 98

Hagelüken C, Buchert M, Stahl H (2005) Stoffströme der Platingruppenmetalle. GDMB-Medienverlag, Clausthal-Zellerfeld, Germany

Hausrath H (1983) Geschichte des deutschen Waldbaus, von seinen Anfängen bis 1850. Schriftenr Inst Forstpolitik Univ Freiburg 3, 428 pp

Ifeu, IAÖ (2006) (Institut für Energie- und Umweltforschung Heidelberg, Öko-Institut für ange-wandte Ökologie, Darmstadt) Beitrag der Abfallwirtschaft zur nachhaltigen Entwicklung in Deutschland-Industrieabfälle. Umweltbundesamt, Dessau, 419 pp

IISD (2002) International Institute for Environment and Development: Breaking New Ground—The MMSD Final Report http://www.iied.org/mmsd/finalreport/index.htm

JCOAL (2007) Utilization of coal ash http://www.jcoal.or.jp/coaltech_en/coalash/ash/02e.html

Krone K, Krüger J, Orbon H, Sommer HW, Vest H (1990) Ökologische Aspekte der Primär- und Sekundäraluminiumerzeugung in der Bundesrepublik Deutschland. Metall 44(6): 559–568

Lurgi AG (1992) Jahresbericht 1991, Frankfurt/Main, Germany

Malenbaum W (1978) World Demand for Raw Materials in 1985 and 2000. E/MJ Mining Infor-mational Service, New York

Massachusetts Institute of Technology (2006) The Future of Geothermal Energy. Boston, MA 367 pp; http://www1.eere.energy.gov/geothermal/future-geothermal.html

Neumann-Mahlkau P (1997) Anthropogenic material flow—a geological factor. Proceedings of 30th International Geological Congress 2 and 3, pp. 61–66

Paschen H, Oertel D, Grünewald R (2003) Möglichkeiten geothermischer Stromerzeugung in Deutschland-Sachstandsbericht. TAB Arbeitsbericht 84, Deutscher Bundestag, Berlin, Germany

Quinkertz R, Rombach G, Liebig D (2001) A scenario to optimise the energy demand of aluminium production depending on recycling quota. Resources, Conservation and Recycling 33: 217–234

Rombach G (2006) Limits of metal recycling. In: von Gleich A, Ayres AU, Gößling-Reisemann S (eds) Sustainable Metal Management. Springer, Dordrecht, pp. 295–312

Schröder D (2004) German Federal Water and Shipping Directorate, personal communication

Schulz I (2006) Ressourcenschutz durch Recycling-Baustoffe—vom Bauabfall zum Sekundär-baustoff. Paper GDMB-Meeting Committee for Construction Materials and Industrial Min-erals, Flechtingen, Germany, 8 June 2006

SRU (Rat von Sachverständigen für Umwelfragen=German Expert Commission for the Envi-ronment) (1994) Umweltgutachten 1994. Deutscher Bundestag 12. Wahlperiode, Drucksache 12/6995, Berlin, Germany

United Nations Environment Programme (1989) Statement of sustainable development: 15th Ses-sion of the governing Council, Governing Council Decision 15/2, May 23, 1989, Annex II, GOAR, 44th Session Supplement, no. 25

Verhoef EV, Gerard PJ, Reuter MA (2004) Process knowledge, system dynamics and metal ecol-ogy. Journal of Industrial Ecology 8(1–2): 23–43

VDEh (2007) Steel Institute Düsseldorf. Blast Furnace Committee, Germany

von Carlowitz HC (1713) Sylvicultura oeconomica oder hauswirtschaftliche Nachricht und natur-mässige Anweisung zur wilden Baum-Zucht. Johann Friedrich Braun, Leipzig, Germany

von Weizsäcker EU, Lovins AB, Lovins LH (1995) Faktor 4—Doppelter Wohlstand–halbierter Naturverbrauch. Droemer Knaur, München, Germany

Wellmer F-W (1998) Lebensdauer und Verfügbarkeit mineralischer Rohstoffe. In: Zemann J (ed) Energievorräte und mineralische Rohstoffe: Wie lange noch? vol. 12. Österreichis-che Akademie der Wissenschaften. Schriftenreihe Erdwissenschaftliche Kommission, Wien, 47–73

Wellmer F-W (2000) The natural resources hierarchy with respect to sustainable development. Poster at Sess.13-1 Mineral Res and Developm. 31st IGC, Rio de Janeiro Brazil 2000, Abstract-CD

Wellmer F-W (2003) Mineral and energy resources: economic factor and motor for research and development. Zeitschrift der Deutschen Geologischen Gesellschaft 154(1): 1–27

Wellmer F-W (2009) Reserves and resources of the geosphere, terms so often misunderstood. Is the life index of reserves of natural resources a guide to the future? Zeitschrift Deutsche Geologische Gesellschaft 159(4): 575–590

Wellmer F-W, Becker-Platen JD (2001) World natural resources policy-focussing on mineral resources. In: MK Tolba Our Fragile World-Challenges and Opportunities for Sustainable

Development, vol. 1. Encyclopedia for Life Support Systems Publishers Co, Ltd., Oxford, UK, 183–207

Wellmer F-W, Becker-Platen JD (2002) Sustainable development and the exploitation of mineral and energy resources: a review. International Journal of Earth Sciences (Geol Rdschau) 91: 723–745

Wellmer F-W, Becker-Platen JD (2007) Global nonfuel mineral resources and sustainability. In: Briskey JA, Schulz KJ (ed) Proceedings for a Workshop on Deposit Modeling, Mineral Resource Assessment, and Their Role in Sustainable Development. U.S. Geol. Surv. Circular 1294, U.S. Geological Survey, Reston, VA, 1–16

Wellmer F-W, Dalheimer M (1999) Trends und Perspektiven der Rohstoffversorgung Deutschlands im 21. Jahrhundert. In Slaby D, Brezinski H (ed) Rohstoffwirtschaft im Prozess der Transformation. Freiberger Forschungshefte 5, Wirtschaftswissensch, 11–52

Wellmer F-W, Kosinowski M (2003) Sustainable development and the use of nonrenewable resources. Geotimes December: 48(12): 14–17

Wellmer F-W, Kosinowski M (2005) A hierarchy of natural resources with respect to sustainable development. Z dt Ges Geowiss 156(2): 247–259

Wellmer F-W, Stein V (1998) Mögliche Ziele nachhaltiger Entwicklungen bei mineralischen Rohstoffen. Erzmetall 51(1): 27–38

Wellmer F-W, Dalheimer M, Wagner M (2007) Economic Evaluations in Exploration. Springer, Berlin Heidelberg New York

Part II
Mining and Sustainable Development

Sustainable Energy and Mineral Resource Extraction and Consumption—Can a Viable Biosphere Be Preserved?

W.G. Ernst

Abstract Modern societies have utilized fossil energy, water, and most other Earth materials at consumption rates far exceeding those of planetary replenishment. The well-known collapse of isolated island communities shows that overexploitation of the environment ultimately ends in disaster. It is now apparent that humanity must reach a steady-state stewardship of the Earth employing efficient, universal mineral resource recovery, recycling, substitution, dematerialization, and conservation. The goal of consumption of renewable resources at or below recharge rates and near-total recycling of non-renewable Earth materials can only be achieved employing universally available, inexpensive energy. However, the Second Law of Thermodynamics dictates that a part of the nonrenewable mineral resource base and most of the spent energy are irretrievably lost as entropy increase. Research-pioneered technological advances leading to the production and ubiquitous availability of environmentally benign, cheap energy will be required in order to reach the sustainable utilization of mineral resources. The principal, virtually unlimited, renewable energy sources appear to be solar and fusion power.

Assuming technical, economic, and political success in achieving universal equity and a comfortable standard of living for the World population, the global ecosystem will be severely impacted by the increased human consumption of Earth materials, reflecting environmental modification attending resource extraction and consumption. The dynamic biospheric equilibrium and ecosystem viability—the carrying capacity of the planet—is deteriorating and is increasingly at risk. Thus, the greatest long-term challenge facing humanity is not global climate change, or even the required transformation to renewable, cheap energy systems, but the necessity of preserving a healthy, sustainable biosphere.

W.G. Ernst (✉)
Department of Geological and Environmental Science, Stanford University, Stanford CA 94305-2115, USA
e-mail: wernst@stanford.edu

J.P. Richards (ed.), *Mining, Society, and a Sustainable World*,
DOI 10.1007/978-3-642-01103-0_6, © Springer-Verlag Berlin Heidelberg 2009

1 Introduction

This paper is an update and revision of an earlier review of mineral + energy resource utilization and global disparities in the standard of living (Ernst 2002). Little has changed in the past 7 years except that environmental degradation has worsened due to the ever-increasing impacts of industrialization, urbanization, and yet more intensive agricultural practices, without an attendant amelioration of world poverty.

Demographers estimate World population, currently slightly over 6.6 billion (U.S. Census Bureau 2008), at 9–10 billion by 2050 (UN 1999; Harrison and Pearce 2000). About 85% of humanity now lives in the Developing Nations, and this proportion is growing. Due to the global information network (e.g., television, the internet, efficient transportation), this is the first worldwide generation to become acutely aware of life styles enjoyed by the inhabitants of the Industrialized Nations. The grossly inequitable distribution of wealth, due partly to resource exploitation (Eggert 2008), is politically destabilizing and must be corrected if we are to avoid global socio-economic conflict between the have and have-not nations.

The question is, can ten billion people be afforded comfortable lives without destroying the carrying capacity and habitability of the planet through exhaustion of the Earth's natural capital? Humans now control a third of both the terrestrial and marine net primary biological production, and our share is increasing. Deleterious environmental impacts include increasing air and water pollution, vastly accelerating loss of biodiversity, ecosystem services, topsoil, fisheries, and tropical rain forests, as well as global warming, sea-level rise, and the increasing severity of tropical cyclones. Disruption of intricate, interactive biological systems is producing some winner species but many more numerous losers.

The Earth's natural capital includes the entire biosphere, consisting of the web of life, its organismal and habitat diversity, and the ecosystem services provided (Field Museum 2008), as well as inorganic mineral resources. This chapter deals principally with naturally occurring solid and liquid Earth materials. The accelerated exploitation of mineral resources and the concomitant environmental degradation have profound implications for future planetary habitability and for the fate of the biosphere, including humanity.

General categories of terrestrial mineral resources include abundant metals, scarce metals, water, soil, building materials, chemicals, carbon-based fossil fuels, and nuclear + renewable energy sources. The extractable portions of these commodities are sequestered in the near-surface environment—some in the oceans and the underlying basaltic crust (Rona 2002), but the vast majority in the continental crust (Craig et al. 2001; NRC 2007a). Among the metals, only silicon (28.2 weight %), aluminum (8.2 weight %), iron (5.6 weight %), magnesium (2.3 weight %), and titanium (0.6 weight %) are present in sufficient quantities in the continental crust to be classified as abundant. All the rest, such as copper, nickel, zinc, silver, and platinum group elements (PGEs) are scarce metals, individually constituting only 0.1 weight % or less (chiefly much less) of the continental crust. Most chemicals, solid, liquid, and gaseous hydrocarbon deposits, and fissionable fuels are comparably scarce. Only supplies of water, a few chemicals, and Earth building materials are plentiful, and in certain regions, nearly limitless.

2 Definitions

Before dealing with the all-important challenge of the sustainable development of the resources sequestered in our planet, we need to clarify a few terms regarding the nature and relative amount of a resource (Einaudi 2000).

A *resource* is the aggregate global inventory of a naturally occurring commodity. For example, iron is an abundant terrestrial element; however, it resides chiefly in the Earth's core as iron-nickel alloy, with lesser amounts sequestered as silicates ± oxides in the mantle, and even less distributed in the oceanic and continental crust. As a practical matter, the only available sources of most minerals reside in the Earth's outer rind (the crust). Here, occurrences of iron ore are sufficiently concentrated (~50 weight % Fe) in sedimentary banded iron formations and in rarer magnetite-rich skarns of the continental crust to be considered as *reserves* (i.e., economically valuable deposits).

A *nonrenewable resource* is present in a finite, fixed amount in the planet, which for all practical purposes means the crust. The amount of iron on Earth is a reflection of the composition of the pre-condensation solar nebula, itself a function of the later stages of old star burnout elsewhere in the galaxy. No iron is currently being generated in the Earth's environment. Crustal iron ore deposits are being formed, but at extremely slow geologic rates comparable to the chemical and mechanical erosion of existing deposits; hence, the net production of iron ore is essentially nil. Moreover, the rate of anthropogenic extraction of iron ore is many orders of magnitude faster than any overall change in the crustal abundance of iron ore deposits.

A *renewable resource* is replenished at a finite rate. The recharge rate may exceed the rate of human usage (e.g., photovoltaic cell and solar-thermal energy production, or wind-turbine electric power generation), whereas in other cases, consumption far exceeds natural renewal (as in the rates of formation and loss of topsoil, groundwater, tropical rain forests, and marine fisheries).

As an example of a renewable resource, the availability of fresh water is a function of the hydrologic cycle (e.g., Schlesinger 1997; Davidson et al. 2002). H_2O is preferentially evaporated from the oceans and precipitated on land, then returned oceanward via surface + groundwater flow, and biospheric evapotranspiration. The scarcity of fresh water is a crucial factor in limiting terrestrial biomass; the rate of biological fixation cannot exceed the rate of recharge for very long without a substantial die-off, and the subsequent re-establishment of a dynamic equilibrium between H_2O availability and the surviving web of life. Groundwater is an important source of fresh water for human use, but is being extracted at rates far exceeding replenishment. The universal availability of safe, inexpensive water is an essential requirement of a sustainable economic system (Bergkamp and Sadoff 2008).

The slow generation and erosion of topsoil is another familiar example of renewable natural capital (Fyfe 1989; Schaetzl and Anderson 2005). Its dispersal in some cases is substantial, such as in pastures and farm tillage, where global soil losses approximate 5.8 cm/100 years (Nearing et al. 2000); in marked contrast, the generation of soil due to physical and chemical weathering of the rocky substrate is extremely slow—28 times slower than agricultural and developmental loss (Wilkinson 2005).

The *World Reserve Life Index* (WRLI) is a ratio that gauges the time to exhaustion of a nonrenewable reserve, assuming the present rate of global consumption. Values of WRLI for selected commodities are presented in Table 1. Slightly more optimistic estimates were provided by Davidson et al. (2002) and Yoder (2002). World reserve life indices often markedly underestimate actual mineral reserve lifetimes because they do not take into account the discovery of new deposits, enhanced recovery techniques, and, through depletion of higher grades of ore, increasing economic attractiveness of lower grade deposits. For instance, based on an assessment of the total complement of copper deposits in the upper 3.3 km of the Earth's crust, Kesler and Wilkinson (2008) have calculated that the present World demand for Cu could be accommodated for 5,500 years. However, the WRLI also ignores hidden costs (i.e., externalities) such as the increasing need for post-extraction environmental cleanup and, most importantly, the rapidly increasing demand for mineral resources by the Developing Nations—a factor that will substantially accelerate the depletion of existing reserves and those of future discoveries. However, as

Table 1 Mineral reserves and global supplies

Metal/mineral	Ore grade (wt %)	World reserve life index
Uranium	0.1–0.2 U_3O_8	65
Iron	30–65 Fe	178
Manganese	25–50 Mn	43
Nickel	1.5 Ni	51
Aluminum	35–50 Al_2O_3	219
Magnesium	70–95 $MgCO_3$	Unlimited
Titanium	20–30 TiO_2	79
Copper	0.2–5.0 Cu	35
Lead	4–8 Pb	20
Zinc	2–4 Zn	19
Silver	0.006 Ag	20
PGE	0.003–0.02 PGEs	190
Phosphate	20–35 P_2O_5	85
Cement	25 clay+sand, 75 limestone	Unlimited
Dimension stone	Marble, granite, basalt, etc.	Unlimited
Oil	1,100 bbl reserves	39
Natural gas	800 bbl equivalent reserves	57
Coal	909 billion metric tons	Centuries

Minerals, Einaudi (2000); oil and gas, Ahlbrandt (2002); coal, British Petroleum (2008).

Table 2 Changes in estimated world reserves of several industrial commodities, 1940–1989 (in millions of tonnes)

Commodity	1940s	1950s	1960s	1970s	1980s
Bauxite	1,605	3,224	11,600	22,700	23,200
Copper	91	124	280	543	566
Lead	31–45	45–54	86	157	120
Zinc	54–70	77–86	106	240	295

After NRC (1996).

Table 3 US scrap recovery of some industrial commodities, 1960–1990, as a percentage of production

Commodity	1960	1970	1980	1990
Aluminum	5	4	11	22
Copper	27	25	28	25
Iron/steel	25	25	15–20	22
Lead	40	37	54	69
Zinc	6	5	6	9

After NRC (1996) and U.S. Geological Survey (2008; http://minerals.usgs.gov/ds/2005/140/ironsteelscrap.pdf).

demonstrated by recent past history, the discovery of new global reserves of some important industrial metals (Table 2), as well as the recycling of scrap metals in the US (Table 3) extends projected lifetimes of these mineral reserves without the prospect of imminent exhaustion (Gordon et al. 2006). Moreover, commodity substitution, technological innovation, and in some cases, dematerialization (i.e., decreasing the quantity of the consumed resource, such as by reduction in size and increased speed of computers, and decreased fuel usage by increased vehicle efficiency) may diminish the threat of doom projected by a finite WRLI (Tilton 2002).

The use of iron is ubiquitous in modern societies (e.g., NRC 2007a). This element is abundant in the continental crust (as are titanium, magnesium, and aluminium), so humankind is unlikely to run out of Fe in the foreseeable future. If usage of other metals is normalized to this element and global consumption is plotted against abundance, as illustrated in Fig. 1, it is clear that for most scarce metals, extraction rates are orders of magnitude greater than for iron; evidently, extractable crustal reserves of metals of low abundance are being rapidly depleted. Consequently, as reserves diminish, either the efficiency of low-cost extraction must increase markedly, or the price of the remaining sought-after metal will increase, in which case deposits of ever lower concentration become profitable to work, thus extending mineral reserves (Skinner 1979).

However, an energy discontinuity—in the form of a major increase in the amount of energy required for beneficiation—exists in the economics of recovery of a scarce resource as its natural geologic concentration diminishes. Where the abundance of the commodity exceeds the solubility of that element in common rock-forming

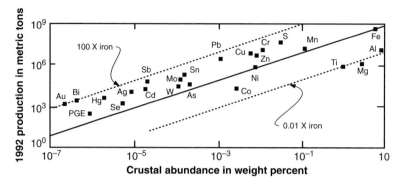

Fig. 1 World annual production of selected metals for 1992 plotted against their abundances in the continental crust, after Skinner (1976) and Einaudi (2000). Considering iron as a reference, elements positioned below the solid line are being extracted more slowly relative to their abundance than is Fe, whereas those lying above the reference iron line are being depleted from the crust at extraction rates faster relative to their abundance than that of Fe

Fig. 2 Energy required to extract a metal such as copper from a host mineral as a function of the metal concentration in the host mineral (e.g., copper sulfide versus copper dispersed in a silicate mineral), after Skinner (1976, 1979)

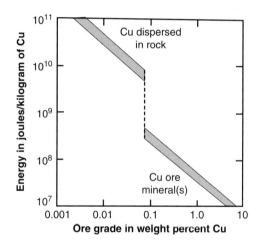

silicates during crystallization, one or more separate ore minerals will form in the deposit as oxides, sulfides, and/or native metals. These ore minerals contain large proportions of the desired constituent; although scattered throughout the rock, ore minerals are readily concentrated by grinding, mechanical and/or chemical separation, and refining (smelting) with the expenditure, in general, of relatively modest amounts of energy. In contrast, in geologic environments where the concentration of an element is less that that necessary to saturate the common rock-forming minerals, such that a metal-rich ore mineral is not formed, the element is dispersed as a trace constituent in the atomic structures of these major phases. Extraction of the metal from silicate host minerals requires several orders of magnitude more energy than is required to liberate the sought-after substance from an ore mineral. This low-grade concentration threshold, illustrated schematically in Fig. 2, is termed the *mineralogical barrier* (Skinner 1976). Clearly, as ore grade declines and reaches this parent-rock compositional threshold, vastly increased quantities of energy will be required to extract further amounts of the metal, and recycling will have to become an increasingly important source of the commodity.

3 Mineral Resource Extraction, Usage, and the Environment

The quarrying of Earth construction materials, mining of ore deposits + coal, and the drilling for and production of oil and natural gas take place in relatively restricted areas compared with the ubiquitous surface modifications wrought by agriculture, forestry, and urban + industrial development (Wolman 2002). Over the 50 years heyday of intense activity, 1930–1980, less than 1% of the U.S. total land area was directly impacted by coal + mineral mining, and by petroleum exploitation (Johnson and Paone 1982). However, topographic alteration and point-source in situ and downstream contamination due to acid mine drainage and hydrocarbon pollution are far more serious than this small areal percentage might suggest, because of the high toxicity of some ore mineral and smelter products, as well as forms of hydrocarbon and oil refinery exhaust. The degradation of resource-exploited lands and those downstream and downwind is well known. Some of this impact is due to industrial activities in general (Freeze 2000), and is only peripherally related to the resource extraction industry.

A conflict between the values of extractable commodities and those of unaltered, pristine lands historically has resulted in the harvesting of natural resources rather than environmental conservation (e.g., World Bank 2003; Ali 2003). This situation partly reflects our economic ability to assess quantitatively and unambiguously the worth of a mineral commodity, but also our inability to determine the financial value of unaltered land containing the natural resource. Until economists are able to place a monetary value on less intensively impacted land and wilderness that is widely accepted as appropriate and fair to all stakeholders, the unequal contest between environmental preservation and resource utilization will continue. Ideally, some of the profits realized from mineral extraction could be used to rehabilitate exploited terrain to near-pristine conditions, providing economic gain, employment, and restoration of the environment (Richards 2002). However, in our present free enterprise system, little incentive exists for a developer to forego extraction or later to undertake remediation of the land, so such goals must be mandated by regional, national, or international regulation (e.g., Khanna 2000).

Markets are competitive, and the total costs of many of our activities generally do not account for externalities such as environmental degradation in the price structure; thus, society is obliged to pay for any remediation. For instance: the price of steel does not include the full downwind impact associated with acid rainfall; the cost of fossil fuel burning fails to capture the costs of global warming and air pollution; and the price of water supplied through groundwater usage typically ignores overdrafting. The public generally defrays these hidden costs through taxation. Instead of maintaining the status quo, consumption problems of this sort could be substantially reduced by assigning the full cost to the activity, and by eliminating counterproductive subsidies (Corson 2002). However, tax incentivization and better stewardship of the Earth depends on a clearer understanding of the externalities by society. Fortunately, attempts to account for such externalities, including cap-and-trade policies, are beginning to distinguish the European Union from most other free-market economies; these efforts may be harbingers of an increasingly global attempt to address hidden costs inherent in natural resource utilization.

4 The Concept of Sustainability

It is instructive to define three contrasting resource-dependant growth scenarios. Consider a bowl of water, to which is added a finite supply of food and a few algae. (1) Exponential algae growth initially occurs, reflecting effectively unlimited sustenance. (2) As time passes, the nutrient resource is drastically reduced due to a greater rate of consumption by the increasing numbers of algae that it must support. The population reaches a maximum, and then declines as food is progressively and rapidly exhausted. Disappearance of the last trace of nutrient signals extinction for the last surviving algae. (3) In contrast, a finite population can be maintained at a particular level if the food supply is periodically or continuously replenished; in this case, the number of algae reaches a dynamic equilibrium controlled by the rate of nutrient replenishment (i.e., the carrying capacity of the system). Thus, the population reaches a sustainable steady state in terms of consumption at no more than the rate of recharge of a renewable resource, and/or the rate of recycling of a nonrenewable resource.

The Sun provides virtually all the energy available to the web of life utilized for driving vital processes, and although finite, the supply of energy is enormous. Chemical and geothermal power represent much smaller and negligible sources of energy for the biosphere, respectively. The Earth's in situ mineral resources are truly limited, however, and are renewed, if at all, on variable time scales. Thus, consumption of mineral resources for abundant as well as scarce commodities must eventually come to a dynamic steady state, through usage at no more than the rate of availability resulting from a combination of natural replenishment and the recycling of existing finite materials. For non-renewable resources, as concentrated deposits are exhausted, lower grade occurrences will become profitable, but will only be recovered with increasing expenditure of energy, especially as the mineralogical barrier is reached. Thus, sought-after resources ultimately must be recovered in part by the recycling of previously worked materials (Gordon et al. 2006). Substitution of more abundant Earth materials for scarcer commodities, and dematerialization (Tilton 2002; Wagner and Wellmer, this book) can ease the transition to sustainable resource usage, but in many cases the process of extracting and refining such substitutes increases environmental degradation.

Because we are dealing with the rate of consumption of Earth materials, we must consider the time over which sustainability is maintained. Self-replicating organisms on Earth began sometime prior to ~3.5 Ga, the age of the oldest yet recorded evidence of Archaebacteria. Since the early Archean, evolving life has maintained a dynamic equilibrium with incoming solar radiation and natural capital derived from the land, sea, and air. The biosphere has been sustained continuously over geologic time, chiefly through the utilization of a range of combinations of solar energy and near-surface, recycled mineral resources. The carrying capacity of the Earth system has evolved, but has not been exceeded, because organisms have only employed readily available recycled mineral resources at and near the surface. This dynamic state of affairs was more-or-less maintained after the appearance of *Homo sapiens* as long as our species remained in a primitive hunter-gatherer mode.

But beginning about 10,000 years ago, climatic excursions lessened appreciably, and settled agriculture was established in the Middle East and in China. This situation allowed for a division of labor, extraction of Earth resources exceeding renewal rates, and the rise of modern civilizations over the course of the past 400–500 generations. Reflecting small numbers of people, low population densities, and relatively simple farming and mining technologies, this borrowing from the future carrying capacity of the planet began slowly and inconspicuously, with a gradual alteration of the land and a progressive loss of forests, topsoil, and biodiversity. As humanity prospered, the increasing population began to exert an ever-greater drain on global mineral and other natural resources. Industrialization vastly enhanced this per capita consumption and environmental impact after about 1800 A.D. (Chesworth 2002; Crosby 2006).

The dynamic equilibrium maintained over the sweep of geologic time between the Earth's crust + hydrosphere + atmosphere and the biosphere has been increasingly altered by the march of civilization. The impacts of war, agriculture, fisheries, urban + rural construction, and anthropogenic resource usage are substantial (Scientific American 1989; Harrison and Pearce 2000). Humanity now controls a profoundly altered global environment in which the health of the biosphere has been severely compromised.

The Brundtland Commission (WCED 1987, p. 8) defined sustainable development in anthropocentric terms, as the ability of humanity "...to ensure that it meets the needs of the present without compromising the ability of future generations to meet their own needs." Or, as stated by Solow (1993, p. 168), "a sustainable path is one that allows every future generation the option of being as well off as its predecessors." Inherent in both statements is an implicit recognition that the dynamic carrying capacity of the planet, like the bowl of water referred to above, is resource limited. In my view, critical to the concept of sustainability—not explicitly stated by either WCED or Solow—is that the process be maintained for a thousand generations, not just for the next generation or two. It is relatively easy to imagine sustainable development over 10–20 years, but intensely sobering to think of it in terms of 10,000 years (Ernst 2002). True sustainability means essentially forever, at least on the time scale of a species such as ours.

5 Energy and the Second Law of Thermodynamics

A commodity such as iron, limestone, or silver can be used, discarded, and dispersed in the environment; then, with the expenditure of sufficient energy, much of it can be recaptured and refined to the desired level of purity. For a nonrenewable, scarce resource, this problem is acute because of our high rate of extraction-depletion, and the elevated energy step function dictated by the mineralogical barrier. Thus, abundant energy is the key to increased mineral resource utilization as ore grades decline over time.

However, unlike other Earth resources, non-renewable energy possess single-use value: the energy content of uranium or coal, for instance, is intrinsic to the substance; once consumed, almost all of the original energy is lost and only a small fraction can be captured for reuse. This is a consequence of the Second Law of Thermodynamics which requires that, during reactions involving transference of energy, an important portion is lost as an increase in entropy (a greater degree of disorder of the system). Equally important, the efficiency of energy conversion is considerably less than the theoretical Second Law maximum value. Consequently, about half to two-thirds of the energy content of the hydrocarbon feed stock used to run an electric power plant is transferred to the surroundings as heat and dispersed gas species, rather than being converted totally into electrical energy.

6 The Unique Role of Energy in Modern Civilization

Widely available, inexpensive energy is the hallmark of industrialized societies. Fossil fuel sources of energy have supplanted human + domesticated animal power and the burning of wood + dung as sources of mechanical power, illumination, heating + cooking, and transportation (Crosby 2006). Rather comparable to American proportions of energy usage, liquid petroleum, coal, and natural gas are first, second, and third in terms of global energy consumption. Together, these three fossil fuels constitute the lion's share of energy usage, with nuclear power and all renewable forms of energy combined making up a distant fourth, as illustrated in Fig. 3. Russia, China, and the United States possess voluminous coal deposits, whereas countries surrounding the Persian Gulf control much of the World's supply of oil. The states of the former Soviet Union have nearly half of the global supply of natural gas. However, these resources are finite, with liquid petroleum and natural gas likely to be virtually exhausted in the next few decades, and coal in several centuries at the present rates of consumption (Table 1).

Unlike other Earth resources, we cannot consume all of the terrestrial complement of a nonrenewable energy resource. This is because energy must be expended in order to extract energy, and it is not economically feasible to spend as much or more energy than the value of that recovered. We may call this obvious concept the *energy barrier*; during extraction, there must be a net gain in the amount of usable energy.

TOTAL PRIMARY ENERGY SUPPLY

The World

Evolution from 1971 to 2005 of World Total Primary Energy Supply* by Fuel (Mtoe)

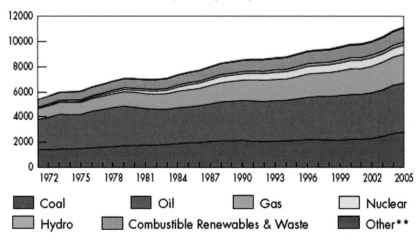

| ■ Coal | ■ Oil | ■ Gas | □ Nuclear |
| ■ Hydro | ■ Combustible Renewables & Waste | | ■ Other** |

1973 and 2005 Fuel Shares of TPES*

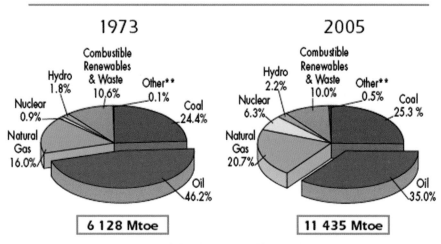

1973

Combustible
Renewables
Hydro   & Waste   Other**
1.8%   10,6%   0.1%
Nuclear   Coal
0.9%   24.4%
Natural
Gas
16.0%
                                     Oil
                                     46.2%

6 128 Mtoe

2005

Combustible
Renewables
Hydro   & Waste   Other**
2.2%   10.0%   0.5%   Coal
Nuclear   25.3 %
6.3%
Natural
Gas
20.7%
                                     Oil
                                     35.0%

11 435 Mtoe

*Excludes electricity and heat trade.
**Other includes geothermal, solar, wind, heat, etc.

Fig. 3 World energy consumption in millions of tons of oil equivalent (Mtoe; modified slightly from the International Energy Agency 2007, p. 6). Utilization is depicted for 1971 and for 2005

In spite of a wide range of estimates regarding the remaining global reserves of carbon-based fuels, they are limited, and sooner or later will be exhausted. Perhaps this is just as well, considering that their usage results in the increased sequestration of CO_2 in the atmosphere and deep sea, and consequently, in long-term global warming and acidification of the oceans (IPCC 2007). So, although fossil fuel resources are large, in order to perpetuate an energy-intensive industrialized society, sustainable utilization of mineral resources in the long run can only be achieved if humanity acquires a ubiquitous, inexpensive, renewable energy source. Research-based technological advances, providing generation and universal availability of low-environmental-impact, cheap energy, will be essential to achieving a state closely approximating Earth resource sustainability. Two sources have the ultimate capability of providing a supply of inexhaustible energy: the Sun and nuclear fusion. The bringing to the market of environmentally relatively benign, abundant power from these or other as yet unanticipated sources of energy will require a substantial increase in the worldwide investment in science and engineering, for such energy sources are not yet within our technological capabilities (Moniz and Kenderdine 2002; Whitesides and Crabtree 2007; NRC 2007b).

Political and technological leadership is needed to provide powerful incentives for the research and technological development of ubiquitous, cheap energy systems, because this work will require a sustained, long-term, interdisciplinary effort. However, by the time the general perception of a serious energy shortfall exists in an ever more populous World, it may be too late to stave off a global increase in the conditions of poverty.

7 World Population and Resource Consumption

The Industrial Revolution ushered in a massive acceleration in the rate of usage of both renewable and non-renewable mineral resources. It also fostered an exponential growth of our species, now more than 6.6 billion people. Because the overall birth rate exceeds the death rate, World population is increasing annually by about 90 million, and will reach 9–10 billion by 2050. Growth is due largely to population age profiles (see Fig. 4), and to birth-rate momentum: half of the people in the Developing Nations are under child-bearing age, so even if each female, on average, has the replacement rate of just 2.1 live births, the global population will continue to increase. Currently, 20% of the World population—that of the Industrialized Nations—uses more than 70% of its natural resources (Doran and Sims 2002). In contrast, the five-and-a-half billion people of the Developing Nations have access to only about 30% of the global resources—and that amount largely because of the rapid industrialization of China (Wagner and Wellmer 2008, this book). With widespread increases in their standard of living, education + empowerment of women, and substantial reduction of poverty, such populations are anticipated to pass through the demographic transition. In this phenomenon, illustrated in Fig. 5, life expectancy increases rapidly due to decreased infant mortality, improved

Fig. 4 Contrasting age
structures of the populations
in (**a**) Developing Nations
and (**b**) Industrialized
Nations, after Harrison and
Pearce (2000). Even with
only modest birth rates,
population momentum will
carry the Developing Nations
to much larger populations; in
contrast, many Industrialized
Nations have virtually
stabilized their populations,
and in some cases, numbers
of people may even gradually
decline, reflecting
less-than-replacement birth
rates

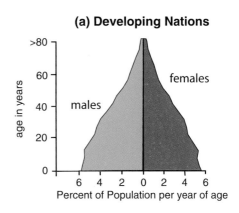

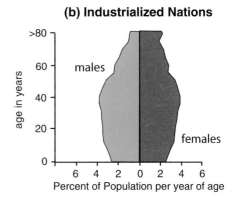

diets, and better health care, followed about a generation later by falling birth rates,
with the population achieving a stable but elevated level.

By 2025, three-fourths of the global population will be concentrated in coastal
zones, most living in the burgeoning, ever-more-interconnected megacities (McNutt
2002; Grimm et al. 2008; Dye 2008). Such coastal environments will become
increasingly vulnerable to tropical storms, land subsidence, and coastal flood-
ing as the effects of global warming become more pronounced (Ernst 2001;
Trenberth 2007). Low-lying coastal environments are also especially prone to out-
breaks of pandemic diseases (Guptill 2001). The Industrialized Nations have sufficient wealth to combat the adverse aspects of climate change in their own countries
(e.g., Mendelsohn and Neumann 1999), but the fate of Developing Nations is much
less certain.

Reflecting spectacular advances in electronic network and telecommunication
technologies, TV and internet communications are global, pervasive, and intrusive.
The people of the Developing Nations (e.g., Fig. 6) are now routinely exposed to

Fig. 5 Schematic (**a**) birth
and death rates and (**b**) sizes
of populations undergoing the
so-called demographic
transition (Stages 1?3), after
Alexander and Ehrlich (2000)

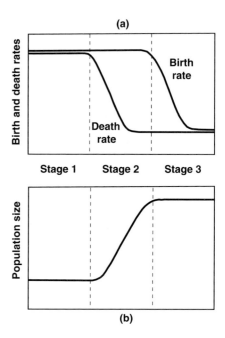

many manifestations of the life styles of the wealthy countries of Europe, North
America, and Japan. Since the rise of civilization, this is the first generation to be so
universally well informed. It cannot fail to occur to inhabitants of the Developing
Nations that the present distribution of wealth is massively unequal, and thus quite
unfair (Galbraith 2002). The inequities are glaring: for example, the World Health
Organization estimates that approximately 43% of the global population lacks adequate sanitation, and 22% do not have clean drinking water (Gleick 1998; NRC
1999; Nordstrom 2002; WHO 2007). Many want and eventually expect to achieve
a standard of living comparable to that enjoyed by people of the Industrialized
Nations. Such a high standard of living, of course, is due partly to the exploitation of limited planetary resources by the Industrialized Nations. Thus, the grossly
disparate distribution of material wealth is a source of serious dissatisfaction among
people of the Developing Nations, and is politically destabilizing. The misery of
poverty begets frustration and in some cases, radicalism.

Humans now control a third of the Earth's net primary biological productivity
(i.e., the global rate of fixation of C, H, O, N, P, and S, and not including an equivalent, steady-state destruction), and our share is increasing (Vitousek et al. 1986;
Rojataczer et al. 2001). The environmental impact is pervasive and predominantly
negative, resulting in increased air + water pollution, a greatly accelerated loss of
biodiversity, an abrupt contraction in ecosystem services, erosion of topsoil, collapse
of marine fisheries, destruction of temperate and tropical rainforests, and global
warming + sea-level rise (e.g., Raven and Williams 2000; Pitman and Jorgensen

Fig. 6 Year 1997 photos of (**a**) men and (**b**) women of the Asmat tribe, Agats, Irian Jaya. Head hunting and cannibalism were practiced in this area into the 1960s, but the Indonesian government has forcibly interdicted such practices. The Asmat people exist in a minimal subsistence environment, but a TV set in the village of Agats provides them with nonstop vistas of the prosperous, Industrialized World. Photos by the author

2002; Raven 2002; IPCC 2007). Attempts to meet the material and quality-of-life aspirations of the teeming populations of the Developing Nations are exacerbating an already serious ecological problem. Damage to the environment inevitably will accompany a vastly increased human consumption of natural capital, resulting in the partial or possibly nearly complete collapse of the Earth's biological systems. Implications for the continued viability of the surviving web of life, and indeed for the future welfare of humanity, are therefore bleak.

   The question is, can all the Earth's people be afforded reasonably comfortable lives without devastating planetary habitability? Failure to address the material needs of people of the Developing Nations is not only morally unjustifiable, it is socio-economically unwise, and politically dangerous (Ernst 2002). However, if the quality-of-life conditions are to be improved substantially for this burgeoning Developing Nation population, resource consumption will be increased by nearly an order of magnitude. Thus, the preservation of a viable biosphere, already under siege due to current resource utilization practices, will become a major global challenge with a very short fuse.

8 An "American World" of Ten Billion People

The United States leads the World in the consumption of most of the Earth's natural resources. Our annual per capita utilization of abundant + scarce metals, water, building materials, chemicals, and fossil fuels is legendary. Non-energy consumption figures used in the following are chiefly from the U.S. Geological Survey (2000) and Tarbuck and Lutgens (2002). On average, every year each American uses 550 kg of iron and steel, 10 kg of copper, 25 kg of aluminium, and 6 kg of manganese. We also utilize 4,100 kg of stone, 3,860 kg of sand and gravel, 360 kg of cement, and 140 kg of phosphate rock. Ignoring saline water extraction, we withdraw an average of 1,280 gal of fresh water off-stream per day, an annual rate in excess of 460,000 gal/capita. These figures simply represent the total US consumption of a commodity divided by the population (272 million in the year 2000).[1] The salient fact is that, as a nation, Americans are major users of scarce and abundant renewable and nonrenewable natural resources. This consumption is viewed as profligate; nevertheless, our standard of living is envied by many people in the Developing Nations. But what if the World population of 2050 were to achieve a level of resource usage approaching our present life style?

   The annual global production of iron and steel from iron ore and scrap exceeds 830 million tonnes. If each of the ten billion inhabitants of the planet were to consume the 550 kg of iron per year that Americans now use, the total demand would be 5.5 billion tonnes—nearly a 7-fold increase in yearly production. A 660+% increase in the level of iron ore mining, however improbable, is conceivable because Fe_2O_3 is widespread and relatively abundant in the continental crust; besides, as is the present practice, reuse of scrap iron would offset some of the increase. However, the amount of recycled scrap is limited, reflecting the more modest volumes of earlier production and its continuing service. Thus the environmental impact of nearly an order of magnitude increase in the mining of iron ore would be intense and far-reaching.

   Now let us consider several of the scarce natural resources, lead, zinc, and PGEs. These metals are essential to modern industrial societies, and are being recovered

---

[1] By early 2008, the American population exceeded 303 million (U.S. Census Bureau 2008) and very likely the utilization of resources has risen commensurately.

for use at much greater rates relative to their natural abundances compared to iron (refer to Fig. 2); US per capita annual consumption of these metals is 6 kg, 5 kg, and 0.8 g respectively. If, by the middle of this century, the global population of ten billion people consumes resources as Americans do now, the continental crust would be obliged every year to produce 60 million tonnes of lead, 50 million tonnes of zinc, and 8,000 tonnes of PGEs yearly, roughly 15, 6, and 3 times the present annual World extraction rates. This might be doable, but at the expense of a more severely impacted environment.

What about the combined per capita average US yearly consumption of stone, sand, gravel, and cement, 8,320 kg? Were the World population of 2050 to employ these ubiquitous construction materials at our present rate, the total annual usage would amount to more than 83 billion tonnes. This represents a mass sufficient to pour a one meter-thick block of concrete over an area slightly in excess of 33,000 km^2, roughly a pavement of dimensions $180 \times 180 \times 0.001$ km—exceeding the sizes of several eastern states (e.g., Maryland + Delaware, or a third of Virginia). This number appears to be manageable until we reflect that such a thick tablet would have to be laid down every year, not just for a generation but for a millennium or more.

A critical limit to growth, especially in arid lands, is fresh water. US per capita usage varies enormously, reflecting marked contrasts in regional geography, climate, and agricultural development. In the East and Midwest, off-stream usage is chiefly for electric power generation, whereas in the Southwest, most is for the irrigation of crops. In terms of the daily off-stream average, we use 1,770 m^3/year (Solley et al. 1998; Hutson et al. 2004). These figures do not include groundwater withdrawals, which average about 20% of the total usage. A global population of ten billion people living in the US style would consume nearly 18,000 km^3 of surface water annually, nearly half of the entire hydrologic cycle return flow of 40,000 km^3 draining from the continents back into the global oceans. Depletion of groundwater is already well advanced in arid regions worldwide, and this problem will worsen due to increased rates of extraction (Alley et al. 2002). Can the land and its wetlands, rivers, lakes, and estuaries sustain this large a draft on the water supply without intensified degradation? Perhaps of even greater importance, the pollution of fresh water by point-source and non-point-source particulates and solutes (toxic chemicals, fertilizers, pesticides, herbicides, sewage, industrial pollutants, etc.) adversely impacts biospheric habitability, and in aggregate is steadily increasing.

Americans currently use about 25% of the Earth's petroleum production. Representing slightly less than 5% of the present global population, this means that by 2050, oil production would have to increase more than five times the present rate of extraction for ten billion people to share our style of consumption. Whether the terrestrial complement of liquid hydrocarbon reserves (already produced, currently available, and likely to be discovered) lies in the 1.8–2.1 trillion barrels range projected by Hubbert (1981) and Deffeyes (2001), or is the 3.0 trillion barrels estimated by Ahlbrandt (2002), this resource is clearly finite (Campbell and Laherrer 1998; NRC 2006; DOE 2008). Some authorities (e.g., Obermiller 1999; Brown 2000) anticipate that World oil production will begin its decline within the current decade. The supply of natural gas is more promising (Ahlbrandt 2002), but is also finite. Moreover, the global demand for petroleum has been rising steadily, and at an accelerating rate. The proportions of various energy sources utilized worldwide are rather similar to those of the US: hence, carbon-based fuels are anticipated to remain the main sources of power for the next few decades (DOE 2001). As World demand exceeds supply, the price of petroleum likely will sky-rocket; if so, marginal oil and gas fields will enjoy new profitability and thus will undergo increased exploitation and extraction. Ultimately, so too will uncon-ventional petroleum deposits such as tar sands and oil shales. But because the Earth's fossil fuel resources (even its massive coal deposits) are finite, consump-tion inevitably will decline over time. For this reason, Developing Nations will not be able to achieve the US hydrocarbon burn rate, and, probably, Americans will not be able to sustain their current per capita use rate either.

Let us assume that the demographic projections are correct, and, further, that today's inequitable distribution of wealth and utilization of natural capital is elimi-nated or much reduced, resulting in an increased standard of living as the Develop-ing Nations transform into Industrialized Nations. The consumption figures derived above are only back-of-the-envelope approximations, but are comparable to an esti-mated overall six-fold increase in utilization of commodities that the International Union of Geology and Geophysics reported to the International Council for Sci-ence (IUGG 2002). Although this increase in standard of living would represent a socio-economical and political triumph of great magnitude, the adverse impact of resource exploitation on the natural environment would likely be immense—an eco-logically undesirable outcome, possibly disastrous in the long run. Assuming abun-dantly available energy, can the planet support anything approaching an American-style consumer population of ten billion people? And if so, for how long?

9 The Achievement of Mineral Resource Sustainability

Modern societies are supported by the extraction of energy, water, and other Earth materials far exceeding renewal rates, limiting the future carrying capacity of the planet. The latter is a function of its aggregate history, but is clearly finite; there-fore, humanity eventually must reach a managed steady state with the resource base and its biospheric life-support system, while preserving environmental quality (Cor-dani 2000; Holdren 2008). Over the grand sweep of geologic time, the evolving biosphere has maintained a dynamic, sustainable interaction with the near-surface environment. The resource excursion supporting civilization (i.e., overdrafting of the resource base) sooner or later will be corrected. It is not a question of whether, but rather the manner in which it will be corrected that is uncertain. Island commu-nities (e.g., Easter Island, Madagascar, Haiti, some islands of the eastern Mediter-ranean) provide sobering examples of cultures that overexploited their local environ-ments through ignorance (Diamond 2005). Judging by the record of human migra-tion and socio-economic evolution of island communities—and continents are just large islands—the future does not appear promising.

Even with heightened technological efficiency of exploration, extraction, and refining, and the ubiquitous recycling of scrap materials, the production and utilization of mineral + metal resources ultimately must encounter the mineralogical barrier for all but the most abundant commodities such as stone, gravel, cement, and probably Fe, Ti, Mg, Al, and possibly Mn. For the scarcer metals, this means that additional quantities of the substance can only be separated from the silicate rocks of the crust through the expenditure of vast amounts of energy. The substitution of alternative materials, and dematerialization of economies seem to be required for future growth. The fresh water supply is renewable, but at a finite rate through the natural hydrologic cycle, and reducing the off-stream usage can only be accomplished through conservation and/or the energy-intensive distillation of non-potable sources, principally seawater (Rona 2002). Fossil fuel reserves are large but finite, and ultimately will dwindle in response to increased consumption by populous countries such as China, India, and Indonesia. Unconventional fossil fuels, such as tar sands, oil shales, natural gas, and coal bed methane constitute important interim energy supplies, but are finite and CO_2 producing (Fouda 1998; George 1998; DOE 2008). Renewable energy sources (biomass, wind turbines, geothermal power, tidal energy, synfuels, hydroelectric, etc.) are currently limited in total global yields, and in aggregate constitute important but, for the foreseeable future, modest sources of power (Perrine and Ernst 1985; Baldwin 2002; Lewis 2007; Stephanopoulos 2007).

Abundant, widely available, inexpensive energy is the linchpin of the Industrialized Nations (NRC 2007b). It stands as the critical enabling factor in addressing problems of non-fuel resource recovery, whether it involves super-efficient exploration, discovery (e.g., drilling to document reserve tonnages), and extraction of new ore deposits, refining of sought-after commodities in low-grade deposits, or enhanced recycling of previously used materials. Because most of the World aspires to, and expects to attain, the standard of living enjoyed by the Industrialized Nations, this will require a virtually unlimited supply of power. Unless it is also very affordable, the energy will not be readily available, so in addition to being ubiquitous, it must be inexpensive. What are the options (Hoffert et al. 2002; U.S. Department of Energy 2008) for long-term sustainability?

Today, interest is being focussed on electrolytic battery-powered vehicles as well as on hydrogen fuel cells (e.g., Ogden 2002; NRC 2004; DOE 2004). Although environmentally benign, both are energy conversion systems, rather than new, primary sources of energy. Electric power supplied to so-called emissionless vehicles is generated at a conventional hydrocarbon- or fission-based power plant; as fossil fuels decline in importance, so will the rate of electric power generation from carbon-based energy sources. Hydrogen fuel cells require a feed stock such as methane, ethanol, or water; hydrocarbons are in limited supply, and the energy required initially to separate molecular hydrogen and oxygen from water exceeds the energy produced by the fuel cell, reflecting operation of the Second Law of Thermodynamics. Fuel cells will have to achieve much greater efficiencies than other forms of energy conversion in order to ameliorate (let alone solve) the growing global energy needs.

Nuclear fission power plants represent an energy bridge toward sustainability as the World passes through the age of carbon. But fissionable mineral resources are also finite, and this form of energy production has proven to be expensive. Most of the public resistance to atomic power is justifiably rooted in the fear of possible technical malfunctions and/or terrorist-inspired sabotage. However, nuclear power technology has advanced considerably over the past several decades (e.g., Marcus and Levin 2002), and no power plants in Western Europe and North America have ever experienced a serious accident comparable to that of Chernobyl in the Ukraine. Presently in the US, it takes about 10^{-15} years from authorization of the plans to build a nuclear power plant, through construction and certification, to the time when its electricity can be switched on to the electric power grid. If we delay the decision to build such generating facilities until consumers need and demand nuclear power, we will wait a long time.

An almost limitless supply of energy, however, is promised by power generated from atomic fusion and/or solar radiation (photovoltaic + solar-thermal), but neither energy source is now economically feasible. Although the Sun's core is fueled by hydrogen fusion, and hydrogen is the main active ingredient in thermonuclear bombs, sustained, controlled fusion for commercial power generation is not yet a reality (European Commission 2004). Photovoltaic (Dupont 2008) and solar heating installations (Sandia National Laboratories 2008) are practical in outer space and in remote terrestrial environments today, but as yet have not achieved widespread adoption because of high costs relative to more competitive forms of energy. Moreover, such energy is not dispatchable—that is, like wind power, which is only available when wind speeds are sufficient, solar can only generate electricity when the Sun shines. Thus storage of energy (i.e., capacitance in the energy system) is required for intervals when power is needed but not produced by these energy sources. Future technological advances in hydrogen fusion and/or photovoltaic power generation, combined with the limitations inherent in other forms of renewable, carbon-, and fission-based systems should eventually result in the commercialization of one or both of these energy sources, or perhaps some other means of providing power not yet envisioned.

Sustainable development of mineral resources in a more equitable World must involve worldwide resource recovery + efficient conservation + dematerialization, while trying to preserve ecosystem integrity. Global exploitation of renewable resources at or below recharge rates, and nearly total recycling of nonrenewable Earth materials can only be achieved, however, employing inexpensive, ubiquitously available energy. Unfortunately, the Second Law of Thermodynamics requires that at least a small portion of the mineral resources and most of the spent energy be irretrievably lost. Thus conservation coupled with basic scientific research-derived technological improvements in utilization of alternative Earth materials and very cheap energy will be absolutely required if humanity is to have a promising future.

Whether or not a World of ten billion people can consume resources approaching the manner in which Americans do now, and still maintain the viability of a more severely impacted and altered web of life, is a daunting question (Skinner 1989; Ernst 2002). A functioning biosphere provides ecosystem services vital to humanity, as well as to other forms of life positioned farther down the food chain (Daily 1997; Daily and Ellison 2002), so for our own self interest, preserving a healthy natural environment is essential if anything approaching sustainability is to be achieved. With a substantially enhanced global research base, scientists and engineers may be capable of supplying the long-term energy needs of the World population (Whitesides and Crabtree 2007). Consequentially, mineral resource sustainability lies within our potential reach. But resource sufficiency is not apt to be achieved in the foreseeable future unless our political, industrial, and technological leaders recognize the serious implications of the looming energy shortfall; they must address this problem by providing strong incentives to conserve and recycle resources and support research in order to develop new, inexhaustible power systems (NRC 2007b).

Real sustainability, however, may ultimately prove to be dependent on the ability of humanity to preserve a viable biosphere and the life services it provides (Balmford et al. 2002). Whether this will be possible or not is uncertain (NRC 1999), and only limited progress has been made so far. Because of population momentum and the ever-increasing consumption of natural resources by both Developing and Industrialized Nations, the next 25–50 years will be absolutely critical in our efforts to achieve a sustained, dynamic equilibrium with the global environment and its biospheric carrying capacity, through conservation, resource substitution, and greatly increased efficiency.

10 Conclusions

Worldwide peace requires the socio-economic evolution toward equitable usage of mineral resources and distribution of wealth. To avoid widespread political unrest, a more even degree of global prosperity must be reached within the next several generations. Of course, all Earth resources, renewable and non-renewable, are present in finite supply. Thus, to achieve stability in mineral resource development, we must immediately increase research in science and technology in order to reach efficient levels in the:

1. Conservation of Earth materials;
2. Recycling of Earth materials;
3. Exploration, extraction, beneficiation, and utilization of Earth materials;
4. Substitution of alternative materials, including dematerialization; and
5. Development and commercialization of universally available, cheap energy.

Political and scientific + technological leadership will be required to promote and achieve these goals. Hopefully, the transition to virtually limitless energy will be accomplished by 2050, and humanity will reach a dynamic equilibrium with the rate of replenishment of renewable resources, and the rate of recycling of non-renewables. However, by 2050, ten billion people consuming resources as intensively as Americans do now would seriously degrade the global environment and biospheric habitability, possibly irreparably. The likelihood is vanishingly small that the Industrialized Nations will purposefully reduce their profligate consumption of mineral resources and energy. Equally implausible is the prospect that equitable resource utilization at a reasonable level of material prosperity can be achieved for ten billion humans in a couple of generations. But, if it is, our real challenge will be to preserve an intact, functioning biosphere. The very habitability of the planet for future life is at risk.

Acknowledgments This paper represents an update of a prior synthesis (Ernst 2002) prepared for an American Geophysical Union symposium entitled "Sustainability of Earth Resources." Stanford University has supported my ongoing research. Brian Skinner, F.W. Wellmer, and Jeremy Richards reviewed this revised, updated manuscript. I thank the above institutions and scientists for their support and constructive feedback.

References

Ahlbrandt, T. S., 2002, Future oil and gas resources of the world: A coming supply crisis? Trans. Amer. Geophys. Union, 83, SM02/html/U32A-02.html

Alexander, S. E., and Ehrlich, P. R., 2000, Population and the environment: pp. 329–345 *in* Ernst, W. G., ed., Earth Systems: Processes and Issues. Cambridge University Press, New York, 566p.

Ali, S. H., 2003, Mining, the Environment, and Indigenous Development Conflicts. University of Arizona Press, Tucson, 254p.

Alley, W. M., Healy, R. W., LaBaugh, J. W., and Reilly, T. E., 2002, Flow and storage in groundwater systems. Science, 296, 1985–1990.

Baldwin, S. F., 2002, Renewable energy: progress and prospects. Physics Today, 55(4), 62–67.

Balmford, A., Bruner, A., Cooper, P., Constanza, R., Farber, S., Green, R. E., Jenkins, M., Jefferiss, P., Jessamy, V., Madden, J., Munro, K., Myers, N., Naeem, S., Paavola, J., Rayment, M., Rosendo, S., Roughgarden, J., Trumper, K., and Turner, R. K., 2002, Economic reasons for conserving wild nature. Science, 297, 950–953.

Bergkamp, G., and Sadoff, C. W., 2008, Water in a sustainable economy: Chapter 8, pp. 107–122 *in* Esty, D.C., ed., State of the World: The Worldwatch Institute, W. W. Norton & Co., New York, 269p.

British Petroleum, 2008 World Coal Reserves: http://wwww.bp.com.sectiongenericarticle

Brown, D., 2000, Bulls and bears duel over supply. American Association of Petroleum Geologists Explorer, May, 12–15.

Campbell, C. J., and Laherrer, J. H., 1998, The end of cheap oil. Scientific American, 278, 78–83.

Chesworth, W., 2002, Sustainability and the end of history. Geotimes, 47(10), 5, 52.

Cordani, U. G., 2000, The role of the earth sciences in a sustainable world. Episodes, 13, 155–162.

Corson, W. H., 2002, Recognizing hidden environmental and social costs and reducing ecological and societal damage through tax, price, and subsidy reform. The Environmentalist, 22, 67–82.

Craig, J. R., Vaughan, D. J., and Skinner, B. J., 2001, Resources of the Earth. 3rd edition, Prentice-Hall, Inc., Upper Saddle River, NJ, 520p.

Crosby, A. W., 2006, Children of the Sun: A History of Humanity's Unappeasable Appetite for Energy. W. W. Norton and Company, New York, 192p.

Daily, G. C., ed., 1997, Nature's Services: Societal Dependence on Natural Ecosystems. Island Press, Washington, DC.

Daily, G. C., and Ellison, K., 2002, The New Economy of Nature. Island Press, Washington, DC, 260p.

Davidson, J. P., Reed, W. E., and Davis, P. M., 2002, Exploring Earth: An Introduction to Physical Geology. 2nd edition, Prentice-Hall, Inc., Upper Saddle River, NJ, 477p.

Deffeyes, K. S., 2001, Hubbert's Peak: The Impending World Oil Shortage. Princeton University Press, Princeton, NJ, 208p.

Diamond, J., 2005, Collapse: How Societies Choose to Fail or Succeed. Viking Press, New York, 575p.

DOE, 2001, International Energy Outlook 2001: Report No. DOE/EIA-0219 (99), U.S. Department of Energy, Washington, DC.

DOE, 2004, The Hydrogen Economy: Opportunities, Costs, Barriers, and R&D Needs: U.S. Department of Energy, Washington, DC, http://www.nap.edu/catalog/10922.html

DOE, 2008, Energy Sources: U.S. Department of Energy, Washington, DC, http://www.energy.gov/energy sources/index.html

Doran, J. W., and Sims, J. T., 2002, Sustaining Earth and its people. Geotimes, American Geological Institute, 47(7), 5.

DuPont, 2008, Science of Photovoltaic Energy: http://www2.dupont.com/Photofoltaics/en_US/science_of/index.html

Dye, C., 2008, Health and urban living. Science, 319, 766–769.

Eggert, R. G., 2008, Trends in mineral economics: editorial perspective, 1986–2006. Resources Policy, 33, 1–3.

Einaudi, M. T, 2000, Mineral resources: assets and liabilities: pp. 346–372 *in* Ernst, W. G. (ed.) Earth Systems: Processes and Issues. Cambridge University Press, New York, 566p.

Ernst, W. G., 2001, The increasing severity of circumpacific natural hazards. International Geology Review, 43, 380–390.

Ernst, W. G., 2002, Global equity and sustainable Earth resource consumption requires super-efficient extraction-conservation-recycling and ubiquitous, inexpensive energy. International Geology Review, 44, 1072–1091.

European Commission, 2004, European Fusion Research Area: http://europa.eu.int/comm/research/rtdinfo_en.html

Field Museum, 2008, Biodiversity and Conservation: The Web of Life, http://www.fieldmuseum.org/biodiversity

Fouda, S. A., 1998, Liquid fuels from natural gas. Scientific American, March 1998, 92–95.

Freeze, R. A., 2000, The Environmental Pendulum. University of California Press, Berkeley, CA, 323p.

Fyfe, W. S., 1989, Soil and global change. Episodes, 12, 249–254.

Galbraith, J. K., 2002, A perfect crime: global inequality. Daedalus, Journal of the American Academy of Arts and Sciences, LV(3), 11–25.

George, R. L., 1998, Mining for oil. Scientific American, 278, 84–85.

Gleick, P. H., 1998, The World's Water 1998–1999. Island Press, Washington, DC.

Gordon, R. B., Bertram, M., and Graedel, T. E., 2006, Metal stocks and sustainability. Proceedings of the National Academy of Sciences, 103, 1209–1214.

Grimm, N. B., Faeth, S. H., Golubiewski, N. E., Redman, C. L., Wu, J. G., Bai, X. M., and Briggs, J. M., 2008, Global change and the ecology of cities. Science, 310, 756–760.

Guptill, S. C., 2001, Disease aftershocks—the health effects of natural disasters. International Geology Review, 43, 419–423.

Harrison, P., and Pearce, F., 2000, AAAS Atlas of Population & Environment: American Association for the Advancement of Science. University of California Press, Berkeley, CA, 204p.

Hoffert, M. I., Caldera, K., Benford, G., Criswell, D. R., Green, C., Herzog, H., Jain, A. K., Kheshgi, H. S., Lackner, K. S., Lewis, J. S., Lightfoot, H. D., Manheimer, W., Mankins, J. C., Mauel, M. E., Perkins, L. J., Schlesinger, M. E., Volk, T., and Wigley, T. M. L., 2002, Advanced technology paths to global climate stability: energy for a greenhouse planet. Science, 298, 981–987.

Holdren, J. P., 2008, Science and technology for sustainable well-being. Science, 319, 424–434.

Hubbert, M. K., 1981, The world's evolving energy system. American Journal of Physics, 49, 1007–1029.

Hutson, S. S., Barber, N. L., Kenny, J. F., Linsey, K. S., Lumia, D. S., and Maupin, M. A., 2004, Estimated use of water in the United States in 2000. U.S. Geological Survey Circular 1268, 52p.

International Energy Agency, 2007, Key World Energy Statistics. Head of Communication and Information Office, Paris, 82p.

IPCC, 2007, Fourth Assessment Report: http://www.ipcc/ch/ipccreports/ar4-syr.htm

IUGG, 2002, Preparation process for the world summit for sustainable development: Chapter 31, Agenda 21. International Union of Geology and Geophysics, Washington, DC, 16p.

Johnson, W., and Paone, J., 1982, Land Utilization and Reclamation in the Mining Industry, 1930–1980: Bureau of Mines Information Circular 8862. Department of the Interior, Washington, DC.

Kesler, S. E., and Wilkinson, B. H., 2008, Earth's copper resources estimated from tectonic diffusion of porphyry copper deposits. Geology, 36, 255–258.

Khanna, T., ed., 2000, Mine Closure and Sustainable Development: World Bank Group, Mining Department. Mining Journal Books, Ltd, London, England, 154p.

Lewis, N. S., 2007, Toward cost-effective solar energy use. Science, 315, 798–801.

Marcus, G. H., and Levin, A. E., 2002, New designs for the nuclear renaissance. Physics Today, 55(4), 54–60.

McNutt, M. C., 2002, Engineering the ocean. Daedalus, Journal of the American Academy of Arts and Sciences, LV(3), 42–54.

Mendelsohn, R., and Neumann, J. E., eds., 1999, The Impact of Climate Change on the United States Economy. Cambridge University Press, New York, 331p.

Moniz, E. J., and Kenderdine, M. A., 2002, Meeting energy challenges: technology and policy. Physics Today, 55(4), 40–46.

Nearing, M. A., Romkens, M. J. M., Norton, L. D., Stott, D. E., Rhoton, F. E., Laflen, J. M., Flanagan, D. C., Alonso, C. V., Bingner, R. A., Dabney, S. M., Doering, O. C., Huang, C. H., McGregor, K. C., and Simon, A., 2000, Measurement and models of soil loss rates. Science, 290, 1300–1301.

Nordstrom, D. K., 2002, Worldwide occurrences of arsenic in ground water. Science, 296, 2143–2144.

NRC, 1996, Mineral Resources and Sustainability: Challenges for Earth Scientists. National Research Council, National Academy of Sciences, Washington, DC, 25p.

NRC, 1999, Our Common Journey: A Transition Toward Sustainability. National Research Council, National Academy of Sciences, Washington, DC, 363p.

NRC, 2004, The Hydrogen Economy: Opportunities, Costs, Barriers, and R&D Needs. National Research Council, National Academy of Sciences, Washington, DC, 256p.

NRC, 2006, Trends in Oil Supply and Demand, Potential for Peaking of Conventional Oil Production, and Possible Mitigation Options. National Research Council, National Academy of Sciences, Washington, DC, 61p.

NRC, 2007a, Minerals, Critical Minerals, and the U.S. Economy. National Research Council, National Academy of Sciences, Washington, DC, 216p.

NRC, 2007b, Rising Above the Gathering Storm: Energizing and Employing America for a Brighter Economic Future. National Research Council, National Academy of Sciences, Washington, DC, 592p.

Obermiller, J., 1999, Historic world oil production: table 10 in Basic Petroleum Data Book: v. 19, sec. 4. American Petroleum Institute, Washington, DC.

Ogden, J. M., 2002, Hydrogen: the fuel of the future? Physics Today, April 2002, 69–75.

Perrine, R. L., and Ernst, W. G., eds., 1985, Ruby Volume III: Energy for Ourselves and Our Posterity. Prentice-Hall, Englewood Cliffs, NJ, 434p.

Pitman, N. C. A., and Jorgensen, P. M., 2002, Estimating the size of the world's threatened flora. Science, 298, 989.

Raven, P. H., 2002, Science, sustainability, and the human prospect. Science, 297, 954–958.

Raven, P. H., and Williams, T., eds., 2000, Nature and Human Society: The Quest for a Sustainable World. National Research Council, National Academy of Sciences, Washington, DC, 644p.

Richards, J. P., 2002, Sustainable development and the minerals industry. Society of Economic Geologists Newsletter, 48, January 2002, 1–12.

Rojataczer, S., Sterling, S. M., and Moore, N. J., 2001, Human appropriation of photosynthesis products. Science, 294, 2549–2552.

Rona, P. A., 2002, Marine Minerals for the 21st Century: Episodes. International Union of Geological Sciences, Delft, The Netherlands, pp. 2–12.

Sandia National Laboratories, 2008, National Solar Thermal Test Facility: http://www.sandia.gov/Renewable_energy/solarthermal/nstttf.html

Schaetzl, R. J., and Anderson, S., 2005, Soils: Genesis and Geomorphology. Cambridge Universty Press, New York, 832p.

Schlesinger, W. H., 1997, Geotimes, 42(2), 44–46.

Scientific American, 1989, Managing Planet Earth. W. H. Freeman and Co., New York, 146p.

Skinner, B. J., 1976, A second iron age ahead? American Scientist, 64, 258–269.

Skinner, B. J., 1979, Earth resources. Proceedings of the National Academy of Sciences, 76, 4212–4217.

Skinner, B. J., 1989, Resources in the 21st century: can supplies meet needs? Episodes, 12, 267–275.

Solley, W. B., Pierce, R. R., and Perlman, H. A., 1998, Estimated Use of Water in the United States in 1995. U.S. Geological Survey Circular 1200, Washington, DC, 71p.

Solow, R., 1993, An almost practical step towards sustainability. Resources Policy, 19(3), 162–172.

Stephanopoulos, G., 2007, Challenges in engineering microbes for biofuels production. Science, 315, 801–804.

Tarbuck, E. J., and Lutgens, F. K., 2002, Earth: An Introduction to Physical Geology, 7th edition. Prentice-Hall, Inc., Upper Saddle River, NJ, 670p.

Tilton, J. E., 2002, On Borrowed Time? Assessing the Threat of Mineral Depletion. Resources for the Future, Washington, DC, 142p.

Trenberth, K. E., 2007, Warmer oceans, stronger hurricanes. Scientific American, 297, 45–51.

U.S. Census Bureau, 2008, http://www.census.gov

U.S. Geological Survey, 2000, Minerals Yearbook: Area Reports: Domestic 2000. U.S. Geological Survey, Washington, DC, 551p.

UN, 1999, World Population Prospects: The 1998 Revision. United Nations Population Division, New York.

Vitousek, P. M., Ehrlich, P. R., Ehrlich, A. H., and Matson, P. A., 1986, Human appropriation of the products of photosynthesis. Bioscience, 36, 368–373.

WCED, 1987, Our Common Future: World Commission on Environment and Development (Brundtland Commission Report). Oxford University Press, New York.

WHO, 2007, The World Health Report 2007—A Safer Future: Global Public Health Security in the 21st Century, http://www.who.int/whr/2007/en/index.html

Whitesides, G. M., and Crabtree, G. W., 2007, Don't forget long-term fundamental research in energy. Science, 315, 796–798.

Wilkinson, B. H., 2005, Humans as geologic agents: a deep-time perspective. Geology, 33, 161–164.

Wolman, M. G., 2002, The human impact: some observations. Proceedings of the American Philosophical Society, 146, 81–98.

World Bank, 2003, Mining and Development: http://www.worldbank.org/ogmc

Yoder, H. S., Jr., 2002, Geology: significant component of new multidisciplinary sciences. Proceedings of the American Philosophical Society, 146, 37–55.

Sustainable Development and Mining—An Exploratory Examination of the Roles of Government and Industry

Arianna Waye, Denise Young, Jeremy P. Richards and Joseph A. Doucet

Abstract A common perception is that the activities of the extractive minerals industry are inconsistent with sustainable development objectives, and that any form of mining is de facto *un*-sustainable. This paper outlines some of the major issues surrounding sustainability in the context of mineral development, and argues that mining can be sustainable if revenues that are derived from mining are collected and used to promote sustainable objectives at both the local (community) and national (or regional) levels.

In order to develop a better understanding of the arguments presented, this study provides a brief empirical investigation of the impact of broad measures of regulation and taxation on the mining sector investment climate, based on data from the Fraser Institute's annual mining surveys, World Bank indices of regulatory activities, and various other sources of information on a variety of country specific socio-economic characteristics. A positive relationship is found between the quality of the regulatory environment and mining companies' perceptions of a country's investment climate, as measured by the Fraser Institute's policy potential index (PPI). However, there is no evidence in the data of a negative relationship between the level of taxation and mining investment climate, except where tax and legislative frameworks are unstable and unpredictable. This allows us to conjecture that there may be scope for increased taxation to deal with local and national sustainability issues. As long as there is political stability and a good regulatory environment, it may be possible to raise taxes to mitigate local mining externalities through community development, or to fund other programs through hypothecated taxes (i.e., taxes raised to fund specific objectives).

A. Waye (✉)
Department of Economics, University of Alberta, Edmonton, AB, T6G 2H4, Canada
e-mail: aewaye@ualberta.ca

J.P. Richards (ed.), *Mining, Society, and a Sustainable World*,
DOI 10.1007/978-3-642-01103-0_7, © Springer-Verlag Berlin Heidelberg 2009

1 Introduction

Applying the concept of sustainable development to the mining industry is fraught with challenges. First, a fundamental incompatibility exists between the non-renewable character of mined resources, and the intuitive understanding of "sustainability" as maintaining a constant level of consumption of those resources across generations. Second, on a practical level, mines can and often do negatively impact the socio-economic wellbeing of nearby communities, contradicting the notion of sustainability as properly meeting the needs of current and future generations.

In light of the above, one approach to sustainable development in the mining sector is to define it in terms of the transformation of natural capital into social or human capital (Peskin 1994; Labonne 2002), thus theoretically providing perpetual benefit streams across generations (Richards 2005). In line with this concept, it is argued here that mining may be sustainable if some of the revenues generated are collected and used to promote sustainable objectives at both the local (community) and national (or regional) levels.

This interpretation of sustainability in mining is of course predicated on a strong government role. Within global markets, the competitive impact of government intervention is an important question. Thus, within the context of broad sustainability goals, it is important for governments to understand the implications of policies that aim to mitigate the negative effects of mining. This paper examines policy decisions that might be used to promote sustainable development. In order to gain a better understanding of the impacts of policy, we use relatively simple statistical methods to analyze empirically the impact of various policy options on overall investment climate. The data are drawn from the Fraser Institute's annual mining surveys, World Bank indices of regulatory activities, and various other sources of information on a variety of socio-economic country characteristics.

2 Sustainable Development and Non-Renewable Resources

The Brundtland report (WCED 1987, p. 8) states that "Humanity has the ability to make development sustainable—to ensure that it meets the needs of current generations without compromising the ability of future generations to meet their needs". Although this may be a useful starting point for addressing sustainability issues, it has long been recognized that there is an inherent ambiguity in this definition. Kates et al. (2005) point out that the vagueness of the Brundlandt definition allows various groups to make contrary claims regarding whether or not a given activity is sustainable. In particular, the relationship between the environment and development is a complex one. Socio-economic development requires inputs from the environment. In order for society to continue to experience improvements in standards of living, the environment must be disturbed. It is therefore incumbent on society to determine the appropriate extent and timing of the development of natural resources, especially non-renewable resources. Decisions must be made in relation to the sustainability of ecological thresholds, human and economic development, and ideally a combination of the two.

Kates et al. (2005) also remind us that sustainable development is not just a local phenomenon, but must also be considered on a global scale. Issues concerning the balance between the environment and human populations re-emerged in the 1970s, as concerns grew about global overpopulation and famine. The United Nations Conference on the Human Environment (held in Stockholm in 1972)[1] addressed the issues of poverty, trade financing, and other matters related to the environment. More recently in 1992, the United Nations Conference on Environment and Development in Rio de Janeiro led to *Agenda 21*, which outlined the actions and international agreements on climate change and biodiversity that were seen as essential for promoting sustainable development (Kates et al. 2005). Ten years later at the World Summit on Sustainable Development in Johannesburg, nations reaffirmed their commitment to sustainable development. However, no working definition or principles of best practice were established, because it quickly became apparent that definitions can vary a great deal depending upon perspective.

Differences in perspectives imply that sustainable development of a nonrenewable resource may be defined differently by each stakeholder: industry, businesses, labourer, society, traditional society, land-owner, and government. To business and industry, it may mean sustaining profits, whereas to a labourer, sustainability may imply keeping mines open to retain jobs. To the various sectors of society, the definition may be more complex, because mines and other resource extraction industries have the potential to cause negative effects on the environment, local human health, and local social wellbeing. On the other hand, mining also generates income and raw materials for society. The various interests of these stakeholder groups are further complicated because they are not independent and commonly overlap. For example, a business owner or a worker may also be a member of the local community.

2.1 Weak Versus Strong Sustainability

The Board on Sustainable Development of the US National Academy of Sciences (USNAS) has attempted to simplify the many concepts and perspectives identified under the broad title of sustainable development. Figure 1 highlights the distinctions between the priorities of various advocates and analysts with respect to sustainable development, while giving some idea of the complex relationship between the two, and the time horizons involved (USNAS 1999).

The possibilities of "what" is to be sustained include nature, life support systems, and community. The areas to be developed include people, the economy, and society. These objectives can sometimes be mutually exclusive and at other times

---

[1]Report of the United Nations Conference on the Human Environment. Last accessed July 29, 2009, at http://www.unep.org/Documents.Multilingual/Default.asp?DocumentID=9

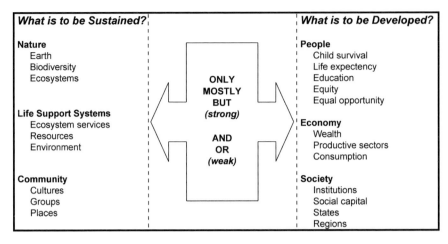

Fig. 1 Defining sustainable development (from USNAS 1999)

complementary (at least to some extent), as indicated by linking *conjunctions* "only, mostly, but, and, or" listed in the center column. Obviously, it is not possible to develop people, the economy, or society without disrupting the environment to some extent.

Some groups advocate that sustaining biodiversity, ecosystems, and ecological thresholds should take priority. This is referred to as the ecological approach to sustainable development (Elliot 1999). Under this view, development would be allowed as long as minimum environmental thresholds and targets are maintained. Under this *strong* approach to sustainability, the environment takes priority in all policy decisions, and biodiversity and ecosystems are not considered substitutes for other forms of capital that might be developed. This definition would seem to be incompatible with any form of non-renewable resource extraction.

On the other hand, the linking words "and, or" suggest a weaker form of sustainable development whereby the environment may or may not be sustained in a strict sense, and it is capital development that is to be sustained. This economic approach is similar to an "optimal savings rule", whereby aggregate levels of productive capital used for economic development are maintained at some constant level. This higher acceptable degree of substitutability between natural and man-made forms of capital is commonly referred to as *weak* sustainable development (Elliot 1999).[2]

Of primary interest for weak sustainable development is that communities are to be sustained and other forms of capital are to be developed. These other forms of capital include human and social capital (institutions, education, and health care in terms of life expectancy and child mortality), as well as financial capital and

---

[2]"Weak" vs. "strong" sustainable development are not indicators of preference; rather, they are definitions.

its distribution (wealth, equity), and physical capital (used for the production of investment and consumption goods). Through some appropriate combination of all of these forms of capital, economic development of local communities, regions, and states can proceed.

2.2 Weak Sustainability and Mining

In the context of the mining industry, and in accordance with the weak definition of sustainable development, measures to mitigate environmental and social damage must be implemented where there are adverse effects from mining. Long-term measures to ensure the economic sustainability of local communities may necessitate investment in economic diversification or, as a last resort, relocation to other areas. At the regional or national level, other forms of capital may be used to sustain development. This implies a differentiation in the definition of, and government responsibilities related to, sustainable development at the local, regional, and national levels.

At the national level, the goal may be to transform mineral wealth into other forms of capital that will create a perpetual benefit stream for both current and future generations. In order for governments to accomplish this, revenues derived from mining can be targeted towards social or environmental objectives. For example, *hypothecated taxes* (taxes collected for specific goals) can be used to achieve various social objectives, such as promoting human development, economic diversification, or protecting the environment.

The discussion above can be summarized in the schematic of local and national sustainable development shown in Fig. 2. Here, sustainable development is presented as a process that is a function of society's decision to sustain natural capital or aggregate capital stocks. The process in Fig. 2 refers to the national level of sustainable development with the exception of the large ellipse, which exemplifies necessary action by local governments to ensure sustainable development at the local level.[3]

Initially, the sustainable development process for non-renewable resources depends upon what society wants to sustain: natural capital (i.e., the environment and the actual resource, which can only be sustained by not mining[4]), or aggregate levels of capital. When a choice is made whereby resources are to be used to sustain human development, societies are saying that natural capital should be transformed into social and physical capital.

---

[3]It should be noted that this model deals only with negative environmental and social externalities. Mining at the local level also provides positive social externalities, such as new transport links that broaden the market for local products, and training that expands the employment opportunities for local workers.

[4]The concept that minerals left in the ground represent a form of capital is only valid if those minerals can at some point be extracted economically. Many argue that no value should be put on such resources until an economic plan for their extraction can be demonstrated.

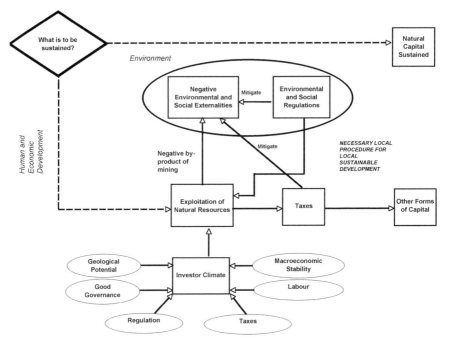

Fig. 2 Local and national sustainable development processes for non-renewable natural resources

In order for resource development to commence, at least in capitalist economies, it is necessary for a project to be profitable, which requires a positive investment climate. A positive investment climate in the mining industry is a function of geological potential, good governance (including clear regulation, stable taxation, and macroeconomic stability), and labour potential (World Bank 2005). However, in order for sustainable development to be achieved, any concomitant negative social and environmental costs must also be mitigated through regulation, or internalized through taxes.

3 Policy Options for Sustainable Development and Mining

Successful sustainable development hinges on the mitigation or internalization of negative social and environmental externalities through regulation or taxation. When implementing sustainable development policy, based on either taxation or regulation, the potential economic and investment consequences must be estimated from the past reactions of industry. These reactions are explored below in relation to specific international regulations and taxation regimes affecting the mining industry.

3.1 Mining Regulation

Regulations are generally put into place in response to market failures, where governments are expected to intervene. In the case of mining, there can be many negative externalities that affect the environment and the socio-economic fabric of communities. In such cases, government regulations can be used to attempt to mitigate market failures by restricting the activities of a mining company. Governments may also need to increase monitoring and legal enforcement within mining communities to deal with social and environmental concerns.

Whereas some regulations apply to mining procedures and environmental standards, thus impacting a company's behaviour, others are designed to ensure that local labourers benefit financially. In Chile, for example, the government has imposed a mandatory profit-sharing scheme, whereby companies are annually required either to share 30% of net profits with their employees, or give them a bonus of 25% of their yearly income (Houde 2005). Similar profit-sharing schemes exist in Mexico and Peru (Houde 2005).

Environmental Impact Assessments (EIAs) are perhaps the most significant example of regulations that affect the mining industry at the local level of development. Environmental impact assessments are used to determine the effects of a proposed activity on the natural and human environment (e.g., human health as a result of environmental impacts). Increasingly, these assessments also consider the socio-economic impacts of development (O'Faircheallaigh 1999). According to Pring and Siegele (2005), and reflecting provisions in the Aarhus Convention of 1998[5] (Davies 2002), the future trend in EIAs will be towards increased public participation in the formulation of laws and best practices. Although public participation should engage the population at large, special care should be taken to include marginalized groups, especially Aboriginal groups, that in the past have commonly not been consulted.

Although developed countries typically have legislation in place to regulate and monitor specific EIA guidelines and recommendations, many developing countries either have minimal legislation, or have substantive EIA legislation but lack the capacity for monitoring and enforcement. In an attempt to respond to this problem, ten of the leading financial institutions in several different countries adopted the Equator Principles in 2003. The Equator Principles "are a voluntary set of guidelines for managing social and environmental issues in project finance."[6] The Equator Principles outline the requirements for obtaining funding for projects costing over US$10 million (Equator Principles 2006). Much of the assessment for funding (especially in non-OECD and low-income OECD countries) is based upon the International Finance Corporation's (IFC) performance standards on social and environmental sustainability (IFC 2006). Equator-Principle financial institutions commit to making the client aware of the content, application, and benefits of applying these

---

[5] http://ec.europa.eu/environment/aarhus/

[6] Quoted from www.equator-principles.com

principles to the proposed projects, and require the client to outline their methods (in the EIA assessment) and actions for adherence to the Principles when seeking (and later continuing) to receive financing.

The voluntary nature and international character of the accord pose some challenges to the implementation of the Equator Principles. However, they have the potential to be remarkably effective because mining companies generally require a significant amount of capital for mine development, which is commonly obtained from these major international banking institutions. Financing from these banks is conditional upon meeting the Equator Principles, and interim reports are monitored carefully as a condition for continued funding. The extent of the requirements in terms of added costs to companies may vary by country, with EIA legislation, or by bank. The IFC outlines environmental strategies that have been included by major financial institutions in the requirements for EIAs under the Equator Principles.[7]

In general, regulation will have a negative (or positive) impact on the investment climate to the extent that it adds to (or reduces) the cost of doing business. In addition, government interventions can fail, worsening investment climate due to poor information, a lack of expertise or resources, credibility gaps, or a lack of public support (World Bank 2005). However, regulations dealing with property rights, which ensure that investments are secure, tend to improve investment climate and are examined further below.

3.2 Mining Taxation

3.2.1 General Principles

Tax revenues from mining operations may be collected for many purposes, including a desire to secure sustainability goals.[8] These revenues can provide resources for meeting the immediate needs of the current generation, for example by the provision of public goods, and also to invest in financial, physical, or social capital in order to provide income or benefit streams for future generations. Several taxation instruments are available to governments, and the choice of tax instruments and rates can have implications for the behaviour of mining companies.

Taxation of mining has a long history, and early rationales were strikingly similar to what would now be termed as sustainability issues. Royalties applied to the mining industry are an application of the "single tax" first suggested by the economist Henry George in 1879 (Petrella 1988).

George believed that actual land was required to create any form of wealth. In addition, he argued that all land was public land, and should be shared equally

---

[7]For more information, see the IFC website at: http://www.ifc.org/ifcext/sustainability.nsf/ Content/EnvSocStandards, the World Bank (2001), and www.equator-principles.com.

[8]Other rationales for collecting tax revenues from mining operations may include: (i) the need to have funds available to moderate foreign exchange flows; and (ii) a need for funds that can be used to pursue macroeconomic policies aimed at smoothing out fluctuations in the business cycle.

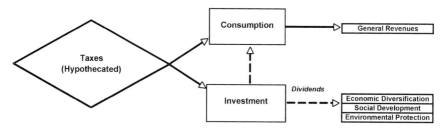

Fig. 3 Transforming mineral wealth into other forms of capital

among current and future citizens. He proposed that land be taxed and that these taxes should be used to promote social goals that promote the greatest common good. This concept is referred to as George's Principle.

A process that collects royalties (or other taxes) for government coffers in exchange for the right to extract and sell non-renewable resources is based on a version of George's Principle, whereby non-renewable resources, instead of land, are the relevant assets that are owned by society. It is the responsibility of government to serve the best interests of society by collecting revenues in exchange for the sale of the natural resources. A related principle is embodied in the 1974 United Nations Resolution 3201 (S-VI) on the Establishment of a New International Economic Order, wherein permanent sovereignty over natural resources is acknowledged.[9]

George's Principle can also be considered a precursor to some of the hypothecated tax funds found around the world today, such as the Alberta Heritage Fund and the Alaska Permanent Fund. The Alberta Heritage fund takes a proportion of revenues from the development of Alberta's oil and gas resources and uses it to promote various objectives such as supporting education (Warrack and Keddie 2002). The Alaska Permanent fund acquires 50% of all mineral royalties, a portion of which is distributed annually to each Alaskan citizen in the form of a dividend (Warrack and Keddie 2002).[10]

At the local, regional, and national levels, depending on the legislative structure of a country, governments can transform mineral wealth into other forms of capital either by consumption or investment (Fig. 3). The results of these policies can have very different results in terms of the capital generated and sustained.

There are two basic channels through which the benefits of hypothecated taxes might be realized. (1) Revenues can be used to support current consumption; that is, the expenditures do not generate another form of capital, but may help to offset any negative environmental, health, or other socio-economic impacts. Funds

---

[9]http://daccess-ods.un.org/acccss.nsf/Get?Open&DS=A/RES/3201(S-VI)&Lang=E

[10]The extent to which the distribution of mineral rents to individuals contributes to local or regional sustainable development depends on the ultimate use to which the funds are put. If the funds are merely spent on immediate consumption, then this could be argued not to represent sustainable development.

are typically spent on new or existing programs, or on capital projects, and are unlikely to generate revenues much into the future because the revenue is spent rather than invested. An exception is expenditure on infrastructure, which may stimulate other investments or social improvements that will bear dividends in the future. (2) Revenues can be invested, with the interest used to fund beneficial programmes rather than spending the principal. The latter type of investment strategy ideally converts non-renewable natural resources into a perpetual benefit stream.

Depending upon a county's level of development, the optimal investment portfolio is likely to vary. More-developed nations might find it optimal to focus on economic diversification and the development of fixed capital, both of which will reap benefits for current and future generations. Developing nations, on the other hand, may need to focus on human capital formation, whereby hypothecated taxes are invested in primary and secondary education as well as primary health care. A skilled labour force is necessary for any form of economic diversification and growth.

It is important to note that hypothecated taxes are *not* a different form of taxation. The tools used to collect tax revenues are similar (e.g., royalty taxes, profit taxes, corporate income taxes, etc.), but the distinction is that the revenues are targetted towards specific objectives rather than being paid into general revenues, with accounting methods in place to enable tracking of expenditures. Without such a system of monetary tracking, it is extremely difficult to measure how revenues from mining (or any other economic activity for that matter) directly contribute to a specific goal such as sustainable development.

3.2.2 Royalty Taxation in Practice

There are different forms of royalty taxes that can be applied to the mining industry, including gross or net income royalties. There are many different definitions of each type of royalty, but the following are the most commonly used. *Unit based royalties* are based upon unit volume or unit weight. *Ad valorem royalties* are based on a mineral's value (whereby the definition of value may vary). A *profit* or *income based royalty* is an ad valorem royalty defined as a percentage of cash income less allowed expenses incurred. In contrast, a *gross income royalty* is also an ad valorem royalty based on a percentage of the gross market, gross sales, net market, or net smelter value of the commodity produced, and is charged whether a profit is made or not. Gross royalties appear to provide maximum benefit to society, but present significant risk to industry, which may stifle investment. In contrast, net royalties payable only on profits are preferred by industry, but may return little value to the government when profits are low, even though mining continues. Table 1 provides an overview of some of the variations that can be found in terms of tax instruments and rates in various jurisdictions.

Investors tend to oppose gross income royalties because they prefer back-end loading of tax payments, which shifts more of the risk onto governments. Profit taxes are seen by industry as offering a lower tax burden in compensation for project

Table 1 Royalties in selected jurisdictions

Country	Royalty type	Royalty rate	Variation
Botswana	Ad valorem (net smelter return)	3–10%	Yes Precious stones: 10% Precious metals: 5% Other minerals or mineral products: 3%
Ghana	Ad valorem (sales revenue)	3–12%	No Same royalty system for all minerals
South Africa	Ad valorem (sales revenue)	Variable	Yes Sliding scale formula for gold Other minerals variable % of either market value or net profit
Zambia	Ad valorem (net smelter return)	2%	No Same royalty system for all minerals
Zimbabwe	N.A.	0%	N.A.
India	Ad valorem or unit-based	0.4–20%	Yes Ad valorem rate or unit-based charge for each mineral
Indonesia (7th Generation COW)	Unit based	N.A.	Yes Unit-based rate for each mineral
Papua New Guinea	Ad valorem	2%	No
Queensland (AU)	Ad valorem or unit-based	2.7% of value, or a variable royalty rate if price exceeds a reference price	Yes Most metallic minerals: 2.7% of value Industrial minerals: $A.25–$A1.00 per tonne Coal: 7%
Argentina	Most provinces have no royalty; others have ad valorem	0–3%	No
Bolivia	Ad valorem, sliding scale based on ratio	1–6% based on sales price position relative to reference prices	Yes
Chile	N.A.	N.A.	N.A.
Peru	Ad valorem, sliding scale-based on annual cumulative sales	0–3% (exported minerals: 1–3%)	No
Arizona (US)	Ad valorem	At least 2%; commissioner to determine rate	Yes
Ontario (CAN)	Profit-based	10%	No

Source: Otto et al. (2006)

risk, because companies do not pay taxes in years where they do not make any profit (such as during project start-up or global downturns in commodity prices). Net income royalties allow companies to recoup their initial costs over the shortest period of time.

It is tempting to draw conclusions concerning the relative competitiveness of a nation's fiscal regime by analyzing their relative royalty rates. However, it is the overall tax burden placed on mining firms that must be considered in order to draw valid conclusions about international investment competition. Effective tax burdens, or average effective tax rates (AETR), are used widely in industry and by governments to assess the overall revenues collected from economic activities. The AETR for a given type of operation, such as a mine, is the amount of total income paid to the government in the form of taxes.[11] It is calculated as the net present value of tax payments (including income taxes, capital taxes, and royalties) divided by the net present value of net income (which is revenues minus costs such as operating expenses, including non-tax fees, capital costs, and amortization).

The impacts of corporate income taxes (CITs) and cost allowances need to be considered when analyzing different royalty regimes and the relative competitiveness of each regime. Standard corporate income taxes are applied in most jurisdictions, and it is the combination of all of the taxes that represents the overall tax burden of industry. Corporate income taxes are attractive from the government's perspective because they require consistent administrative practices and legal frameworks. In general, capital allowances for capital intensive projects mostly permit 100% expensing of exploration and development costs. Other deductions from income taxes are generally for operating costs, depreciation, continued exploration expenses, environmental protection, land reclamation, and community development. The more write-offs granted, the lower the present tax base and, therefore, the lower the current revenues collected. This may shift the time profile of taxes, but does not necessarily lead to an overall reduction in tax payments.

Governments commonly walk a fine line between generating tax revenues from industrial activities such as mining, and encouraging capital investment and resulting economic development. Given that companies have many choices regarding where to invest, there is competition between jurisdictions and a resulting convergence of tax rates. According to Brewer (2005), in 2002, most jurisdictions that had been taxing outside the average range have moved towards the average. However, because rates of convergence differ, a wide variation in tax rates persists across jurisdictions, as shown in Table 2 (Brewer et al. 2006). Companies operating in Brazil seem to be paying the most in taxes (as percent of net income on marginal operations) for both the 10 and 25% internal rate of return cases, whereas companies in Canada and the US in general seem to pay the least.

---

[11]Note that tax rates may vary by commodity in many countries (Table 1), complicating these comparisons.

Table 2 Average effective tax rates at 10 and 25% rate of return

Jurisdiction	10%	25%
Canada-Quebec	17.7	34.2
Canada-Newfoundland	26.0	29.6
Canada-Manitoba	26.2	
US-Nevada	28.2	26.1
US-Alaska	28.4	28.5
Canada-Ontario	28.6	42.9
Canada-British Columbia	30.4	38.7
Chile	32.1	32.0
Venezuela	35.7	
Papua New Guinea	37.3	
Indonesia	43.1	37.8
Mexico	46.3	44.3
Australia	48.2	37.6
Brazil	54.1	49.9

Source: Brewer et al. (2006)

3.2.3 Taxation to Mitigate Local Negative Social and Environmental Externalities Resulting from Mining

Governments have the option of: (1) taxing mining companies an extra amount which could be used to compensate local communities; or (2) requiring that the mining company directly provide infrastructure or other socio-economic or environmental services to the community (which is financially equivalent from the standpoint of the cost to the firm). According to a Pareto efficiency criterion, whereby no-one is made worse off as a result of economic activity, if the people negatively affected by mining are adequately compensated, then negative externalities are mitigated. However, unless taxes are hypothecated, with the funds directly allocated to mitigation projects, it is difficult to ensure that the extra taxes will in fact positively impact local sustainability.

4 Sustainability Measures and Industry Investment: The Bottom Line

Investment decisions in the mining industry commonly relate to projects with time horizons of 10–20 years or more, with large capital costs borne up front and revenues spread over time. Therefore, uncertainty and risk are critical factors in the investment decisions of mining companies, and include geographical, market, and political risks, in addition to industry-specific risks. Mineral commodity prices are extremely volatile, and producers are therefore forced to make investment decisions without knowing future prices. This results in uncertainty in production, specifically surrounding potential cut-off grades, the quantities of ore to be extracted, and resulting operating costs. Mines require large amounts of capital to be invested up front, leaving them at risk if prices drop dramatically or if business or fiscal conditions (including tax rates) or regulations change.

An evaluation of the feasibility of a mining project will involve net present value (NPV) calculations, where the estimated costs per unit of production are compared to expected revenues over the life of the project. The cost per unit of production depends upon accessibility (infrastructure), ore quality, input costs, taxes, etc. It is also important to evaluate the likelihood of, and sensitivity of the economics of the project to, changes in these costs (i.e., risk), because increased costs will reduce the expected rate of return on a project, making firms less willing to invest. If investment has already taken place and policies change in an unfavourable direction, companies will be less inclined to invest again or to expand. That is, government policies generally have impacts on a company's "bottom line". To the extent that economic benefits outweigh social costs, governments will prefer policies that provide a favourable investment climate.

Different companies' bottom lines may be of different magnitudes due to the fact that they all have different capabilities and strategies. However, they all come to a decision about where to invest based upon maximizing expected profits, which are in turn influenced by location-specific costs and risks. In addition to the cost factors mentioned above, location-specific risks include: regional geological potential, security of tenure, physical security, and corruption. Additional risk factors arise from the potential for *changes* in: government regulations and laws, the tax regime, political stability, macroeconomic stability, labour markets, and the quality of financial markets (World Bank 2005; Otto et al. 2006). Governments play a key role in shaping investment climate through many of these factors, such as taxation, security of property rights, government regulation, political stability, and control of corruption. Governments also have high levels of control over labour and financial markets, as well as over levels of infrastructure (both physical and human capital). All of these factors shape the investment decision by positively or negatively affecting a company's bottom line.

Many authors have provided similar lists of factors that are likely to be most important in terms of investor decisions in the mining sector. For example, according to Pritchard (2005), the most important requirements for a positive investment climate are the ability to sell in local and foreign markets, to obtain payment in a currency that is convertible, to be able to convert payments at a favourable exchange rate, and to be able to service loans and pay taxes. Other factors that are important are security of tenure, a stable tax regime, consistent and timely processes for obtaining government approvals for development and maintenance of mining rights, the ability to obtain equipment from local and foreign suppliers, and the freedom to engage local and foreign labour (Pritchard 2005).

According to Williams (2005), four of the five most important aspects for investment in the mining industry are directly related to the *types* of regulation: nature of exploration and mining rights, procedure for access to rights, security of title and tenure, and maintaining operating rights and obligations.

Otto (2005) ranked investment decision factors prior to exploration as follows (rankings at the mining stage are shown in parentheses):

1. Geological potential (N/A)
2. Security of tenure (1)
3. Ability to repatriate profits (2; with measure of profit, 3)
4. Consistency of policy (9)
5. Country has management control (7)
6. Mineral ownership (10)
7. Foreign exchange regulation (6)
8. Stability of mining/exploration terms (4)
9. Ability to predict tax liability (5)
10. Ability to predetermine environmental liability (8)

All of these authors (Otto 2005; Pritchard 2005; Williams 2005) seem to agree that security of tenure, stability of the terms of regulations, and the ability to repatriate profits are key factors for investors. Actual taxation rates do not appear at the top of any of their lists, although predictability of tax burdens is important.

Obviously, there are many factors that come into play when it comes to investment decisions. Some of these, such as taxes and regulation, are under the control of the government. In practice, governments need to strike a balance between ensuring sufficient mining sector investment to generate revenues and benefits from mineral development, while implementing sustainability measures for the good of the local, regional, and national communities. An empirical analysis of the impacts of policy measures on industry perceptions and behaviour can provide important information regarding the suitability of establishing policies, especially if they will make the jurisdiction relatively more or less expensive compared with other jurisdictions in terms of dollar costs or risk exposure.

5 An Empirical Model of Investor Reactions to Government Policy

The available tools that governments have for addressing sustainability issues in mining fall into two major categories: regulation and taxation. That is, they must change the regulatory environment and/or introduce (and re-allocate the revenues from) taxes in order to alter the behaviour of mining firms, and/or make their own investments in social and physical capital. It is of interest to policy-makers to see the extent to which the regulatory and tax tools that they can use to address sustainability issues impact the desirability of mining investment in their jurisdictions, controlling for other factors. To this end, an empirical model is presented below that attempts to shed light on these impacts.

Ideally, information on actual investment decisions, more specifically current exploration data, would be used to examine the impacts of government policy measures. Because this type of data is not readily available, we use a proxy for investment climate that is drawn from the Fraser Institute's annual mining surveys spanning 2000–2006 (Fraser Institute 2000; Fredricksen 2001, 2002, 2004; McMahon

2005, 2006; McMahon and Melham 2007). In 2006, for example, the Fraser Institute (McMahon and Melham 2007) surveyed 3,000 exploration, development, and mining consulting companies around the world and received responses from 333 of them. The companies that participated in the survey spent US$644 million on exploration in 2005, and US$1.02 billion in 2006. Thus, survey respondents represent 12.6% (US$5.1 billion) of total global exploration budgets in 2005, and 14.5% (US$7.13 billion) in 2006, as reported by the Metals Economics Group.[12,13] Based on the responses to the survey, the Fraser Institute constructs an annual index of perceived investment climate for several countries.

More specifically, the Fraser Institute's Policy Potential Index (PPI) is a "composite index, measuring the overall policy attractiveness of the 65 jurisdictions in the survey" (McMahon and Melham 2007, p. 5). The PPI is normalized to a maximum score of 100, representing a jurisdiction that ranked first in every category, whereas a zero score would represent a jurisdiction that ranked last in every category. For example, in 2005/2006, the highest score was 93.1 (Manitoba) and the lowest was 2.4 (Zimbabwe).

Use of this proxy enables construction of a panel of data covering the 2000–2006 period. The explanatory variables included in the data set have been chosen based upon the theoretical groundwork explored above, as well as informal discussions with individuals at Natural Resources Canada (Minerals and Metals Department), the Mining Association Canada, and Falconbridge Ltd. (now Xstrata plc.) in the summer of 2006. Of primary interest are the impacts of regulation and taxes on investor perceptions of the attractiveness of operating in a particular jurisdiction.

Including taxes directly as an explanatory variable in our model proves to be problematic for two reasons. Firstly, tax regimes are multi-faceted, with a variety of types of taxes and tax rates in place in a given jurisdiction at any point in time (see, for example, Otto et al. 2006, for detailed information on the tax systems in several jurisdictions). One way to summarize the overall tax regime for any given country is to calculate an effective tax rate that takes into consideration the impacts of the entire gamut of taxes that are paid by mining firms. Average effective tax rate (AETR) figures are occasionally published for a few countries, but are not available on a widespread basis (across countries or time). Also, tax rates are commonly held constant within a given jurisdiction over long periods of time. The lack of availability of data for many countries combined with the lack of variation across time poses statistical problems for measuring the impact of taxes on investor perceptions. Therefore, we are only able to provide indirect empirical evidence regarding the impacts of tax regimes on investor perceptions.

Our strategy for indirectly examining the effects of mining tax regimes is to include "fixed effect" country-specific intercepts in our model, and to examine their relationship with a variety of country-specific tax-system features. These country-specific intercepts will capture the impacts of conditions not explicitly entered into

---

[12]http://www.metalseconomics.com/default.htm

[13]These figures indicate that respondents may include a disproportionate number of (i) junior exploration companies and (ii) North American-based companies.

the model, such as tax regime characteristics. Once these country-specific intercepts are estimated, they can be examined for patterns across countries that might be related to various country-specific tax features, such as tax rates and the complexity of the tax system. This strategy is discussed in more detail below.

As with taxation, regulatory structures within any jurisdiction are complex. Ideally, we would like to focus on regulatory structures that are specific to the mining sector. Such data are not available, but the World Bank does provide a set of indices on a variety of aspects of the regulatory and political climates for a large number of countries. These "good governance" variables include a regulatory quality index that covers factors such as the prevalence of "market-unfriendly" policies, and perceptions of burdens imposed by excessive regulation. There are also indices of voice and accountability, political stability, government effectiveness, rule of law, and control of corruption. The availability of these data allow us to explore, albeit imperfectly given the fact that these variables are not specific to the mining sector, the impact of regulatory factors directly within our model. More detailed descriptions of the individual indices can be found in Appendix 1.

Investor perceptions of the attractiveness of a jurisdiction are likely also to be affected by a variety of other factors. Many of these are controlled for by including country-specific geological and economic characteristics in the model. Geological potential obviously plays an important role in determining the extent of mining investment opportunities within a country. However, for a number of reasons, geological potential is also very difficult to measure. It is conventional in the literature to include mineral exports as a proxy for the geological potential of a country (Sachs and Warner 2000). This variable may also capture the level of development of the mining sector, especially related to the amount of mining-related infrastructure (e.g., transportation facilities) already in place in the jurisdiction.[14] The general level of economic development and size of the economy are captured through each country's level of GDP per capita. Uncertainty in a country's economic environment is measured through the rate of change of its currency exchange rate (home currency per US dollar), the domestic inflation rate, and the rate of change of local wages. Finally, a set of year dummies is included in the model to control for variables that vary across years but not countries. Among other things, these might include the rate of interest faced by mining firms and world commodity prices. See Appendix 1 for detailed variable definitions and sources.

Before proceeding to the formal model, it is useful to examine summary statistics for the variables used in our regression analysis (Table 3).[15] It can be seen that the average PPI is slightly above 50 with a standard deviation of 20.5, and ranges from

---

[14]One drawback of the World Bank measure used in this study is that it does not capture the entire gamut of mining exports (for example, processed gold, diamonds, and precious stones are omitted, whereas some scrap metals are included). Also note that any export measure will not fully capture resource sector potential because an extracted resource may be used domestically.

[15]Due to missing values for some variables for various countries and years, a total of 100 observations over the 2000–2006 period are available for model estimation purposes. The summary statistics only include these complete observations. Any observations with missing data are left out of our model and summary statistics.

Table 3 Summary statistics

Variable	Mean	Standard deviation	Minimum	Maximum
PPI (Fraser Institute)	51.15	20.5	8	94
Metal and ore exports (%)	12.07	15.83	0	71.93
Change in exchange rate (%)	8.62	66.66	−28.23	627.26
GDP per capita (USD)	10994.7	13816.69	383.89	48604.2
Inflation rate (%)	−0.08	2.8	−24.55	10.5
Change in wages (%)	0.58	5.73	−30	15.6
Voice and accountability index	0.35	0.85	−1.66	1.58
Political stability/no violence index	−0.12	0.88	−1.94	1.61
Government effectiveness index	0.41	1.04	−1.12	2.13
Regulatory quality index	0.34	0.96	−2.27	1.82
Rule of law index	0.19	1.09	−1.66	1.96
Control of corruption index	0.27	1.2	−1.29	2.51

a low of 8 (Zimbabwe) to a high of 94 (Ireland).[16] Exports of minerals as a per-centage of merchandise exports averaged 12.07%; these values tend to be lower for diversified, developed countries, whereas developing countries, especially in Africa, tend to have higher rates of mineral exports as a percentage of total merchandise exports.

The average annual percentage change in the exchange rate is a 8.62% increase, with a maximum of 627% for Zimbabwe between 2003 and 2004. GDP per capita averages about US$11000, with the lowest being found in Ghana and the highest in Ireland.

The good governance variables *generally* range from a low of –2.3 to a high of 2.5. The lowest sample values for the governance variables in our data set were found in countries such as China, Indonesia, and Zimbabwe. The highest values corresponded to countries such as Finland, New Zealand, and Canada.

Based on this data set, we estimate the following relationship

$$Y_{it} = \alpha_i + x_{it}\beta + \gamma_t + \varepsilon_{it} \tag{1}$$

where: Y = measure of investment climate (PPI); α_i = the fixed effect intercept for country i; x_{it} = the set of independent variables (regulation variables and controls); β = a vector of slope parameters to be estimated; γ_t = time effects; and ϵ_{it} = a random error term.

Given that not all countries are included in the Fraser Institute survey every year, and some of the variables such as the World Bank indices are not available for each year for some countries, the model is estimated based on an unbalanced panel of data, using fixed effect panel methods.[17]

---

[16]See Appendix 3 for more details.

[17]Fixed effects panel methods estimate regression models for panel data, allowing for unobservable differences across countries to be captured by the fixed effects coefficients. The data are unbalanced in our sample, meaning there the number of observations per country varies.

6 Results

The model includes many of the variables deemed to be important for explaining investment climate in the mining industry. Economic variables (GDP, and the annual percentage change in the exchange and wage rates), good governance indicators (voice and accountability, political stability, regulatory quality, rule of law, and control of corruption), as well as a proxy for geological potential (exports) and time-specific variables for the years 2000–2006, were all deemed to be potentially valuable in explaining variations in investment climate.

The results for two versions of the model are presented in Table 4. In Model A, all of our explanatory variables are included, and it is found that the only

Table 4 Fixed effects model regression results

	Variable	Model A	Model B
Macroeconomic variables and geological potential	Mineral exports	−0.668	
		(0.553)	
	Exchange rate (% change)	−0.002	
		(0.025)	
	Inflation rate	0.173	
		(0.564)	
	GDP per capita	0.001*	0.001
		(0.001)	(0.001)
	Wages (% change)	−0.169	
		(0.327)	
Good governance variables	Voice and accountability	0.451	
		(12.654)	
	Political stability	12.318	
		(8.427)	
	Government effectiveness	1.349	
		(17.471)	
	Regulatory quality	18.772*	10.371
		(10.42)	(6.928)
	Rule of law	−40.017	
		(24.095)	
	Control of corruption	5.181	
		(14.79)	
Regression diagnostics	LM test for heteroskedasticity	0.011	0.101
	Adjusted R^2	0.742	0.755
	Log-likelihood	−344.55	−349.75
	N	100	100
Wald tests for joint significance	GDP per capita and regulatory quality	7.658**	7.313**
	Remaining macro and geological variables	1.893	
	Remaining good governance variables	4.19	

Notes: (1) Fixed country and time effects estimated but not reported, (2) values in parentheses are standard errors, and (3) ∗ denotes significance at the 10% level; ∗∗ denotes significance at the 5% level.

individually significant variables are GDP per capita and the Regulatory Quality Index. Because Wald tests of joint significance indicate that (1) the other macro/geological variables and (2) the remaining good governance variables are not significant, they are dropped from Model B. In both models, Regulatory Quality and GDP per capita are jointly significant in determining the mining investment climate at the 5% level.

In terms of basic diagnostics, in both models there is no evidence of heteroskedasticity in the error terms, and adjusted R^2 values for the models are high, indicating that about 75% of the variation in the dependent variable is explained by the model.[18]

The factors of primary interest in our study are those that relate to regulation and taxation. Regulatory Quality, one of the two significant variables in the two models, has a positive coefficient, indicating that improved regulatory quality should improve investor perceptions with respect to a country, because good regulations lower the idiosyncratic risk associated with investments.

6.1 Fixed Effects Coefficient and the Impact of Taxation

Taxation is one of the key factors expected to influence mining investment. As explained above, it is not possible to look at the impacts of taxation directly in our model. However, an examination of the correlation of the country-specific intercepts with available information on taxation may provide some indication of whether or not taxation is in fact related to investor perceptions. To this end, we examine Spearman rank correlations between the fixed effects coefficients and a set of tax variables taken from the World Bank's Doing Business Project database: the profit tax rate, the labour tax rate, the average rate of other taxes, the total tax rate (the sum of the first three measures), the number of tax payments required, and the number of hours that must be devoted to complying with the country's tax

---

[18]It is possible that the high R^2 values could result partially from endogeneity between the dependent variable PPI and the good governance variables. Both the Fraser Institute Index and the World Bank Indexes are constructed from surveys of knowledgeable parties and stakeholders. The World Bank data sources mostly consist of surveys of companies, individuals, and the assessments of commercial risk rating agencies, non-governmental organizations, and other multilateral aid agencies. The Fraser Institute surveys mining company executives. As a result, although the variables are not identical, they are likely to have underlying factors in common. An examination of sample correlations indicates that although the Fraser Institute Index is correlated with the good governance variables (the correlations are in the 0.58–0.67 range), these correlations are small in comparison to the correlations among the good governance variables (in the 0.74–0.98 range). From a statistical perspective, the most problematic possibility would be the case where the Fraser Institute Index values influence (or cause) the values of the World Bank Indexes. It is much more likely, however, that the causality runs the other way and that it is (the underlying factors captured in) the World Bank Indexes that influence the perceptions of the Fraser Institute survey respondents.

requirements.[19] The first four tax variables provide a proxy for the magnitude of taxes in the various countries, while the other two provide proxies for the complexity of the tax system.

By definition, countries with higher fixed effects coefficients are those countries that, due to factors not explicitly controlled for in the model, are more attractive to the respondents to the Fraser Institute survey. The basic idea behind the strategy of comparing these two measures of various components of the tax system is that if taxation is one of these uncaptured factors that matters to firms, there should be a relationship between the sizes of the fixed effects coefficients and the size of the tax burden that firms face in that country. Spearman rank correlations are used to see if there is any significant correlation between the ordering of the fixed effects coefficients with the ordering of the tax variables (both from smallest to largest). These correlations are calculated for Models A and B, and are presented in Table 5.

Given that it is best to control for as many other factors as possible when calculating the fixed effects coefficients, the fixed effects from Model A should be most informative because they have been purged of the impacts of all of the potentially important explanatory variables included in the model. The fixed effects coefficients from Model B, on the other hand, will include any impact that the omitted macro/geological and good governance variables might have on mining companies' perceptions regarding a particular country. It turns out that none of the Spearman rank correlations reported in Table 5 for Model A are significant. Only one of the correlations for Model B is significant, with the counter-intuitive result that higher fixed effects coefficients are positively correlated with the number of hours required to comply with tax regulations. In general, the model results do not provide any strong (indirect) evidence that either tax rates or the complexity of the tax system have an impact on mining companies' perceptions as captured through the Fraser Institute Index.

Table 5 Spearman rank correlations between fixed effects coefficients and tax measures

	Model A	Model B
Profit tax	−0.02518	−0.29392
Labour tax	−0.12753	0.22222
Other tax	0.01587	0.08758
Total tax	−0.0509	0.05857
Tax payments (number)	−0.16092	0.22113
Time required (hours)	0.06294	0.57635

Note: For $N=28$ (number of countries in sample), the Spearman rank correlation is significantly different from zero if its absolute value is greater than 0.377.

---

[19]Ideally, mining-specific tax rates would be used for this exercise. However, the number of countries covered in cross-country comparisons of mining taxes is too small to allow for any meaningful analysis.

7 Discussion of Empirical Findings

Our examination of the factors that impact the perceptions of mining companies regarding investment in particular jurisdictions might be viewed as an extension of the development literature analyzing which factors and institutions facilitate development and economic growth. Economic growth, which is closely linked to a positive investment climate, is significantly affected by the institutions present within a given jurisdiction. Markets rely on the enforcement of contracts, which in turn relies on strong regulatory and legal systems. In order to function well, financial systems rely on a nation's central bank monetary policies with respect to inflation, exchange rates, and interest rates. In addition, governments need to be competent, maintain order, and control corruption to ensure well-functioning markets (Perkins et al. 2006). Economic growth is driven by investment, and therefore by investor perceptions as to whether or not the environment is conducive to investment and profit maximization.

We find that there are two major factors that have a positive impact on mining companies' perceptions regarding the suitability of investing in a particular jurisdiction. There is a decided preference for jurisdictions where the economy is already fairly well developed (as measured by per capita GDP) and for jurisdictions with a high quality regulatory system in place. There is little evidence in this particular (and limited) data set of any significant impact of absolute taxation rate on mining companies' perceptions.

7.1 Regulation and Its Contribution to Investment Climate

The results regarding the importance of a high quality regulatory framework are consistent with mainstream views. According to Perkins et al. (2006), one of the most important issues that must be addressed to ensure efficient market behaviour is that effective property rights are established. For property rights to be meaningful, they must be exclusive and clearly defined. Otherwise, the property may be considered a public good, free for anyone to use. If the good is clearly defined, but not exclusive, there may be issues with externalities, whereby bystanders bear a cost as a result of owner decisions (e.g., pollution). In order for owners and management to have the appropriate incentives, it is also necessary for these rights to be transferable, as well as secure for definite periods of time. These issues surrounding how private goods are defined, and the extent to which they are secure and transferable, depend upon the quality of regulation in a country. A related issue surrounding property rights is that they must be enforceable through a well established legal system. Governments have a great deal of influence over the extent to which property rights are defined, monitored, enforced, and regulated.

Security of title is an important consideration for investors. There are many cases where governments have expropriated and nationalized a specific reserve or industry. A recent example from Canada was the Windy Craggy copper-cobalt deposit in the Tatshenshini Area of northwestern British Columbia, which was first discovered in 1958 by prospector J. McDougall. The deposit was later being explored in the 1980s and 1990s by Geddes Resources Ltd., who had submitted a development proposal to the government of British Columbia. In 1992, after the company had already invested $48 million on exploration, the government of British Columbia under premier Harcourt suspended review of the project, and subsequently in 1993 announced that the region around the proposed mine would be preserved as a wilderness provincial park (Day and Affum 1995). The Federal government endorsed the region as a UN Heritage site in December of the same year. According to Webster (1998), "This 'taking' without fair process and without good reason has led mining company managers, who are responsible to their shareholders, to shift the bulk of their exploration planning and budgets out of BC. This has resulted in much of Canada's exploration and mining expertise finding its way abroad—to areas as far away as the former Soviet Union, Latin America, China, and sub-Saharan Africa." It can be argued that a government that sets aside property as a UN Heritage site is interested in sustainability. However, the point to be made here is that, regardless of the motivation for a change in regulation, such changes can have impacts on investment patterns.

Regulations that enhance the security of land titles and stability of society are generally expected to improve the investment climate (World Bank 2005). Government regulations can enhance the investment climate by improving levels of certainty and minimizing risks in the investment process, such as by improving the liquidity of the asset. For example, a new model of mining regulation related to security of tenure in Latin America has had a very positive effect on investment climate.[20] Chile, Colombia, Venezuela, Peru, and Bolivia all have single concessions, which guarantee transferable, long-term mineral rights following exploration with a single payment (Ossa 2005). Exploration rights are guaranteed under a "first come, first served" basis in a new map staking procedure (title registry) that prohibits overlap (Williams 2005). A single payment is required to claim mineral rights to exploration *and* expropriation (whereas most other countries require a dual license for these activities). In addition, existing mining rights are transferable and can be bought from the owner (Williams 2005). Failure to meet annual payments results in fines or suspended operations depending on remediation, but *not* by cancelling title. In these countries, mine operators can decide the cut-off grade, optimal level of production, and the appropriate process to obtain mineral output. In addition, firms are free to export with no exchange restrictions or profit sharing requirements (Williams 2005). Latin America is on the forefront of an innovative way of minimizing risks to investors without governments taking on an unreasonable burden. Investors see these regulations on the process of obtaining and maintaining rights as very reasonable policies, and conducive to investment.

---

[20] In contrast, in many other countries, the procedure for maintaining rights can be quite burdensome because many operational aspects are legislated, such as cut-off grade, production levels, processes used, etc. If conditions are not met, then property rights can be assumed by the government.

Policies related to the *acquisition and transfer of land rights* following exploration can have a significant impact on investment climate through increased levels of risk. Exploration costs are high, and if investors are not guaranteed rights when they find resources, or if rights are conditional on a set of unreasonable regulations, the jurisdiction will not be viewed as a favourable place to invest. One policy that positively affects investment climate is a retention licence, which allows companies to hold exploration periods for a number of years until the reserve is proven economically feasible (Williams 2005). This policy extends the time that a firm can maintain rights without developing or producing from a site. Such licenses have been applied with positive effects in Tanzania, Botswana, Namibia, and a few other African countries.

7.2 Taxation and Its Contribution to Investment Climate

The patterns of the fixed effects coefficients provide some indirect evidence regarding how industry responds to taxation. The lack of strong correlations between the relative rankings of taxation measures and the country-specific fixed effects coefficients invites conjectures that taxation may not be a strong determinant of investment climate. This result is in general accordance with the ranking of factors deemed to be significant in investment decisions in the mining industry according to Otto (2005), Pritchard (2005), and Williams (2005). That is, other factors, such as the level of the development of the economy and the regulatory institutions in place, may dominate in terms of pivotal factors that are taken into consideration by investors. These will remain conjectures in the absence of good, accessible data across jurisdictions and time on actual investment decisions made by individual firms.

7.3 Limitations of the Empirical Results

The analysis presented above is limited for several reasons. Firstly, tax rates are often held constant within a jurisdiction over long periods of time; the lack of data for many countries combined with this lack of variation across time poses statistical problems for measuring the impact of taxes on investor perceptions. Until improvements are made in either country or government reporting of taxation information, it will be difficult to draw further conclusions.

To this end, the Global Reporting Initiative (GRI) may help.[21] The GRI has developed a sustainability reporting framework, one of the major components of which is a set of performance indicators that includes the total taxation payment to individual governments. However, this sort of information is not yet widely

---

[21] http://www.globalreporting.org

available, and so it is currently very difficult to analyze taxation issues within and especially between countries.

Furthermore, our investment data from the Fraser Institute survey consist of measures of the stated perceptions of a subset of mining companies regarding the attractiveness of various jurisdictions. This sample may be biased, for example towards junior companies in the exploration business rather than to majors who undertake the bulk of investment. The sample may also be biased towards North American companies, and gold miners and explorers. Furthermore, what people say and what they do do not always perfectly coincide. Ideally, information on actual investment decisions (specifically current exploration data) would be used to examine the impacts of government policy measures, but this type of data is not readily available. The drawbacks associated with the currently available data point to the need for an increased availability of detailed taxation and investment data in order to verify (or refute) the tentative conclusions that can be drawn from this study regarding the impacts of regulation and taxation on the mining investment climate.

8 Conclusions

In theory, both regulation and taxation can be used as policy instruments by governments that are attempting to pursue sustainable development strategies based on the wealth created by mining. Our empirical results, although tentative, lead us to suggest that there may be scope for increased taxation to deal with local and national sustainability issues. As long as there is a good regulatory environment, it may be possible to raise taxes to mitigate local mining externalities or to fund other programs through hypothecated taxes, without seriously damaging investor confidence.

Minerals are non-renewable resources, such that, if they are mined, the actual resources cannot be sustained. However, through appropriate use of regulation, taxation, accounting, and investment instruments, the benefits that are derived from mining may be sustainable. This weak form of sustainable development aims to sustain communities at the local level and sustain aggregate capital levels at the national or regional level (depending upon who is constitutionally the owner of the resources). At the local level, taxation and a Pareto-improving reallocation of funds may provide a workable way to deal with negative social and environmental externalities that can result from mining activities. Nationally, revenues derived from taxation can be earmarked for a variety of social objectives such as promoting human development, investments aimed at transforming natural capital into other more sustainable forms that will benefit the present as well as future generations, economic diversification, and protection of the environment. Sustainable development based on the wealth generated from mining may thus be feasible through the use of taxation and appropriate regulation.

Acknowledgments This research was supported by a Research Development Initiatives grant from the Social Sciences and Humanities Research Council of Canada, for which we are grateful. We thank Simon Handelsman for editorial handling, and two anonymous reviewers for their constructive comments that greatly improved the manuscript.

Appendix 1: Variable Definitions

(*Definitions taken directly from source*)

Geological Potential and Macroeconomic Variables

Metal and Exports: Ore and metal exports as a percentage of merchandise exports. Components may not sum to 100% because of unclassified trade. Ores and metals comprise the commodities in SITC sections 27 (crude fertilizer, minerals excluding coal, petroleum, and precious stones); 28 (metalliferous ores, scrap); and 68 (non-ferrous metals). From the World Bank staff estimates from the COMTRADE database maintained by the United Nations Statistics Division.

Exchange Rate: Official exchange rate refers to the exchange rate determined by national authorities or to the rate determined in the legally sanctioned exchange market. It is calculated as an annual average based on monthly averages (local currency units relative to the US dollar). From the IMF at: www.imf.org/external/pubs/ft/weo/2006/02/data/download.aspx

GDP Per Capita (constant 2000 US$): GDP per capita is gross domestic product divided by midyear population. GDP is the sum of gross value added by all resident producers in the economy plus any product taxes and minus any subsidies not included in the value of the products. It is calculated without making deductions for depreciation of fabricated assets or for depletion and degradation of natural resources. Data are in constant US dollars. From World Bank World Development Indicators WDI Online data set: www.worldbank.org/data/onlinedatabases/onlinedatabases.html

Inflation: Inflation as measured by the consumer price index reflects the annual percentage change in the cost to the average consumer of acquiring a fixed basket of goods and services that may be fixed or changed at specified intervals, such as yearly. The Laspeyres formula is generally used. From the Political Risk Services group, available at: www.prsgroup.com

Percentage Change in Wages: Annual percentage change in real wages. From the Political Risk Services group, available at: www.prsgroup.com

Good Governance Variables

Good Governance Indicators: The units in which governance is measured follow a normal distribution with a mean of zero and a standard deviation of one in each period. This implies that virtually all scores lie between −2.5 and 2.5, with higher scores corresponding to better outcomes. This also implies that our aggregate estimates convey no information about trends in global averages of governance, but they are of course informative about changes in individual countries' relative positions over time. These boundaries correspond to the 0.005 and 0.995 percentiles of the standard normal distribution. For a handful of cases, individual country ratings can exceed these boundaries when scores from individual data sources are particularly high or low. From: www.worldbank.org/wbi/governance/govmatters6/. Data source for all "Good Governance Indicators": http://info.worldbank.org/governance/wgi2007/

Voice and Accountability: This variable includes a number of indicators measuring various aspects of the political process, civil liberties, and political rights, measuring the extent to which citizens of a country are able to participate in the selection of governments.

Political Stability: Political Stability and Absence of Violence combines several indicators that measure perceptions of the likelihood that the government in power will be destabilized or overthrown by possibly unconstitutional and/or violent means, including domestic violence and terrorism.

Government Effectiveness: This variable combines responses on the quality of public service provision, the quality of the bureaucracy, the competence of civil servants, the independence of the civil service from political pressures, and the credibility of the government's commitment to policies.

Regulatory Quality: Regulatory Quality instead focuses more on the policies themselves, including measures of the incidence of market-unfriendly policies such as price controls or inadequate bank supervision, as well as perceptions of the burdens imposed by excessive regulation in areas such as foreign trade and business development.

Rule of Law: Rule of Law includes several indicators that measure the extent to which agents have confidence in and abide by the rules of society. These include perceptions of the incidence of crime, the effectiveness and predictability of the judiciary, and the enforceability of contracts.

Control of Corruption: Control of Corruption is a measure of the extent of corruption, conventionally defined as the exercise of public power for private gain. It is based on scores of variables from polls of experts and surveys.

Taxation Measures

All taxation data are from the World Banks's "Doing Business" Project. The project is described at: http://www.doingbusiness.org/

The data were downloaded from http://www.doingbusiness.org/features/taxes.aspx. Variable definitions are from www.doingbusiness.org/MethodologySurveys/PayingTaxes.aspx

Profit Tax: Amount of taxes on profits paid by the business as a percentage of commercial profits.

Labour Tax and Contributions: Amount of taxes and mandatory contributions on labour paid by the business as a percentage of commercial profits. This amount includes mandatory social security contributions paid by the employer both to public and private entities, as well as other taxes or contributions related to employing workers.

Other Tax: Property taxes, turnover taxes, and other small taxes (such as municipal fees and vehicle and fuel taxes).

Total Tax Rate: This variable measures the amount of taxes payable by the business in the second year of operation, expressed as a share of commercial profits. Doing Business 2007 (World Bank 2007) reports tax rates for fiscal year 2005. The total amount of taxes is the sum of all the different taxes payable after accounting for deductions and exemptions. The taxes withheld (such as sales tax or value added tax) but not paid by the company are excluded. The taxes included can be divided into five categories: profit or corporate income tax, social security contributions and other labor taxes paid by the employer, property taxes, turnover taxes, and other small taxes (such as municipal fees and vehicle and fuel taxes).

Tax Payments Indicator: Reflects the total number of taxes paid, the method of payment, the frequency of payment, and the number of agencies involved for this standardized case during the second year of operation. It includes payments made by the company on consumption taxes such as sales tax or value added tax. These taxes are traditionally withheld on behalf of the consumer. The number of payments takes into account electronic filing. Where full electronic filing is allowed, the tax is counted as paid once a year even if the payment is more frequent.

Time Required: Time is recorded in hours per year. The indicator measures the time to prepare, file and pay (or withhold) 3 major types of taxes and contributions: the corporate income tax, value added or sales tax, and labor taxes, including payroll taxes and social contributions. Preparation time includes the time to collect all information necessary to compute the tax payable. If separate accounting books must be kept for tax purposes—or separate calculations made—the time associated with these processes is included. This extra time is included only if the regular accounting work is not enough to fulfill the tax accounting requirements. Filing time includes the time to complete all necessary tax forms and make all necessary calculations. Payment time is the hours needed to make the payment online or at the tax office. Where taxes and contributions are paid in person, the time includes delays while waiting.

Appendix 2: PPI Dataset

PPI: The Policy Potential Index (PPI) is a composite index, measuring the overall policy attractiveness of the jurisdictions in the survey. The PPI is normalized to maximum score of 100. A jurisdiction that ranks first in every policy area would have a score of 100; one that scored last in every category would have a score of 0. For example, in 2005/2006, because no nation scored first in all categories or last in all, the highest score was 93.1 (Manitoba), while the lowest score was 2.9 (Zimbabwe).

For countries that were broken down by sub-national jurisdictions in the Fraser Institute's mining survey (Australia, Canada, USA), a weighted average was used using the percentage of national total metallic minerals mined in the province or state as the weights. Production years were chosen to be the lower of the set (i.e., 2001/2002 was chosen to be 2001). Data were obtained from NRCan, USGS, and AUSSTAT websites.

The values of production used for calculating the weighted PPI (PW) for Australia, Canada, and the USA were obtained from:

Canada: production measured as the value of metallic plus non-metallic minerals; Data source: mmsd1.mms.nrcan.gc.ca/mmsd/production/ production_e.asp

USA: production measured as value of non-fuel mineral production; Data source: minerals.usgs.gov/minerals/pubs/commodity/statistical_summary/ index.html#myb

Australia: production measured as value of metallic minerals in 2000, and value of metallic minerals plus industrial minerals for 2002–2005; Data source: www.abs.gov.au/AUSSTATS/abs@.nsf/DetailsPage/8415.02004-05?OpenDocument

Appendix 3: Fraser Institute Policy Potential Index (PPI)

Country	2000	2001	2002	2003	2004	2005	2006
Argentina	63	65	54	58	44	62	41
Australia	61	78	75	75	73	70	75
Bolivia		64	70	57	20	24	9
Brazil	71	75	64	79	47	66	51
Canada	56	64	68	65	67	70	72
Chile	87	85	85	85	74	87	64
China		28	38	50	49	40	28
Ecuador					38	34	30
Finland					62	67	62
Ghana		49	45	47	60	61	45
India			26	42	68	35	32
Indonesia		27	19	23	12	22	23
Ireland				72	94	67	47
Kazakhstan		21	24	38	30	35	15

(continued).

(continued)

Country	2000	2001	2002	2003	2004	2005	2006
Mexico	79	70	71	63	71	84	64
New Zealand			42	57	60	40	52
Papua New Guinea	33				25	12	14
Peru	75	69	67	61	46	38	30
Philippines		33	29	20	24	18	14
Russia		20	23	35	17	23	16
South Africa	51	45	47	43	32	45	29
Spain					78	60	71
Sweden					64	56	66
Turkey				57	55	62	52
USA	54	57	56	47	61	58	65
Venezuela		50	44	34	21	13	5
Zambia					38	24	31
Zimbabwe		22	20	26	8	3	2

Note: This table includes all reported PPI; however, not all PPI are included in the analysis due to missing values in independent variables (incomplete observations).

Websites

1. GRI website, last accessed September 7, 2007 at http://www.globalreporting. org/AboutGRI/WhoWeAre/
2. Canada production measured as value of metallic plus non-metallic minerals information taken from: http://mmsd1.mms.nrcan.gc.ca/mmsd/production/production_e.asp
3. US production measured as value of non-fuel mineral production. Taken from: http://www.minerals.usgs.gov/minerals/pubs/commodity/statistical_summary/index.html#myb
4. Australia production measures taken from: http://www.abs.gov.au/AUSSTATS/abs@.nsf/DetailsPage/8415.02004-05
5. IMF data: http://www.imf.org/external/pubs/ft/weo/2006/02/data/download.aspx
6. WDI indicators: http://publications.worldbank.org/WDI
7. http://www.metalseconomics.com/default.htm, Last accessed September 3, 2007
8. Political Risk Services group, available at: www.prsgroup.com
9. International Finance Corporation contribution to the Equator Principles is available at: http://www.ifc.org/ifccxt/sustainability.nsf/Content/EnvSocStandards

References

Brewer K (2005) Trends and directions in mining taxation in the 2000s. In: Bastida E, Wälde TW, Warden-Fernandez J (eds) International and Comparative Mineral Law and Policy: Trends and Prospects. Kluwer Academic Publishing, Hague, the Netherlands, pp. 517–530

Brewer K, Bergevin G, McCutcheon B (2006) Government approaches to mineral policy and taxation. Natural Resources Canada (NRCan) Minerals and Metals Sector, Ottawa. Retrieved September 20, 2007, from http://mmsd1.mms.nrcan.gc.ca/efab/docs/GovtApproches_e.pdf

Davies P (2002) Public participation, the Aarhus convention, and the European community. In: Zillman DN, Lucas AR, Pring G (eds) Human Rights in Natural Resource Development. Oxford University Press, Oxford, pp. 155–185

Day JC, Affum J (1995) Windy Craggy: Institutions and stakeholders. Resources Policy 21:21–26

Elliot J (1999) An Introduction to Sustainable Development, 3rd edn. Routledge, New York

Equator-Principles (2006) A financial industry benchmark for determining, assessing and managing social & environmental risk in project financing. Retrieved October 4, 2007, from http://www.equator-principles.com/documents/Equator_Principles.pdf

Fraser Institute (2000) Annual Survey of Mining Companies. Last accessed July 29, 2009, at http://oldfraser.lexi.net/publications/surveys/2000mining/s2_execsummary.html

Fredricksen L (2001) Fraser institute annual survey of mining companies 2001/2002. Last accessed September 3, 2007, at http://www.fraserinstitute.org/commerce.web/product_files/MiningSurvey2001pt1.pdf

Fredricksen L (2002) Fraser institute annual survey of mining companies 2002/2003. Last accessed September 3, 2007, at http://www.fraserinstitute.org/commerce.web/product_files/MiningSurvey2002.pdf

Fredricksen L (2004) Fraser institute annual survey of mining companies 2003/2004. Last accessed September 3, 2007, at http://www.fraserinstitute.org/commerce.web/product_files/MiningSurvey2003.pdf

Houde S (2005) International Perspective of the Tax Burden for Mining. Natural Resources Canada (NRCan) Minerals and Metals Sector, Ottawa

IFC (International Finance Corporation) (2006) Policy and Performance Standards on Social and Environmental Sustainability. International Finance Corporation, Washington, DC

Kates R, Parris T, Leiserowitz A (2005) What is sustainable development? Goals, indicators, values and practice. Environment: Science and Policy for Sustainable Development 47(3):8–21

Labonne B (2002) Commentary: Harnessing mining for poverty reduction, especially in Africa. Natural Resources Forum 26:69–73

McMahon F (2005) Fraser institute annual survey of mining companies 2004/2005. Last accessed September 3, 2007, at http://www.fraserinstitute.org/commerce.web/product_files/MiningSurvey2004.pdf

McMahon F (2006) Fraser institute annual survey of mining companies 2005/2006. Last accessed September 3, 2007, at http://www.fraserinstitute.org/commerce.web/product_files/MiningSurvey2005.pdf

McMahon F, Melham A (2007) Fraser institute annual survey of mining companies 2006/2007. Last accessed July 29, 2009, at http://www.fraserinstitute.org/commerce.web/product_files/MiningSurvey2006.pdf

O'Faircheallaigh C (1999) Making social impact assessment count: A negotiation-based approach for indigenous peoples. Society & Natural Resources 12:63–80

Ossa J (2005) Recent developments in latin American mining legislations. In: Bastida E, Wälde TW, Warden-Fernandez J (eds) International and Comparative Mineral Law and Policy: Trends and Prospects. Kluwer Academic Publishing, Hague, the Netherlands, pp. 759–768

Otto J (2005) Security of mineral tenure: Time-limits. In: Bastida E, Wälde TW, Warden-Fernandez J (eds) International and Comparative Mineral Law and Policy: Trends and Prospects. Kluwer Academic Publishing, Hague, the Netherlands, pp. 353–372

Otto J, Andrews C, Cawood F, Doggett M, Guj P, Stermole F, Stermole J, Tilton J (2006) Mining Royalties: A Global Study of Their Impact on Investors, Government, and Civil Society. World Bank, Washington, DC.

Perkins D, Radelet S, Lindauer D (2006) States and Markets, Economics of Development, 6th edn. Norton and Company, New York

Peskin HM (1994) Sustainable resource accounting. In: Assigning Economic Value to Natural Resources. National Research Council (U.S.). Commission on Geosciences, Environment, and Resources. National Academy Press, Washington, DC, pp. 59–66

Petrella F (1988) Henry George and the classical scientific research program: The economics of Republican millennialism. American Journal of Economics and Sociology 47: 239–256

Pring G, Siegele L (2005) International law and mineral resources development. In: Bastida E, Wälde TW, Warden-Fernandez J (eds) International and Comparative Mineral Law and Policy: Trends and Prospects. Kluwer Academic Publishing, Hague, the Netherlands, pp. 127–146

Pritchard R (2005) Safeguards for foreign investment in mining. In: Bastida E, Wälde TW, Warden-Fernandez J (eds) International and Comparative Mineral Law and Policy: Trends and Prospects. Kluwer Academic Publishing, Hague, the Netherlands, pp. 73–97

Richards JP (2005) The role of minerals in sustainable human development. In: Marker BR, Petterson MG, McEvoy F, Stephenson MH (eds) Sustainable Minerals Operations in the Developing World. Geological Society, London, Special Publication 250, pp. 25–34

Sachs JD, Warner AM (2000) Natural resource abundance and economic growth. In: Meier GM, Rauch JE (eds) Leading Issues in Economic Development. Oxford University Press, Oxford, pp. 161–167

US National Academy of Sciences (USNAS) (1999) A Transition Toward Sustainability. National Academy Press, Washington, DC

Warrack A, Keddie R (2001) Alberta Heritage Fund vs. Alaska Permanent Fund: A comparative analysis. Proceedings of the International Conference on Management, Xi'an China, 5–7, May, 2001, taken from http://www.apfc.org/iceimages/library/Warrack_Alberatafund.pdf

Warrack A, Keddie R (2002) Natural Resource Trust Funds: A Comparsion of Alberta and Alaska Resource Funds, last accessed July 29, 2009 at http://www.business.ualberta.ca/cibs/publications/IB%20pdf/72.pdf

Webster M (1998) The Windy Craggy experience. Fraser Forum, January 1998, last accessed October 4, 2007 at http://oldfraser.lexi.net/publications/forum/1998/january/cover_story.html

Williams J (2005) The Latin American mining law model. In: Bastida E, Wälde TW, Warden-Fernandez J (eds) International and Comparative Mineral Law and Policy: Trends and Prospects. Kluwer Academic Publishing, Hague, the Netherlands, pp. 741–757

World Bank (2001) Making Sustainable Commitments: An Environmental Strategy for the World Bank. World Bank, Washington, DC

World Bank (2005) World Development Report: A Better Investment Climate for Everyone. World Bank, Washington, DC

World Bank (2007) Doing Business 2007: Comparing Regulation in 175 Economies. World Bank, Washington, DC

World Commission on Environment and Development (WCED) (1987) Our Common Future, Report of the World Commission on Environment and Development (the Brundtland Commission). Oxford University Press, Oxford

Natural Resource-Based Sustainable Development Using a Cluster Approach

Indira Singh and Jim Evans

Abstract Natural resources have provided a livelihood to many regions and nations, and a platform for some to achieve sustainable growth. In the global economy, natural resources have the potential to contribute to sustainable development goals through the development of local suppliers of inputs, services, and products, and global supply chains that encourage innovation and new products and services. The presence of physical clusters of related industries and organizations has the potential to accelerate local sustainable development, and the evolution of virtual clusters holds the potential for achieving more global sustainable development. A series of questions is proposed to help develop a framework for achieving sustainable development goals. It is contended that there are powerful agents that will shape and affect this journey.

1 Introduction

Throughout the history of mankind, nations and regions have used natural resources to transform their cultures and create economic prosperity for their people. Natural resources, in the manner they are extracted, manufactured, and used, have generally had a beneficial impact on human society, its culture, and its standard of living. But there has been a downside: the past two hundred years of the industrial age has seen a considerable degradation of our environment. We urgently need to reform our wealth-creation processes so that they are sustainable in the context of both the environment and societal needs.

Sustainable development is about economic activity in an environmental and social context:

I. Singh (✉)
Ministry of Northern Development and Mines, Thunder Bay, ON, P7E 6S7, Canada
e-mail: Indira.singh1@ontario.ca

J.P. Richards (ed.), *Mining, Society, and a Sustainable World*,
DOI 10.1007/978-3-642-01103-0_8, © Springer-Verlag Berlin Heidelberg 2009

Sustainable development seeks to meet the needs and aspirations of the present without compromising the ability to meet those of the future. Far from requiring the cessation of economic growth, it recognizes that the problems of poverty and underdevelopment cannot be solved unless we have a new era of growth... (United Nations 1987, p. 11.)

Sustainable development is maintaining a delicate balance between the human need to improve lifestyles and feelings of well-being on one hand, and preserving the natural resources and ecosystems on which we and future generations depend. (The Global Development Research Center; gdrc.org/sustdev/definitions.html)

From these two definitions one can glean three key elements of sustainable development: economic growth; environmental stewardship; and corporate social responsibility. Economic growth is crucial for the continued generation of public revenues, enabling social responsibility and a commitment to environmental stewardship. Social stewardship and environmental stewardship both require policies, programs, and practices for minimizing the adverse impacts of resource development. Economic growth, social responsibility, and protecting the environment are not separate challenges; they are linked. Historically, economic growth has been associated with damage to the environment because it puts increased pressure on environmental resources. However, governments and companies are showing an increasing realization that the growth of economies must remain firmly attached to their ecological roots, and ensure that these roots are protected and nurtured so that they can support future growth. The economy is not just about the production of wealth, and ecology is not just about the protection of nature. They are both important for improving the lot of humankind. Corporate social responsibility requires principles of community collaboration, a respect for local people's values, and stakeholder participation in resource development. The sharing of economic benefits generated by operations, and taking the opportunity to develop social and institutional structures, are encouraged. Another dimension of corporate social responsibility is to respect human rights and to adhere to ethical business practices. Companies need to take measures to ensure that sustainable development is high on their corporate agenda.

The concept of sustainable development provides a framework for the integration of economic, social, and environmental policies, and the development of implementation strategies. Pursuit of sustainable development goals requires new ways of thinking about economic development, patterns of behaviour, and value criteria. There is a need to explore approaches, practices, and possibilities that can be used to leverage natural resources towards achieving sustainable development goals.

Natural resources are commonly divided into two major categories: regenerative resources such as trees, plants, soil, water, and air; and extracted resources such as minerals and fossil fuels. These extracted resources can be further sub-divided into those consumed as energy, typically hydrocarbons, and those that are used to create manufactured products, typically minerals. Because minerals are generally capable of being recovered and reused, they can be considered, provided that we make the appropriate decisions to avoid degradation, comparable to regenerative resources.

During the past two hundred years we have witnessed the development of an industrial economy which, through innovation and specialization, has dramatically increased productivity and quality of life. But the underlying model of "take, make, consume, and waste", associated with an industrial economy has also degraded the environment through waste and pollution, and has increased the gap between the rich and the poor. This has not, therefore, been sustainable development.

The supply chain of an industrial economy starts with the harvesting and extraction of natural resource inputs for processing into goods and services, which are then consumed, with any by-products being discarded at the end of their life (recycling still only captures a small proportion of consumed materials). Issues like the depletion of resources and climate change are increasingly bringing attention to the fact that the present supply chain model is unsustainable.

The natural resource industries at the beginning of the industrial economy supply chain typically include agriculture, fisheries, forestry, and hydrocarbon and mineral extraction. Agriculture, fisheries, and forestry, and their underlying environments, naturally self-renew (unless harvested to exhaustion). Alternative approaches to the harvesting and use of regenerative resources are clearly capable of leading to a sustainable future.

However, the extraction of hydrocarbons and minerals is a more complex matter. Extracted hydrocarbons are generally consumed to make energy, which is then used to produce new outcomes such as more advanced products, motion, or modified states. The consumption of hydrocarbons may well be beneficial to human society in the near term, but hydrocarbons are a finite resource, and their consumption threatens the environment.

Minerals are also extracted, and may be alloyed into more exotic forms to respond to the needs of more developed societies, but are seldom completely consumed. Minerals are primarily lost by being discarded, wasted, or through degradation over time (e.g., Ayres et al. 2003). Resource sustainability, therefore, requires more sophisticated supply chain strategies and public policies to protect the capacity to regenerate, limit the harmful consequences of consumption, and increase preservation through recycling of enduring resources like minerals, while also encouraging and enabling benefits to society.

So the question remains: "How can natural resources create sustainable development?" This represents a critical starting point to a much larger re-evaluation of economic activity and its environmental and social context (e.g., Fairbanks and Lindsay 1997).

2 Shifting Trends in the Global Economy

Older-style industrial economies tend to have a strong association with the extractive and manufacturing activities of the past two hundred years. The evolution of processes and products in these industrial economies typically has a long lead time, and innovation and improvements take place over years and decades. Newer, post-industrial economies tend to work with knowledge and have a services orientation. Such economies have a strong reputation for rapid and agile computer-aided product and service development, and, because they suffer fewer physical limitations, are typically associated with a continuing stream of rapid product and service innovations that take place in days, weeks, and months. But the reality is that there is a considerable coexistence of industrial and post-industrial economies in present-day supply chains, with much of the post-industrial creativity and innovation being highly dependent on industrial economy inputs.

The older style industrial economy supply chains, which invariably started with the extraction of natural resources, were seen as a direct input into wealth creation. The development of increasingly advanced manufacturing techniques for manipulating and combining natural resources, substantially increased productivity and resulted in increasingly more sophisticated products. These advances both encouraged and enabled the creation of knowledge-based creative societies. For example, the wealth gained from the gold and silver rushes of California and Nevada built San Francisco and the surrounding Bay area.

Post-industrial economies that work with knowledge as their "natural resource" are less reliant on extracted resources, and instead create wealth through universities and other centres of innovation, frequently in partnerships, which develop products and services based on advanced and frequently complex theories and concepts. But these "virtual"[1] economies continue to have old economy linkages. They remain directly dependent on natural resources for their energy and the infrastructures that enable the processing and sharing of knowledge. The energy is consumed, but may be renewable, and the infrastructures (computers and networks) are used without being consumed.

Participants in the industrial economy are also capable of a transition from dependence on resources as a commodity, to knowledge-based product and service economies, through more sophisticated approaches to natural resource harvesting and extraction, and the creation of value-added products and services that appeal to consumers. This transition, in turn, leads to the protection and enhancement of the community's quality of life, itself a component of sustainability. This next generation of natural resource harvesting and extraction "specialists" also utilizes knowledge and services from the new economy.

Some see natural resources as a "curse" (e.g., Sachs and Warner 1995) and argue that nations and regions whose production and exports are based solely on comparative advantages—be they natural resources, location, and/or climate—in the long run remain poorer as measured by their ability to buy goods and services in their respective countries (their purchasing power parity), as shown in Fig. 1 (from Fairbanks and Lindsay 1997). This graph shows that a rich natural resource endowment has the potential to make government, industry, and citizens complacent and heavily reliant on the export of raw materials. Failures or absence of public policy (the nation state is typically the "owner" of the resource) lead to a lack of development

---

[1]"Virtual" implies of the digital or electronic world. A service or product exists as software or a website without necessarily having a physical base or material existence. Information-sharing and decision-making processes are therefore released from physical constraints using computer technology and communications networks.

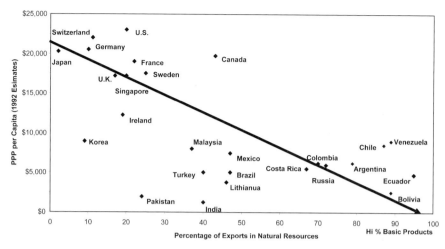

Fig. 1 Percentage of nation's exports in basic products versus purchasing power parity (PPP) per capita, 1992. The correlation coefficient determines the relationship between the two properties: coefficients closer to 1 or –1 indicate that variations in one variable are perfectly explained by variations in the other. The correlation between purchasing power parity and exports in natural resources is –0.66 (*dashed line*), which suggests a strong correlation between these parameters. Note: PPP is standard in this presentation as opposed to GNP because it is an estimation of exchange rate changes based on keeping prices of goods in different countries fairly similar by offsetting inflation differentials with changes in the currency exchange rates. Diagram redrawn from Fairbanks and Lindsay (1997); data from United Nations (1994)

of local supply chains and the stagnation of related industries, resulting in a lack "second generation" industries. Resource-based industries then compete solely on the export at the lowest cost of raw materials, with a consequent risk of environmental and societal degradation, and a further risk of reinforcing failed public policy.

Although the rent or revenues from the natural resources provide a basic livelihood to citizens of resource-based economies, there is little capacity for innovation. In contrast, resource-poor developing countries commonly explore alternative economic development opportunities to support the standard of living of their citizens. Denmark, Japan, the Netherlands, South Korea, and Taiwan, because of their shortage of natural resources, have increasingly based their exports on processed and manufactured goods and services. Research has shown that "nations who export manufactured goods are wealthier than those that export simple materials; the market pays a premium for knowledge imbedded in manufactured goods" (Gutierrez et al. 1997, p. 69).

Our contention is that natural resources themselves are neither a curse nor a panacea for creating sustainable development. What they do offer is a comparative advantage, which can be used to create competitive advantage through development of what Ramos (1998) calls "production complexes or clusters" (Ramos 1998, p. 105), which in turn have been shown to create sustainable economic growth in the developed regions and nations (Singh, 2001).

2.1 Industry Clusters

Groupings of related suppliers of inputs, services, equipment, know-how, and expertise have led to the formation of mature clusters such as the Finnish Forestry Cluster, and Latin America's emerging clusters around oilseed (Argentina) and mining (Chile). Ramos (1998) suggested that the capacity of regions rich in natural resources to reduce reliance on the export of raw materials is through the development of home grown suppliers and supporting institutions such as research and education.

In this context, an explanation of what clusters are and why they are important is warranted. Clusters are a group of industries whose linkages mutually reinforce and enhance their competitive advantage. Clusters are composed of diverse groups, including primary, related secondary, and supporting industries, universities, research centres, training institutions, and government agencies. They represent consumers of each others' products, competitors, partners, suppliers, or research and development sources. Porter (1998b) described clusters as the building blocks of a productive, innovative economy. A cluster is more than a single industry making a single product.

Clusters have been shown to drive innovation and increase competitiveness by their geographic concentration. Proximity has the following effects: it helps bring partners together, providing opportunities to network and share expertise; it enables firms to exchange knowledge and tacit information that cannot be codified; it provides shorter lines of communication between suppliers and end users; and it accelerates development of new generations of products which are normally tested first in the home market.

Provided that appropriate public policy frameworks are in place, there are four phases through which nations and regions develop strategies to transition from the simple export of low-value raw materials to higher value activities, and achieve prosperity (Ramos 1998). In the first phase, natural resources are extracted and exported with minimal local processing, and machinery and equipment are imported. In the second phase, processing and export activities are initiated, and a start is made on import substitution, with local production of some inputs and equipment. In the third phase, a cluster begins to export some of the goods and services that it originally began to produce for import substitution purposes. And finally, in the fourth phase, all types of goods and services are produced locally and are exported, and a trade balance is struck between imports and exports.

The experiences of developed economies such as Sweden, Finland, and the United States suggest that promoting clusters of related and supporting industries, such as manufacturers and suppliers of goods and services, can assist in achieving sustained growth. Examples of successful clusters that grew out of and around natural resources shows that there are numerous possibilities for catalysing economic potential. Many of today's developed countries have used natural resources as a platform to develop clusters of related and supporting industries, services, and products, which in turn create local employment, increase revenue, and attract investment. There is a general consensus that sustainable economic growth is indispensible to all forms of human progress (i.e., health, education, infrastructure). The wealth created by economic diversification can be used to invest in promoting environmental best practices, sharing benefits, and to support the distribution of wealth close to resource areas.

The performance of any industry sector depends on the macro- and micro-economic foundations and the cultural values of nations and regions, not on the mere abundance of natural resources. According to Harvard Professor Michael Porter, renowned for his pioneer work on competitiveness and cluster theory[2]:

> A nation can be prosperous and productive in virtually any field. What matters is how a nation competes, not what industry it competes in... we must stop thinking that traditional industries are bad and that the nation must move into high tech (Porter 1998b, p. 3).

Porter (2005) has argued that competitiveness is not a "zero-sum game", but that all countries can become more competitive because all countries can become more productive. There is an almost unlimited amount of human need that countries are competing to serve; therefore, if productivity goes up, more of those needs can be served at the same cost.

Figures 2, 3, and 4 provide faceted perspectives on cluster relationships, linkages, and opportunities for value-added products, services, and technologies. These are the local adaptations of generally applicable cluster representations. These figures

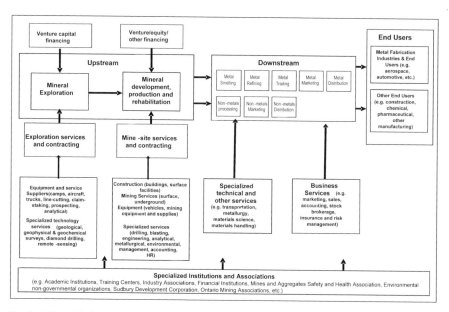

Fig. 2 Mineral industry cluster—schematic view. Diagram design adapted from Porter (1998b), with content from the Ontario Ministry of Northern Development and Mines

---

[2]For more information on Porter's academic work on cluster theory and competitiveness see www.drfd.hbs.edu

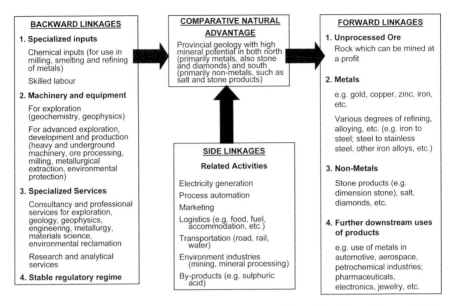

Fig. 3 Mineral industry cluster and its linkages. Diagram design adapted from Ramos (1998), with content from the Ontario Ministry of Northern Development and Mines

help the development of a comprehensive understanding of what is really going on, to create strategic maps that enable whole systems to be appreciated, and to catalyze the numerous possibilities associated with their sustainable development.

3 Examples of Natural Resource-Based Clusters

The following is a selection of examples of clusters that grew around natural resources. They serve to illustrate how natural resources can be used to support sustainable development by building the capacity of local suppliers of equipment and services, and the export of value-added products.

3.1 The Italian Ceramic Tile Cluster

Italian companies are the world leaders in producing and exporting ceramic tiles. Local red clay deposits (a comparative advantage) helped Italian equipment manufacturers develop and export equipment and develop expertise to work with imported white clay. Sixty percent of the Italian tile manufacturers are located in the Sassualo area. The geographic concentration of these manufacturers encouraged the establishment of many other supporting companies. The contributing success factors for the Italian ceramic tile industry are: sophisticated and demanding local buyers and customers; intense rivalry among local buyers; strong and unique distribution

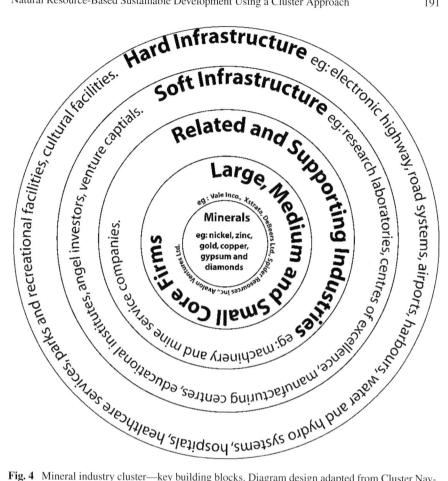

Fig. 4 Mineral industry cluster—key building blocks. Diagram design adapted from Cluster Navigator Ltd., with content developed by Ontario Ministry of Northern Development and Mines

channels; private ownership of the companies in the sector; highly developed local machinery manufacturers and suppliers; a number of supporting industries; specialized consulting companies; the strong presence of the Tile Industry Association; the involvement of the University of Bologna; and, most importantly, geographic concentration.

3.2 The Finnish Forestry Cluster

The Finnish forestry cluster is thoroughly developed. It generates 25% of the country's exports, and supplies 40% of the world plywood market and 25% of the world's market for all types of cardboard and paper. The Finnish forestry cluster also includes a machinery export industry for all phases of activities. The cluster continues to increase its proportion of value-added products, and maintains its competitiveness not only through its natural comparative advantage, but increasingly through continual improvements in productivity. Recently, the Finnish Forestry Cluster outsourced its bulk products (pulp and paper) processing operations closer to fast growing markets such as Southeast Asia and South America, thus creating a "Finland-originated" international cluster (Ramos 1998).

3.3 The Swedish Forest Industry Cluster

According to Dan Sjögren, project manager at Swedish Agency for Innovation System:

> ...in future, the challenges of the forest industry will be to advance ahead the supply chain, to produce a wider range of product (i.e. bio-refinery – green chemistry using 100% of raw material) and to develop energy efficient processes. (D. Sjögren, July 26, 2008, personal communication.)

The Swedish Forest Industry Cluster consists of a forestry and forest products industry, businesses, suppliers, packaging and processing industries, and university and research institutes and services including engineering, IT, chemicals, and logistics. Together, the Swedish Forest Industry Cluster makes a significant contribution to Sweden's economy.

In 2004 the Swedish Forest Industry accounted for 12% of the country's exports, 20–25% of its industrial investments, and 3% of GDP. The Swedish Forest Industry Cluster is one of the best examples of how a renewable natural resource, such as forest timber, can create wealth for a nation by striking a balance between environmental stewardship, economic sustainability, and social well being. Because the forest and forest products industries are dispersed throughout the country, the forest industry is of great importance to its regional economies, indicating progress in overcoming the limitations of physical clusters. The industry includes sawmills, and pulp and paper mills in 250 towns and villages in Sweden, often in areas where other types of employment are scarce. In many cases, these mills are essential for the maintenance of a variety of service functions in these sparsely populated rural areas.

3.4 The Norwegian Marine Cluster

Norway accounts for 10% of the world's seaborne transportation. Its support industries and internal suppliers are the main reason for this success. This cluster includes the maritime transport industry, fisheries, and shipyards that can build the most highly specialized tankers. These enterprises produce and export machinery and equipment for fishing and ship building industries on an international scale (Porter 1998b).

3.5 The Dutch Flower Cluster

This cluster has created competitive advantage in the growth, production, and marketing of flowers throughout the European market. The Netherlands continues to produce flowers, but it also imports them for re-export. The comparative advantage lies in the development of new varieties, and the competitive advantage lies in marketing its products outside the country and the development of a transportation hub specializing in perishable products (Porter 1998a).

3.6 The Argentine Oil Seed Cluster

This cluster generates 25% of Argentina's exports, and has recently experienced extremely rapid growth. An important processing industry has grown from converting agricultural inputs, such as soybeans and sunflower seeds, into oils and oil by-products. A domestic industry has also been established which produces 90% of the machinery required by the processing plants. The international competitiveness of Argentina's oil seed cluster demonstrates success in penetrating the most demanding markets (Ramos 1998).

3.7 The Chilean Mining Cluster

Over 25% of the world's copper ore reserves are estimated to occur in Chile, giving the country a strong comparative advantage. An industrial cluster has grown around the extraction and processing of copper ore, such that the country internally supplies two-thirds of the inputs, 40% of the equipment, 70% of the engineering services, and accounts for 60% of the costs of input machinery and engineering services for goods and services produced by related industries. At the governance and macro-economic level, the most important innovation has been to deposit the fiscal surpluses generated by the extraction of copper ore into an Economic and Social Stabilization fund (Ramos 1998).

3.8 The Ontario Mineral Industry Cluster

Mining has many important links to other industries and sectors in the economy, including geology, geophysics, geochemistry, remote sensing, technology, mine construction, legal services, banking, transportation, and environmental management. Together these constitute a cluster of mining-related industries and organizations. The Ontario Mineral Industry Cluster (www.omicc.ca) includes exploration companies, major mine operators, suppliers, financial institutions, education and training centres, research and development institutions and trade associations, and other allied industries (Fig. 4). It contributes to the wealth-creation and well-being of communities, especially of those in close proximity to the natural resource base, and provides opportunities to develop and export value-added products and services in highly efficient and advanced environmental, recycling, and rehabilitation technologies, and energy efficiency just to name a few (Singh 2005a).

4 The Benefits of Clusters

In all the above-mentioned examples, the real long-term benefits come not just from the presence of natural resources themselves (forests, mines, flowers, seeds, oceans, etc.), but from the development of associated activities, often subcontracted, that occur around them. These include the development and export of related machinery, equipment, services, and complex value-added products.

A further benefit is that the global marketplaces that form following cluster development tend to be relatively immune to the economic cycles of the host countries. Economic activity first evolves around natural resources, but then survives their depletion. For example, the Finnish mining industry was the basis for today's global mining equipment and machinery manufacturing industry.

In almost all successful clusters, more specialized universities, training programs, and regulatory frameworks, and close collaboration among educational institutions, research institutions, and industry, have played an important part in strengthening and intensifying linkages and competitiveness, promoting value-added industries, and maintaining environmental and societal standards and opportunities. For example: Holland has premiere research institutions in the cultivation, packaging, and shipping of flowers; and in Argentina, strong and ongoing increases in crop yields are largely due to the services offered by Argentine research centres and the University of Buenos Aires.

A report by the Economic Commission for Africa (Dales et al. 2004) provides a comprehensive analysis of the economic importance of clustering and shows that "The development of small-scale mining clusters contributes to the alleviation of poverty on a local scale and has potential to enhance regional development through linkages." (Dales et al. 2004, p. 147.)

4.1 Focus on the Ontario Mineral Industry Cluster

The minerals industry in Ontario has transformed itself into a modern, innovative, and highly technologically intensive cluster of related industries, organizations, and businesses (www.omicc.ca). Jobs in the mining industry have been, and continue to be, among the highest paid of all industries in Ontario, and productivity, as measured by GDP per hour per capita, is more than 50% above the all industry average (Ontario Mining Association 2006a).

Regional offices of major mining companies, such as Xstrata Nickel and Vale Inco, are located around the Sudbury copper-nickel mines, and the mining industry is the main regional employer of Aboriginal people. Exploration and mine development projects provide business and employment opportunities, especially to the remote communities in northern Ontario. For example, De Beers Canada's Victor Mine near Attawapiskat, James Bay, Ontario, has made the City of Timmins a major procurement centre for this almost C$1 billion capital project (Ontario Mining Association 2006a).

Dispersal of mining activities has created employment and growth opportunities throughout the province. The City of Toronto is a major north American centre for business expertise. The Toronto Stock Exchange (TSX) and the TSX Venture Exchange are home to approximately 3,600 issuers with a market captialization in excess of C$1.5 trillion. Some 1,200 of the TSX issuers are mining exploration companies. This global mining franchise promotes economic development in Ontario and supports a wide range of highly skilled, high paying jobs.

A vibrant cluster of related and supporting industries has emerged in the Sudbury–North Bay–Timmins triangle. These companies provide specialized equipment, services, and expertise to the local market, as well as exporting equipment and services to other mining jurisdictions around the world (Singh 2005b).

A critical mass of research organizations works closely with industry to conduct applied research for the development of new services and breakthrough technologies, which gives the local minerals industry a competitive advantage. Similarly, throughout Ontario, colleges and universities offer specialized education and training programs ranging from mining technology to satellite imagery. A Federated School of Mines, a centre that provides specialized education and training and a single-point access, meets the needs of industry's increasing demand for a highly trained workforce.

The Ontario government, in collaboration with the mining industry, is proactively and constantly striving to clean up environmental legacy issues of the past, and to improve the way it operates to leave less of a mark on the land in future, to prevent pollution, reduce energy consumption, and encourage the greater recycling of metals. The mining industry in Ontario also makes a significant investment in environmental protection and improvement.[3] There is an increased awareness of the need to mitigate environmental impacts, and corporate social responsibility is taken seriously by all involved—government, communities, and the mineral industry.

The mining and minerals industry in Ontario takes proactive steps to build capacity in the communities and provide business and employment opportunities. For example, De Beers Canada Inc. has taken concerted positive steps to reach out and employ Aboriginal people in its operations, and is committed to sustainable development, as noted by Jeremy Wyeth, Vice President, Victor Project:

> De Beers is committed to the concept of sustainable development and we intend to go to all lengths possible, not simply to comply with legislation, but practice proper environment

---

[3] The mining industry in Ontario now spends approximately $85 million annually on environmental protection, environmental improvement, and pollution prevention. In the early 1990s, $1 billion was spent on sulphur dioxide abatement programs in Sudbury alone.

management and continue open communication with the governments, employees, local communities and the public. (Ontario Mining Association 2006b, p. 17.)

Environmental stewardship is taken seriously by both industry and government, as evidenced by Xstrata's ranking in the Dow Jones Sustainability Index as the Global Sector Leader for Basic Resources (www.sustainability-indexes.com) for the second year in a row in 2008, and the Rehabilited Mine Site Repository (www.omicc.ca/rehabilitatedsites.php). The mining industry recognizes that in addition to its environmental responsibility, it also has social and economic responsibilities. These three areas of responsibility are the pillars of a right to operate in Ontario.

5 The Contribution of Clusters to Sustainable Development

Geography, culture, and innovative capacity, the stability and maturity of democratic institutions, and public policy frameworks are the primary factors that can influence the approach to sustainable development. Other variables, such as public education and awareness, corporate social responsibility, developmental agency success with capacity building, and how developed countries take their moral obligation towards reducing poverty, and leadership by example, will also have an impact on how sustainable development goals are achieved.

Natural resources, by definition, are found "in place", and the initial harvesting or extraction must be local. This appears to be a natural advantage for local development, but so often the local benefits have been limited and the environmental consequences have been disastrous. An inability to turn natural resources into new local industries is commonly associated with a lack of public policy, and frequently manifests itself as a failure of governance or mismanagement. In contrast, with appropriate public policies in place, natural resources can create opportunities for sustainable economic development close to the resource. For example, the Ekati diamond mine in the Canadian Northwest Territories purchases goods and services from more than 400 firms, most of which are located in the Northwest Territories (Natural Resources Canada 2000).

Mining activity in Canada and Ontario is spread out, and therefore a significant number of communities benefit from mines. For example, there are 185 mining-related communities in Canada, of which 88 have a reliance on this industry of 50% or greater, and 97 have a reliance of 30–49%. The economy of these communities depends either on local mining activity or on metal-processing plants (Natural Resources Canada 2001).

The further enhancement of world economies through ongoing cluster development will be a function of the capacity of virtual collaboration,[4] which enables

---

[4]"Virtual collaboration" involves using computer software and services, together with powerful, high speed networks, to support rich and complex information exchanges and decision-making processes, which enable economic growth and cultural enrichment beyond physical limitations. Examples of enablers include: blogs, wikis, web services, and webinars.

extended supply chains of participants to complement physical collaborations, as evidenced by the evolving global reach of the Finnish Forestry Cluster, the Swedish Forest Products Cluster, and Dutch Flower Cluster. By utilizing a combination of physical and virtual collaboration to more substantially integrate participants from the developed and developing economies into the same supply chain, the economic potential of natural resources can be further leveraged towards sustainable development.

According to Harrison and Huntington (2000), culture is a significant determinant of a region's and nation's capacity to achieve economic growth and prosperity. Consider the examples of Ghana and South Korea. In the early 1960s, both Ghana and South Korea had similar economies. Forty years later South Korea has become an industrial power with one of the largest economies in the world, whereas Ghana has achieved little growth. "Undoubtedly, many factors played a role, but is seemed. . . that culture had to be a large part of the explanation. South Koreans valued thrift, investment, hard work, education, organization and discipline. Ghanaians had different values. In short, culture counts." (Harrison and Huntington 2000, p. xiii.)

Many governments and societies lack an awareness that innovation and new technology are key aspects of the resource industry, as well as the new technology industries. Innovation linkages with the resource industry in resource-based nations is critical in the management of economic risk, and can reduce losses due to increased costs of extraction, labor, and capital, market fluctuations, resource depletion, and the emergence of lower cost competitors, disruptive new technologies, or material substitutions.

The Conference Board of Canada (2001, p. 7) noted that "Innovation is increasingly seen as a national imperative to achieving success in the global, knowledge-based economy. Canadians are now recognizing that they will be able to reach their sustainable development goals only by generating better ideas and implementing innovative products, services and processes." The capacity for innovation in the resource sectors will be increased by the expansion from physical to virtual collaboration, and this will accelerate the development of global supply chains.

The Conference Board of Canada also noted that the transition to sustainable development needs to be driven by all involved: industry, government, public, research and development institutions, and individuals. In particular, "Government has an important role to play in facilitating the integration of the economic, social and environmental objectives of all stakeholders—and this is a sustainable development role." (Conference Board of Canada 2001, p. 23.) A case in point is the Government of Ontario's establishment of the Ontario Mineral Industry Cluster Council, where the Ontario Ministry of Northern Development and Mines plays a "facilitator role" by bringing all stakeholders and partners of the cluster together to leverage the province's rich mineral endowment to achieve sustainable development balanced with economic growth and environmental stewardship.

The boom in metal and petroleum prices is generating new national wealth in the 21st century in Russia, the Persian Gulf states, Venezuela, Brazil, Chile, and elsewhere. Countries need to rethink how they will reinvest this wealth into long-term sustainability. Norway is an example of how its society has benefited from petroleum income from the North Sea. Finland is another example of a country that has reinvested mineral wealth into specialization in mineral processing technologies.

Governments need to commit funds and people to major initiatives that will benefit the development of specific technologies and business development in their jurisdictions. An excellent example is the national commitment to energy technology innovation (wind, solar, fuel cells, etc.) in some European countries. As a result of national commitments these countries are attracting scientists, engineers, and companies. These examples strongly argue for governments to promote the development of business and research clusters (Singh, 2003).

The experience of developed countries shows that there is a strong correlation between a stable standard of living, and positive societal and personal values. Public policy frameworks therefore need to influence a multiplicity of issues related to economic development, social equity, social justice, democracy, and civic sense. Variables, such as the distribution of wealth, reducing the gender gap, and dispelling perceptions about natural resources being a curse, all contribute to the potential to create and sustain economic growth.

6 A Proposed Pathway for Sustainable Development

> No single blueprint for sustainability will be found, as economic and social systems and ecological conditions differ widely across the globe. Each nation will have to work out its own concrete policy implications. (United Nations 1987, p. 11.)

Each nation or region needs to find its own specific pathway for a transition from high reliance on the export of natural resources, to sustainable development through the development of clusters of supporting and related industries based on accumulated know-how and expertise. However, "Sustainable development is not an automatic result of favourable factor conditions and the functioning of markets... It requires the purposeful collective action and an adequate coordination of public and private initiatives. All this stems from a shared strategic vision among all agents in the region regarding how to build a better future, and to build this strategic vision should be a central goal of any cluster project." (Dales et al. 2004, p. 145). A program to improve the competitiveness for resource-based economies is proposed here to guide that journey. It should be commissioned to conduct qualitative and quantitative analyses and gather insights and information based on the following characteristics:

- What are the major barriers to meaningful downstream and upstream processing?
- What are the necessary governance and public policy "levers" or tools, and how can they be put to best use?
- What are the growth opportunities?
- What are the growth inhibiting factors?
- What are the generic technologies that can migrate to other industries?

- What mechanisms are in place for representatives of and players in different sectors to collaborate?
- Is there inter-firm participation and co-operation (i.e., between major firms, suppliers, educational institutes, etc.)?
- Are benefits shared with people near the resources, especially Aboriginals?
- Who are the present and potential customers?
- What is imported versus exported, and what is the ratio of imports to exports?
- How can information, communication, and technology-based tools be used to improve both collaboration and competitiveness?
- What lessons can be learned and best practices adopted to achieve a balance between the three main pillars of sustainable development: the economy, society, and the environment? The "supply chain" behaviour of the old economy failed to reinvest wealth gained into the local economy. This remains a challenge.

It is recommended that an umbrella body be created to represent government at all levels, the private sector, associations, communities, universities, and non-governmental organizations, to develop and organize a collaborative plan. This could involve a network that engages new businesses, ideas and innovation, and draws lesson from the case histories of other cluster ventures. Such collaborative frameworks or clusters would create plans, ideas, directions, and opportunities to reinvest the wealth created into building new businesses and better local opportunities.

7 Conclusions

Natural resources have provided a source of livelihood to many parts of the world, and a comparative advantage that many developed countries have used to establish a stable standard of living for their citizens. Natural resources, even in modern society, remain indispensible. But going forward, economic growth, environmental stewardship, and corporate social responsibility must be better entwined, to create a viable sustainable development agenda.

Natural resources have the potential to stimulate both local and global economic growth, but on their own they do not guarantee a competitive advantage. Resource-based economies have achieved sustained economic growth through the development of locally based suppliers of inputs, services, and products, commonly referred to as clusters. Successful examples of resource-based clusters support this statement.

The achievement of sustainable development goals depends on the ability to harness many primary and secondary agents. Primary agents include geography, culture, innovative capacity, stability, and maturity of democratic institution and public policy frameworks. Secondary agents include public education and awareness, corporate social responsibility, developmental agency success with capacity building, how developed countries take their moral obligation towards reducing poverty, and leading by example. Exactly how these primary and secondary factors are orchestrated will have a profound impact on how sustainable development goals can be achieved. Clusters provide a vehicle for doing so.

Given the local mix of agents and mechanisms for their orchestration, each region's and nation's capacity to achieve sustainable development goals, and the pace at which they do so, will vary. But in all cases, the pursuit of sustainable development goals will require a new orientation, new approaches, and a new way of thinking about managing natural resources and achieving prosperity.

Acknowledgments This article has benefited from the comments of Pentti Noras. The authors would also like to thank Arnaud Segla for his technical assistance.

Disclaimer: The views and opinions expressed herein are those of the authors, and not necessarily those of the Ontario Ministry of Northern Development and Mines.

References

Ayres R.U., Ayres L.W., Rade I. (2003) The Life Cycle of Copper, Its Co-Products and Byproducts. Kluwer Academic Publishers, Dordrecht, the Netherlands

Conference Board of Canada (2001) Investing in Innovation in the Resource Sector—Industry Needs, Barriers and Opportunities for Action.

Dales D., Walker M., Black P., Botha A., and Mtegha H. (2004) Minerals Cluster Policy Study in Africa: Pilot Studies of South Africa and Mozambique. Economic Commission for Africa, Addis Ababa, 158p.

Fairbanks M., Lindsay S. (1997) Plowing the Sea, Nurturing the Hidden Sources of Growth in the Developing World. Harvard Business School Press, Boston, MA.

Gutierrez R., Martinez C., Sfeir-Younis A., Fairbanks M., Lindsay S., Holden P., Brugger E. (1997) Challenges for the New Millennium in Latin America. Unpublished report.

Harrison L.E., Huntington S.P. (eds) (2000) Culture Matters—How Values Shape Human Progress. Basic Books, a member of the Perseus Books Group, New York.

Natural Resources Canada (2000) Canadian Suppliers of Mining Goods and Services: Links Between Canadian Mining Companies and Selected Sectors of the Canadian Economy. Ministry of Public Works and Government Services Canada. Catalogue no. M37-48/2000E

Natural Resources Canada (2001) Mining-Reliant Communities. Online document at http://atlas.nrcan.gc.ca/site/english/maps/economic/rdc2001/rdcmin/1

Ontario Mining Association (2006a) Ontario Mining: A High-Tech Productivity Powerhouse Economic Contribution Study, Ontario Mining Association Report

Ontario Mining Association (2006b) Toward Greener Footprints 2006, Report on Significant Environmental Achievements

Porter M.E. (1998a) Cluster and the New Economics of Competition. Harvard Business Review, November–December 1998

Porter M.E. (1998b) Address at the Wellington Town Hall, New Zealand, in November 10th 1998. http://www.clusternavigators.com/parter_paper.htm

Porter M.E. (2005) Defining Competitiveness—A Zero Sum Game, in Global Competitiveness Report 2005–2006: Interview with Michael Porter. World Economic Forum, September 2005

Ramos J. (1998) A development strategy founded on natural resource-based production clusters. Economic Commission for Latin America and Caribbean Review

Sachs J.D., Warner A.M. (1995) Natural resource abundance and economic growth. NBER Working Paper No. 5398. National Bureau of Economic Research, Cambridge, MA

Singh I. (2001) Natural resource-based clusters in the new economy: Theory and Reality. Paper presented at the 4th Annual International Conference of the Competitiveness Institute, Tucson, AZ. http://www.omicc.ca

Singh I. (2003) Can government catalyze cluster? Examples of government actions. Paper presented at 6th Annual International Conference of the Competitiveness Institute, Gothenburg, Sweden. http://www.omicc.ca

Singh I. (2005a) Poised to compete globally: The Ontario mineral industry cluster is in position—ready to propel Ontario's mining industry forward. Canadian Mining Magazine, Summer: 10–12

Singh I. (2005b) Achieving higher value-added in Ontario's mineral industry cluster. CIM/ICM Bulletin, November: 12–14

United Nations (1987) Our Common Future, Report of the World Commission on Environment and Development—Our Common Future, (the Brundtland Commission). Oxford University Press, Oxford

United Nations (1994) UN SITC Trade Statistics Data Revision 2, World Bank, World Tables 1994

Part III
Mining and the Environment

Improving Environmental Performance in the Minerals Supply Chain Using a Life-Cycle Approach: The Role of Fuel and Lubricant Suppliers in Enabling Sustainable Development

Turlough F. Guerin

Abstract Suppliers have a pivotal role in enabling the mining and minerals industry to achieve their goals for sustainable development and demonstrating corporate responsibility. Reputable suppliers know their products and services and the industry, and are often well placed to have unique knowledge of these in relation to their customers' business needs. A survey of professionals from the mining and minerals industry was conducted, revealing factors limiting the influence of suppliers on a mining company's move toward sustainable development and specifically their environmental performance. Factors that affect a mining company's ability to engage with suppliers and the reasons why these companies believe suppliers are important to the achievement of their goals for sustainable development are also identified and discussed. This aspect of sustainable development in the mining and minerals industry has not been studied extensively. Petroleum hydrocarbon suppliers, in particular, affect a mine's goals for sustainable development because of the extensive reach of petroleum hydrocarbon products into the mining and minerals product life-cycle, their impact on operational efficiencies, cost, and mine viability, and their potential for leaving negative environmental as well as safety legacies. The petroleum hydrocarbon life-cycle is a framework that enables structured engagement between supplier and customer on a range of sustainable development issues because it is an example of an input into the mining industry that affects the entire mining and minerals processing value chain. The life-cycle starts with supply of fuels, lubricants, speciality chemicals, and services to the mine, through plant operation and maintenance, transport of mined products, and ship loading, and finally to mineral processing and other downstream value-adding. Eco-efficiency opportunities in this life-cycle are the main focus of this chapter. There are barriers within the mining industry to leveraging suppliers' capabilities which have to be overcome before the industry will realise the full benefits from such supplier engagement.

T.F. Guerin (✉)
Telstra Corporation Limited, Melbourne 3000, Australia
e-mail: turlough.guerin@hotmail.com

The views presented in this chapter are those of the author and do not necessarily reflect those of his employer, Telstra Corporation Limited.

1 Introduction

As companies compete for market share, they are increasingly focussing on their core competencies to become customer-centric. This involves, among many other elements of a business transformation, reducing costs, which inevitably involves looking to the supply chain to increase efficiencies and enhance the value created. Companies in all sectors of industry are increasingly being required by their stakeholder groups to state where their raw materials are coming from, and take action over and above this recognition and disclosure, to influence the supply chain to improve business as well as environmental performance.

Strategic supply chain management has been recognised in the business and management literature for many years as a critical element of any business planning process. However, it is only in the past two decades that environmental performance has been recognised to be of strategic importance in the supply chain (Lloyd 1994; Mehta 1994; Fiksel 1995; Lamming and Hampson 1996; Anonymous 1997, 1999; Tyler 1997; Christensen 2002; Hagelaar et al. 2004; Lutz 2005). Retailers and manufacturers, particularly in the automotive and electronics industries, have been leading progress in the greening of supply chains (Lamming and Hampson 1996; Anonymous 1997, 1999; Eskew 1999; Christensen 2002; Lutz 2005; Rao and Holt 2005; Barton 2006; Ellinor 2007; Ryu and Eyuboglu 2007; Simpson et al. 2007). Business's awareness of cleaner production (or eco-efficiency) and its uptake has helped drive this change (Altham and Guerin 2005). However, there are fewer published studies that explicitly describe the role of suppliers to the mining and minerals processing industry in greening its supply chain (Robinson et al. 1995; Enever and Robertson 1998; Guerin et al. 2004; Guerin 2006a, b).

Globally, sustainable development principles relevant to the mining industry were adopted by the International Council for Mining and Metallurgy (ICMM) in May 2003. ICMM member companies, which include the world's largest mining and minerals processing companies, have pledged to report on their progress in implementing these principles and these are being adopted internationally. In Australia, the Minerals Council of Australia (MCA) has developed a framework for sustainable development for member companies, which is based on these principles. The MCA's framework, which is called Enduring Value, was released in October 2004. This framework recognises the role that suppliers play in the transition of mining companies to a sustainable future (Anonymous 2008; Tables 1 and 2).

Two of the elements of this framework focus explicitly on how mining and minerals processing companies who commit to the framework (referred to as Signatories) are to work with suppliers.

A survey of the international mining industry (Lane and Danielson 2001), though several years old, recognised that sustainable development, in the context of mining, included the following:

Table 1 International guiding principles of sustainable development in the mining and minerals processing industry directly relevant to the supply chain

Organisation	Principles and/or signatory commitments	Source
The International Council on Mining and Minerals (ICMM)	ICMM has developed and published ten principles in relation to sustainable development in the mining and minerals processing industry. The following sub-set of seven principles have relevance to improving environmental performance in the supply chain:	www.icmm.com
	• Implement/maintain ethical business practices and sound systems of corporate governance. • Integrate sustainable development considerations within the corporate decision-making process. • Implement risk management strategies based on valid data and sound science. • Seek continual improvement of environmental performance. • Facilitate/encourage responsible product design, use, reuse, recycling, and disposal of products. • Contribute to the social, economic, and institutional development of the communities in which the industry operates. • Implement effective and transparent engagement, communication, and independently verified reporting arrangements with stakeholders.	
The Global Mining Initiative (GMI)	The GMI has set out the following principles from the Mining Minerals and Sustainable Development (MMSD) study: • Minimise waste and environmental damage along the whole of the supply chain. • Ensure transparency through providing all stakeholders with access to relevant and accurate information.	www.iied.org

- Impact on lives of people in the local communities in which mining communities operate;
- Interaction and consultation with local communities, particularly regarding the economic and social impacts of mining;
- Impacts on the environment where mining occurs.

The same survey demonstrated a general trend that there was a widening of an organisation's perceived area of responsibility in relation to sustainable development. The survey also showed that the mining and minerals processing industry has seen the emergence of many specialist contractors, for example in areas such as earth moving and maintenance, and that it is normal that business-critical activities are being outsourced. It indicated that while mining and minerals processing

Table 2 The Australian adaptation and development of guiding principles of sustainable development in the mining and minerals processing industry directly relevant to the supply chain[a]

Minerals Council of Australia (MCA) *Enduring Value* framework—Overview	*Enduring Value* will apply to all exploration, mining and minerals processing activities of Signatories, wherever they operate. It will also apply to the relevant activities of contractors engaged by the Signatories to undertake such activities. In addition, signatory companies will strongly encourage application of *Enduring Value* to operations in which they hold a non-controlling interest and to other supply chain partnerships. When referring to *Enduring Value*, Signatories will be transparent in identifying those aspects of their business that are covered by their Signature. For mining companies, this may entail identifying relevant operations. Commitment to *Enduring Value* brings with it a number of obligations. In summary, these are:
	• Progressive implementation of the International Council on Mining and Metals (ICMM) Principles and Elements;
	• Public reporting of site level performance, on a minimum annual basis, with reporting metrics self-selected from the Global Reporting Initiative (GRI), the GRI Mining and Metals Sector Supplement, or self-developed; and
	• Assessment of the systems used to manage key operational risks (using either internal or external assessment as appropriate).
Minerals Council of Australia (MCA) Enduring Value framework— Implementation guidance for Element 2.4	Element 2.4: "Encourage customers, business partners and suppliers of goods and services to adopt principles and practices that are comparable to our own."
	• Implement a procurement policy that includes sustainable development performance outcomes in key contracts;
	• Promote product stewardship initiatives throughout the supply chain through partnerships with contractors, suppliers and customers;
	• Encourage customers, contractors, suppliers, and business partners to adopt sustainable development policies and practices;
	• Establish "suppliers of choice" which include sustainable development criteria, such as the role of local employment, service, and supply to foster local economies.

Table 2 (continued)

Minerals Council of Australia (MCA) Enduring Value framework— Implementation guidance for Element 5.1	Element 5.1: "Implement a management system focused on continual improvement of all aspects of operations that could have a significant impact on the health and safety of our own employees, those of contractors and communities where we operate." • Implement an occupational and community health management system consistent with recognised quality standards that includes: • Control of hazards/risks of activities, products, and services over which the organisation has control, including the activities, products, and services of contractors and suppliers; • Identified management structures, responsibilities, resources, training, awareness, and competencies; • A communication system that includes employees and other interested parties, and provides for the relevant and timely reporting of performance; • Involve employees and other relevant stakeholders in auditing management systems and in management reviews.
Minerals Council of Australia (MCA) Enduring Value framework— Implementation guidance for Element 8.2	Element 8.2: "Conduct or support research and innovation that promotes the use of products and technologies that are safe and efficient in their use of energy, natural resources and other materials." • Where appropriate support research to improve eco-efficiency of production processes and products; • Review and innovate to reduce waste through cleaner production processes recycling and reuse of materials; • Review usage and innovate to improve efficiency in the use of energy and water; • Take other users' present and future requirements into account, including air and water quality and environmental flows of water; • Involve suppliers in identifying opportunities to reduce energy consumption or use renewable sources to reduce production of greenhouse gases and other emissions; • Where feasible, collaborate in industrial ecology activities to develop synergies in resource usage.

[a] These were obtained from the Enduring Value framework documents available at www.minerals.org.au

companies have experience in dealing with contractors, they are not familiar with using their influence over suppliers. Of the 32 companies surveyed, 78 and 59% required specific environmental standards to be met by contractors and suppliers, respectively. Respondents exhibited a strong interest in ensuring that local suppliers are used in their operations, and 90% of respondents stated that engaging with stakeholders effectively was one of the top five economic issues of concern to their company (Lane and Danielson 2001).

1.1 A Definition of Sustainable Development

A challenge for mining companies is defining what sustainable development means at an operational level (Azapagic 2004; Guerin et al. 2004). This includes the extent to which the minerals value chain is included within the scope of a mining and minerals processing business, and therefore to what extent suppliers are "within scope". There are many stakeholders for any one mining operation, and also numerous approaches available to a mining operation for assessing the impact of suppliers, one of their major stakeholder groups. In the context of this chapter, sustainable development is defined in relation to the wider business impacts of mining within the mining and minerals processing supply chain. Sustainable development, should, by its implication, encompass impacts throughout the supply chain both from a product as well as an input perspective. As referred to in the previous section, this has not been extensively studied in the literature. An explanation for this is the emphasis on the direct impacts of the mining industry, which are material in and of itself, even without considering the wider supply chain impacts. As the triple bottom line performance of mining and minerals processing companies improves, it could be expected that there will be a refocus of attention on the inputs into the industry. This chapter focuses on this input side of the supply chain.

The working definition of sustainable development in this chapter is framed in the concept of stewardship. It is implementing the industry's commitment to taking direct responsibility for its production, including inputs and processes, and a shared responsibility with customers, suppliers, and end users to ensure that all outputs are produced, consumed, and disposed of in an environmentally and socially responsible way.

1.2 Suppliers' Role in Supporting Sustainable Development in the Minerals Industry

Suppliers have traditionally been viewed as integral to the normal operation of mining companies (Enever and Robertson 1998). With the advent of heightened stakeholder awareness of environmental and social impacts of a mine, the role of suppliers is coming into sharper focus as an important contributor to both a mine's liability and opportunity for contributing to sustainable development. There are several specific drivers emerging both from within and external to the mining industry, which are influencing suppliers to recognise and embrace their role in assisting the minerals industry to work towards sustainable development (Enever and Robertson 1998; Blowfield 2000; Halme et al. 2007). These include:

- Formal recognition by the industry of a supplier's role in assisting mining companies in working towards sustainable development (e.g., ICMM Principles, and MCA Framework for Sustainable Development; Tables 1 and 2);
- Enhanced recognition by the industry that extended producer responsibility applies to products supplied to industry as raw materials, as well as the mineral products purchased as a result of mining (i.e., product stewardship);
- The business need for suppliers themselves to be more competitive. This is driving product and service differentiation in the mining industry marketplace through social, environmental, and financial performance;
- Recognition by responsible corporations, including both suppliers and mining companies, that their activities, products, and services interact with and affect the broader environment and the communities in which they do business; and
- The business needs of the mining industry to identify materially important eco-efficiency gains across their business, and their recognition that suppliers can help drive these types of improvements.

There are two main mechanisms by which suppliers can affect a mining customer's operations. These are through indirect or direct supply-chain leverage, which was highlighted in a presentation made to the forestry and timber industry in Australia (Guerin et al. 2003), but which is equally applicable to any industry. Indirect mechanisms include engaging with industry groups (common to supplier and customer) to assist in moving the entire industry forward, such as by development of industry and professional standards, frameworks, or codes of practise, and direct mechanisms through the supplier's unique understanding of their product and/or service, their life-cycle, and nature of risks and opportunities in relation to their customer's business.

There are numerous suppliers for any mining company or mining operation. These include product, service, and people suppliers covering every aspect of the minerals value chain. Table 3 provides examples of generic supplier groups, suppliers active in the industry, and the types of products and services they can provide. It also provides a description of the niche leverage that each supplier grouping can exert in support of their mining customer's sustainable development performance. This paper addresses these direct mechanisms.

1.3 Purpose and Scope

This chapter describes ways in which suppliers to the mining and minerals processing industry can support its move towards sustainable development. It also describes

Table 3 Role of suppliers in influencing environmental performance of the mining value chain

Supplier's industry	Services or product provided	Niche value-add to customer	Example of supplier[a]
Electricity supplier	Electricity (and commonly natural gas) supply	Provide carbon offset programs for customers; renewable energy offerings	Origin Energy: AGL (in Australia)
Explosives	Explosives and related services	Technologies to increase blast efficiency and reduce environmental impacts; provide expertise in engaging with and managing neighbour relationship	Orica, Akzo Nobel
Facility managers	Building and facility management	Identify and incorporate environment-related key performance indicators into mining contracts; identify and drive initiatives to reduce water and energy use	Transfield, Spotless and United Group Services (UGS) (all operating in Australia)
Fuel supplier	Fuel supply, distribution, and related services	Provide biofuels, carbon-offset fuels, and low particulate/low emissions fuels; advice on fuel efficient driving	Shell, BP, ExxonMobil, Caltex
Labour hire	Temporary staff hire services	Providing staff with environmental skills, awareness training programs for staff	Skilled Group (Australia); local indigenous labour hirers, Adecco, Exalt

Table 3 (continued)

Supplier's industry	Services or product provided	Niche value-add to customer	Example of supplier[a]
Lubricant supplier	Lubricant supply, distribution, and related services	Biodegradable lubricants alternatives; advice and services on lubricant life extension; life cycle management of lubricants	Fuchs, Shell, Castrol
Mining contractors	Mining operations services	Identify and incorporate environment-related key performance indicators into mining contracts	Roche Bros. Mining (Australia), Kaipara (New Zealand)
Telecommunications	Voice, data, internet access, dedicated networks for mining operations	Travel substitution such as high definition video conferencing; telemetry solutions for remote real-time monitoring	Telstra, Bell Canada, Vodaphone, Verizon
Waste management contractors	Total waste management services	Identify and incorporate environment-related key performance indicators into mining contracts; advice to mine on waste prevention strategies; implement and drive waste reduction initiatives across the mine	Thiess; Veolia; Transpacific Industries (Asia Pacific region)

[a] Provision of a supplier's name does not imply that they provide the niche value-added services, or that it is being endorsed by the author.

the perceived barriers within mining companies to harnessing the opportunities presented by their suppliers, and how these barriers may be overcome. The chapter has been prepared to help mine personnel and procurement and operations managers to leverage greater value from a mine's relationship with its suppliers, particularly those supplying petroleum hydrocarbons (fuels and lubricants). The chapter does not address the issues of suppliers engaging the services of people from communities in which their mining customers are operating.

2 A Qualitative Survey of Sustainability and the Minerals Industry Supply Chain

2.1 Background

Australia is a major player in the global mining industry. The expansion of the industry over the past 5–10 years has led to a large investment to support this growth, particularly in Western Australia, the nation's richest source of minerals. The flow-on effect from this expansion has been widespread across the Australian economy and society, with large increases in wages (in the mining sector) and house prices in Western Australia. In 2006/2007, mining contributed to 8% of Australia's GDP, employed 127,500 people directly, and 200,000 people indirectly (including suppliers). It also represented 26% of Australia's total capital investment, and contributed exports totalling A\$91.3 Bn. Suppliers have been major stakeholders in and beneficiaries of this industry's wealth, and therefore have had an important role in influencing and shaping the Australian industry's transition to sustainable development.

2.2 Survey Purpose

During 2006, a qualitative survey was conducted by the author of a segment of the Australian mining industry to identify views, opinions, and examples of the types of suppliers providing products and services to the mining industry. The purposes of the survey were to investigate what mining companies perceive the role of suppliers to be in their supply chain, and to identify any barriers that suppliers should be aware of that could negatively affect the role that they play. It was also conducted to establish a baseline of the wider minerals industry to understand the potential leverage that exists among suppliers to help meet its own objectives for environmental performance, and to work towards sustainable development goals. Specifically, it was anticipated that the survey would generate qualitative data and anecdotal evidence that suppliers contribute to the sustainability of the minerals supply chain and how they make this contribution.

2.3 Survey Method

The survey asked a series of multiple-choice and open-ended response questions. To develop the questions, the author engaged several marketing managers from supplier organisations and procurement managers from mining companies, as well as other corporate environmental managers. Twenty two professionals in the Australian mining industry and their suppliers, were surveyed using an online survey delivery program. These individuals were colleagues known to the author. Although this was a relatively small sample, the major industry sectors represented in the survey included diversified mining and minerals processing companies, metal producers, exploration, and energy companies (51% of respondents). The majority of the respondents (65%) were mining and minerals processing and consulting companies. Other respondents included researchers, academics, government organisations, and suppliers to the mining and minerals processing companies. Suppliers and other support organisations to the mining sector were also represented. The majority of roles represented were consultants, contracting and procurement, and community engagement staff.

2.4 Survey Findings

Energy, chemicals, telecommunications, and equipment were, not surprisingly, the most important supplied products and services to the mining industry (Fig. 1).

The most important finding was that when mining companies engaged with suppliers, greater than 50% of the respondents indicated that the most effective interactions occurred when:

- The supplier understood the needs of the business and tailored its approach accordingly;
- The supplier could demonstrate how its own commitment to environmental management and sustainable development would benefit the mining operation;
- The supplier created value for the mining operation by reducing costs and providing an improved solution (compared to existing solutions); and
- The supplier knew the life-cycle impacts of its own goods and/or services on the mining operation's business.

These findings show the high expectations that mining operations have of their suppliers, and they provide useful guidance for suppliers aspiring to work for the mining industry.

Secondly, the survey explored the major barriers identified to maximising the role of suppliers in, and leveraging their contribution and influence to, enhancing a mining company's strategy for working towards sustainable development. These barriers include limited engagement between a mining company's contracting and

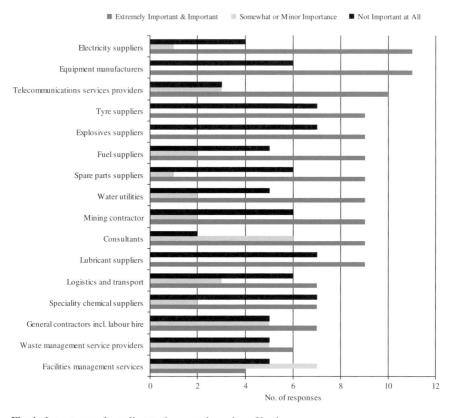

Fig. 1 Importance of suppliers to the normal running of business

procurement staff and environmental teams, and suppliers having limited under-standing of their mining company customers' operations and business needs. Others barriers included mining companies having limited financial and human resources for engaging with suppliers, and the hurdles presented by preferred vendor status or similar programs in effect at mining operations (Fig. 2).

Respondents indicated that the most important action a supplier could take to improve a mining company's drive towards a more sustainable future was demon-strating the supplier's own commitment in these areas. These are described in Table 4.

These findings reflect the largely cultural issues of maximising the value obtained from supplier relationships and in particular resistance to change. These findings underscore the importance of effective relationships between suppliers and mining companies such that there is fruitful exchange of ideas, innovations and relevant information to address problems or to identify opportunities for improvement. Pre-ferred vendor status can work well, though less so when prices start to increase (without corresponding value increase). Such programs can also promote the

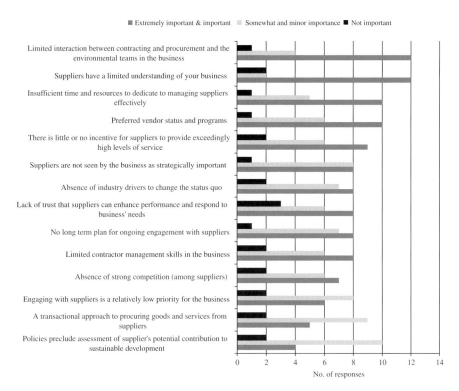

Fig. 2 Barriers to maximising suppliers' influence

Table 4 What is the most important action that a supplier to your business could do to improve your organisation's commitment and transition to sustainable development?

Improve their own sustainability performance

Understand the business of their customers

Make sustainable development part of the selling proposition and ensure the proposition is cast at the customer audience as they may not be experts in the field

Provide products that are energy efficient, have a limited impact on the environment, and are socially responsible

Be proactive in promoting to customers the sustainability aspects of their products and services as awareness is a key issue

Demonstrate their commitment to customers and an understanding and alignment with customer's needs and aspirations

---

status quo and limit innovation (such as increasing environmental performance) in contracts.

The third major finding identified from the survey was the barriers that limited mining companies from further engaging with their suppliers. These were a lack of time (for this particular activity), absence of commitment from senior mine management to such engagement, uncertainty of outcomes from such engagement, and a lack of interest from suppliers to such an engagement. Other barriers include the perceived increase in cost from such engagement, and that such engagement is unusual (i.e., not standard business practise or part of management culture) for mining operations (Fig. 3).

When respondents were asked what made their interaction with suppliers effective (using examples), 86% of respondents indicated that it was extremely important or important to them that the suppliers understood the needs of the business and tailored their approach accordingly, and that the suppliers could demonstrate how their own commitment to environmental management and sustainable development could benefit the business (Fig. 4).

For each of the following statements, approximately three quarters of respondents stated it was extremely important or important that the supplier:

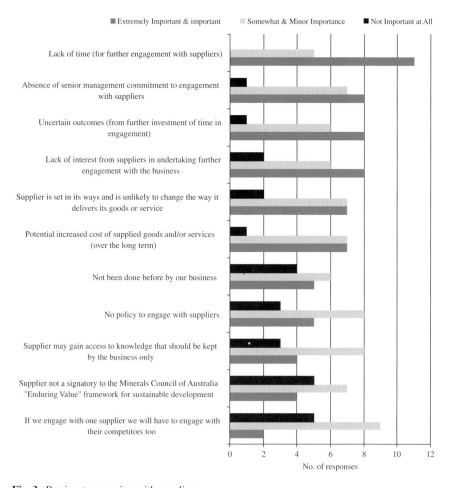

Fig. 3 Barriers to engaging with suppliers

Fig. 4 Importance of engaging with suppliers

- Provided the services required but then went beyond what was expected of them (in terms of delivering their products and services);
- Provided the goods and/or services required and at a competitive price;
- Helped reduce procurement costs (i.e., the process of procurement);
- Created value for the business by reducing costs and providing an improved solution (compared to that existing); and
- Knew the life-cycle impacts of its goods and/or services on the business.

These characteristics provide useful criteria for how to select suppliers or supply chain partners. They are aspirational attributes for any supplier to the minerals industry.

This survey indicated fuel suppliers were perceived as important to the mining and minerals industry, and the remainder of this paper focuses on their role, along with that of lubricant suppliers, in providing a life-cycle examination of how a supplier impacts a customer's business.

3 Overview of the Supplied Petroleum Hydrocarbon Life-Cycle

The downstream oil industry is a major supplier to the mining and minerals industry globally as demonstrated in the survey described in the previous section and also in other industry reports (Guerin et al. 2004; Guerin 2006a, b). Suppliers in this industry interface with many parts of the mining and minerals production process.

A useful way of understanding the inter-connections between a supplier and a mining or minerals processing operation (customer) is through the supplied product and service life-cycle across the mine operation. The life-cycle approach enables petroleum hydrocarbons for example to be tracked from the point of supply through to their end-of-life (Fig. 5).

Petroleum hydrocarbons are used through the entire mining and minerals production process and can generate impacts along the minerals processing value chain.

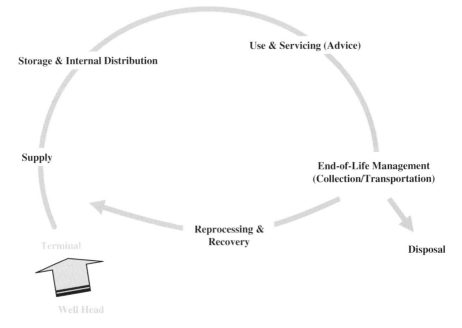

Fig. 5 Lubricant life-cycle

Typically, the most recognised impacts of petroleum hydrocarbon usage are negative (Table 5), often associated with pollution.

A life-cycle analysis approach enables the full benefits of closer engagement with petroleum hydrocarbon suppliers to be identified. It also facilitates development of subsequent joint action plans to address problems and explore opportunities. Many of the opportunities can be positive, and the minerals industry can reap the benefits if they are aware of and invest in the relationships with their petroleum hydrocarbon supplier(s). Similarly, credible and professional suppliers can add depth to the level of services or quantity and range of products they supply to a mine, including those that bring environmental and social as well as financial benefits.

The life-cycle approach provides a useful model for mapping supplier-customer relationships and therefore marketing opportunities across the range of businesses in the minerals industry. Engagement between suppliers and their customers should start at the tendering stage, continue during contract management, and remain in ongoing interactions. Suppliers can also be a catalyst in other parts of the mining operation, in addition to product supply and procurement, such as health, safety, and the environment (HS & E), to initiate activities in the customer organisation to enhance environmental performance of the mine.

The following four areas of a mining company's business (in relation to petroleum hydrocarbons) are affected by and need to be considered by mining operations for effective petroleum hydrocarbon management. These are also stages of the petroleum hydrocarbon life-cycle, and illustrate where a mining operation will need to proactively manage hydrocarbons (Fig. 5):

- Supply and procurement;
- Storage and internal distribution;
- Product use and servicing; and
- End-of-life management.

These stages are discussed in order. Selected examples of how suppliers of petroleum hydrocarbon can interact with mining companies are provided at each of the stages of the life-cycle.

3.1 Supply and Procurement

During procurement and supply of petroleum hydrocarbons, there is an opportunity for the supplier to consider supply transaction options, and understand the types of products and/or related services needed at the mine. There is also the opportunity to review existing supply arrangements, that will help improve delivery, and reduce costs, and to provide environmentally-preferred products where these are available. This stage is critical in driving change in the mining and minerals industry. In the future, there will also be increasing pressure on suppliers as well as lubricant users as to the types of products used in particular applications. At this stage of

Table 5 Environmental impacts as a consequence of equipment maintenance and supplied petroleum hydrocarbons to the mining industry

Aspect of maintenance program[a]	Description of impact or potential impact[b]
Liquid petroleum hydrocarbons[c]	Groundwater contamination from spillages and leaks[d]
	Soil and surface water contamination from spillages and leaks[d]
	Air pollution (from uncontained volatile components if present)
	Disposal without reuse or energy recovery is unsustainable
Solvents	Air pollution—uncontained vapours could adversely affect human health
	Disposal without reuse or energy recovery is unsustainable
	Groundwater contamination from spillages and leaks[d]
	Surface soil contamination from spillages and leaks[d]
Waste grease	Soil and groundwater contamination from petroleum hydrocarbons and metals impregnated in grease[d]
	Disposal without reuse or energy recovery is unsustainable.
Detergents	Toxicity to numerous aquatic organisms; emulsification of petroleum hydrocarbons (and reduced waste oil recovery)
Sediment (suspended solids)	Surface water contamination from spillages and leaks[d]; blockages cause flow-on effects in oily wastewater management systems

Table 5 (continued)

Aspect of maintenance program[a]	Description of impact or potential impact[b]
Contaminated rags, paper, protective equipment, and gaskets[e]	Excessive use and wastage is unsustainable
Used oil filters, burst hydraulic hoses, empty fuel, oil and grease containers (e.g., drums)	Often landfilled which is unsustainable. Petroleum hydrocarbons within these materials have the potential to impact soil and groundwater (i.e., act as contamination sources)
Energy usage (powered equipment)	High levels of use can be unsustainable if not linked into an energy generation system on site
Water usage (for washing vehicles prior to maintenance)	High levels of use are unsustainable (may reduce quantity and quality of water supplies for other uses, e.g., availability for local communities near mine)

[a] Including contaminants, co-contaminants, or waste types.
[b] Note that these are qualified as "potential" impacts because these wastes will not always have a negative impact on the environment (i.e., if they are managed and disposed of properly).
[c] These include fuels and engine, hydraulic, gearbox, differential, and steering oils, off-specification fuels and oils, cutting fluids, and oily wastewater.
[d] Soil and groundwater contamination can add significantly to the environmental liability of a mining operation, and site investigations, risk assessments and remediation programs may be required to address these at closure or as part of the divestment of the operation.
[e] Also include oil-contaminated plastics, speciality chemicals, leather and cloth (e.g., gloves).

the life-cycle suppliers can help mining companies achieve their sustainable development goals through:

- Product design and development, supporting research and optimisation of product selection, and offering product options and alternatives to conventional products;
- Product procurement transactions and supply chain leverage.

3.1.1 Product Design and Development, and Offering Product Options and Alternatives

Petroleum hydrocarbon suppliers invest resources into developing and producing new products for their customers. This is reflected in the financial commitments made by large oil companies into product research. Examples of this include the manufacturing and supply of low emission fuel products. Low sulphur diesel (50 ppm) is now being produced at refineries in Australia, and benzene reduction units are currently being installed at Australian refineries to produce low benzene petrol (1 ppm). These required maximum concentrations for sulphur and benzene will be reduced even further as fuel regulations continue to become more stringent. Biofuels development will be of increasing importance as the price of crude oil continues to increase and the demand for these products increase from larger users such as mining and other heavy industries. Formulations that allow for longer storage life and that do not cause engine power to be significantly reduced are also needed by the mining industry.

Petroleum hydrocarbon suppliers can provide alternatives to the conventional range of products currently being offered to the mining industry. Though cost is critical, customers of lubricants increasingly want to exercise their ability to choose options when purchasing products, including options related to environmental performance. For example, biodegradable lubricants are preferable for applications where there are acute risks from mining or operations in environmentally-sensitive areas such as during exploration and at ship-loading facilities (Battersby et al. 2003).

3.1.2 Product Procurement Transactions

Petroleum hydrocarbon suppliers can use their purchasing power to secure supply arrangements with specialised chemical manufacturers or suppliers. This includes third party supply of specialist products which can be procured at lower cost than can be achieved by the end user (mine), such as specialist greases, fluids, coolants, and solvents. The benefits of this also include reduced administration to the mine, and the fact that it places responsibility for security of supply of these specialist products with the petroleum hydrocarbon supplier.

Suppliers can also assist a mining operation's overall environmental program in the development of environmental management plans (EMPs) for supplied products, which can be negotiated at the contract stage of the procurement process. Some of the larger mining operations in Australia are now stipulating that EMPs be prepared by major fuel suppliers that are supplying product to their operations. EMPs should highlight the risks and controls in place in relation to the supplied product or service (including its transport, storage, and handling); this increases the assurances that the mining company has identified and is controlling these risks.

3.2 Storage and Internal Distribution

The second stage of the life cycle relates to facility design and layout, which influences the placement of supplied product in relation to the operational needs of the mine. Storage and internal distribution issues can have a significant impact on the potential or likelihood of environmental contamination from products such as from leaks, particularly those undetected, which can result in additional costs to the mine. These costs can be incurred during the normal life of the mine, or will be realised at mine closure if no action is taken during normal operations. Strategic capital investment in appropriate storage and internal distribution facilities ultimately reduces the long-term financial liability for a mine, because it can eliminate or reduce environmental contamination from product losses.

Elements of the product storage and internal distribution stage of the life-cycle, where suppliers can help mining companies achieve their goals for sustainable development, include the following:

- Ensuring facility design meets construction standards appropriate for the petroleum hydrocarbon and chemical tanks and infrastructure present at the mine.
- Optimising fuel and lubricant delivery across an operation to ensure the lowest cost and safest way of keeping the mobile (i.e., portable or transportable) plant running.
- Identifying and assessing compliance of chemical storage areas to dangerous goods standards (for packaged products).
- Stock reconciliation to account for product flows into and across a mine or a series of mines.
- Testing of infrastructure (asset) integrity to prevent and minimise stored product losses.

3.2.1 Meeting Design Standards

Fuel and lubricant storage distribution, and dispensing facilities must be designed and built to meet minimum engineering standards. There are standards that cover issues such as materials, tank and pipe configurations, electrical, safety, and environmental issues. In Australia, one of the main standards is Australia Standard (AS) 1940:2004, that describes the requirements for storage of non-flammable liquids such as diesel and lubricants. Petroleum hydrocarbon suppliers have expertise in auditing and redesigning, rebuilding and/or repairing such facilities because they are continually working with fuel and lubricant infrastructure at their own facilities. They also have extensive experience in applying these standards because they audit and manage their own facilities. Suppliers are in a position to offer focused auditing capabilities to their customers, and to know which fuel and lubricants standards will be applicable to the mine. There are a range of other industry standards as well as those for handling flammable goods, and for construction of fuel and lubricant storage and dispensing facilities.

3.2.2 Optimising Product Delivery

As a mine expands, and the location of the mined ore body changes relative to the mine's fixed infrastructure, so does the mining operation's need for the supply and dispensing of fuels and lubricants. For a mine to optimise the delivery of fuels and lubricants in the mine, it requires extensive knowledge of transport and distribution logistics. This will ensure that capital is not wasted on infrastructure that could become redundant as a result of inappropriate placement of fuel or lubricant delivery infrastructure. Minimising the amount of time required for refuelling and maintenance ensures loss of productivity is kept as low as possible as well as environmental impacts.

3.2.3 Stock Reconciliation Solutions

Fuel and lubricant stock reconciliation systems include, for example, simple mechanical measurement (i.e., dipping) of tanks, reconciliation of flow meters on a regular basis across a single mine, and more complex network-level (i.e., across multiple sites) leak detection systems that have data collection, statistical analyses, and red-flag reporting mechanisms. Reporting from stock reconciliation systems identifies where stock control practices are inadequate, and identifies tanks, or users of mobile and fixed mechanical plant (i.e., machinery), that have or contribute to unusually high product losses. Such systems are particularly important for underground product storage facilities. Inventory control and monitoring systems are a relatively small investment that can reduce environmental testing and remediation costs in the long-term, and are the only effective and preventative mechanisms for monitoring leaking underground storage systems. Stock reconciliation systems can also enable better control of fuel management data for more effective reporting and reconciliation of greenhouse gas emissions.

3.2.4 Asset Integrity Testing

Asset integrity testing is the assessment of petroleum storage, distribution and dispensing equipment, and other facilities for product leaks. Mining operations and other facilities that handle fuel and lubricants are required to conduct integrity testing on their assets at specified time intervals. Ten-year test intervals are common in many jurisdictions. Asset testing can include positive and negative pressure testing systems, which can measure the loss of pressure or vacuum in the product storage or distribution system over time to determine the presence and extent of leaks. Asset integrity testing should form the first stage of assessing the risks associated with the storage and dispensing of fuel at a mine site. It is not uncommon to find that asset integrity testing reveals that a proportion of underground storage structures (pipes and tanks) are leaking at a facility. A recent study of a commercial fuel network supplying the road transport industry revealed that there were approximately 10% of sites that reported failures using vacuum testing of all underground storage and distribution assets across the network. The presence of a failure from asset integrity testing, when using the common vacuum testing approach, indicates that there is air ingress and/or a crack or hole in the infrastructure. Where an asset failure has occurred, the concrete or surface overlying the underground asset will have to be removed to examine and identify the reason for failure. Apparent asset failures (i.e., reporting false positives during the vacuum test) may simply be a loose collar on a pipe or loose pipe fittings, and not necessarily a hole in a tank or pipe.

3.3 Product Use and Servicing

Petroleum hydrocarbon products can and should be managed, during their working life, to ensure that they do what they are supposed to do during this time. The third stage of the life-cycle examines the impacts of the supplied petroleum hydrocarbon product(s) on fixed and mobile mine plant components and how these products can be best serviced to extend their own as well as the plant's life. Fluids selection across a mine's fixed and mobile plant can include consolidating the range of grades of lubricants being used. Consolidation itself can reduce the range of products and containers stored (and ultimately disposed of) at a mine site, which can enhance waste management. But more importantly, fluid selection can have a dramatic impact on the eco-efficiency of mining equipment. Suppliers can work with mining companies in the product use and servicing stage to achieve the mine's goals for sustainable development, in the following ways:

- Reviewing the mine's maintenance strategy to enhance reliability of mobile and fixed mine plant.
- Recommending the use of energy efficient-lubricants for high-friction applications.
- Managing lubricant cleanliness to maximise lubricant and plant life.
- Developing lubricant laundering (i.e., cleaning) as an option to extend the useful life of lubricants.

3.3.1 Reviewing the Mine's Maintenance Strategy to Enhance Reliability of Mobile and Fixed Mine Plant

Further examples from this stage of the life-cycle are the contributions suppliers can make to maintenance strategies. These should include planning for maintenance activities, monitoring and analysing maintenance costs, establishing targets for maintenance performance (in particular percentage downtime) and establishing preventative maintenance programs. Preventative maintenance is an area where considerable cost savings may exist for a mine, particularly because the numbers and sizes of fixed and mobile plants can be large. An important part of any preventative maintenance program is to have predictive tools to define equipment defects as early as possible. Early detection of a defect allows for better failure analysis to improve the equipment's service life performance. It also assists in identifying the true problem rather than a symptom of the problem. In many cases, what we see as the failed component is a symptom of what the true cause of the failure was. To determine the causes of failure, fuel and lubricant suppliers can provide preventative maintenance services as a means of extending both product and plant life at customer sites using thermographic techniques and condition monitoring programs that involve lubricant analysis and diagnosis. Such programs help prevent plant breakdowns, while at the same time delivering business and environmental benefits through lower operating and capital costs, and reducing rates of waste oil generation (Pearson 2004; Mercer 2005; Garvey 2006; West 2006). For example, infrared thermography has been found to be a valuable tool in the mining industry. It can survey equipment at a mine, including electrical distribution systems, pumping systems, piping systems, exchangers, process fired heaters, and many other types of equipment. Infrared thermography can assist in finding the underlying true cause of failure. It is seen as a predictive tool that supports other predictive technologies, such as vibration analysis and compression analysis. One primary advantage is that it is faster than many of the existing techniques in identifying and detecting a problem. It has the ability to find defects before a secondary catastrophic failure occurs. A technician can view many pieces of mechanical equipment very quickly to determine if a possible problem exists. Various petroleum hydrocarbon suppliers are providing preventative maintenance strategies, that often package the solutions together for clients (Messenger et al. 2004a, b).

3.3.2 Recommending the Use of Energy-Efficient Lubricants for High-Friction Applications

Energy-efficient lubricants have a niche role in enhancing plant performance. By switching to synthetic lubricants, the most common examples of energy-efficient lubricants, a mine can improve both efficiency of plant energy use, and environmental performance. For example, synthetic lubricants have long been recognised for their benefits compared to conventional mineral oil-based lubricants for increasing oil service life, reducing wear, system deposits, and improved viscosity/temperature behaviour. They are not used widely and this is primarily because of their cost.

One benefit that has been largely overlooked is that of energy savings. Because the mining industry must focus on reducing CO_2 emissions, there is great benefit to using products, that contribute to reduced energy consumption, even though the drop in CO_2 emissions may be minor compared with emissions generated by a mine. One application is the lubrication of worm gears, which have unique requirements relative to lubrication of standard helical or spur gears. In particular, the high degree of sliding contact in worm gears generates considerable friction. These types

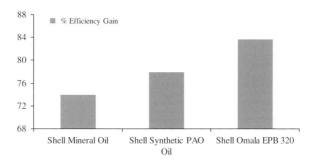

Fig. 6 Energy efficiency gains over conventional mineral oil from application of selected synthetic lubricants in high friction environments

of gears have numerous applications across processing plants, and in particular, in gear boxes and stationary equipment. Efficiencies with worm gearboxes are often as low as 70% which is a result of the high loads and high friction in transferring energy through such gear configurations. Highly polar polyalkylene glycol (PAG)-based worm gear lubricants, such as Shell Tivela S oils, are ideal for lubrication of steel-on-bronze worm gears, because they can lower the friction in boundary lubrication and thereby reduce inefficiency. Maximising efficiency also means that less power is lost in friction and converted into heat. In the David Brown Radicon efficiency worm gear test, in which the input and output torque of the gearbox is used to determine the efficiency, the PAG-based Shell Tivela S demonstrated 15.8% improvement in energy efficiency relative to a mineral oil based product. Based on new PAG-technology, the Shell Tivela S offers benefits of 9% energy savings relative to older PAG-based worm gear lubricants. Relative to a polyalphaolefin (PAO)-based fluid, the energy saving is 11%. In terms of cost savings, for an operation running 50 gearboxes for 168 h/week (50 weeks per annum), with an average power output of 7.4 kW and electricity at a cost of $AUD0.14/kW/h, the savings relative to a mineral oil (with gearbox efficiency 74.1%), in using a PAG based lubricant amounts to $AUD82K per annum (Guerin et al. 2004). These savings will increase as the price of electricity continues to rise at the current high rates (10% per annum, or more).

In the same worm gear test, the benefits of energy efficiency for a biodegradable gear oil, Shell Omala EPB, are shown (Fig. 6). For the gearbox running on Shell Omala EPB, a 12% efficiency gain over a mineral oil-based gear oil (of the same viscosity grade) and 7% over a PAO-based gear oil were obtained. The switch to Shell Omala EPB in such an application would also bring the additional environmental benefits of a biodegradable lubricant (Battersby et al. 2003; Guerin et al. 2004) and could be considered for any environmentally-sensitive application such as on a wharf, jetty, or ship loader.

3.3.3 Managing Lubricant Cleanliness to Maximise Lubricant and Plant Life

Another example of product use and servicing is managing lubricant contamination. The impact on heavy vehicles from contaminated lubricants can be extremely costly due to lost productivity, increased maintenance, and spare parts costs. There are many risks associated with lubricant contamination, especially where dirt, road grime, and dust are abundant. As far as particulate matter is concerned, how much is considered too much, and how will this contaminant impact on a machine's life? The impact of lubricant contamination will depend on the hardness, volume, and size of the contaminating material. Harder materials such as silica, bauxite, and iron ore will cause accelerated abrasive wear, whereas softer materials such as talc and coal can cause build-up in oil ways and tooth roots that can lead to failure (Carlin et al. 2003). Any size and number of particles in a lubricant can cause problems; however, larger particles tend to fall to the base of the plant's fluid reservoir. The smaller particles remain suspended and are pumped into bearings and other critical working components. To prevent this problem, rather than relying entirely on oil filters, it is critical that plant and product container breathers are kept clean (Fig. 7).

The most common lubricant and fluids (including brake, hydraulic, steering fluids) contamination sources and causes include:

- The mechanical seal on metal drums working loose during abnormal transportation conditions, releasing metal particulates and causing drum varnish to flake into the oil.
- Bulky plastic product containers having breathers that allow the product to breathe, but that leave it exposed to atmospheric contamination.
- Bulk lubricant transport systems, which are used to administer lubricant products to equipment in the field and can contain residue from previous loads and/or dust particles.
- On-site practices designed to make life "easier" for on-site personnel who handle lubricants. For example, such as leaving a grease hopper lid open so that truck drivers can monitor grease levels also leaves the product open to the elements and increases risk of product contamination from dust.

Several studies have dealt with this issue in greater detail (e.g., Huth 1975; Pavlat 1984; Rakic 2004).

3.3.4 Developing Lubricant Laundering as an Option to Extend the Useful Life of Lubricants

A final example is a technology called lubricant laundering, which brings the benefit of reduced costs in purchase of new lubricants (Messenger et al. 2004a, b). Lubricant laundering is the refurbishing or cleaning of a lubricant so it can be reused as a lubricant. This process can also result in fewer oil changes which means less used oil to manage. Applications of this technology in the Oceania region are limited at the current time because of the relatively high capital cost for the equipment, and the labour required to handle and manage the laundering operation. Lubricant laundering offers the potential for a mine to reduce its lubricant purchase costs; however, it should be viewed as only one of a number of strategies to help extend

Fig. 7 Uncovered (*top left*), contaminated (*top right*), open (*bottom left*) breathers, and sloppy dispensing activities can lead to lubricant contamination

the life of the supplied petroleum hydrocarbon at a mine (Neadle 1994; Messenger et al. 2004a, b)

3.4 End-of-Life Management

The final stage of the petroleum hydrocarbon life-cycle is managing the supplied product at the end of its useful life. Although various technologies and strategies can extend the life of supplied product, lubricants eventually become ineffective and need to be managed as either wastes or a feedstock for energy recovery purposes. Suppliers can help mining companies achieve their goals for sustainability during the end-of-life management stage of the life-cycle, including the following:

- Product packaging and stewardship;
- Used oil collection and management;
- Management of maintenance wastes;
- Fuel and lubricant infrastructure management; and
- Fuel and lubricant disaster and spill management.

3.4.1 Product Packaging and Stewardship

Providing an outlet for off-site removal of used oil and oil containers is an ongoing challenge for both mining operations and packed product fluid suppliers. Consolidation of supplied pack sizes into a single size, e.g., 18 L (where packed product is required at a mine) and switching to bulk lubricants (where possible), are ways in which suppliers can assist a mining operation. In Australia, an environmental and economic review of lubricant pack size consolidation in the 10–20 L pack size range was conducted by the author in 2003 (unpublished). The results demonstrated that the plastics recycling industry in Australia, while technically able to reprocess the volume of containers produced as a result of the mining industry's consumption, is at the stage of maturity such that the costs for reprocessing of used oil container plastic is too high to provide a cost-effective and an equitable take-back service for all mining and/or industry customers. In Australia, the introduction of a National Packaging Covenant may help provide an incentive for the lubricant and specialist chemical supply industry to provide the most environmentally-preferred and cost effective packaging solutions for their industry to supply the mining industry. The National Packaging Covenant puts the onus on suppliers of products with containers and packaging to demonstrate how they will reduce the environmental burden of the product packaging they supply, and particularly so as this covenant is now legislated at the state-level across Australia.

There is no consistent enforcement of pollution laws in the Australian mining industry to drive the prevention of disposal of used oil containers at mining operations. Hence, pollution continues to occur at mining operations from empty fluid containers and from other operational wastes.

3.4.2 Used Oil Collection and Management

The used oil management industry benefits greatly from the mining industry because of the large used oil volumes generated by mining and mineral processing. For example, the volume of used oil collected in Australia is approximately 500 ML annually. Of this volume, approximately 50 ML is generated by the mining industry. Used oil handlers provide a network of collection services reaching most locations in Australia, including remote mining areas. These suppliers often work with each other; as sub-contractors to other used oil handlers, depending on the region. There are however, a wide range of quality standards to which these suppliers work to, and this has meant there are varying levels of service quality provided to the mining industry. There are no specific legislated standards to which these suppliers have to work, with the exception of AS 1940:2004; this standard regulates the storage and handling of dangerous goods and has been enforced in many Australia jurisdictions by state governments. The author recently reviewed and audited all major used oil handling facilities in Australia (unpublished report). Some facilities are certified to ISO 14001, but many of the facilities have poor housekeeping practises. Overall, these facilities are improving due to the increased levels of competition, largely due to the Australian federal government's initiative to implement legislation that maximises the value of the used oil resource. Used oils are reprocessed back into base oils at various reprocessing facilities across Australia, including most capital cities.

3.4.3 Management of Maintenance Wastes

A further example is the management of maintenance wastes (Table 6).

Such waste, which includes used petroleum hydrocarbons, poses a significant challenge to the mining industry (Guerin 2002). If maintenance activities are not conducted effectively so as to minimise losses of petroleum hydrocarbon wastes to the environment, they can lead to significant long-term environmental liabilities from soil and ground water contamination. Typically, many older mining operations (i.e., those established for >20 years) do not manage their maintenance wastes effectively, based on a survey previously published by the author (Guerin 2002). Good housekeeping in maintenance areas is critical to prevent soil and groundwater contamination; such house keeping includes for example proper waste segregation and storage of drums and wastes (Fig. 8).

Petroleum hydrocarbon suppliers often have the capability or the supply chain influence to provide wide-ranging services that improve management of maintenance activities at a mine. Petroleum hydrocarbon suppliers can assist by:

- Auditing maintenance waste streams;
- Advising on process improvements to reduce volumes and types of maintenance wastes; and
- Advising on life-cycle management of maintenance wastes from prevention through to treatment.

Table 6 Sources of wastes from maintenance operations in the minerals industry [a]

Stage of mining process	Practice or specific site location	Type and source of waste
Exploration	Drill maintenance areas	Spillages and leakages of oils, grease, and degreasers during maintenance to drilling rigs
	Drill mast maintenance areas	Grease and oil sand blasted from mast frame before overhaul maintenance and re-painting is carried out
	Drilling operations	Drilling muds with ores containing hydrocarbons
Mine	Shovels, excavators, scrapers, backhoes, wheel loaders, and bucket loaders	Waste oil from oil changes to mine equipment, spillages from breakdown maintenance, blown hydraulic hoses, spillages from refuelling, maintaining oil and grease levels on field equipment; empty drums and used protective clothing
Maintenance	Wash down areas	Wash down of mobile equipment, effluent containing oils, diesel, grease, detergents and soil
	Heavy vehicle equipment servicing	Oil and filter changes on mobile equipment, waste grease containers, blown hydraulic hoses, used protective clothing and lead acid batteries; waste tyres; worn brake pads; solvent for engine parts cleaner; plastic drums; waste coolant, brake and transmission fluid
	Light vehicle servicing	Oil and filter changes on mobile equipment, waste grease containers and lead acid batteries; tyre bay wastes; general waste around car ramps; worn brake pads; solvent brake cleaner; solvent for engine parts cleaner; plastic drums; waste coolant, brake fluid and transmission fluid
	Servicing pits	Spillage during vehicle servicing, regular greasing and cleaning out of sludge pits; used protective clothing
	Workshop floors	Spillage onto workshop floor during maintenance and repairs, and leakage and spillage from oil storage area and from wash down practices
	Oily wastewater separators	Incorrectly designed or poorly maintained equipment
	Oil filter draining	Spillages around collection vessel

Table 6 (continued)

Stage of mining process	Practice or specific site location	Type and source of waste
	Waste oil storage	Spillages during storage and transfers
	Workshop drain cleaning	Sludge (from build-up)
	Compressor sheds	Oil changes, leakages, compressor clean down, water/oil drainage from filters and air receiver, wash down of concrete floor
	Drum storage areas	Leaks/spills from drums, wash down of concrete floors and drum cleaning
	Fuel supply depots and infrastructure	Leaks of diesel and gasoline (on- or off-site), sometimes from underground supply pipework; refuelling leaks, (overflows and broken seals); surface water run-off
	Oil supply bays	Spills during filling of storage tanks, filling of vehicles and mobile tankers
	Equipment refuelling	Spillage (overfilling) during refuelling of equipment and servicing trucks, leaking pumps and blown hoses
Upstream (or primary) processing	Processing plants	Oil changes on scrubbers, screens and conveyor belts; grease
Ore shipment/transport	Crusher areas	Dust suppression foam; grease and oils
	Stackers, reclaimers, conveyors, train load out areas	Grease and leaked oil, particularly hydraulic fluid
Downstream processing	Milling, smelting, refining, preparation for sale	Metals and minerals; petroleum hydrocarbons spills, soil and groundwater contamination in particular from lubricants, cutting fluids, and hydraulic fluids

[a]This is a comprehensive listing of potential sources and types of wastes observed at mining operations during site visits by the author.

Fig. 8 Petroleum hydrocarbon wastes including used filters (*top left*) should be segregated from other types of wastes. Unbunded storage or ineffectual bunding (*top right*) should be rectified to meet appropriate industry standards. Spill absorbent is applied to a workshop oil loss (*bottom left*) and improper storage of used oil containers at a lube bay (*bottom right*)

These areas have been dealt with extensively elsewhere and are not discussed further in this chapter (Guerin et al. 1994; Guerin 2002).

3.4.4 Fuel and Lubricant Infrastructure Management

A further example of end-of-life management of petroleum hydrocarbons is managing aging fuel and lubricant infrastructure and assets. This stage of the life-cycle poses the single biggest financial risk to mining companies from petroleum hydrocarbons. Therefore, the procurement of cost effective and technically proficient environmental consultants and civil contractors to adequately delineate and remediate contaminated soil and groundwater, is critical. Furthermore, to minimise the amount of remediation needed, asset integrity testing should also be carried out. Petroleum hydrocarbon suppliers in Australia and the wider Asia Pacific (Oceania) region have developed testing specifications that consultants and contractors are required to use for soil and groundwater assessment and remediation. Such an approach specifies the expected outcome or objective of each phase of the assessment is for the mining operation, enables standardised consultant and contractor performance, produces consistent results across operations and countries, and minimises wasted efforts in ineffective environmental assessments. Petroleum hydrocarbon suppliers, specifically their environmental and remediation teams, therefore have expertise that may help mining companies ensure that any work done by external parties to delineate contamination for improved environmental management and closure planning is effective, and properly estimated and executed. These resources can be drawn upon by the mine procurement staff as part of the fuel and lubricant supply agreements.

3.4.5 Fuel and Lubricant Spill Management

Fuel and lubricant suppliers usually have specialised expertise and dedicated resources for the management of petroleum hydrocarbon spills. Mining operations can harness this expertise by having their suppliers review on-site spill or disaster management plans, purchasing spill kits or related products, and engaging them for co-ordinated disaster management training. These services can be included in supply agreements or procured out of scope of the supply agreement. Suppliers can also assist audits, develop schedules for testing disaster management programs, and identify resources that are needed to ensure effective planning.

3.5 An Example of Engagement Between a Fuel and Lubricant Supplier and an Australian Mining Company

3.5.1 Background and Rationale

Incorporating environmental considerations into the minerals supply chain, such as those opportunities described in the preceding sections, requires a willingness for deliberate engagement between the supplier and the mine through focused planning sessions. These planning sessions need to be a collaborative effort by both parties; otherwise, if driven by the supplier only, they could be perceived as simply a move by the supplier to elicit the provision of more product purchases or services into the mine. This section reports an example of how the life-cycle approach was successfully used by a coal mining company to engage with their fuel and lubricant supplier and improve the environmental performance of its operations.

The mine produces approximately 10 Mt of coal annually and is located in the central coastal region of eastern Australia. It has approximately 10 satellite mines within approximately 200 km from the mine's head offices, and is of strategic importance to the electricity generation industry on the east coast of Australia. The mine's diesel fuel consumption is approximately 50 ML per year. Lubricant usage is estimated at approximately 2 ML per year.

3.5.2 Engagement Process

Both the supplier and mining operations agreed that there could be value in running an engagement session to systematically identify ways to improve the value each party would obtain from the relationship. The existing contract did not preclude such an engagement process. A half day engagement session was co-ordinated by the supplier on the mine's site. Participants in the engagement session were the:

- Procurement Manager from mine
- Two Operations Personnel from mine
- Environmental Officer from mine
- Key Account Manager from supplier
- Environmental Adviser from supplier
- Alternative Fuels Manager from supplier
- National Mining Marketing Co-ordinator from supplier

Approximately half an hour was dedicated to brainstorming each of the 4 stages of the supplied lubricant and fuel product life-cycle (Fig. 5). The remaining 2 h were used to review and clarify each of the ideas raised and prioritise these as actions. Actions were ranked according to whether they were high, medium or low priority (as agreed by both parties), and accountabilities for the high priority actions were set.

3.5.3 Engagement Outcomes

Table 7 lists the outcomes from the engagement session.

For each of the stages of the life-cycle, the most important issues were captured and key actions to address these issues were identified. For the supply stages of the life-cycle, alternative fuels and guaranteed supply of existing fuels were of most concern. For storage and distribution, checking compliance to relevant standards

Table 7 Outcomes from an engagement session between a petroleum hydrocarbon supplier and a mining customer in Australia

Stage of hydrocarbon life-cycle	Issue	Description of ideas and issues raised in engagement session	Priority (L/M/H)	Solution option/actions: (C) = Customer and (S) = Supplier to action
1. Supply and procurement	Alternative fuel	There is a need to identify bio-diesel sources available to the mine and whether it is feasible for the mine. The following issues were raised: • Impact on power, odour, impact on emissions (levels and quality), compliance to evolving laws/regulations. • Impact on emissions is critical issue. • Particulates were identified as critical issues. • Whether or not alternative fuels would lower engine life.	H	• Conduct trials on biofuels (C). • Regulatory changes to be monitored (C). • Any new alternative fuels to have testing prior to supply to assess the impact on emissions (S) as well as performance impact assessed (S).
	Guarantee of supply	Security of supply/availability of product: The following issues were raised: • Consolidating the purchasing of other chemicals with petroleum hydrocarbons. • Identified the importance of understanding what back-up plans are there to ensure security of fuel and lubricant supply to the customer in future.	H	Priority of this issue noted
	Delivery	The following issues were raised: • Identify if road transport is the best/only way of supplying product to the sites • Ascertain the supply footprint (i.e., total environmental cost of supplying of product). • Acknowledged that damaged drums were leading to environmental risks across the sites. • Identified need to ascertain if fuel unloading is AS 1940:2004 compliant.	M	Noted. C and S agreed that these would require attention in the contract

Table 7 (continued)

Stage of hydrocarbon life-cycle	Issue	Description of ideas and issues raised in engagement session	Priority (L/M/H)	Solution option/actions: (C) = Customer and (S) = Supplier to action
	Procurement process	Mine acknowledged the need to influence suppliers (using new safety, environmental regulations and controls). The following issues were raised: • Minimise paper (required to have fuels and lubes delivered). • Complexity of procurement process (electronic data interface—Quadrem).	M	Investigate introduction of Quadrem—an electronic trading medium (C & S)
2. Storage and internal distribution	Compliance	The mine identified the need to ascertain if: • there was scope within the fuel supply contract for conducting integrity testing of tanker and checking compliance AS 1940:2004. • the supplier was auditing the fuel facilities at the mine (was not known by the mine if this was the case).	H	• Supplier has an advisory role (S). • External audits required every 3 years according to the mine's own auditing requirements(C). • Internal environmental compliance, including latest version of AS 1940:2004 (C).
	Product handling	Safety and Environmental risk of moving product: A solution could be in piping product • Minimal handling of products should be a goal at the mine. • HSE compliance with regards to moving product (e.g., hazardous). • There is a need to provide training for key personnel at mine sites on dangerous goods management. • Piping long wall fluid to the underground mine (i.e., getting product to mine face) could reduce manual handling of 20 L containers.	H	• 20 L oil containers to be transported in self-bunded pallets (C & S). • Move to bulk (to develop pay back scenario, need to undertake a benefit cost analysis(C & S). • Investigate training options for machine operators (C & S).

Table 7 (continued)

Stage of hydrocarbon life-cycle	Issue	Description of ideas and issues raised in engagement session	Priority (L/M/H)	Solution option/actions: (C) = Customer and (S) = Supplier to action
	Storage facilities	Mine requested information on best methods for storage including bulk tanks, efficiency of product use on site, underground storage for bulk lubricants, portable underground diesel tanks, and increased availability of fleet	M	Noted. Supplier to provide best practise information (S)
3. Use and servicing	Environmental contamination	Contamination of the environment from petroleum hydrocarbons • Minimising spills and wastage. • Spillages from use and handling on site. • Due diligence and contaminated groundwater/soil. • Liability arising from contamination. • Emulsion: Where does it go? Could go to groundwater and soil. • How much soil and groundwater contamination is at the customer sites?	M-H	Noted

Table 7 (continued)

Stage of hydrocarbon life-cycle	Issue	Description of ideas and issues raised in engagement session	Priority (L/M/H)	Solution option/actions: (C) = Customer and (S) = Supplier to action
	Product contamination	On site contamination of lubricants and other fluids • Wrong product going into a machine. • Coal contamination in lubricants. • Product cleanliness is a high priority in constricted underground environment. • Have lubricant cleanliness assessment been conducted? • Breathers on bulk and 1,000 L storage—have they been assessed?	M-H	• Upgrade workshops to bulk (C). • Training for mine staff in minimising lubricant contamination (C & S). • Minimise the number of lubricant grades (C & S). • Reduce lubrication contamination points (e.g., breathers). • Storage facilities upgrade (use best practise) (C). • Drums colours should be such that it is more difficult for operators to use incorrect fluids (C & S). • Mine wanted to know if lubricants could be laundered (S). • How can water be prevented from entering into oil in machines (S)?
	Product rationalisation	Reducing number of products on site • Lubricant survey to determine best fit of products to plant. • Can number of products used be streamlined? • Product identification: wide range of oils and lubricants on site—need materials and MSDS register updates • Does customer have a strategic maintenance program and does it include lubricant types, lube longevity studies, and oil change out frequency assessment?	M	Supplier lubricant engineer to update and/or complete site lube surveys and identify opportunities to rationalise grades (S)

Table 7 (continued)

Stage of hydrocarbon life-cycle	Issue	Description of ideas and issues raised in engagement session	Priority (L/M/H)	Solution option/actions: (C) = Customer and (S) = Supplier to action
4. End of life management	Used oil	• The mine does not have any reconciliation process for determining the mass balance of lubricants across the mine. • A single waste management contractor for collection and disposal of waste oils, drums, spilled fuel, filters, oil rags, hydrocarbon waste, and grease should be considered.	H	*This links to waste removal and storage (below)*
	Drum and container disposal (waste removal)	Removal and management of containers: segregation of hydrocarbon waste and general waste • Disposal of used filters. • Separation/sorting of waste products is an issue for the mine. • Drum disposal is an issue (consolidation to bulk pods could reduce this problem). • Drum crusher has potential for safety issues at the mine. • 20 L drum and filter draining in workshop causing contamination of oil.	M	• Mass balance assessment and reporting needed at mine (C & S). • Supplier to consider certifying used oil handling contractors (S). • Oil water separator at mine: investigate upgrading (C). • Have a single waste removal contractor (C) • Drums: Segregate used drums from general waste, remove drum crusher, reduce drums by going to bulk (C) • Drums: Supplier to supply contact details for recyclers: Sims Metal (S). • Used oil filters requires disposal (C & S). • Has customer done a waste management audit (C)?
	Waste storage	Dedicated used oil storage • Mixing of oil contaminated waste with other wastes. • Used hydrocarbon tanks that can be used to collect waste oil spills on site using customer personnel.	M	Noted

and laws and product handling were the most important issues. For the use and servicing stage of the life-cycle, contamination of the environment and of product (i.e., of lubricants) was the most important issue, and for the end-of-life management stage, waste oil was the most important issue. There were several issues that were raised by the mine that related to the operation of the fuel and lubricants contract. These included: the mine inquiring as to which party was responsible for auditing the fuel and lubricant facilities; discussion regarding improving manual handling of 20 L drums at the mine; and the need for training of operators particularly in relation to product handing and cleanliness. In summary, the outcomes were identification of a range of actions, by both the supplier and the mine, to address the highest priority issues. Actions were followed through as part of normal engagement between the mine and supplier.

4 Barriers to Optimising the Contribution of Suppliers

The chapter highlights opportunities where petroleum hydrocarbon suppliers can work with mining companies to help them work towards a sustainable development agenda, including the provision of specific products and services that add significant environmental value (Table 8).

However, there are barriers to getting such initiatives embedded at mining operations as has been identified earlier in the chapter. There are technical barriers to implementing supplier-driven environmental improvements in the mining supply chain. However, the most difficult barriers, are those relating to changing culture in both the supplier and mining customer businesses. Mining companies need to appreciate and value suppliers, products, and/or services that will or can contribute materially to the achievement of their goals for long-term, sustainable operation of their mines. Such an appreciation is reflected in collaboration between suppliers and contractors that includes the engagement planning sessions described previously. While collaboration can be considered the driving force behind effective supply chain management, there is still limited evidence that companies have truly capitalised on its potential. This is a challenge for the mining industry, which in this regard is considerably behind other industries such as the food, automotive, and electronics industries.

There are other barriers that will limit the implementation of examples described in the preceding sections of this chapter. The lack of an understanding by suppliers that their long term commitment to a mining operation is critical. Such commitment will require ongoing relationship management and a 2-way commitment to improve the value provided back to the supplier and through to the mining operation. The challenge is for the supplier to remain engaged and not lose interest or margin. Also the perceived benefits of stakeholder engagement along the supplied product life-cycle can be intangible, with only limited direct evidence of impact on financial performance. The challenge for suppliers is to demonstrate the financial value in all the offerings provided to the mining company, in addition to the benefits that help the mining company become more sustainable.

Table 8 Specific opportunities to enhance end-of-life management of petroleum hydrocarbon wastes at mine sites[a]

Strategy	Examples of technologies, innovations or practise(s) which can be provided by suppliers
Prevention	Strategies to improve maintenance planning and scheduling (e.g., preventative maintenance to enhance oil life in plant and reduce wear)
	Segregation of wastes as close to the point of origin as possible (e.g., keeping contaminated waste oils separate and contaminated grease from uncontaminated grease)
	Quick release couplings on storage vessels for facilitating loading and unloading
	Use of vacuum pumps to evacuate engines, gearboxes, differentials, and fixed plant through access (drain) points
	Spill containment pallets that hold 2 × 205 L or 4 × 205 L drums for temporary storage
	Handling equipment for empty and full 205 L drums for improved ergonomics/efficiency
	Use product tanks that have integral secondary containment, vacuum generator pump system, vacuum gauges (for transfers to collection vessel), bunding (i.e., to capture leaks or spillages), and colour coding to minimise mixing products
	Keeping product storage areas away from water runoff
	Efficient (manual) refilling of engines with oil to minimise wastage from spillages
Reduction and minimisation	Automatic dispensing of grease to specified lubrication points on plant to minimise losses and increase efficiency of use
	Use of bulk containers for purchase of chemicals and petroleum hydrocarbons
	Mobile oil draining systems (i.e., tanks on wheels) for workshops
	205 L drum funnels for improving handling of waste oils to minimise splashing including lockable funnels for improved security for ensuring segregation
	Oil transfers using pumps e.g., vacuum pumps on trucks to avoid manual handling

Table 8 (continued)

Strategy	Examples of technologies, innovations or practise(s) which can be provided by suppliers
Reuse, recycling and recovery	On-site oil filtering and oil laundering
	Reuse of waste oil in explosives production (e.g., ammonium nitrate fuel oil or ANFO)
	Reuse of waste gear oil in lower value applications such as chain oil
	Blending used oil into fuel (e.g., using waste oil as a diesel extender for use in burners or in diesel engines and industrial and space heaters, where appropriate/allowable)
	Reprocessing (off-site) of used oil to produce base oil (hydraulic and gear oils)
Treatment and disposal	Separation and recovery of oil from oily wastewater
	Treatment of oily wastewater using biotreators
	Use of composting for treatment of biodegradable solid and semi-solid wastes containing petroleum hydrocarbons (such as sludges)
	Use of biodegradable cellulose fibre for absorbing spills

[a] These technologies are now available commercially and are commonly used in Australia and other developed countries.

Integrating life-cycle considerations into the purchasing process at a mine requires a commercial decision by the supplier to provide the necessary resources and linkages with its mining customer to ensure the value of both products and services is delivered. This should involve sharing of information between environmental managers or their equivalent between each organisation.

Corporate and or mine procurement groups and other critical decision makers within mining companies may not recognise their role in implementing the environmental and sustainable development goals of their company in a commercial context. Procurement staff require training and awareness of these issues. Key performance indicators need to be set by senior mine management to emphasise the importance of environmental or sustainable development concerns or values in purchasing decisions.

It is this last point that presents a challenge to the conventional negotiation process, and is currently the major hurdle to mining companies obtaining potential value from petroleum hydrocarbon suppliers. It is also this point where there are considerable opportunities for improvement. Much work is yet to be done to educate supply and procurement staff to the business value of close engagement with suppliers. Both suppliers and mine staff need to take time to broaden the relationship between their organisations so that opportunities to improve the environmental performance of the supply chain can be explored more comprehensively for their particular mining operation.

5 Conclusions and Recommendations

In relation to sustainable development in the mining industry, suppliers have a pivotal, though often unrecognised role in enabling mining industries to achieve their goals. Suppliers can enhance sustainable mining practises by helping operations become more efficient in their use of supplied products and input resources; this leads to improved business as well as environmental performance. In particular, petroleum hydrocarbon suppliers can help drive mining company performance to reduce fuel and lubricant use and therefore costs, to optimise the value gained from the supplied products, and to extend the life of products, and to ensure used products are recycled or reprocessed efficiently. Suppliers have specialised knowledge, and there are numerous examples of how they can assist mining customers throughout the life of the supplied petroleum hydrocarbons. A survey we conducted underscores the important perceived role of suppliers to the mining industry in relation to sustainable development and corporate responsibility objectives.

Opportunities for improving petroleum hydrocarbon management at each stage of the life-cycle will require close interaction between the supplier and the responsible mining operations and corporate personnel. To ascertain the potential benefits from any or all of the above stages of the petroleum hydrocarbon life-cycle at a mine, focused assessment across a mine is the first step. While there are numerous opportunities for a mining operation to enhance its own move towards sustainable development in relation to petroleum hydrocarbons, there are real barriers entrenched in the way in which suppliers are currently engaged. These barriers will need to be overcome before the benefits of closer engagement with suppliers, to sustainable development will be realised.

The author's recommendations for overcoming these barriers are:

- Increase the time planned for and dedicated to engagement between suppliers and mining companies. This should be linked with a focused engagement session between each major supplier and mining company to identify opportunities.
- Establish key performance indicators for supply and procurement personnel to ensure they are systematically exploring opportunities to incorporate environmental improvements into purchasing decisions.
- Proactive engagement needs to be undertaken by suppliers to ensure that mining companies are aware of the opportunities for improving environmental performance of their supply chain.

A relatively small investment in time and resources to engage with suppliers through engagement and planning sessions can provide a useful platform for identifying, discussing, and developing joint actions to address issues that directly affect a mining operation's objectives for sustainable development and corporate responsibility.

References

Altham J, Guerin TF (2005) Cleaner production. In: Rajaram V, Dutta S, Parameswaran K, editors. Sustainable Mining Practices. A.A. Balkema Publishers (Francis & Taylor Group Plc), London, pp. 93–120

Anonymous (1997) Green purchasing in need of further boost. Supply Management 2(9): 9

Anonymous (1999) Green actions pay, managers told. Supply Management 4(7): 42

Anonymous (2008) Minerals Council of Australia. From www.minerals.org.au

Azapagic A (2004) Developing a framework for sustainable development indicators for the mining and minerals industry. Journal of Cleaner Production 12(6): 639–662

Barton A (2006) Turn the supply chain green. Supply Management 11(25): 15

Battersby N, Greenall S, Gustafsson G (2003) Field Trials of Ecologically Acceptable Hydraulic Fluids in Sweden. Eighth Scandinavian International Conference on Fluid Power, Tampere, Finland

Blowfield M (2000) Ethical sourcing: A contribution to sustainability or a diversion? Sustainable Development 8(4): 191–200

Carlin P, Messenger A, Guerin T (2003) Why lubricant cleanliness is so important. Beneath The Surface – The Customer Magazine from Shell Global Mining September: 3–4

Christensen L (2002) The environment and its impact on the supply chain. International Journal of Retail & Distribution Management 30(11/12): 571

Ellinor R (2007) Costing the earth. Supply Management 12(2): 24

Enever J, Robertson AC (1998) Role of equipment suppliers and mining consultants in the mining cycle. Proceedings of the 1998 Annual Conference on Mining Cycle, AusIMM, April 19–23, 1998. Mount Isa, Australia, p. 247

Eskew ML (1999) Profiting through environmental supply chain management. Executive Speeches 14(1): 5

Fiksel J (1995) How to green your supply chain. Environment Today 6(2): 29–30

Garvey R (2006) Preventing downtime hinges on knowledge of lubricant condition. Pulp and Paper 80(11): 48

Guerin TF (2002) Heavy equipment maintenance wastes and environmental management in the mining industry. Journal of Environmental Management 66(2): 185–199

Guerin TF (2006a) Realising minerals theories "Down Under". Mining Environmental Management March: 12–16

Guerin TF (2006b) A survey of sustainable development initiatives in the Australian mining and minerals industry. Minerals & Energy – Raw Materials Report 20(3): 11–44

Guerin TF, Guerzoni F, Mullarky G (2003) The Role of a Major Supplier to the Australian Forestry and Timber Industry. Future Forests & Timber, Abacus Management Pty Ltd, Sydney

Guerin TF, Turner O, Tsiklieris J (2004) Moving towards sustainable development in the minerals industry – The role of a major supplier. Proceedings of the Australian Institute of Mining & Metallurgy, New Zealand Branch, Nelson, New Zealand, pp. 136–143

Guerin TF, Rhodes SH, Leiner C, Hammerschmid K, Roden S, McAllister PJ, Peck PC, Kelley BC (1994) Management and treatment of wastes from maintenance operations in the mining industry. In: Hargreaves A and Montegner J editors. Maintenance in the Mining and Metallurgical Industries. The Australasian Institute of Mining and Metallurgy and The University of Wollongong, Wollongong, pp. 255–266

Hagelaar G, van der Vorst J, Willem JM (2004) Organising life-cycles in supply chains: Linking environmental performance to managerial designs. Greener Management International 45: 27

Halme M, Anttonen M, Kuisma M, Kontoniemi N, Heino E (2007) Business models for material efficiency services: Conceptualization and application. Ecological Economics 63(1): 126

Huth WJ (1975) In-plant extension of lubricant service life from the lubricant aspect. Lubrication Engineering 31(2): 65

Lamming R, Hampson J (1996) The environment as a supply chain management issue. British Journal of Management 7: S45

Lane G, Danielson L (2001) Mining and Minerals Sustainability Survey. PriceWaterhouseCoopers and MMSD, London, England, p. 38

Lloyd M (1994) How green are my suppliers? – Buying environmental risk. Purchasing & Supply Management [PSU]: 36–39

Lutz P (2005) Rhetoric and reality of corporate greening: A view from the supply chain management function. Business Strategy and the Environment 14(2): 123–139

Mehta SK (1994) Environmental concerns in the supply chain. Purchasing & Supply Management: 26

Mercer M (2005) Lubricant service program expanded. Diesel and Gas Turbine Worldwide 37(2): 34

Messenger A, Carlin P, Guerin TF (2004a) Born to run. World Mining Equipment March: 50–53

Messenger A, Guerin TF, Carlin P (2004b) Lubricant laundering – The cost effective option. Beneath the Surface – The Customer Magazine from Shell Global Mining January: 2–3

Neadle DJ (1994) Lubricants recycling. Industrial Lubrication and Tribology 46(4): 5–7

Pavlat M (1984) Extending paper machine bearing life with silt control filtration. Practical Lubrication & Maintenance 7: 26

Pearson C (2004) Condition monitoring with thermography: Training, certification and accreditation. Insight: Non-Destructive Testing and Condition Monitoring 46(3): 164–165

Rakic R (2004) The influence of lubricants on cam failure. Tribology International 37(5): 365–373

Rao P, Holt D (2005) Do green supply chains lead to competitiveness and economic performance? International Journal of Operations & Production Management 25(9): 898–916

Robinson GJ, Hagan TN, Tucker AJ (1995) Mine owner and explosives supplier. Partners in controlling the environmental effects of blasting. AusIMM Annual Conference – Technical Proceedings, Newcastle, Australia, Australasian Inst of Mining & Metallurgy, Carlton, Australia, 23–25 March, pp. 263–269

Ryu S, Eyuboglu N (2007) The environment and its impact on satisfaction with supplier perfor-
    mance: An investigation of the mediating effects of control mechanisms from the perspective
    of the manufacturer in the U.S.A. Industrial Marketing Management 36(4): 458–469
Simpson D, Power D, Samson D (2007) Greening the automotive supply chain: A relationship
    perspective. International Journal of Operations & Production Management 27(1): 28–48
Tyler G (1997) Blueprint for green supplies. Supply Management 2(7): 36–38
West K (2006) Red spells danger. Plant Engineer (London) 50(2): 22

Global Trends in Mine Reclamation and Closure Regulation

James M. Otto

Abstract This chapter examines global trends in mine reclamation and closure regulation. It addresses the following main topics: identification of key issues; statutory requirements and voluntary guidelines; implementation during the various phases of mining; and financial assurance requirements, methods, and costs. Statutory examples are provided from a number of jurisdictions to illustrate different approaches. Best practice recommendations are offered.

1 Introduction

Every mine will close. This may be a result of the deposit being depleted or because economic or social factors make it no longer viable to mine. In prior times, neither the company nor the host government concentrated much effort on the closure process, and the result has been, in many jurisdictions, a substantial legacy of abandoned mine sites that pose health, safety, environmental, and visual blight challenges for present and future generations.

The expectations of stakeholders, including mining shareholders, mine management, employees, affected communities, regulators, consumers, and watchdog non-governmental organizations, regarding mining have evolved over the past several decades, and these expectations have led to a paradigm shift in how the closure process is typically implemented. Governments and the mining industry increasingly recognize the need of mines to obtain and maintain a social licence to operate, and maintaining such licence for the industry, over the longer term, means handling

J.M. Otto (✉)
Mineral Policy, Law and Economics, Boulder, CO 80305, USA
e-mail: jim.otto@comcast.net

An earlier version of this paper was presented and published in conference proceedings as J. Otto, "Global Trends in Mine Reclamation and Closure Regulation," *International Mining and Oil and Gas Law, Development, and Investment*, Paper No. 9A, Page 9A-1 (Rocky Mt. Min. L. Fdn. 2007).

reclamation and closure in a socially and economically acceptable manner. The term "social license to operate" means generally the acceptance of the mining activity by those stakeholders who have a significant influence on the ability of the mine to operate.

This chapter examines global trends in mine reclamation and closure regulation. It begins by identifying the key policy issues that have driven the development of such regulation and then proceeds to examine these issues from regulatory and economic perspectives. Examples are provided of various regulatory approaches and trends are described. Finally, recommendations are offered.

2 Reclamation and Closure Policy Issues

The complexity of reclamation and closure has increased over time. Fifty years ago, at closure, a mining company might focus on removal of equipment for salvage or sale, collection of accounts receivable, laying off its workforce, and extinguishing its legal title. As environmental awareness increased, attention was paid to reclamation, initially with regard to pollution containment and safety but increasingly focusing on ecologic system (i.e., flora, fauna, water, etc.) restoration. More recently, social and economic issues are being considered, particularly where mines have substantial economic linkages to the local and/or national economy. These latter issues range from employee termination compensation to implementation of sustainable community development programs. Other emerging closure issues include:

- Restoration of the mine site so as to maximize future economic potential rather than to restore a marginal ecosystem;
- At the local level, transfers of assets to communities, such as infrastructure, fire equipment, ambulances and so forth; and
- At a national level, attention to macro-revenue stabilization and intergenerational trust schemes, and the means to repay government infrastructure debt formerly serviced with mine user fees.

The particular issues faced by any one mine will be in some ways unique to that mine dependent on the nature of its operation and how it fits into the physical, ecological, social, and economic landscape. The importance of each issue will vary amongst stakeholders. Table 1 lists sample reclamation and closure issues that may be of particular interest to different stakeholders.

The terms reclamation and closure are not synonymous. Reclamation is the process whereby a mine's landform and ecology are altered to achieve a planned state. Closure includes actions such as the physical shutdown of the mine and the host of activities, such as final reclamation, equipment removal, community disengagement, employee severance, debt settlement, and so forth that occur when the company determines that it will no longer mine the property.

Table 1 Sample reclamation and closure issues

Reclamation/closure issue	Stakeholder
Profit maximization (cost minimization)	Company
Termination of liability	
Social licence to operate in the future	
Termination benefits	Employee
Retraining	
Future status of pension	
Future economic viability	Community
Economic diversification	
Unemployment	
Environment	
Debt repayments	
Recurrent costs previously paid by the mine	
Transfer of infrastructure and facilities	
Safe site	National government
Ecological recovery	
Landform transformation	
Revenue stabilization	
Intergenerational benefit and cost distribution	
Macro-economic linkages	
Infrastructure debt repayment (ports, rail, water, roads)	

3 Timing of Reclamation and Closure Activities

Whether required by regulation or not, planning for reclamation and closure commences before a mine is built. Engineering and fiscal feasibility models are used to analyze the economic viability of the mine and to optimize it. Key parameters are looked at, such as mine size and life. The mine's life, and year of closure, will be determined, although for many mines, that life may be later extended. The mining method will be carefully studied and a mine design chosen.

Today, this process will also take into account reclamation and closure. For example, where should tailings dumps be sited so as to avoid costly future acid drainage problems? Should lime be added to tailings on an ongoing basis to reduce future acid generation or will a permanent water treatment plant be needed? Where will stripped top-soil be segregated and stored for future topping applications? In addition, the scheduling of reclamation activities throughout the life of the mine (progressive reclamation) will result in less onerous final closure costs, and must be planned at the outset. Mine reclamation and closure has now become an integral part of the mine design process.

It is common for a mine's initially planned life to be extended through progressive additions to reserves as a result of additional brown-field exploration, or technology improvements that allow lower grades of mineralization that were previously sub-economic to become economic to mine. Thus, ideal planning for reclamation and closure is evolutionary, commencing with feasibility and adapting as circumstances change throughout the life of the mine. For example, waste dumps and tailings are usually located close to a mine to minimize transport costs, but land utilized for disposal may later prove to hold economic reserves. Additionally, what was once considered waste may become economic to exploit as prices go up or new recovery technologies emerge.

Some mines are amenable to on-going reclamation as mining leaves one part of the deposit to work another part (for example, strip mining of coal or mineral sands), while others will incur most reclamation costs when mining ceases (for example, an open-pit mine). Companies are also aware that even in circumstances where reclamation and closure regulatory requirements are minimal or absent at mine startup, this may not be the case at the time the mine closes. Retrofitting a mine to comply with new requirements can be more costly than building the mine initially with forethought for reclamation and closure. Today's mining engineers graduate having learned that good mine design includes engineering plans and cost estimates for reclamation and closure.

Not every mining company is concerned about its social licence to operate, and particularly amongst smaller companies, many of whom operate marginally economic mines, reclamation may be considered to be a non-essential part of their business. Even among large companies with highly profitable mines, every mine manager strives annually to minimize costs and non-revenue generating activities, such as reclamation and plans for closure, and the manager may delay implementation of reclamation anticipating that that burden can be borne by his successor manager. To ensure that mines do not become abandoned without adequate planning processes, design, and implementation, almost all jurisdictions now impose reclamation and closure regulatory requirements that must be approved before permits and licenses to mine are issued. These requirements commonly take the form of statutory or administrative law but can be imposed through negotiated mining agreements. Additionally, the mining industry, through various associations, has developed its own voluntary guidelines.

4 Regulatory Systems

Given the wide range of issues involved with reclamation and closure it is common to see a number of different laws applied to these activities. In jurisdictions that supplement their codified laws with ad hoc mining agreements, as is common in Africa and some Asian jurisdictions, detailed requirements can be brought together in a holistic, unified approach. In jurisdictions without such agreements, reclamation and closure requirements usually arise from three primary sources: labor law, mining and environmental law, and good practice guidelines.

4.1 Labor Law Issues

When a mine closes or approaches closure, workers will no longer be required and laws relating to the labor force may impose substantial obligations on the employer. These obligations may include advance termination notice requirements, as well as the need to provide termination benefits. In some jurisdictions, for example many jurisdictions in Europe, advance notice periods can be substantial, as can termination benefits. Termination benefits can range from payment to a worker of a lump-sum based on a multiple of his monthly salary (e.g., an amount based on his or her length of employment), a transition period of health insurance coverage, and so forth. Because these costs will occur after the mine is no longer generating revenues, workers' termination benefits may be at risk if the company does not have other sources of revenue, has not made real rather than mere accounting provision for these costs, or declares bankruptcy. In contrast, in many developing jurisdictions, termination obligations imposed by labor law are minimal or absent.

Statutory requirements to provide training so that laid-off workers can find work in another employment field are rare but have been a feature in the closure of some state-owned mines (for example, coal mine closures in Poland, Ukraine, and the United Kingdom; Egorova and Otto 1998). Some companies voluntarily provide laid-off workers with such training or assist them with locating new employment, whether or not this is required by the regulatory system. For mines that provide a pension scheme to their workers, another issue is how that scheme will be handled once the mine closes.

If a mine is major employer in the area, closure-induced unemployment impacts can be great and are not limited to the immediate work force. For every worker employed by a mine more than one job is created in the wider economy, and this is commonly referred to as the employment multiplier effect. Depending on both the nature of a mine and the local economy, the employment multiplier effect may be small or large. Schodde (2006) reports that in data drawn from 16 studies, the employment multiplier effect for mines ranged from a low of 1.39 to a high of 3.34 at the local level (i.e., for every 1 worker employed by a mine, 1.39–3.34 additional jobs were created in the local economy), and was slightly higher at the national level. In a study of the Chilean mining industry, Aroca (2000) found that the employment multiplier effect ranged from just over 1–1.7 for older mines, and could be as high as 5.7 for new mines that use a high level of contract labor or who outsource services done internally at older mines. There is no standardized approach used in computing employment multiplier effects, and the above numbers should be used with caution. However, regardless of the calculation methodology, in almost any mining situation, for every worker employed by the mine, other employment will be generated in the wider economy.

Many mines worldwide are increasingly relying on casual or subcontracted labor and services. In India, where the practice of hiring "casual" laborers rather than permanent employees is widespread, a government mining policy study reported that the ratio of direct to indirect employment in the mining sector is 1:10 (Government of India 2006). Labor law-related mine closure requirements do not extend to indirect labor impacts, although mitigation of such impacts may be done as part of a voluntary or mandatory community sustainable development effort. Depending on the level of mandated termination benefits, substantial savings can be realized by mines that operate with casual and sub-contracted labor. Where labor unions are strong or regulatory restrictions are in place, this practice may be minimal, but in other jurisdictions it is a growing trend and an item for political discussion. Regardless of whether a mine operates with only its own employees or uses subcontracted labor, at closure, unemployment impacts can be large where the local economy is dependent on wages paid directly or indirectly through mine-created wealth.

4.2 Mining and Environmental Law Issues

Almost all jurisdictions that have substantial mining industries impose statutory reclamation and closure obligations. The general nature of these obligations is usually similar, although the means to impose the obligations differ. A key policy issue is a determination of which government ministry or department (hereafter referred to as agency) will have responsibility for reclamation oversight, with the usual choice being either the agency responsible for mines or the one responsible for environmental matters. If the objective of reclamation was solely land form stabilization and safety (for example, the sealing of shafts), then the mining agency would be the obvious best choice. However, most jurisdictions also seek some degree of environmental stabilization, and have concerns about issues such as ecological restoration and maintenance of water quality, topics where the environmental agency may be a better choice.

If the regulatory responsibility is with the agency responsible for mining, reclamation obligations are usually outlined in the mining law and in regulations associated with the mining law. One benefit of this approach is that the mining agency will have a pool of experts (geologists, mining engineers, mineral economists) familiar with mining. In contrast, one of the challenges for jurisdictions that impose reclamation obligations through a law regulated by an environmental agency is that requisite internal expertise regarding mining may be lacking. For example, staff employed by the agency responsible for the environment may be unable to estimate the severity of future acid generation and drainage from tailings impoundments, or the amount of financial assurance required to ensure re-contouring of the mined landform. Some jurisdictions have special regulations pertaining to mine reclamation that are part of the environmental law, whereas others have only general environmental requirements that do not specifically address mines.

Dual agency responsibility for reclamation is not rare, although most legal systems in theory strive to place statutory oversight of a regulatory issue with a single agency. A typical example of dual regulatory authority is found in Mozambique. There, under the mining environment regulation[1] that was brought into force

---

[1] Mining Environmental Regulation, Decree No. 26/2004 of 20 August.

through its Council of Ministers in 2004, both the minister responsible for mines and the minister responsible for environment are given respective powers and duties relating to reclamation and closure. A mining company is required to obtain an environmental licence based on an environmental plan prepared in conformance with the environmental mining regulations (including a reclamation plan and provision for a reclamation financial assurance), which is approved and issued by the minister responsible for the environment, and the company cannot commence mining until it obtains such an environmental licence. Failure to comply with the terms of the environmental licence (the approved environmental/reclamation plan) can lead to cancellation of the mining license by the minister responsible for the mining code.

Another dual agency approach is where one agency is given the regulatory authority for reclamation, although that agency is required to consult on certain matters with other agencies.

The means to ensure that reclamation and closure are done in an approved manner vary, and enforcement provisions in the following extract from the British Colombia mining law are typical.

> (8) If the owner, agent, manager or permittee fails to perform and complete the program for reclamation or comply with the conditions of the permit to the satisfaction of the chief inspector, the chief inspector, after giving notice to remedy the failure, may do one or more of the following:
> (a)   order the owner, agent, manager or permittee to stop the mining operation;
> (b)   apply all or part of the security toward payment of the cost of the work required to be performed or completed;
> (c)   close the mine;
> (d)   cancel the permit.[2]

4.3 Good Practice: Voluntary and Industry Guidelines

While many governments have acted to impose reclamation and closure requirements, in some jurisdictions such requirements are still absent or are stated generally without adequate means to encourage compliance. Recognizing that industry's "social licence to operate" could be imperilled if mining companies continue to leave behind mines that are abandoned in such a way as to impose health, safety, and aesthetic concerns, some companies have adopted internal policies and various industry organizations have worked to create voluntary guidelines. Being internal or voluntary, such guidelines are regarded as "soft laws" lacking the means for enforcing violations, but they do demonstrate the increasing willingness of industry to adhere to standards of good practice.

---

[2]British Colombia, Canada, Article 10(8), Mines Act [RSBC 1996] Chapter 293.

4.3.1 Corporate Policies

Many of the world's larger mining companies have internal policies regarding reclamation and closure which they implement at all their operations. Newmont Mining Corporation's (2007) policy position is typical:

> Each Newmont operation will develop, during the design phase, and implement closure and reclamation plans that provide for long-term environmental stability and suitable post-mining beneficial land-uses. . . . Closure and reclamation are the completion of the life cycle of a mining operation. The post closure environmental condition and beneficial land uses of Newmont sites is the Company's lasting legacy. Consideration of closure and reclamation must occur during the design of a project and must be included as an integral component during the life of the operation. The operations will be managed during the life of each facility in a manner consistent with full implementation of the closure and reclamation plan. The cost of reclamation and closure must also be included in all front-end project evaluations. Adequate financial provisioning for closure and reclamation will be provided for all operations.

Some major mining companies, such as Vale (Vale 2003) and BHP Billiton (BHP Billiton 2004), have developed detailed internal closure guideline procedures. The comprehensive guide used by BHP Billiton outlines procedures that begin with exploration and are ongoing to post-closure. The company's closure plans for each major operation are updated every 3 years, including reclamation cost estimates. However, not all large mining companies have strong internal reclamation and closure policies, and many smaller companies have none.

4.3.2 Industry Organization Guidelines

In many jurisdictions there exist national mining associations whose members include the most important mining companies that operate there. To belong to some of these associations, members are expected to voluntarily comply with the policies and principles that the association develops. During the 1990s many of these organizations developed principles that dealt with reclamation and closure issues; others did not. One of the most active of these associations is the Mining Association of Canada. This association has adopted general principles that include commitments that commence with exploration and continue to closure, but like many such organizations, their general principles lack specific statements on reclamation and closure commitments:

> We will demonstrate leadership worldwide by:
> - Involving communities of interest in the design and implementation of our Towards Sustainable Mining initiative;
> - Proactively seeking, engaging and supporting dialogue regarding our operations;
> - Fostering leadership throughout our companies to achieve sustainable resource stewardship wherever we operate;
> - Conducting all facets of our business with excellence, transparency and accountability;
> - Protecting the health and safety of our employees, contractors and communities;
> - Contributing to global initiatives to promote the production, use and recycling of metals and minerals in a safe and environmentally responsible manner;

- Seeking to minimize the impact of our operations on the environment and biodiversity, through all stages of development, from exploration to closure;
- Working with our communities of interest to address legacy issues, such as orphaned and abandoned mines ...
  (Mining Association of Canada 2004)

Some national mining organizations which had previously adopted or were working on reclamation and closure principles for their members have instead decided to become members themselves of the International Council on Mining and Metals (ICMM). At the start of 2008, ICMM members included 18 of the world's largest mining and metal companies, and 30 national mining and global commodities associations. Like the Mining Association of Canada, ICMM's stated objectives and principles address issues related to reclamation and closure, but lack the level of detail that would provide a global operational standard. For example, with regard to reclamation and closure, members are committed to a number of sub-principles including the following:

- Contribute to community development from project development through closure in collaboration with host communities and their representative.
- Rehabilitate land disturbed or occupied by operations in accordance with appropriate post-mining land uses.[3]

Work by the ICMM continues and brings together thinking and research from all around the globe. Its goals and principles are general in nature, and may remain that way, but they might also evolve into detailed industry standards. For example, ICCM has been investigating ways in which companies can plan for reclamation and closure costs, and has stated:

As outlined in one of our guiding sustainable development principles, our members are committed to designing and planning all operations so that adequate resources are available to close them down properly. Consequently, ICMM members view the provision of financial assurance as an important aspect of the mining industry's commitment to sustainable development... Our work is meant to address the issues related to financial assurance policies. The need for financial assurance is clear, but choosing the right approach needs careful consideration. Policies that meet environmental objectives can and should also be compatible with a healthy investment climate and be financially efficient. We hope our work will help both the industry and government achieve that.[4]

The organization has completed a major financial assurances study, and work such as this may eventually evolve into an industry standard and act to influence not only its members but also government regulators. The primary weakness in a voluntary principles approach that is not backed by standards is that it is difficult to determine whether a member is in compliance or not. Additionally, if a member is found to not be in compliance, enforcement mechanisms are lacking other than removal from the membership roster. Ultimately, voluntary standards may transition to hard law regulations imposed by host jurisdictions.

---

[3]Position statements supporting ICMM guiding principle numbers 6 and 9, ICMM www.icmm. com, accessed on June 23, 2008.

[4]ICMM www.icmm.com, accessed on June 23, 2008.

4.3.3 The Equator Principles: Lending Covenants

While most voluntary approaches to closure principles and obligations arise from organizations consisting of mining companies, policies of affiliated industries can also have an appreciable impact. An example is the Equator Principles. These are a set of principles originally adopted by 10 major banks in 2003. In 2006, the principles were amended, and today there are over 45 Equator Principles Financial Institutions (EPFI). These include many of the financial institutions that lend or provide financial services to the mining industry. The key part of the principles is a commitment by each EPFI that it will not lend to a borrower (e.g., mining company) that does not abide by the principles. The commitment reads:

> These Principles are intended to serve as a common baseline and framework for the implementation by each EPFI of its own internal social and environmental policies, procedures and standards related to its project financing activities. We will not provide loans to projects where the borrower will not or is unable to comply with our respective social and environmental policies and procedures that implement the Equator Principles.[5]

The principles address a wide range of environmental and social issues. What makes them effective is the requirement that the participating lender incorporate covenants in the lending contractual documents that require the borrowing company to do certain things. The covenants section of the general principles reads:

> Principle 8: Covenants
> An important strength of the Principles is the incorporation of covenants linked to compliance. For Category A and B projects,[6] the borrower will covenant in financing documentation:
> a)   to comply with all relevant host country social and environmental laws, regulations and permits in all material respects;
> b)   to comply with the AP [Action Plan] (where applicable) during the construction and operation of the project in all material respects;
> c)   to provide periodic reports in a format agreed with EPFIs (with the frequency of these reports proportionate to the severity of impacts, or as required by law, but not less than annually), prepared by in-house staff or third party experts, that i) document compliance with the AP (where applicable), and ii) provide representation of compliance with relevant local, state and host country social and environmental laws, regulations and permits; and
> d)   to decommission the facilities, where applicable and appropriate, in accordance with an agreed decommissioning plan.[7]

From a reclamation and closure perspective, the requirement for EPFIs to impose contractual requirements on borrowers to implement an action plan (which implements a wide range of environmental and social objectives) and a decommissioning plan is unprecedented, and clearly indicates that mining companies that do not

---

[5] www.equator-principles.com, accessed on February 26, 2007.

[6] *Category A*—Projects with potential significant adverse social or environmental impacts that are diverse, irreversible or unprecedented; *Category B*—Projects with potential limited adverse social or environmental impacts that are few in number, generally site-specific, largely reversible and readily addressed through mitigation measures.

[7] www.equator-principles.com, accessed on February 26, 2007.

comply will find that their funding options are severely constrained. The adoption of the Equator Principles by so many of the world's banks is a major step in institutionalizing this reclamation and closure-related requirement as international law. The Equator Principles are relatively new, and as they evolve may become more stringent. For example, the requirement 'to comply with all relevant host country social and environmental laws, regulations and permits in all material respects' has little meaning in a jurisdiction where such laws are absent or weak.

5 Timing

Mining consists of various phases, commencing with exploration, then transitioning to development (construction), exploitation (mining), closure, and post closure. All these phases may be subject to reclamation/closure regulation.

5.1 Exploration Phase

Most jurisdictions do not address reclamation issues during the exploration stage, and if they do, obligations are most often levied on the exploration concession holder through the mining law and its regulations, rather than under the environmental law. The concession holder may be required, for example, to cap and plug all drill holes, to contour and replace lost vegetation disturbed by temporary roads, and disassemble and remove evidence of camps.

However, this simple approach is not always the case and an increasing number of jurisdictions require that some type of environmental mitigation plan, including reclamation measures, be prepared by the company and be approved by the government. The following extract is from the Philippine mining regulations:

> Applicants for Exploration Permits . . . shall submit to the Bureau . . . an EWP [Environmental Work Program] . . . detailing the environmental impact control and rehabilitation activities proposed during the exploration period including the costs to enable sufficient financial resources to be allocated to meet the environmental and rehabilitation commitments. . . . The EWP shall provide a description of the expected and considered acceptable impacts and shall set out the environmental protection and enhancement strategies based on best practice in environmental management in mineral exploration. It shall include a statement on post-exploration land use potential for various types of disturbed land and extend to the completion of the commitments in the rehabilitation of the disturbed land in a technically, socially and environmentally competent manner. The program shall be based on acceptable, practical and achievable options and demonstrated practice. Finally, the program shall include implementation schedules, system of environmental compliance guarantees, monitoring, reporting and cost provisions.[8]

---

[8] Section 168 Environmental Work Program (EWP), DENR Administrative Order No. 96-40 Series of 1996 of December 19, 1996, Revised Implementing Rules and Regulations of Republic Act No. 7942, otherwise known as the "Philippine Mining Act of 1995".

One of the challenges in requiring an exploration phase reclamation plan, either as a separate plan or as part of an environmental protection plan, is that one rarely knows where, and the extent to which, future disturbances will take place. Exploration is progressive, and commonly it is only late in the exploration concession term that a discovery is made and environment and landform altering activities, such as diamond drilling, commence. It is probably for this reason that most jurisdictions still achieve exploration reclamation objectives by merely mandating that certain types of disturbances named in the mining law be addressed by the explorer, rather than requiring some sort of formal reclamation plan for each exploration concession area.

Although governmental regulation may be lacking in many jurisdictions, voluntary good practices are increasingly being applied. An example is the Environmental Excellence in Exploration (e3)[9] program launched by the Prospectors and Developers Association of Canada in 2003. This web-based compilation of worldwide good practice gives guidance for environmental and community engagement activities during the exploration phase. It is estimated to now be in use by over 2,000 practitioners representing over 80% of the world's exploration expenditures, and has been translated into French, Spanish, and Portuguese, with Russian and Chinese (Mandarin) in progress. It represents a huge body of knowledge on mitigation and remediation at the exploration stage.

5.2 Development Phase

In most jurisdictions, regulatory requirements concerning reclamation and closure begin during the development phase. The development phase for the purposes of this paper encompasses activities including feasibility, mine planning and construction prior to the commencement of commercial mineral production. In many jurisdictions it is immediately before or during the development phase that the government requires the preparation of mine and environmental mitigation plans. Some jurisdictions integrate reclamation and closure requirements into one or both of these plans, or can require a stand-alone reclamation plan. Likewise, a closure plan can be associated with a mine plan, environmental mitigation plan, or reclamation plan, or can be a separate requirement.

A concern of the private sector is the extent to which the government exerts control. Does the company need to submit such plans for informational purposes, or does the government need to approve such plans? If approval is required, how long does approval take? To what extent can the government require changes to a plan? Can the company proceed with development before a plan is approved? What recourse does the company have should a plan be rejected? These are sensitive issues because they address a balancing of sovereign power and responsibility with the need for mines to operate in a predictable regulatory environment. In jurisdictions

---

[9]http://www.pdac.ca, accessed on June 17, 2008.

with well established mining industries and a track record of administrative procedures, company fears of arbitrary or delayed decisions will be lower than in nations
with a newly emerging industry. Additionally, where government agencies have little experience with mining, or generally lack capacity and knowledge of matters
such as the environment, reclamation, sustainable development, and so forth, policy
makers may be hesitant to implement legal requirements, or officers may hesitate to
grant approvals.

5.2.1 Uniformity or Discrimination?

Not every mineral sector project is identical, and while some are similar, there is
generally great diversity. For example, an artisanal miner or a small quarry will have
very different environmental and closure impacts than a world-class porphyry copper mine. The former will not be able to bear a large and involved reclamation and
closure study and plan, whereas the later will have the ability to fund a very comprehensive approach including a detailed environmental impact assessment. Ideally,
every mine regardless of size should be able to clean up after it is done, or it should
not be permitted to open. However, the reality is that in many developing jurisdictions large segments of the mining industry, for example artisanal and small scale
mining and quarrying, operate without any legal authorization. The question then for
regulatory policy-makers is: to what extent should reclamation and closure requirements, including financial assurances, be uniform or different for various types and
scales of operations?

One approach is for the mining law to differentiate between different types and
scales of mining, either on the basis of the concession type or its physical parameters. In many jurisdictions that have legitimized their artisanal miners, such miners
do not file any sort of plan for approval. Instead they are simply prohibited from
certain actions, or are required by law to take certain actions. The following excerpt
from the Eritrean mining act is typical of how governments may impose different
reclamation requirements depending on the licence type:

30.   Health, Safety end Environmental Protection

. . .

4)   Prior to expiration or termination of the license, the licensee shall fill, close, block
or otherwise render safe all tunnels, pits and other installations of a potentially dangerous
nature.

5)   The holder of a mining license shall progressively restore or reclaim the land covered
by the license and, if applicable, a lease so that, prior to termination of the license, the area
has been completely restored or reclaimed for beneficial future use, except if such progressive restoration or reclamation is not feasible as determined by the Licensing Authority in
writing or the Licensing Authority approves otherwise.

6)   The holder of an artisanal mining license shall take all environmental protection measures commensurate to his operations; in particular he shall fill pits and plant trees and shall
not be allowed to use mercury or similar materials in his operation.[10]

---

[10]Eritrea, Proclamation No. 68/1995, A Proclamation to Promote the Development of Mineral
Resources.

In Eritrea, there is no requirement for a mine to submit a reclamation plan for government approval regardless of scale of operation. However, if a company would like to enter into a mining agreement with the government, additional requirements are imposed in the mining agreement. This is a common practice throughout much of Africa.

> 10.3 *Reclamation of Mining License area*. The Company shall reclaim area disturbed by Agreement Operations under a Mining License in accordance with the Environmental Management and Rehabilitation Program Plan approved pursuant to Article ... of this Agreement before the expiry of that Mining Licence.
> 10.4 *Financial assurances*. The Company is obligated to provide and maintain amounts of financial assurances in the types as approved in the Environmental Management and Rehabilitation Program Plan pursuant to Article ....[11]

Another example of discrimination based on scale of operation is found in Brazil, where lesser requirements apply to small mines:

> The Brazilian Environmental Policy (BEP) is the responsibility of the Ministry of the Environment (Ministério de Meio Ambiente, MMA) and is executed at three levels: federal, state, and municipal. ... Environmental licenses required for all mining activities are managed at the federal level. The environmental legislation applied to mining is basically consolidated in the following environmental requirements: environmental impact study (EIA), environmental licensing (LA), and plan for recovery of degraded areas (PRAD). An EIA applies to mining projects of any mineral substance. The LA is mandatory for installing, expanding, and operating any mining activity under the systems of mining concession or licensing. A PRAD requires suitable technical solutions to rehabilitate the soil and other aspects of the environment that might be degraded by mining operations. In recognition that an EIA can represent a substantial financial burden for smaller project, a company can undertake a less detailed form of EIA called "Environmental Diagnostic Report". This report is submitted to the Environmental State Council (Conselho Estadual do Meio Ambiente, CONSEMA), which has the authority to waive the need of a full EIA.[12]

Discrimination can also be based on the type of mine. In the United States, for example, closure requirements for hard-rock mines differ from those for coal mines.

5.2.2 Preparation of the Reclamation and Closure Plan

In most jurisdictions it is the responsibility of the mining company to prepare any required reclamation and closure plan. However, in some jurisdictions there is a concern that the preparation of such plans, which are intended to minimize environmental and social costs, by a bottom-line – oriented mining company is less than an optimal approach. Some jurisdictions thus require, particularly where the reclamation/closure plan is part of an environmental mitigation plan, that the plan be prepared by a third party who may or may not need to be approved by the government.

Another issue is to what extent third party stakeholders will be involved in reclamation and closure planning. In many jurisdictions, both the mining company and

---

[11] Eritrea, Model Mining Agreement 2006.

[12] Brazilian environmental and reclamation requirements are summarized at www.diagem.com/en/1032/index.php, accessed on February 19, 2007.

government have respective roles, but as the concept of sustainable development takes hold, other interested parties are increasingly being involved. Such involvement can be voluntary or mandatory. The two examples below come from Nigeria and Queensland, Australia:

> Community Development Agreement must have certain provisions. The Company shall negotiate with the host community the terms of the Community Development Agreement, and such agreement shall include at least the following provisions: ... the obligations of the Company with regard to the host community including but not necessarily limited to: ... consult with the community in the development of a Mining Operation closure plan that seeks to prepare the community for the eventual closure of the Company's Mining Operations.[13]

> Stakeholders such as the background landowner, immediate neighbors, employees, government agencies and any other persons directly affected by the operation and closure of the mine should be involved throughout the planning process. The most effective way of involving external stakeholders is through the establishment of a forum such as a closure committee or group that meets formally or informally to discuss issues such as are appropriate. This will ensure that members of the community and government agencies are kept abreast of the future plans for the mine thereby providing for a transparent process.[14]

Although many jurisdictions are moving to adopt sustainable development objectives, including addressing the impacts of closure on local communities, many have yet to implement these policies at the statutory level. For example, in India, the National Mineral Policy specifically mentions rehabilitation and closure and emphasizes the need for the process to take into account human rehabilitation:

> Whenever mine closure becomes necessary, it should be orderly and systematic and so planned as to help the workers and the dependent community rehabilitate themselves without undue hardship.[15]

While the Indian mining act and regulations requires that every mine have an approved progressive mine closure plan and a final mine closure plan, which in part address rehabilitation and reclamation, the implementing rules do not indicate that the plan is required to take into account sustainable development issues. There is also no requirement to involve third party stakeholders.[16]

5.3 Exploitation Phase

Once a mine is up and running, impacts on the environment, landscape, communities, and economy will increase. Some impacts will be foreseeable and will have been planned for, but others may emerge as the project develops. In many jurisdictions it is required that various plans that relate to reclamation and closure be updated either on a periodic basis, for example every 3 years, or when an event

---

[13] Nigeria, Article 12 Model Mineral Agreement 2006.

[14] Guidelines for Mine Closure Planning in Queensland, Queensland Mining Council, 2001, p. 6.

[15] India, National Mineral Policy, §7.15.

[16] India, MCR§22(5)(va) as defined in MCDR§23A, MCDR§23B.

occurs, such as an expansion or material change in the mining method or plan. It is usually during this phase that financial assurances need to be provided and maintained, and these are discussed below. Many mines will undertake progressive reclamation work during this phase, whether required by regulation to do so or not.

5.4 Closure Phase

Regulatory requirements for closure usually provide for a notice period and for certain minimum statutory requirements to be met, in addition to specific requirements levied under a closure or reclamation plan. The following provision from the Mongolian mining act is typical:

> Article 41. Requirements for closure of a mine
> 41.1.   Before closure of a mine, mining license holders shall take preparatory measures pursuant to regulations of the specialized inspection institution to protect the health and safety of local residents. License holders shall inform the specialized inspection institution by an official letter that the mine shall be closed in whole or in part, at least one year prior to any such closure, and the following measures must be implemented during the preparatory period:
>> 41.1.1.   to take all necessary measures to ensure safe use of the mining site and mining claim for non-mining purposes and to protect the environment;
>> 41.1.2.   to seal safely and fence off the parts of the mining area that may be dangerous during the use of the area for other purposes;
>> 41.1.3.   to remove all machinery, equipment and other property from the mining area except as permitted by local administrative bodies or the specialized inspection institution.
> 41.2.   Mining license holders shall prepare a detailed map on an appropriate scale showing dangerous or potentially dangerous areas created by mining operations, shall submit the map to the specialized inspection institution and the local Governor, and shall place the necessary warnings and markings in the vicinity of the mining claim.[17]

Most jurisdictions with a legacy of abandoned mines now require formally submitted and approved reclamation and closure plans prior to granting mining licences, which impose legal obligations on the mining company. Others, however, simply use general requirements such as in the Mongolian example.

5.5 Post-Closure Phase

Many reclamation/closure regulatory systems do not deal with the post-closure phase of the operation. Once closure has taken place and the mine is reclaimed, the mining concession is cancelled and the company is released from further obligations and, in some cases, liability arising from statutory law. In other jurisdictions, liability may be ongoing but would arise from tort actions rather than through liability imposed under the mining or environmental law. To avoid tort actions, companies

---

[17]Mongolia, Minerals Law, as revised 2006.

operating in this type of legal environment may chose a business structure where the entity holding the mining concessions ceases to exist after the mining concession ends. Some regulatory systems impose on-going open-ended obligations.

For some types of mines, substantial environmental impacts can be generated long after the mine ceases production. Such impacts are commonly associated with contaminated water, which may be acidic and/or laden with toxic metals, or toxic dust leaving the site. For example, it was common in the past to place tailings impoundments conveniently at the bottom of a valley, but here they may be subject to natural runoff and erosion. More modern practices, where practical, place acid generating or toxic tailings away from run-off. It may be necessary to put into place and to maintain water control systems, such as diversion or collection ditches, water-impervious caps to keep water away from tailings, or to treat water exiting the site, and such measures may need to be maintained for decades. The challenge for both the regulator and the company is how to structure a way to ensure that money is available on an on-going basis for these post-closure activities. Such costs can be substantial, and without adequate regulatory obligations the government may be left with paying recurrent costs. For instance, Pring (2001) reported that construction of a plant to treat heavy metal-laden water run-off from the Argo Tunnel in Colorado (a tunnel used in the past by several mines to supply a mill) cost USD4.7 million in 1998, and its annual operating costs are about $1.1 million, of which the US Environmental Protection Agency pays 90% and State of Colorado 10%.

6 Financial Assurances

6.1 Why Are Financial Assurances Required?

There a number of ways in which governments can influence a company to comply with reclamation and closure requirements, but some of these ways are not effective in some circumstances. For example, under many mining laws, failure to comply with statutory obligations can lead to the imposition of financial penalties (fines), administrative or judicial orders for specific performance, or cancellation of the mining concession. Such measures can be effective in encouraging or forcing on-going reclamation efforts at a profit-making mine, but may be ineffective in the event that the company is under financial duress, is dissolved, or declares bankruptcy. When a mine closes, mineral revenues will cease, but for many types of mines this is the time when substantial reclamation and closure costs will be incurred.

To ensure that reclamation and closure costs will be covered, many jurisdictions require that some form of financial assurance be provided by mining companies. For example, the Queensland Government (2003, p. 1) states in its guidelines that the purpose of a financial assurance is "To ensure funds are available to the government should the company default on environmental requirements or become bankrupt." Governments, particularly those with a legacy of abandoned mines or an active environmental movement, increasingly require financial assurances. In some jurisdictions the requirement may simply allow the company to make an accounting provision for these costs, but the clear trend is to require a more secure and certain form of assurance.

Mining is quite diverse and a type of assurance that works well for one type of mine, for instance a tin dredge that continually backfills as it mines, may be ill-suited for another type, such as an open-pit copper mine. There are many approaches to financial assurances, including methods that are intended to provide a protected "working fund" from which reclamation costs are charged on an ongoing basis, and "insurance" type methods where, in the case of company default, a sum of money becomes available to the government for reclamation purposes. The two approaches are not mutually exclusive and can be used in combination.

In the remainder of this section, financial assurance methods are introduced, then such methods are assessed to provide a generalized comparison of their respective strengths, weaknesses, and opportunity costs (e.g., their costs taking into account the time value of money).

6.2 Method for Determining the Amount

The method for determining the amount of financial assurance that must be provided by a mining company for a project varies widely from jurisdiction to jurisdiction. Two prevalent systems are assurances based on mine area, and assurances based on cost estimates.

In the some states within the United States, for certain types of mines the financial assurance amount is determined by multiplying the area occupied by the mine times a statutory amount per unit area.[18] The advantage of this method is that it is relatively easy to administer, and does not require a detailed reclamation/closure cost estimate. Obviously, its weakness is that the assurance amount will be higher than necessary for some mines and too low for others. If too low and the mine does not meet its reclamation/closure obligations, the government may be left without adequate funds to reclaim the mine site.

The second method, where the amount of financial assurance is based on an estimate of the closure and reclamation cost for the operation, is potentially less risky for governments but imposes a higher administrative burden on government and additional costs on the mining company, who will need to bear the expense of preparing the cost estimate. While this author knows of no study comparing the number of jurisdictions using one method or the other, many, but not all, developing jurisdictions tend to use the cost-estimate based system.

A key part of any financial assurance system is how much financial assurance needs to be provided at any given time in the project's life. This is important because

---

[18]Pring (2001), in a study of abandoned mines in the United States, reported that reclamation and closure for hard-rock mines is dealt with primarily at the state level in the US federal system, and each state is free to tailor its own closure-cleanup program or have none at all.

many forms of financial assurances have costs associated with them that are in proportion to amount of the assurance. The amount that would be required to meet reclamation and closure costs in, say, the 5th year of a mining project will most likely be much less than in its 15th year. Should the financial assurance then be the amount based on costs in the 15th year or should the amount be adjusted taking into account the cost-time profile of the mine?

Different regulatory systems approach this question in different ways. To illustrate this, the following six figures have been prepared, with each figure showing the identical cost to meet reclamation and closure costs if the mine were to close in any year over a 15 year period. Each figure then shows the amount of financial assurance required in each year by a different regulatory approach. The figures are intended as conceptual aids such as might be useful to policymakers, and do not reflect any actual mine. In a robust economic comparison of various financial assurance methods for an individual mine, factors such as the time value of money, cost escalation, technological advancements, and other factors affecting reclamation costs can be built into more accurate predictive models. However, in this author's experience, policy makers are wary of complex approaches involving time value of money concepts and predictions about cost escalation, and are most comfortable with conceptual aids such as these six figures.

In practice, a mine's life and reclamation costs may, and probably will, change over time, raising the additional issue of how financial assurance requirements may also need to be amended over the life of a mine. Some jurisdictions provide that the amount of financial assurance, regardless of the method used to allocate it over time, must be adjusted periodically.

The most costly approach for a company is usually when the amount of assurance required in any year is equal to the estimated final closure cost or total closure cost (Fig. 1).

In an ideal system, the amount of financial assurance required at any point in time would exactly match the amount required to meet reclamation and closure costs should the mine close at that point in time (Fig. 2). This method, however,

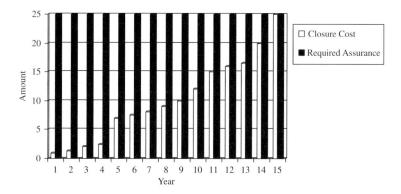

Fig. 1 Total (or final) cost upfront assurance method

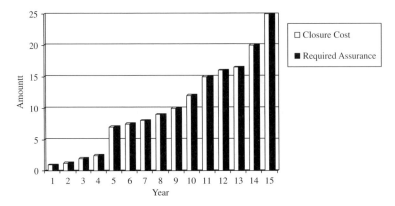

Fig. 2 Annual readjustment assurance method

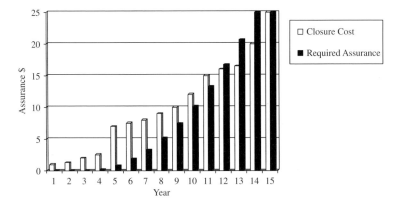

Fig. 3 Fixed schedule assurance method

requires that cost estimates must be prepared and reviewed annually, and this is a heavy burden for the mining companies, and if the country has many mines, for the government.

A variety of methods that "adjust" the amount of assurance required are used by governments. For example, in Quebec, Canada, the amount of financial assurance required in any year is set according to a fixed schedule of percentages of the estimated closure cost (Fig. 3).

A simple method is to set the level of financial assurance required at any point in time as the total cost divided by a set time period that applies to all mines. For example, as shown in Fig. 4, the amount of assurance required increases by 1/10th of the total closure cost each year for 10 years.

Another method is where the amount of assurance required increases by a factor of 1 divided by the mine life times the total estimated closure cost (Fig. 5). The challenge for this method is that the mine life may change, and if the period becomes suddenly shorter, there may be a shortfall.

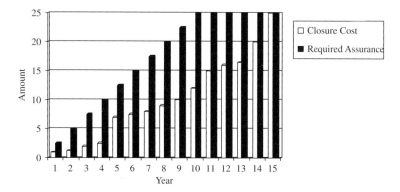

Fig. 4 Equal annual increase over set time period assurance method

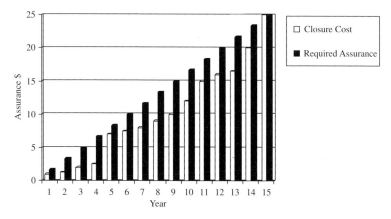

Fig. 5 Equal annual assurance method based on life of mine

A popular method is to require that the company periodically submit a revised reclamation/closure plan, including a cost estimate. The amount of assurance required in any year equals the estimated closure cost if the mine were to close at the end of the current planning period. Figure 6 illustrates this approach for a 5 year revision schedule.

6.3 Which Party Determines Reclamation and Closure Costs?

Regulatory systems differ in regard to which party is responsible for estimating the amount of reclamation and closure costs. As was previously mentioned, in some jurisdictions these costs are not relevant because either no financial assurance is required or such amount is based simply on the land area occupied by a type of mine. For other jurisdictions where the amount of assurance is based on estimated reclamation and closure costs, there may be a concern that a mining company may

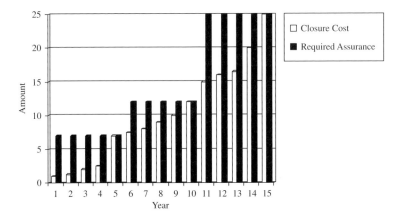

Fig. 6 Periodic readjustment assurance method

underestimate these costs in order to minimize the amount of the financial assurance required. For this reason, in some jurisdictions the reclamation/closure plan, often as part of an environmental management plan, must be prepared by a qualified third party. However, in most jurisdictions, the reclamation and closure plans, along with their cost estimates, are prepared by the company but are subject to approval by the relevant government authority. Where the company's estimate is found to be too low, some jurisdictions allow the authority to determine the amount of required assurance.

Another important policy consideration is whether the cost estimate should be based on the assumption that the company will do the reclamation, or the cost that third parties would charge to do it in the event that the company does not. Cost estimates based on the company doing the work are almost always less than estimates based on a third party effort. It is best practice policy to impose a system that bases reclamation cost estimates upon third party costs. In this way, if the company goes under, a third party can come in and close the site out satisfactorily.

6.4 What Forms of Financial Assurance Are Used?

There are many forms that financial assurances can take. Many governments offer a number of choices for companies to choose from, or allow companies to propose a method which may be approved at the discretion of a government officer on a project by project basis. The most commonly used forms include those listed in the financial assurances allowed by the Arizona (USA) government (Kuipers 2000, p. II.8):

Any or a combination of:

(1) Surety bond
(2) Certificate of deposit

Table 2 Allowed assurance methods in selected US states

| Type of assurance | Is form of assurance allowable in this state? | | | | | |
	Arizona	California	Colorado	Nevada	New Mexico	Utah
Surety	Yes	Yes	Yes	Yes	Yes	Yes
Letter of credit	Yes	Yes	Yes	Yes	Yes	Yes
Cash and certificates of deposit	Yes	Yes	Yes		Yes	Yes
Trust funds and deeds	Yes	Yes	Yes	Yes		
Corporate guarantee	Yes		Yes	Yes	Yes	Yes
Other	Yes	Yes	Yes	Yes	Yes	Yes

Source: Derived from Kuipers (2000, p. I-2, Table 1.1.1).

(3) Trust fund with pay-in period
(4) Letter of credit
(5) Insurance policy
(6) Certificate of self insurance
(7) Cash deposited with State Treasurer
(8) Evidence of ability to meet corporate financial test or corporate guarantees as provided by 40 Code of Federal Regulations 264.143(f)
(9) Annuities
(10) Additional financial assurance mechanisms that are acceptable to the Inspector.

Not all US states allow such a broad range of assurance methods, but as indicated in Table 2, many do.

Some jurisdictions, such as some in the US, provide a list of specific types of financial assurances that are acceptable, while other jurisdictions provide a government official with the authority to determine whether a form is acceptable or not. The approach used in New South Wales, Australia, provides a great deal of flexibility, including cash deposits, other kinds of assurances approved by the minister in regulations or guidelines, or other types allowed by the mines department on a case-by-case basis. The relevant provision states:

28 Security to be lodged in respect of mineral claim
For the purposes of section 190(4)(b) of the Act, the security to be lodged with the mining registrar:
(a) must be:
(i) in the form of cash, or
(ii) in the form of a security instrument of a kind approved by the Minister, being an instrument issued by an authorized deposit-taking institution, or
(iii) in any other form that the Director-General may approve, and
(b) must be of an amount that the Director-General may determine.[19]

---

[19] New South Wales, Australia, Mining Regulation 2003.

The most common forms of financial assurance allowed by governments are described below, followed by an indication of their relative cost to a mining company.

6.4.1 Cash Deposits

Most jurisdictions with substantial mining allow a financial assurance in the form of a cash deposit, deposited with a government department or, more commonly, in an interest-bearing account with a commercial bank in the name of the government. Advantages and disadvantages of this approach include:

Advantages to government:

- Very low risk;
- Easy to administer;
- Immediately accessible;
- Interest earned helps offset inflationary impacts.

Disadvantages to company:

- Very high opportunity cost;
- Will make project less economically viable because economic measures such as IRR and NPV will be lower;
- Potential for abuse by corrupt government officials who access the money.

Regulatory systems differ on how earnings from cash deposits are handled. Some systems require that the interest earned on the deposit stays with the deposit to offset the effects of inflation thus reducing the need to readjust the amount periodically.

6.4.2 Certificate of Deposit

A certificate of deposit is similar in most ways to a cash deposit. The company purchases a time deposit instrument in the name of government, usually with a maturity period of 1 year, which automatically renews on maturity unless part or all is withdrawn on maturity by the government. Advantages and disadvantages include:

Advantages to government:

- Very low risk;
- Easy to administer;
- Accessible upon maturity;
- Interest earned helps offset inflationary impacts;
- Easy to increase or decrease the amount of assurance.

Disadvantages to company:

- Very high opportunity cost (ties up money);
- Will make project less economically viable because economic measures such as IRR and NPV will be lower.

The following example is from the Quebec, Canada, mining regulations:

115. The person ... shall submit a guarantee ... in one of the following forms...
...
(3) Guaranteed investment certificates or term deposit certificates, in Canadian dollars, *issued on behalf of the Minister of Finance* by a bank.... Such certificates shall have a term of at least 12 months, shall be *automatically renewable* until the issue of a certificate of release provided for in section ... of the Act and shall not include any reservation in respect of redemption during its term.[20]

6.4.3 Irrevocable Letter of Credit (LOC)

The usual attributes of an irrevocable letter of credit are as follows:

- It is an agreement between a bank and a mining company whereby the bank will pay cash funds to the government under the terms of the letter of credit.
- The bank is required to honor government claims in compliance with the terms of the letter of credit.
- Any changes to the letter of credit must be agreed by all parties.
- It is issued for 1 year and is automatically extended if the company satisfies an annual financial review by the bank.
- If the bank decides not to extend the LOC, the government is notified and has an option to draw down the full value of the LOC.

Advantages and disadvantages include:

Advantages to government

- Low risk;
- Easy to administer;
- Accessible upon breach.

Advantages/disadvantages to company:

- Low cost for most companies (some may not be able to obtain it);
- Bank fees are negotiable; will be based on creditworthiness of the company, usually less than 1% of face value; the company may be required to provide an indemnity or guarantee;
- Reduces the mining company's borrowing capacity.

---

[20]Quebec, Mining Act SQ 1965, c.34, as amended.

The following extract is from the Ontario, Canada, mining act:

145(1) The financial assurance required as part of a closure plan shall be in one of the following forms...:

....

2. A letter of credit from a bank named in Schedule I to the Bank Act (Canada)[21]

6.4.4 Performance/Surety Bond

The usual attributes of a performance/security bond are as follows:

- It is an agreement between an insurance company (or bank) and a mining company whereby the insurance company (or bank) will pay cash funds to the government under the terms of the bond.
- If issued by an insurance company it may be called a surety bond.
- The government is the beneficiary under the bond.
- Any changes to the bond must be agreed by all parties.
- It issued for a defined time period; may be for any length of time; and is automatically extended if the company satisfies a financial review by the insurance company (or bank).
- If the insurance company (or bank) decides not to extend the bond, the government is notified and has an option to draw down the full value of the bond.

One drawback of a performance or surety bond approach is that not all companies can obtain such a bond. Availability may depend on a mining company's creditworthiness. In some instances, a mining company may not qualify unless, if a subsidiary, the parent company provides the bank a guarantee, or the company deposits cash equal to the maximum amount guaranteed by the bond. Advantages and disadvantages include:

Advantages to government:

- Low risk;
- Easy to administer;
- Accessible upon breach;
- Can be for life of the mine; however, the amount of the bond may need to be adjusted from time to time.

Advantages/disadvantages to company:

- Low cost for most companies: fees are negotiable; will be based on creditworthiness of company; usually less than 2% of face value; the company may be required to provide an indemnity or guarantee;
- Reduces the company's borrowing capacity.

---

[21]Ontario, Mining Act R.S.O. 1190, c.M-14, as amended.

The Australian state of Queensland has issued an official guideline, "Financial Assurance for Mining Activities: Appendix C", that contains a standardized two page Bond agreement acceptable to the State.[22] Such a standardized sample agreement is not common among governments, but can be a useful means to let the company know what the government's expectations are.

6.4.5 Insurance Policy

Insurance policies to cover reclamation and closure obligations are relatively new, but have been obtained by some mining companies in the United States. Other jurisdictions list insurance as an acceptable means of meeting financial assurance obligations. In its simplest form, premiums are paid by the mining company to the insurance company, which may then invest them as it sees fit, and claims are paid out to the government by the insurance company. The insurance company may enter into arrangements with other insurance companies to spread the risk (reinsurance). In some instances, the insurance company may be owned by the mining company or by an affiliate of the mining company. This type of "captive policy" acts to segregate the closure risk internally within the company.

The following case study description of a more complex approach is provided by Miller (2005, p. 54):

> The package is a combination of three main components: a conventional surety bond, accumulation of cash within the policy, and insurance protection for overruns and for changing requirements. The insurance company works with the operator's best estimate of reclamation cost (normally lower than the amount of security regulators would want). In addition, the insurer "optimizes the approach by a combination of pre-funding and credit evaluation." ... Each year's reclamation cost is estimated, taking inflation into account. A discount is applied, depending on the operator's credit rating. The adjusted net present value (NPV) is the basis for the annual payment. AIG [the insurance company] guarantees the reclamation. From the funds deposited, AIG issues the required security bonds to government. And AIG pays the actual reclamation costs. At the end of the project, if there is a surplus in the account, it goes back to the operator. If a deficit, AIG pays. Term of the policy is for 15–30 years (30 maximum). After the fifth year the operator can commute the fund and withdraw if there are better options.

Advantages and disadvantages include:

Advantages:

- Less expensive early in the project than a cash or certificate of deposit type trust fund;
- Less administration than is required for many forms of trust fund;
- Insurance premiums may be a tax-deductible expense;
- Amount of insurance can be changed to reflect changed circumstances.

---

[22] www.env.qld.gov.au/environmental_management/land/mining/guidelines/

Disadvantages:

- Not available or tested in most jurisdictions;
- Some companies may not be acceptable to an insurer, or costs may be too high to be practical;
- Insured must pay insurance brokerage fees.

6.4.6 Trust and Reclamation Funds

There are many types of trust funds and their requirements and attributes can vary from one jurisdiction to another. Attributes of a trust fund may include:

- The purpose is to build over time a fund of money and invested assets that will be used for rehabilitation and other stated purposes.
- A trust fund is established and the mining company pays amounts into the fund on a predetermined schedule.
- The beneficiary may be the mining company, the government, or typically both.
- The trust fund is held, invested, and administered by a third party trustee (such as a bank or trust company).
- The trustee manages the fund: approves disbursements back to the company for activities in an approved plan. Although a third party trustee is desirable, other management possibilities exist such as a Board or Committee with government, company, and community representation.
- The fund is subject to periodic audit.
- After closure and reclamation obligations have been met, excess funds go to the company, or are held and disbursed for post-closure costs (such as acid drainage control).
- Not all jurisdictions have laws governing the creation of trusts.

A fund that is established and where the disbursing party is the government instead of a third party trustee is sometimes referred to as a reclamation or rehabilitation fund. Advantages and disadvantages of such funds include:

Advantages/disadvantages to government:

- Low risk: cash or assets accumulate;
- Administration: time-consuming;
- Easily understood by the public: high visibility;
- Can be used even after the company is released from the closure obligation.

Advantages/disadvantages to company:

- Payments can be scheduled over time;

- May be costly depending on whether payments into the fund coincide closely with disbursements out;
- If linked to an approved plan and administered by a neutral third party trustee, risks are low;
- Transparent;
- Administration can be time-consuming;
- Not all civil law jurisdictions have laws governing the creation of trusts.

Unlike some other financial assurance types, some forms of trust and reclamation funds provide an actual account that is drawn upon to pay the company, or a third party, to undertake rehabilitation and closure activities. This is in contrast to most other financial assurances that are there only to cover the possibility that the mining company will not fulfill its legal obligations. In the event of a company failure, the likelihood is great that the money in the trust will be used for its original purpose; other types of assurances may simply be paid out to government, with a lower likelihood that they will be spent as originally intended (e.g., this year, hospitals are a greater priority). Three examples of reclamation funds are provided. The first example is a third party trustee type of fund, and comes from the Quebec (Canada) Mining Regulations[23]:

115 The person . . . shall submit a guarantee to the Minister in one of the following forms. . .
. . .(7) a trust constituted in accordance with the provisions of the Civil Code . . . meeting the following requirements
    (a) The purpose of the trust is to ensure completion of the work provided in the rehabilitation and restoration plan;
    (b) The Minister of Finance and the [mining company] are the beneficiaries of the trust;
    (c) The trust is a bank. . .;
    (d) The trust patrimony is comprised only of sums in cash, or of bonds or certificates. . .
116 In the case of a trust, interest yielded by the trust patrimony belongs to the trust. Interest kept as part of the trust patrimony shall not be used as payment of the guarantee.

The current trend in some jurisdictions to require more than just environmental protection, and physical site reclamation is exemplified by the second example, which is an excerpt from the Philippines mining law where a trust fund is also used to address social issues.[24]

Section 69 Environmental Protection
Every contractor shall undertake an environmental protection and enhancement program covering the period of the mineral agreement or permit. Such environmental program shall be incorporated in the work program which the contractor or permittee shall submit as an accompanying document to the application for a mineral agreement or permit. The work program shall include not only plans relative to mining operations but also to rehabilitation, regeneration, revegetation and reforestation of mineralized areas, slope stabilization of mined-out and tailings covered areas, aquaculture, watershed development and water conservation; and socioeconomic development.

---

[23] Quebec Mining Regulations c.M-13.1, r.2.
[24] Philippine Mining Act of 1995 Republic Act No. 7942.

Section 71 Rehabilitation
Contractors and permittees shall technically and biologically rehabilitate the excavated,
mined-out, tailings covered and disturbed areas to the condition of environmental safety,
as may be provided in the implementing rules and regulations of this Act. A mine rehabilitation fund shall be created, based on the contractor's approved work program, and shall be
deposited as a trust fund in a government depository bank and used for physical and social
rehabilitation of areas and communities affected by mining activities and for research on
the social, technical and preventive aspects of rehabilitation. Failure to fulfill the above
obligation shall mean immediate suspension or closure of the mining activities of the con-
tractor/permittee concerned.

The third example, which does not have a third party trustee, is from British
Columbia, Canada:[25]

10 (1) Before starting any work in, on or about a mine, the owner, agent, manager or any
other person must hold a permit issued by the chief inspector and, as part of the application
for the permit, there must be filed with an inspector a plan outlining the details of the
proposed work and a program for the conservation of cultural heritage resources and for the
protection and reclamation of the land, watercourses and cultural heritage resources affected
by the mine, including the information, particulars and maps established by the regulations
or the code.
(2) Despite subsection (1), if the chief inspector is satisfied that, because of the nature of
the proposed work, it is not necessary to obtain a permit, the chief inspector may exempt in
writing the owner, agent or manager from the requirement to comply with this section with
respect to the proposed work.
(3) If the chief inspector considers the application for a permit is satisfactory the chief
inspector may issue the permit, and the permit may contain conditions that the chief inspector considers necessary.
(4) The chief inspector may, as a condition of issuing a permit under subsection (3), require
that the owner, agent, manager or permittee give security in the amount and form, and
subject to conditions, specified by the chief inspector
(a) for mine reclamation, and
(b) to provide for protection of, and mitigation of damage to, watercourses and cultural
heritage resources affected by the mine.
(5) If required by the chief inspector, the owner, agent, manager or permittee, in each year,
must deposit security in an amount and form satisfactory to the chief inspector so that,
together with the deposit under subsection (4) and calculated over the estimated life of the
mine, there will be money necessary to perform and carry out properly
(a) all the conditions of the permit relating to the matters referred to in subsection (4) at the
proper time, and
(b) all the orders and directions of the chief inspector or an inspector respecting the fulfill-
ment of the conditions relating to the matters referred to in subsection (4).
(6) The owner, agent, manager or permittee, or an inspector, may apply to the chief inspector
for a revision of the conditions or an extension of the term of a permit issued under this
section, and the chief inspector may revise the conditions or extend the term, as the case
may be.
(7) For the purposes of subsection (6), if the chief inspector considers it necessary, the chief
inspector may impose additional conditions or changes in the existing conditions, including
changes to the security required or the term of the permit, with or without an application
under this section. . . .
Mine reclamation fund

---

[25] British Columbia, Mines Act [RSBC 1996] Chapter 293.

12 (1) In this section, "**fund**" means the mine reclamation fund.

(2) The Lieutenant Governor in Council may, by regulation, establish a fund to be known as the mine reclamation fund into which must be paid security, that is in the form of money, given by the owner, agent or manager of a mine under section 10.

(3) Money received from an owner, agent or a manager must be credited to a separate account in the fund in the name of the mine.

(4) The minister may requisition payments from an account in the fund

(a) to refund money and interest earned on it to the owner, agent or manager of a mine from time to time if in the opinion of the chief inspector it is no longer required for mine reclamation and protection of, and mitigation of damage to, land and watercourses affected by the mine, or

(b) to pay for the cost of work required under section 10 (8) (b).

Miller (2005) reports for the ICMM that, from an industry perspective, the principles in the two following Tables 3 and 4 should be taken into consideration by government when designing trust fund type requirements.

6.4.7 Balance Sheet Test

Some jurisdictions apply a "balance sheet" test, and if a mining company meets that test it is exempted from normal financial assurance requirements. Such a test may be used in place of financial assurances or used to qualify for a reduction of the amount of financial assurance required. The policy assumption is that if a company meets defined tests of financial strength, the probability that it will not comply with reclamation and closure requirements is small and no (or a reduced) financial assurance is required. Tests of financial strength vary from nation to nation and may include measures based on net worth, working capital, bond rating, level and the value of other domestic assets. Usually companies meeting a balance sheet test must submit annual audited reports to confirm that their financial status has not degraded, and if it has, they may need to provide a financial assurance. Advantages and disadvantages of a balance sheet test approach include:

Advantages/disadvantages to government:

- Simple to administer;
- A good fit for many large companies, but not suitable for small companies;
- A company can go from great strength to great weakness very quickly, thus creating a risk for the government that the company will not be able to meet expenditure requirements;
- If a balance sheet test is not met after mining commences, the company may be unable to provide another form of assurance.

Advantages/disadvantages to company:

- Very low cost; just annual reporting.

Table 3 Criteria for the efficient design of a fund

Issue	Industry recommended principle
Site-specific basis for fund	Each mine should be assessed individually and the security required should reflect the costs and risks associated with reclaiming that site
Basis for cost estimates	Estimated costs should be based on careful engineering and technical studies accompanied by formal risk assessments to take into account the probabilities and consequences of alternative scenarios
Responsible management of reclamation	The design of the fund should encourage mining companies to manage their reclamation programs in an active and responsible manner, in order to control costs and to develop innovative technical solutions to reclamation challenges
Similarity to pension fund	The principles for setting up a fund should be similar to those used to establish a pension fund
Investment policy	Investment policy should permit investments that optimize the risk-return ratio, bearing in mind that the fund is a long-term investment
Investment manager	The fund should be managed by an investment manager selected by the company. The company should at the same time have the option of managing the fund internally with reasonable guidelines, as with a pension fund
Monitoring legislation	Legislation modelled on pension statutes or other similar legislation can be used to monitor performance of the fund and to ensure compliance with investment policy
Choice of financing mechanism	As justified by the circumstances, a company should have the option to determine which government-authorized financing mechanism (or combination of mechanisms) represents efficient use of the company's capital
Expenses deductible for tax	Where a government-mandated mine reclamation fund is required, payments into the fund should be allowed as a deductible expense at the time they are made for purposes of income tax and mining taxes
Fund income sheltered from tax	Income generated by a fund should be tax-sheltered until withdrawn
Investment management fees	All investment management costs should be financed from the proceeds of the fund
Fund trustee	An independent third party, such as a trust company, is an acceptable trustee of a fund
Sole government control	The mining industry is opposed to the government having sole control over the management of investments in a fund

Source: Adapted from Miller (2005, Table A1).

The following example of balance sheet test approach is taken from the Ontario (Canada) mining regulation:[26]

---

[26]Ontario Regulation 240/00, amended to O. Reg 282/03.

Table 4 Guidelines for the review and audit of a fund

Issue	Industry recommended principle
Site-specific basis for fund	Each mine should be assessed individually and the security required should reflect the costs and risks associated with reclaiming that site
Basis for cost estimates	Estimated costs should be based on careful engineering and technical studies accompanied by formal risk assessments to take into account the probabilities and consequences of alternative scenarios
Periodic review or audit	A periodic review or audit of activities of a fund is necessary to ensure appropriate disbursement and use of funds pursuant to the approved decommissioning plan
Scope of audit	An audit would include the preparation of financial statements and a technical review of work performed. It should also include, where applicable, a reassessment of reclamation requirements and funding contributions
Conduct of audit	An appropriate panel should be engaged to undertake the review and audit, using technical, engineering, legal and actuarial expertise
Frequency	A review should be held with a stated frequency, which could be from 3 to 5 years, or more frequently if deemed desirable by the government or the company
Disposition of surplus funds	Any surplus funds determined by a review should be returned, net of appropriate tax adjustments, to the company

Source: Adapted from Miller (2005, Table A2).

16(1) If a proponent's credit rating meets or exceeds two of the following credit ratings from the stated credit rating services, the proponent complies with the corporate financial test . . .

. . . A3 from Moody's . . . A1 from Standard and Poors. . .

(4) If, as a result of a downgrading. . ., the proponent no longer complies with the corporate financial test under . . . (1), the proponent shall . . . provide the Director with financial assurance in the form and in the amount identified in. . .

6.4.8 Third Party Guarantee

Another approach to financial assurances is where a third party guarantees to pay for the mining company's reclamation and closure costs. The form of such guarantee can vary, such as in a financial guarantee or an indemnity agreement. This method allows, for example, a parent or affiliate of the mining company to fulfill the financial assurance obligation. The parent or affiliate may have funds, or meet financial tests, that may satisfy the financial assurance requirement. Usually in such cases, the government obtains the guarantee directly from the parent or affiliate, or has legal recourse against them.

Advantages and disadvantages of a third party guarantee:

Advantages/disadvantages to government:

- Practical solution where the mining company on its own cannot raise the financial assurance;
- Requires administrative oversight of an additional party;
- May require third party to obtain bond or LOC;
- Question of fees between affiliated parties may arise (may affect taxes).

Advantages/disadvantages to company:

- Do not have to deal with an "outside" entity;
- Low cost.

6.4.9 Comparison of Assurance Method Costs

It is not always apparent to government and companies what the economic implications are of using one form of financial assurance versus another. This is in part due to the related concepts of opportunity cost and the time value of money. When a company has funds tied up in a financial assurance that it is not immediately expending to meet its reclamation and closure obligations, it is not able to invest that money in other economic opportunities, and the company will thus forgo returns that could have been made had those funds been available for investing. Additionally, if funds are tied up in the early years of the project rather than later in the project, then this has an impact on project economics because of the time value of money.

To gain an appreciation of the direct and opportunity costs to a company posed by different assurance methods, the author has used a copper mine model to roughly estimate the cost of five assurance methods. The model used has been reported widely in previous studies published by the author, and is perhaps best known for its use in comparative taxation studies (Otto et al. 2000). The mine model assumes a 20 year mine life and an actual closure/reclamation cost expenditure of US$25 million in the last year of the 20 year mine life. Five systems are assessed including: cash or time deposit, bond, letter of credit, trust or reclamation fund, and financial test. Table 5 summarizes the direct and opportunity costs under each type of financial assurance. A discount rate of 12 percent was used as the basis for the time value of money.

Table 5 Cost impact of five assurance methods (million $)

	Financial assurance method				
	Cash deposit	Performance bond	Letter of credit	Reclamation fund	Balance sheet financial test
Closure cost	25	25	25	25	25
Fees for assurance	Nil	10	5	Nil	0
Lost earnings (opportunity cost)	115	18	9	45	0
Total cost	140	53	39	70	25

Note: Assumptions include as follows:

Cash or time deposit method:
- One time payment at startup: $25 million.
- Deductible for income tax at time of payment.
- Earnings on the deposit: accrue at 3% and are returned to the company.
- Company can earn 12% on its usual investments.

Performance bond:
- Amount of required bond: $25 million.
- Annual insurance/banker's fee: 2% of bond.
- Annual fee: deductible for income tax.
- Company can earn 12% on its usual investments.

Letter of credit:
- Amount of required bond: $25 million.
- Annual letter of credit fee: 1% of credit.
- Annual fee: deductible for income tax.
- Company can earn 12% on its usual investments.

Trust or reclamation fund:
- Amount of required bond: $25 million.
- Annual deposit to fund: equal annual installments ($25m/mine life).
- Earnings on the deposit: accrue at 3% and are returned to the company.
- Annual deposit: deductible for income tax.
- Company can earn 12% on its usual investments.

7 Summary and Recommendations

7.1 Reclamation and Closure Obligations

Almost all jurisdictions today impose reclamation and closure obligations on mines, and, while the nature of such obligations vary, in most jurisdictions they can be summarized as follows:

The objectives of mine closure planning are:

- to reduce or eliminate adverse environmental effects once the mine ceases operations
- to establish physical and biological conditions which meet regulatory requirements

– to insure the closed mine does not pose an unacceptable risk to public health and safety.[27]

In some jurisdictions, reclamation and closure obligations now also take into consideration objectives relating to sustainable development, local communities, and social impacts. As pressure continues to be applied on mines to mine their product not only in a safe but also an environmentally sound way, it can be anticipated that more governments will increasingly move from sustainable development policy to sustainable development-related statutory and contractual obligations. In the future, reclamation and closure obligations will expand in many jurisdictions to meet objectives that will work toward the mining industry maintaining a social licence to operate. As such obligations come into force, implementation costs will increase and methods for their estimation will move from ad hoc approaches to approaches grounded in voluntary or hard-law procedures.

7.2 Stakeholder Needs

To meet the needs of key stakeholders the following best practice recommendations are offered:

- Rules and procedures regulating reclamation and closure should be clear and unambiguous.
- Governments should publish guidelines in addition to statutory requirements.
- The legal framework for reclamation and closure should evolve in a predictable manner to allow for innovation and changed circumstances.
- Reclamation and closure rules will in most cases need to exempt artisanal mining operations.
- A reclamation/closure plan should be part of the mine plan, or be tied to it and evolve with it.
- Clear lines of authority and oversight should be assigned, and closely coordinated.
- The system adopted should be practical and not exceed the government's administrative capacity.
- Financial assurance should be required and should be:

  • Practical; and
  • Affordable for most prospective mines (attention must be paid to the costs of various options and to timing the amount of assurance to be provided).

- A variety of financial assurance options should be allowed to accommodate:

---

[27]Guidelines for Mine Closure Planning in Queensland, Queensland Mining Council, 2001, p. 6.

- Different reclamation scenarios depending on mine type, location, size, and impact; artisanal miners should be exempt;
- The capacity of companies to qualify for various forms of financial assurance.

- Companies should be allowed to propose the form of financial assurance, selecting from a government approved set of options: the government should have the power to agree or not to the proposed method. The types of financial assurance allowed should include at least the following options:

  - Cash account
  - Time deposits
  - Letter of credit
  - Performance bond
  - Balance sheet test
  - Trust or reclamation fund
  - Third party guarantee
  - Other methods acceptable to the Director of Mines/Environment.

- The amount of the financial assurance should be defined through the reclamation/closure plan process.
- The level of financial assurance required in any 1 year should depend, to the extent practical, on the cost to implement the plan for the current planning period (this will guarantee a satisfactory level of assurance without imposing unnecessary costs on the company).
- The amount of financial assurance required should change over time to reflect on-going reclamation efforts, changes in mine capacity, or changes in technology (for example, to require a reclamation/closure plan update every 5 years, or at any time that a change takes place that would cause a need for a new plan).

References

Aroca P (2000) Diversification and Development in Local Economies Based on Mining Sector: The Case of the Chilean II Region. UNCTAD: Proceedings of the Regional Workshop on Growth and Diversification in Mineral Economies, Cape Town 7–9 November 2000

BHP Billiton (2004) Closure Standard, Issue 1.0 July 2004. www.bhpbilliton.com, accessed February 26, 2007

Egorova V, Otto J (1998) Restructuring the Ukrainian coal industry. Resources Policy 24:157–166

Government of India (2006) National Mineral Policy. New Delhi: Report of the Planning Commission, Government of India

Kuipers J (2000). Hardrock Reclamation Bonding Practices in the Western United States. Boulder: National Wildlife Federation

Miller CG (2005) Financial Assurance for Mine Closure and Reclamation. International Council on Mining and Metals, available at www.icmm.com, accessed on June 23, 2007

Mining Association of Canada (2004) Towards Sustainable Mining Guiding Principles 2004. www.mining.ca, accessed on February 19, 2007

Newmont Mining Corporation (2007) Environmental Policy. www.newmont.com/en/social/
    policy/environment/index.asp, accessed on February 26, 2007

Otto J, Beraun M, Cordes J (2000) Global Mining Taxation Comparative Study, 2nd edition.
    Golden: Colorado School of Mines and distributed the UNCTAD and the World Bank

Pring G (2001) The Role of Government and the Private Sector in Reclaiming Abandoned Mines:
    Lessons from the U.S. Experience. Tokyo: Metal Mining Agency of Japan

Queensland Government (2003) Guideline 17 – Financial assurance for mining activities.
    Queensland Environmental Protection Agency

Schodde R (2006) The Role of World-Class Mines in Wealth Creation. Society of Economic
    Geologists, Special Publication 12:71–90

Vale (2003) Guide to Mine Closures. www.cvrd.com.br, accessed February 26, 2007

Environmental Liability in the Mining Sector: Prospects for Sustainable Development in the Democratic Republic of the Congo

Marie Mazalto

While protecting the environment is expensive, the cost of doing nothing will be much more expensive.

(Annan 2002)

Abstract The return of international financial institutions to the Democratic Republic of the Congo (DRC), after many years of dictatorship and war, occurred in 2001. Since 2002, the transition government, in collaboration with financial backers, has implemented a series of large-scale reforms, beginning with the adoption of a new Mining Law, conceived to encourage foreign direct investment in the country. During this process, the mining sector has experienced a new cycle of growth, and the spectre has arisen of substantial environmental liabilities with consequences for the entire country. This chapter raises the question of environmental challenges for the country in terms of sustainable development. This leads to the thorny issue of who is responsible for solving this problem: the state, private companies, and/or international financial institutions? These issues are approached using the results of an environmental audit of Gécamines (a state-owned mining company) financed by the World Bank Trust Fund, and conducted by Canadian consultant group SNC-Lavalin. The audit was submitted to the Bank and government of the DRC for review in April 2003, and was made public in 2004.

M. Mazalto (✉)
Groupe de Recherche sur les Activités minières en Afrique (GRAMA), Université du Québec à Montréal (UQAM), Montréal, QC, H3C 3P8, Canada
e-mail: marie.mazalto@cirad.fr

J.P. Richards (ed.), *Mining, Society, and a Sustainable World*,
DOI 10.1007/978-3-642-01103-0_11, © Springer-Verlag Berlin Heidelberg 2009

1 Introduction

In February 2007, the United Nations Development Programme (UNDP)[1] and the United Nations Environment Programme (UNEP), at their meeting in Nairobi, together set up the Poverty and Environmental Facility to contribute to poverty reduction by preserving the environment. The objective of this initiative was to help countries of the South integrate "healthy environmental management" into their policies to fight against poverty. UNDP administrator Kermel Davis declared at this meeting that "Eliminating poverty and hunger and protecting the environment are inseparable" (UNDP 2007, p. 1). This reflected the international community's adoption of the UN Millennium Development Goals to "integrate the principles of sustainable development into country policies and programmes; reverse loss of environmental resource."[2] This action added to a growing list of international codes of conduct drawn up for investors (Mazalto and Campbell 2004), within a logic of voluntary compliance.

From the beginning of the 1990s, international financial institutions (IFIs) elaborated the first safeguard measures in development project financing, aimed at minimising the impacts of certain projects on communities and the environment. States and investors were required to provide social and environment assessments as a pre-condition for development projects. In 2006, the World Bank Group (WBG) revised an internal list of very rigorous performance norms, among the strictest in the world, with regard to the social and environmental viability of projects to be financed (International Finance Corporation 2007).

The beginning of the first decade of 2000 was marked increased questioning of companies' social responsibility, initiated by the IFIs[3] and taken up by the international community, regional development banks, and certain private banks. Countries receiving aid were to adopt the triple imperatives of sustainable development: economic development, social development, and preservation of the environment. In cases where pressure groups opposed large-scale investment projects for environmental reasons, the community of donors sets itself the objective of reconciling these two essential dimensions.

---

[1]This chapter contains numerous abbreviations, particularly those of the plethora of agencies in the DRC involved in the mining sector. To help the reader, all abbreviations are listed in an Annex at the end of the article. In the case of the DRC, I present the function in English in the text, followed, when appropriate, by the abbreviation (in French). The full French titles are provided alphabetically (by the abbreviations) in the Annex. I have also chosen, for reasons of transparency, to indicate each time that a quotation is translated into English ("our translation") even if this is frequently obvious to the reader.

[2]http://www.undp.org/mdg/goal7.shtml

[3]In October 2002, a small number of banks convened in London, together with the World Bank Group's International Finance Corporation (IFC), to discuss these issues. The banks present decided jointly to try and develop a banking industry framework for addressing environmental and social risks in project financing. This resulted in the Equator Principles (www.equator-principles.com), which were announced in Washington, DC, on June 4, 2003.

Raw materials produced by the extractive sector are a key factor for sustainable growth in industrialised, emerging and developing economies. They are a particularly valuable asset for sustaining growth and reducing poverty in many of the poorest countries in the world. It is in our common global interest that resource wealth be used responsibly so as to help reduce poverty, prevent conflicts and improve sustainability of resource production and supply. (G8 2007)

In this chapter, I critically examine this mobilization of principles, which express a development perspective that requires companies to address specified social and environmental conditions. To do this I use a case study: the process of reform of the legal and institutional frameworks of the mining sector in the Democratic Republic of the Congo (DRC), initiated in 2001. The conditions under which environmental measures have been elaborated and put into practice in the DRC raise questions about the potential impact of externally-imposed policies.

This choice of focus is based on several grounds. Firstly, the mining sector poses important challenges. Within the extractive sector, mining operations are readily recognized over and above all other activities for their negative impacts on ecosystems: the scars are highly visible and geographically concentrated. The impact of the forestry, agriculture, fishing, and chemical industries, on the other hand, are more diffuse. The mining industry directly generates deforestation, erosion, soil degradation, toxic discharges, dust emissions into the atmosphere, and pressure on water resources: these are all phenomena intimately related to the presence of mines and other industries. If industrial mining and artisanal (or small-scale) mining[4] represent problems of a different scale, both are still crucial to consider with respect to the environment. Such impacts directly affect the local population's living conditions, most critically in poor regions where survival depends on access to natural resources. There are therefore major challenges in reconciling the objectives of sustainable development and the promotion of the mining sector.[5]

Together with South Africa, the DRC is one of the great mining nations of sub-Saharan Africa, and possesses 36% of the world's cobalt reserves, between 25 and 65% of its tantalum reserves, 7% of its tin, and 25% of its diamonds (Berke et al.

---

[4] I use the term artisanal mining and small-scale mining interchangeably throughout this text. Artisanal miners include men, women, and children, working in a variety of roles such as "creuseurs" (diggers), screeners, store-keepers, dealers, transporters, traffickers, intermediaries, and guards. In the Mining Code, the legislature has preferred the term artisanal miners, which is less pejorative than diggers. Section 1, point 21: Artisanal Exploitation, reads: "Any activity by means of which a person of Congolese nationality carries out extraction and concentration of mineral substances using artisanal tools, methods and processes, within an artisanal exploitation area limited in terms of surface area and depth up to a maximum of thirty metres." Point 22: Small-scale mining exploitation, reads: "Any activity by means of which a person carries out permanent small-scale exploitation, requiring a minimum amount of fixed installations, by using semi-industrial or industrial processes, after a deposit has been found."

[5] "The sources of growth more inclusive in the short term are mining, agriculture and energy. To enjoy these sectors one needs to develop infrastructure, and donors can also play an important and catalytic role in these sectors". Final press statement by the director general of the World Bank, Ngozi Okonjo Iweala, Kinshasa—July 25, 2008 (author's translation): http://www.forumrdc.org/index.php?option=com_content&task=view&id=399&Itemid=2

2007). This mining wealth is today considered by the donors such as the IFIs as the DRC's principal lever of development.

This focus on the DRC is also motivated by the fact that this country has been one of the first (since 1982) to adopt structural adjustment policies promoted by the IFIs. The country was subjected to an economic policy of liberalization and social austerity, as well as suffering the consequences of predatory regimes. Two decades later, according to the UN Human Development Indicator (UNDP 2007), the DRC is today one of the ten poorest countries in the world, with 70% of the population living below the absolute poverty threshold on incomes hardly exceeding 30 cents (US) per day. Several years after the IFIs withdrew from the country in 1991 for reasons that were essentially political, the process of reform in the mining sector amounts to a new attempt to re-align development programmes based on an export economy.

To what extent does this reform, promoted by the financial institutions in the name of the fight against poverty (Ravallion 1995) and the Millennium Development Goals, help the DRC mining sector to achieve sustainable development, in both its social and environmental dimensions?

Studying the reform process is central to understanding how environmental norms have penetrated the mining sector through the pressure of donors, the principal investors, and the international community. It seems that the origin, orientations, context of adoption, and application of the legal and institutional frameworks of the reforms are major obstacles to positive changes in practices, especially for the environmental and social dimensions contained in the new mining norms and Mining Regulations.

To develop the analysis, this chapter consists of three parts. The first presents an analysis of the environmental norms contained in the law (the Mining Code of 2002) and the Mining Regulations of 2003. The second part focuses on the reform of institutions responsible for the application of environmental standards. The third part examines the current environmental and social challenges in the mining sector, starting with the environmental liability inherited from intensive exploitation of the country's mining resources. The chapter concludes with a reflection on the potential for considering the whole issue of environmental and social consequences in new forms of governance in the mining sector.

2 The Process of Strengthening Environmental Norms in the DRC Mining Sector

2.1 Aligning Environmental Norms With International Standards

Since the 1980s, mainly through development programmes, the poorest countries that possess important mineral reserves have been encouraged to revitalize their national economies through mining. During the 1990s, and increasingly from the first years of the new millenium onwards (thanks to the recovery of world market prices of minerals, until recently), donors have consolidated this macroeconomic approach. They have developed a new framework of intervention aimed at achieving "good governance", which is gaining momentum in countries of the South. A so-called "selective" approach has been implemented by the IFIs that makes their funding conditional upon the potential for reform, as well as the performance of the poorest and most indebted countries. Respecting associated norms—of transparency, the fight against corruption, the participation of populations, decentralization, and privatization—are presented as a *sine qua non* condition for guaranteeing their economic competitiveness in the world market-place.

The first level of intervention, largely institutional, is at the national level. The IFIs introduce this first in the framework of development programmes. The process leads to a rapid and profound reconfiguration of the legal framework, the role of the state and the institutional environment.

Thus, sectoral environmental norms (in the mining sector, forestry, water-resources, etc.) became part of the national legislation in countries of the South from the end of the 1990s. Tied to a whole new set of development standards aimed at introducing "good governance" (Campbell 2005), the social and environmental dimension was considered in the context of providing security for investments, strengthening the financial performance of projects, and, more generally, reinforcing the mining sector. This approach reflected the conceptual framework within which the reform was being proposed: "The quality of a country's environmental management system is becoming a key asset in the competition for foreign direct investment" (Weber-Fahr et al. 2002, p. 1).

The second level of adoption of social and environmental standards is more informal, operating at the transnational or supranational levels. These standards are developed to strengthen "corporate social responsibility" (Laforce 2006), and have been drawn up, promoted, and legitimized by international organisations, the banks, and multinational corporations (e.g., OECD, World Bank, UN). Intended to limit the international legal constraints placed on investors, these standards are typically presented as codes of conduct, which multinational corporations adopt voluntarily ("soft law"). Within this regulatory system, companies are encouraged to be self-regulating. The main function of these codes is to control their practices, generate their own credibility, and, in return, strengthen the legitimacy of the economic actors operating in the mining sector. In the context of development programmes, the IFIs encourage the adoption of these codes of conduct, to the extent of conditioning the loans they provide to multinational corporations upon the adoption of these transnational standards (Lawrence et al. 2006).

Donors therefore identify several dimensions that must be satisfied in drawing up their lending policies: "The loan conditions would include 'environmental and social' covenants consistent with the World Bank Group's operational directives on the environment, resettlement and indigenous peoples" (Onorato et al. 1998, p. 4).

In fact, since the 1990s, all projects obtaining IFI financing have been required, in the framework of safeguard policies, to submit a social and environmental impact study. Adopted by the OECD, the international community has committed itself to developing "indicators of their progress in applying national and infra-national policies with respect to the environment" (OCDE and Ministres de l'Environnement 2004, p. 4).

David Szablowski refers to these codes of conduct as "certification mechanisms", tied to the imperatives of economic, social, and environmental performance (Szablowski 2007). He analyses this phenomenon according to the "proliferation of transnational legal ordering" (Szablowski 2007) which directly affects the modes of regulation and the "criteria of legitimacy" of laws (Habermas 1978). For Habermas, the legitimacy of the law and the social integration of norms depend on institutionalized legal procedures. But the process of the transnationalization of norms appears to reflect a very different logic: such codes seem to be conceived *by and for* the international economic actors. Their existence is justified by the potential weakness of many states to impose their own norms; hence, the superposition of legal, national, and transnational orders. Imported by the economic actors, the transnational norms are presented as a supplementary guarantee to ensure that good practices will prevail, even in zones said to be under "weak governance". In this way, they are considered as instruments that should allow the countries of the South to make their transition to a state of law.

This overlapping of legal, national and transnational regimes clearly generates risks of normative conflicts. It is very likely that certain transnational codes of conduct establish development standards that differ from, or are even incompatible with, norms prevailing in the host country. Moreover, contrary to national laws, such transnational codes are quite often used to give legitimacy to mining activities by declaring a commitment to their principles. Yet the control mechanisms put in place and the reports provided are often inadequate, and provide no guarantee that the standards will really be applied. As Szablowski has emphasized, we are witnessing a "transfer of power" (Szablowski 2007) from the authority of the state towards a new form of counter-power held by the multinationals. On this point, the generalization of "control mechanisms", on the basis of a company's environmental self-regulation, opens up a series of questions about the new forms of regulation in the mining sector. The absence of, or ambiguity concerning, the accountability, or the relatively high risk of a mismatch between international commitments as these are reflected in practices observed at the local level, create risks of "drift" that need to be underlined. The study of the implementation mechanisms of national policies in the DRC mining sector enables us to understand better the dynamics involved.

2.2 Strengthening Environmental Norms in Mining Legislation in the DRC

Reflecting its commitment to the international scene, the DRC signed the Convention on Biological Diversity on 12 June 1992, and ratified it on 15 September 1994. This engagement led to the adoption of the Congolese National Plan of Environmental Action (PNAE), a Biodiversity Coordinating Unit, and, at the level of the

National Council of NGOs, a Technical Resources Unit for Environmental Issues. A study on biodiversity published by Centre d'Echange d'Informations gives a relatively complete picture of the natural resources in the country (CEI 2002). One is struck, however, by the fact that the report does not assign more importance to the impacts of mining on the environment. The experts merely mention these in passing as though they were not a real problem, whereas they are in fact central for Congolese biodiversity and ecosystems.[6]

The 2002 process of Mining Law reform, which attempts to integrate the new environmental norms,[7] is an echo of the DRC's international commitments at the national level. In the absence of an environmental law (Secrétaire Général des Mines 2008),[8] these norms are included in sectoral legislation. Arising from the reform process initiated by the IFIs, references to commitments to protect the environment are found throughout the various orientation documents for national development: the Emergency Multisectoral Programme for Partnership and the Transition and Revival of the Economy (PMPTR), the Emergency Multisectoral Programme for Reconstruction and Rehabilitation (PMURR), the Strategy for Poverty Reduction Documents (DSRP), and the National Programme for Environmental Action (PNAE).

In the mining sector, the DRC has therefore opted for sectoral regulation, with clauses concerning social and environmental dimensions included in all phases of mining.[9] Therefore, the new Mining Law of 2002 (RD Congo 2002) and the Mining Regulations of 2003 (RD Congo 2003a) differ from the old legislation dating from 1981, in which environmental dimensions were absent (Mutombo 1999). The new legislation incorporates environmental obligations[10] for the holder of mining rights and the State, for each category and cycle in the exploitation of a mine. To this end, the law provides a framework according to environmental norms for the four phases characteristic of all mining projects: prospecting, mineral extraction, mineral processing, and mine closure. The legal and institutional framework of the mining sector, from now on, possesses instruments to make development sustainable: the respect and promotion of the environment, and contributions to the improvement of social conditions for the population.

---

[6]A major study, entitled "The study of the impacts of mining operations on biodiversity" was conducted between 1999 and 2003, mandated by the Ministry Environment, Nature Conservation and Tourism. The final report is not currently available.

[7]Mining Code, Protection of the Environment, articles 202–218.

[8]There is a project for framework legislation on the environment, dated July 2000, which has been updated in 2008, but this has not been adopted by the Congolese parliament. This bill has been heavily criticized within the Mining Ministry, which objects to the failure to take account of existing sectoral legislation.

[9]The Ministry of Environment and Conservation of Nature is involved in the preservation and rehabilitation of mining sites. It works with the Ministry of Mines through the Standing Committee Evaluation (CPE) in the protection and rehabilitation of sites affected by mining and quarrying.

[10]Provisions for inspections and monitoring are set out in Article 15a of the Mining Code, and articles 444–449 of the Mining Regulations.

These norms answer a dual concern from a technical and management perspective (Campbell 1997): respect for the environment and promotion of social development are now placed alongside new norms, such as the fight against corruption and the need for transparency, decentralisation, and participation by the population. These measures are aimed at strengthening management performance as well as the performance of investment projects, and are intended in part to reassure investors concerned about the limits of their responsibilities. These norms therefore have to provide reliable and stable frameworks, guaranteed by competent and efficient public institutions (Weber-Fahr et al. 2002). The whole gamut of environmental measures contained in the Mining Regulations is presented below under three headings, each corresponding to a specific cycle in a mining project: (1) exploration and the start-up of operations; (2) the operating stage; and (3) site closure.

2.2.1 Impact Studies and Environmental Management Plans

The following studies and plans are required prior to permitting for exploration and mining activities:

- A Mitigation and Rehabilitation Plan (MRP) is required in order to apply for mineral or quarry exploration rights.
- An Environmental Impact Study (EIS) and an Environmental Management Plan of the Project (EMPP) are required to receive authorisation for exploitation.[11]
- Prior to any exploration activities, an Environmental Adjustment Plan (PAE) must be filed by the owner within six months of receiving authorisation for exploration. For exploitation activities, the Environmental Adjustment Plan must be filed within twelve months of authorisation.[12]
- "The holder of an artisanal miner's card must comply with the regulations regarding safety, health, use of water and the protection of the environment which apply to his mining activity, in accordance with the regulations in force. He must compensate the farmers for any damage caused by his activity. The Mining Regulations set forth the conditions of execution of the regulations regarding public safety, public health and the environment".[13]

2.2.2 Environmental Protection Measures

Several measures and control mechanisms are in place to guarantee the actual application of prevention and attenuation legislation foreseen in the EIS, EMPP, MRP, and PAE:

---

[11] Mining code, articles 64–85.

[12] Article 408 in Mining Regulations.

[13] Article 112 of the Mining Code, and Annex V in the Mining Regulations, article 416 for Saesscams' mandates.

- An annual report on activities by the title-holder.[14]
- An independent environmental audit, conducted once every two years, and paid for by the title-holder.[15]
- Inspections conducted by the Directorate for the Protection of the Mining Environment (DPEM).[16]
- Site monitoring, sampling, and analysis at times decided by the DPEM or any other agency to which it delegates this task.
- The title-holder can be released from any environmental obligations to the state by satisfying the requirements for mine closure through an Environmental Closure Audit.[17]

2.2.3 Financial Resources for Environmental Protection

The following provisions are made for environmental protection:

- The title-holder of mining rights must make a provision for site rehabilitation. The maximum amount to be allocated for this provision is equal to 0.5% of the turnover for the tax year during which it is made.[18]
- Confiscation of funds for rehabilitation of the site: if, on completion of the exploration or exploitation work, the holder of a mining or quarry licence does not voluntarily execute the obligations agreed to in the EMPP or MRP, at the request of the Mines Authority, a court shall order the confiscation of the funds set aside for rehabilitation by the title-holder.[19]
- Ten percent of the cost of obtaining an artisanal mining permit is set aside to rehabilitate artisanal mining zones.[20]

From the foregoing it can be seen that social and environmental dimensions are strongly inter-related in the Congolese Mining Code and Mining regulations. For instance, the Environmental Management Plan of the Project (EMPP) must, among other things, allow for:

---

[14] Articles 445 and 458 in the Mining Regulations.

[15] For more details see article 459 and 461 of the Mining Regulations. For example, in article 459 of the Mining Regulations: "Every two years from the date of approval of the Environmental Impact Assessment of the original draft, the holder of a mining or quarrying is required to implement at its own expense an audit by an approved office of environmental study other than that which prepared the Environmental Impact Assessment of the project or the Environmental Management Plan Project".

[16] Articles 447 and 461 in the Mining Regulations.

[17] Article 258, Provision for site rehabilitation, in the Mining Code, and 472 in the Mining Regulations.

[18] Article 258 in the Mining Code.

[19] Article 294 in the Mining Code.

[20] Article 417 in the Mining Regulations.

Improving the well-being of local communities by establishing economic and social development and foreseeing the compensation of populations in case of displacement of where
they live;
    Reducing to an acceptable level the harmful effects of the mining operation on the atmo-
sphere, water sources and rivers;
    Reducing the harmful effects of the mining operation or of quarries such as explosions,
noise, dust, etc. on the activities of human and animal populations which live in the surrounding area. (Our translation, RD Congo 2002.)

The environmental impact study itself consists of detailed research, principally
descriptive, of the state of the ecosystem prior to exploitation: flora and fauna, topo-
graphical soils, air quality, and surface and subsurface water. It allows for a survey
of resources that could be affected by the mining operation, and envisages taking
measures "to protect the environment, eliminating or reducing different forms of
pollution and the reconstitution of sites as well as verifying the effectiveness envisaged by these measures" (our translation, RD Congo 2002). Most studies have been
conducted by one of the twelve consulting agencies accredited by the Mining Ministry and paid for by the companies. These studies are then evaluated by the Directorate for the Protection of the Mining Environment (DPEM). The law also requires
each title-holder to submit an annual report to the DPEM on mining, attenuation,
and rehabilitation activities.

Nevertheless, the legal framework for environmental and social impacts of mining activities foresees terminating these activities in cases of negligence. In fact, if
it could be proven that certain activities had had deleterious effects "on the environment, public health, and security", the State could legally sanction the title-holder
with an immediate suspension of activities "in proportion to the gravity of the fault
committed and its effect on the environment, public health, and security" (our translation, RD Congo 2002). The law is based on the principle of the title-holder's full
legal obligations for damages caused by site occupation. Moreover, the law obliges
investors to deposit a financial security fund.[21] This reserve fund, based on to the
"polluter pays" principle, guarantees that the investor can satisfy its environmental
obligations both during and upon the completion of mining activities.

Unfortunately, several obstacles impede the application of these environmental
measures contained in the Mining Code and Regulations. The first is the lack of
means and expertise available to the DPEM to oversee the performance of companies. The second obstacle arises from the absence of state representation in most of
the country's mining provinces. Moreover, it seems that the Mining Ministry contributes, in part, to an approximate application of the Mining Code and Regulations.
There is thus, currently, no public watch dog to verify company-provided data or to
analyse their annual impact studies and company reports.

There is a dual problem here. Firstly, there is no independent expertise to evaluate
the private sector's results, thus weakening the State's ability to control the sector.
Secondly, the government seems to lack the means and political will to carry out

---

[21] Article 566 in the Mining Regulations.

this role, so what appraisal is done is principally undertaken in the capital, Kinshasa, with little direct contact with the actual miners and mining regions.

From the above it can be seen that the legislative framework for social and environmental responsibility in the mining sector has been clearly set out, but implementation remains a problem. In the next section, I examine the practical application of these measures.

2.3 Overhaul of the Institutional Framework: Conditions for Applying Environmental Norms

2.3.1 A Process of Modernizing Mining Institutions

The process of overhauling the legal and regulatory framework of the Congolese mining sector has been accompanied by a major programme of institutional reform (Fig. 1). Several specialized agencies have been created, or modernized. A more or less important part of their mandates, varying between institutions, concerns the environmental and social dimensions contained in the Mining Law. Certain environmental perspectives, complementary to the legal dispositions, are contained in the terms of reference of the Mining Plan. This plan is an orientation document,

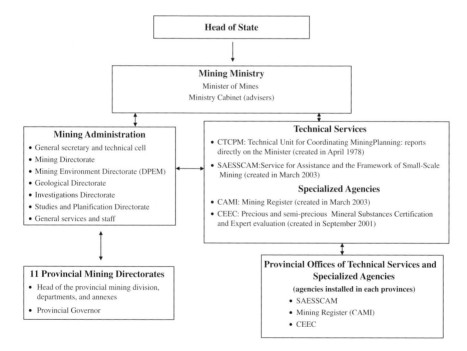

Fig. 1 The institutional landscape: the mining sector in the Democratic Republic of the Congo

currently still being elaborated. In terms of the environmental problems, it reads: "Environmental protection: for this, a general plan of sustainable development will be drawn up" (CTCPM 2006); however, no details are provided on the content or modalities of the application of such a plan.

Environmental protection depends almost exclusively on a specialized institution (set up in 2003) within the reform framework: the Directorate for the Protection of the Mining Environment (DPEM), which answers to the Mining Ministry. The DPEM therefore carries out its obligations in coordination with the state agencies responsible for mining management. The DPEM has broad mandates:

- Setting pollution limits;
- Developing the procedure for impact studies;
- Collecting basic data on the state of the environment in the mining area;
- Ensuring that companies respect environmental norms;
- Supervising the conduct of environmental impact studies.[22]

The Technical Unit for Coordinating Mining Planning (CTCPM), the Geological Directorate, and the Directorate of Mines contribute to formulating the Environmental Impact Study (EIS) and the Environmental Management Plan of the Project (EMPP), by way of the Permanent Committee for Environmental Evaluation (CPE). The Central and Provincial Mining Register (CAMI) also participates in these two stages.

The Service for Assistance and Framework for Small-Scale Mining (SAESS-CAM).[23] which is responsible for the artisanal sector, has been subordinated to the Mining Ministry since 2003. SAESSCAM is in charge of training programmes for small-scale miners (providing internships and education) and is mandated to generate awareness of environmental problems. It is also represented on the Permanent Commission for Environmental Assessment, composed of 13 members. A permanent committee, created in the context of the reform, acts as coordinator between different ministries, and its members perform the final stage of informing the EISs and EMPPs. According to the Mining code the final decision on the environmental viability of a mining project is made by the DPEM directorate, but in fact the Commission's recommendations are decisive.

To obtain an artisanal exploitation permit, small-scale miners are supposed to make annual contributions to a Rehabilitation Fund, but few of them make such payments, so the Fund is effectively empty. Moreover, the local authority charged with managing the Rehabilitation Fund for artisanal mining operations is unspecified. In addition, in the absence of recognized boundaries of artisanal zones between

---

[22] Article 12 in the Mining Regulations.

[23] Previously known as the Inter-ministerial Committee for Small Scale Mining (CISSCAM), set up in November 1999.

2002 and 2007,[24] there is still a very little evaluation of the environmental impacts of small-scale production. This situation is highly problematic.

2.3.2 Reforming Institutional Frameworks and Practices

It is clear that, at all levels of the mining hierarchy, environmental policies are still only being partially applied. Several factors explain why this is the case. Firstly, the State's central and provincial services are functioning without adequate means to ensure that operations actually perform to the expected standards. Lacking adequate transport, expertise, equipment, and supplies, the directorate is unable to enforce the environmental standards contained in the law (Secrétaire Général des Mines 2005). In the mining regions, state bureaucrats, who are typically poorly or irregularly paid, tend to be complacent towards investors, while corruption is reported to continue to dominate the institutional landscape (Global Witness 2006).

It is also true that the crisis of the 1990s in the DRC contributed to the almost complete degradation of public mining expertise. To this was added the lack of material and human resources, which encouraged disregard of the law and recourse to private actors in the sector to compensate up for the State's shortcomings. In such a context, three options can be distinguished:

Firstly, the State could have used accredited consulting companies to apply environmental controls in the mining sector. This option would legitimize a form of State privatization, by delegating control of important functions to private subcontractors. Moreover, the costs incurred by such sub-contracting would have been a significant proportion of the meagre budget within which the institutions were operating; this option was therefore little used. The Mining Ministry grants accreditation to Congolese consulting companies specializing in the environmental aspects of mining, but a number of officials in the ministry pointed out the lack of expertise of most of these consultants, many of whom who also work for the mining companies.

The second option would have meant disregarding certain environmental provisions in the Mining Law, or bypassing the law entirely.

The third option would have been for investors themselves to finance the controls to be conducted by public agencies. Mining Ministry officials informed us that there were very few environmental controls carried out by State agencies at the mining sites. Lacking the means to do this, the State representatives are dependent on companies to pay for their travel expenses to verify that state-imposed controls are adopted.

If reviews of what is actually happening on mining sites are almost non-existent, how are the environmental studies, which are supposedly preconditions for granting mining rights, handled in Kinshasa? The mining service responsible for environmental protection is first charged with verifying the environmental viability of a project, but does the lack of capacity to perform this task slow down or block the processing of mining rights? An analysis of records considered by the DPEM in

---

[24]In 2008, the Mining Ministry created 42 specific artisanal areas in the main mining provinces.

2006 shows that 265 of the 281 submitted applications received positive responses, 16 companies were required to provide further information, but none was refused.[25] In other words, the lack of human and material resources does not block the process of granting mining rights—in fact, one could even propose the opposite, and that the environmental assessment stage is merely an administrative formality, with few constraints being placed on future investors.

Moreover, in Kinshasa, as in the provinces, public administration is confined to offices in a pitiful state of repair. Indeed, it was private companies that paid for the rebuilding of the headquarters of the Kinshasa Mining Directorate after an arson attack in 2006. A close reading of the 2006 annual activities report of the Mining Environmental Protection Directorate (Muhindo Songe Luyeye 2007) reveals that the DPEM identified substantial material, personnel, and training needs.[26] The activity report concluded that, under the present circumstances, state services cannot enforce the environmental norms contained in the Mining Law. Yet, if we refer to the text itself, the Mining Regulations specify which proportion of state receipts should be assigned to specialized mining services to ensure their ability to function properly. For instance, the Mining Register (CAMI), an institution playing a central role in regulating the sector, should be benefiting from the annual surface rights payments (i.e., the amounts paid by the companies for surface access according to the area of their concessions). Similarly, the Mining Environmental Protection Directorate should receive 6% of these revenues (RD Congo 2003a), yet this clause of the ordinance is not being applied by the central government.

It should be emphasized that in the last few years, the DPEM has attempted to initiate some changes. For example, in 2008, it prepared a list of specifications on "measures to maintain workers' security and environmental protection during research and exploitation phases" (our translation). If this restrictive measure is adopted by the ministry, it will enable the government to define national standards that were absent from the 2002 Mining Law. However, such initiatives could well fall foul of the current lack of political will or means available to the government to implement them.

While the DPEM has concentrated its activities on industrial-scale production, another institution, SAESSCAM (which was also created within the reform framework), is responsible for the artisanal sector (RD Congo 2003b). This administratively-autonomous technical service is mandated to provide the framework for small-scale mining activities. A recent study conducted in Katanga (Sakata et al. 2007) identifies several weaknesses in SAESSCAM, including:

---

[25] Here I draw no distinction between the Mitigation and Rehabilitation Plan (MRP), the Environmental Impact Study (EIS), Environmental Recommendations of the Environmental Adjustment Plan (PAE), and Environmental Management Plan of the Project (EMPP).

[26] In 2006 and 2008, the World Bank provided a certain amount of training in environmental expertise. The training was described to us by Mining Ministry officials as significant but insufficient, too theoretical, and, in the end, poorly adapted to the government employees' training needs.

- Weak logistical capacity;
- Lack of financial resources;
- Lack of financial security for employees (irregular and arbitrary payments replace salaries);
- Inappropriate personnel in critical posts; and
- Several cases of dysfunction in carrying out the mandate assigned to SAESSCAM.

SAESSCAM's role is particularly important because, according to the World Bank, the artisanal sector represents approximately 90% of mining production in the DRC (World Bank 2007). In fact, since the collapse of the industrial sector, small-scale mining has been attracting a very large number of informal workers. The World Bank has estimated that between 500,000 and 2,000,000 people, including women and children, are engaged in this informal sector across the whole national territory (Banque mondiale 2008). The Human Development Indicators are lowest precisely in these mining zones. It is clear that this unregulated mining activity is an important factor in the destruction of ecosystems, a factor contributing to the impoverishment of local populations. Erosion, dust, the discharge of polluting products onto the ground and in water systems, deforestation, and the abandoning of sites without corrective measures are among the principal impacts that have been identified. Moreover, these mining populations typically live in temporary encampments erected close to, or even directly within, the mining sites themselves. The degradation of the environment creates important public health consequences.[27] In its training and monitoring mandate, SAESSCAM could be called on to play a role in increasing awareness and educating miners to help them grasp these problematic environmental issues and the resulting public health consequences. The miners' families are concerned with the lack of health services and schools close to the mining sites. In such an environment, they consider that their children's futures are permanently compromised.

It seems that the low likelihood of putting into practice the environmental measures contained in the Mining Code and the Mining Ordinance is a more general reflection of the deficiencies observed in the application of the Mining Law as a whole. Therefore, one of the principal challenges today is to provide a minimum guarantee of the application of the law adopted in 2002. Is this shortfall due to a lack of means available to the central State, and/or a lack of political will? To all appearances, the Congolese government seems, here as elsewhere, to be vacillating between two functions that sometimes appear incompatible: the State as *promoter* of the sector, and the State as *regulator* of the sector.

More generally, such a proposition opens up the question of the compatibility between the norms contained in the legal framework and how the Congolese administration functions. It would seem that, lacking the mechanisms to apply them, the

---

[27] In mining camps, children are often affected by malaria and diarrhea, which can be killer diseases if left untreated.

introduction of the new norms has not been sufficient to generate reforms in practices, and this is the case at all levels of the State system. The State lacks the capacity to enforce its own legislation, a problem that is not unique to DRC, but a widely-recognized limitation on sustainable development.

Such findings raise questions about the mining reform in its totality. It would seem that the legal and institutional instruments elaborated by the donors are, in the end, poorly adapted to the Congolese context, both in their orientations and in the strategy for applying them. This line of questioning in fact goes beyond concern about the State's capacity to exercise its "regal" functions. It needs to take into account the time-lag between attempts to graft international standards of governance onto existing structures, the instruments that are adopted, and the extent of challenges posed by the context of the DRC: environmental liability, dysfunction of the State in a country that is continental in its dimensions, the heterogeneity of the problems caused by mining, major insecurity in the east, etc.

The norms contained in the Mining Law correspond to the introduction of an Environmental Management System (SGE), based essentially on mining projects' social and environmental performance goals. This approach illustrates the way the World Bank encourages mining countries to adopt transnational legal norms in their legislation while poorly taking account of each country's specific conditions. Moreover, environmental issues are strongly interlinked and have inter-sectoral aspects with agriculture, forestry, mining, land ownership, and water resources. That is why the definition of environmental norms must result from a participatory process involving civil society and government. No-one is more aware than the Congolese of the challenges that face their country.

By creating new legal and institutional frameworks, the strategy that has been adopted is instead introducing a climate that encourages the return of foreign direct investments to the DRC, and introducing "good corporate practices" as a guarantee of sustainable development. This situation is widely found in a Congolese context in which the state is unable to conduct its own monitoring activities. Thus, companies' compliance with practices that respect the environment is often limited to a commitment in principle, but with few consequences for non-compliance in reality.

The second issue concerns the limitations in a company's responsibilities. As noted by Bruno Sarrasin, "This model of development is based on a formula by which growth in exports contributes to reducing poverty, while also protecting the biodiversity" (Sarrazin 2005, p. 801). Such a perspective tends to propose a "natural" association between economic growth, the fight against poverty, and protection of the environment, which could in part explain why donors typically fail to acknowledge the limited potential of the State to appropriate, and hence apply, new environmental norms. In fact, it seems important to emphasize that introducing and rapidly aligning environmental standards with international norms is important but does not guarantee that the State will be able to apply its own laws in its territory.

In the DRC, the new mining institutions are relatively inoperative, and are therefore only effective at a marginal level. Moreover, the massive arrival of mining investors—there are estimated to be more than 300 in Katanga alone—does not enable the State to apply real control over all the operators' environmental impacts. A management model that confers on companies the responsibility for social development and environmental protection (in place of public control) clearly has its limitations. Can the application of a national policy to protect the environment in a sustained way be replaced by environmental mining standards involving voluntary practices by mining companies?

Some elements of the answer can be found in reports that document the environmental situation in the Congolese mining sector, as discussed below. These cast light not only on the complexity but also the political dimension of governance (Campbell 2001), and the associated environmental stakes.

3 Environmental Liabilities in the Mining Sector in the DRC: Social Implications

3.1 Impacts of Mining Activities

The failings of the mining administration in the DRC are well known, but the impacts of mining activities are little known or poorly documented.

One reason for this, proposed in a report prepared for the World Bank, is that: "It might seem surrealistic to talk of the environment in a country where the basic needs are not satisfied, and where the Human Development Indicators such as life-expectancy and infant mortality are among the worst on the planet." (SNC-Lavalin International 2004 p. 1, our translation).

In fact, in some of the regions where industrial-scale mining has developed—either dating from before or after independence—it is only recently that the extent of negative environmental impacts is beginning to be recognized.

However, it is difficult to obtain data documenting the nature and extent of this deterioration, making it difficult to formulate any kind of response. Public expertise is almost nonexistent, and when expertise is available it is usually limited to projects controlled by private mining interests, that is, by investors or donors. Therefore, even if some data do exist, with few exceptions, they are rarely published (Banque Africaine de Développement 2007). To make up for this chronic lack of information, a number of NGOs have attempted to acquire more wide-ranging mining expertise.

In 2002, European Commission (EC) experts concluded that "there are no exhaustive reliable data on the pollution situation caused by mining activities in Katanga and, still less, for the country as a whole" (Vande Weghe et al. 2005). Yet in the rural areas they point to the "relatively catastrophic" effects of mining activities on the levels of pollution in the soil, air, and surface waters due to discharges of acids, solvents, cyanide, and heavy metals.

A more systematic analysis of Katanga province provides an illustration of the extent of environmental impacts from industrial and artisanal mining. The industrial-scale activities raise two questions: Where does the responsibility lie for managing environmental liabilities, and what are the impacts of current methods of mineral extraction and concentration? Commonly, these companies employ so-called "illegal" subcontracted labour forces, without recognizing the *creuseurs* as company employees. In the absence of any State regulation, artisanal activity is expanding, without concern for the associated social and environmental stakes (Réseau Ressources Naturelles 2007).

3.2 An Environmental Inventory of Katanga: Impacts of the Mining Industry

The failure of Gécamines (Générale des Carrières et des Mines) at the end of the 1980s led to a halt in investments in Katanga. Even though the enterprise had a "division of the environment to combat pollution", its activities were minimal, mainly limited to taking and analyzing samples from certain zones designated as "at risk". No substantive preventive or corrective measures were adopted by this under-financed division. Gécamines' activities nevertheless had, and continue to have, important impacts on the province's environment. Environmental degradation is also the result of a massive influx of people seeking work, putting substantial pressure on the region's natural resources. For example, in Katanga, the production of charcoal is contributing directly to the rapid deforestation in some zones. While the environment was beginning to emerge as a sensitive issue internationally, in Katanga no further investment was committed to maintain and improve facilities or equipment.

> In Katanga, and particularly in Kolwezi, Likasi, Kipushi and Lubumbashi, the heavy concentration of mining concessions (Gécamines, Forrest, etc.), quarries, and industries involved in refining metal (copper, cobalt, zinc, uranium, etc.) is responsible for the total degradation of the soil and of the ecosystems (. . .) and the loss of large areas of agricultural land which will be impossible to rehabilitate. (Vande Weghe et al. 2005, p. 45.)

It is a fact that the arrival, or return, of the World Bank and of the major mining companies to the DRC in the new millenium led to a real awareness of the extent of the environmental problems. But concrete measures to correct these problems have been slow in coming. Six years after the beginning of the reform process, many mining companies present in Katanga have yet to adopt either an Environmental Management Plan of the Project (EMPP), or an Environmental Adjustment Plan (PAE) (Muhindo Songe Luyeye 2007). Clearly, the State as "promoter" is today more inclined to promote prospecting activities that might identify new sites to offer to investors, than to assess the environmental impacts of mining activities, or to mobilize partners for mine-site rehabilitation.

In such a context, the role of donors and investors is not inconsequential, not least because the environmental standards adopted in the Congolese mining sector were largely instigated by the World Bank with the intention that they would be adapted to investors' needs. In 2002, the World Bank commissioned a keynote study, more widely known by the name of the consultancy group that produced it: SNC Lavalin International. The initiative was undertaken at the beginning of the reform process, when a mission was sent to Katanga, the principal industrial-mining region in sub-Saharan Africa. Katanga is part of the African Copper Belt, which extends into Zambia, and is host to extensive copper and cobalt mining operations once dominated by the state-owned company Gécamines. In 2002 and 2003, SNC Lavalin's experts inspected 39 sites in the mining districts of Lubumbashi, Likasi, and Kolwezi, and assessed environmental performance at 32 of these sites. The conclusions contained in a preliminary report are incontrovertible[28]: the environmental situation at the sites visited, and in the region in general, is described as highly problematic:

> Of a total of 32 sites assessed, 13 (41%) are considered as presenting Priority 1 environmental problems (requiring immediate corrective measures), and 10 sites (31%) present moderate, Priority 2 problems (requiring corrective measures in the short term). Finally, 9 sites are considered as presenting few or no environmental problems and require few or no corrective measures (Priorities 3 and 4). (SNC-Lavalin International 2004, p. IV.)

The authors of the report noted the striking absence of recent data that would permit detailed documentation of the extent of environmental consequences of mining activity in Katanga. Consequently, they indicated their inability to "distinguish which of the different forms of contamination are most problematic" (SNC-Lavalin International 2004). As an illustration, five cases are described below, which are representative of the recurring environmental problems on the sites studied in Katanga.

3.2.1 Kipushi

The Kipushi mine is a Gécamines concession first exploited in 1926, and closed down in 1993 due to lack of funds. In the late 1980s, this copper and zinc mine produced up to 143,000 t of zinc and 43,000 t of copper per year.

The mine's tailings were released into the Kipushi river valley, covering an area of about 240 hectares over a distance of 2.5 kilometres. The river retained some 38 t of residues (0.02% of zinc and 1% of flotation reagents). Tailings spread to within a few metres of the Zambian border.

> It is considered to contain 38 million metric tons of residues with copper concentrations of 0.2% and zinc of 1% and reagents used in flotation. Beyond the contamination of the water circulating on the surface of the residues and which flows into the Kafubu river, the existing basins are an important source of dust that is blown towards the community by the predominant winds. These discharges contain arsenic, cadmium, lead and zinc. (SNC-Lavalin International 2004, our translation, p. 50.)

In 2003, the town of Kipushi, located on the Zambian border, had a population of approximately 174,000 people. Serious public health problems such as pulmonary and ocular irritations have been identified in people living close to the discharge sites and in the town itself. Even more seriously, repeated cholera epidemics have broken out in Kipushi and other mining towns such as Likasi, which are directly attributable to the poor quality of the drinking water (WHO et al. 2008). The dust deposits also

---

[28]At the time of writing, no final version of the report was available.

affect the vegetation and animals close to the discharge basins. This discharge site was classified by the World Bank experts as Priority 1, and is therefore considered to require immediate corrective measures. Yet, since the report was issued in 2003, no significant measures have been undertaken.

In a study financed by the Katanga Natural Resources Network (Réseau Ressources Naturelles 2007), the pan-Congolese NGO revealed that water contamination largely exceeded international norms in the area around the Kipushi river. Only detailed epidemiological studies will permit a precise analysis of the effects on the health of such concentrations.

In 2007, American Mineral Fields Inc. took over the concession (with 51% of the holdings) in a joint-venture with Gécamines (49%). How will the mine's operators deal with the question of environmental liabilities? Even if the problem of apportioning responsibilities between new and previous title-holders is not unique to the DRC, given the extent of the environmental damage, it is worth drawing attention to question, particularly because the Congolese State does not seem to be in a strong position to negotiate with foreign investors that it is at the same time trying to reassure about the security of their investments.

3.2.2 Shinkolobwe

The Shinkolobwe uranium mine is located near the town of Likasi in Kambove territory, in the district of Haut Katanga. The mine is best known for supplying the uranium used in making the atomic bombs dropped in Hiroshima and Nagasaki. From 1921 to 1959 it was the flagship of the Congolese mining industry, but today is a good example of the mining environmental liabilities that are still mortgaging the development of certain regions. In 1960, underground operations in the mine were halted and the mine shafts were closed and sealed with concrete. However, these measures did not stop the site from being the object of "illegal" exploitation since the 1990s by artisanal miners. There have been hundreds, even thousands, of *creuseurs*, the majority of whom are women and children, looking for copper and cobalt (e.g., heterogenite, a cobalt oxide mineral).

The 2003 SNC-Lavalin report notes the presence of highly radioactive slag, which neighbouring populations commonly use as landfill:

> At some of the factories visited that were treating the Shinkolobwe minerals, it was observed that the residues were sometimes used as filler for low-lying land or for route maintenance, a factor encouraging the dispersion of the material. (SNC-Lavalin International 2004, p. 89, our translation.)

In 2004, following the collapse of a mine-shaft that cost the lives of several people, the UN commissioned a number of inquiries (Pasche 2004a, b), and the site was officially closed by presidential decree on 28 January 2004:

> The tables show high concentrations of cobalt, nickel, copper and uranium for most of the soils and sediments. Cobalt, nickel and copper concentrations up to 1–8% were found. The concentrations of some heavy metals in water samples have been compared to the WHO guidelines for drinking water. Except for uranium, all the concentrations were lower than

the WHO guidelines for drinking water. Water sample Shinkolobwe S6 contained more than 20 times higher uranium concentrations than the WHO guideline. (Pasche 2004b, p. 17)

A radiological report confirmed contamination of the site by radioactive substances, and noted the severe risk that this posed for the "diggers" (Waggitt 2004). UN experts who visited the site reported:

The Group did not observe any artisan mining on the day of its visit. However, several of the individuals interviewed, including the police agent and his assistants, did state that artisan mining is an ongoing activity at Shinkolobwe. (De Rivero 2006, p. 32.)

Finally, the experts responsible for five reports—on public health, the environment, artisanal workers, and the humanitarian and radiological situations—all concluded that immediate closure of the site was necessary to stop anyone from entering it. They noted the need for eventual decontamination of the site, and recommended further examination of the problems and vigilance to ensure there would be no access to mine. However, mining appears to have continued into 2007, when Katanga authorities launched an enquiry into the dumping of 18 t of copper minerals with a high concentration of uranium into the Mura River, close to Likasi. According to the International Atomic Energy Agency, minerals from Shinkolobwe were continuing to contribute to illegal uranium exports and arms trafficking (Africa Intelligence Mining 2007). Beyond the site itself, no corrective measures have been undertaken to remediate the environmental and social impacts generated by the dispersion of dangerous minerals in residential zones.

3.2.3 The Chemaf (Chemicals of Africa) Hydro-Metallurgical Plant, Lubumbashi

In contrast to most of the sites that were inspected by the SNC-Lavalin experts in 2003, the hydro-metallurgical plant near the Tshimilemba neighbourhood of Lubumbashi, owned by Chemaf (a subsidiary of Indian company Shalina), was still in construction but partly operational. It was intended to produce 1,000 t of cobalt carbonates and 1,200 t of copper carbonates per year. Despite the fact that it is a new plant, the site is already classified as Priority 2, that is, requiring short-term corrective measures. The reasons mentioned in the report for this classification concern principally the plans to discharge untreated effluent (barren solution and sodium hydroxide) into the environment, as well as the conditions under which certain potentially polluting substances (iron cake) were stored. It was observed that a treatment basin was being constructed to hold the discharges. At the time of the mission, effluents were being directly dispensed into the environment without any prior treatment.

In 2007, a delegation of national parliamentarians publicly denounced the contamination of drinking water by the Chemaf plant (Radio Okapi 2007). Untreated discharges and industrial waste-water were running through trenches, and the surrounding vegetation was burned by the acids. These accusations were prompty refuted by the company. The Environment Minister at the time, Didace Pembe, supported the company's position, but groups representing civil society have for several years been demanding corrective measures (Nouvelle Dynamique Syndicale 2004; Ligue pour la Bonne Gouvernance 2005).

In September 2008, Chemaf was again accused of polluting the water supply by Tshamilemba habitants who live in close proximity to the factory, and who suffer from skin and respiratory diseases (Le Potentiel 2008). In reply to these various accusations, Chemaf financed a number of community projects (e.g., health, culture, education, and infrastructure development) and made this widely known. This situation illustrates the inadequacy of policies based on corporate social responsibility (i.e., the company did not undertake these corrective measures voluntarily), and the need for stronger State involvement in protecting the population and controlling company activities (Droits et Démocratie 2007).

3.2.4 The Lubumbashi Electrical Smelter

The Lubumbashi Electrical Smelter (FEL) is a pyrometallurgical factory owned by Gécamines. This type of factory is characterized by high-temperature water discharges. In the case of FEL, the water released is between 40°C and 60°C higher than the temperature of the ambient water courses. The liquid discharges also contain Cu, Co, Cd and Ge, and contribute to the contamination of the Lubumbashi River. Atmospheric emissions from the factory's chimney are also high, and contain sulfur dioxide and metals. SNC-Lavalin classified this as a Priority 3 site, and recommended the construction of a treatment plant and the capture and treatment of flue gases (SNC-Lavalin International 2004). It is not known whether these corrective measures have been adopted.

The examples described above are considered to be just the tip of the iceberg, below which lies a legacy of decades of industrial mining operations that had little regard for environmental impacts. This history puts a heavy mortgage on future strategies for sustainable development in the mining sector.

3.3 The Central and Eastern Mining Provinces: Informal Small-Scale Production and Impacts on the Ecosystem

In the east of the country, the mining problems take a different, but no less complex, form. Even before the two wars broke out in 1996 and 1998, the artisanal mining of diamonds, tin, gold, and coltan dominated the sector. It is very hard to find documentation on the quantities of minerals produced because the majority of the operating sites were being exploited "illegally" by small-scale miners. Certain NGOs, working directly with these miners, have nevertheless attempted to document the social and environmental conditions of these sites (Mutabazi and Sanganyi 2007). The artisanal activities are known to be an important cause of soil and water pollution, provoking land degradation (erosion) due to intense pitting and trenching: "The most serious problems in the small-scale mining industry are environmental in nature" (Hilson 2002, p. 869).

In north and south Kivu, small-scale mining is known to be strongly linked with economic and political dimensions of conflicts (Global Witness 2005; De Failly 2007; De Jonghe and Berck 2007; Hocquard 2007; Garret 2008). In fact, human and environmental impacts of mining sector are closely connected. As illustration, studies by the Dian Fossey Foundation (Armstrong et al. 2003) examined the impact of coltan (colombite-tantalite) mining in and around the Kahuzi-Biega National Park (NPKB) in South Kivu. The results illustrate the devastating effects of the increased prices of this mineral during the years 2000 and 2002. In 1982, the price of tanta-lite fell to US$34–44/kg, and then climbed back up to US$65/kg. As of February–March 2000 and up to the end of that year, the tantalite price soared progressively, reaching ~US$600/kg (De Failly 2001). There were at the time estimated to be 12,000 miners illegally present in the park, many fleeing the famine and extreme violence (kidnapping of children, rape, stealing harvests) committed in their vil-lages. All categories of the population—women, children, government employees, and soldiers—turned to exploiting the coltan in the Park (Redmond 2001). The impacts on the environment of this massive human presence have been described by the Dian Fossey Foundation as disastrous. The miners and soldiers killed and ate protected species of wild animals (elephants, gorillas, etc.) and caused rapid defor-estation of certain zones. Pollution of water courses and soil erosion were further consequences of mining activities (Armstrong and D'Souza 2003). Epidemiological research revealed high contamination levels among coltan workers in North Kivu due to the presence of naturally occurring radioactive materials (Mustapha et al. 2007).

In Western Kasai, a province located in the centre of the country, industrial-scale mining has almost completed given way to artisanal exploitation. The principal con-cession of the Société Minière de Bakwanga (MIBA) conceals numerous abandoned and unsealed mineshafts invaded by the *creuseurs* (Maninga and Mbikayi 2007). In most of the mining zones, the solid, gaseous, and liquid discharges resulting from intensive exploitation have made soils sterile (Vande Weghe and Franssen 2005). Soil erosion affects the whole town of Mbuji-Mayi and its surrounding area. East-ern Kasai used to produce large amounts of maize, peanuts, palm oil, and beans, but since the 1980s, agriculture has been abandoned for mining. Without an agricultural policy, numerous tracts of land are now being dug up in search of minerals and are no longer cultivated.

In this region, as in Katanga, the withdrawal of the State from the mining sector is evident from the decay of infrastructure that used to be maintained by the mining companies as part of their social policies. The decline of State hospitals and absence of health centres close to the artisanal working sites increases the risks associated the *creuseurs'* and their families' lifestyles: water-born diseases, HIV-Aids, accidents, dust, pollution, etc.

Despite these drawbacks, artisanal mining is a major source of job creation for rural populations, which are among the poorest in the country, and cannot therefore simply be ignored. Currently, the lack of any production framework by the State ser-vices significantly negatively affects the social and environmental conditions which artisanal miners and their families are experiencing. The 2002 Mining Law gave recognition to small-scale mining, but, in the absence of any effective public policies, the contribution of artisanal mining to "poverty reduction" is extremely limited.

4 Conclusions

The DRC is today in a period of major restructuring of its mining sector. Both the environmental liabilities and the environmental impacts of ongoing mining operation are still poorly documented, and are minimally taken into account, either by the government or the mining companies, and still less by artisanal miners. Yet the few studies that have been done illustrate that industrial-scale and artisanal activities are contributing to the rapid and sometimes irreversible degradation of ecosystems. In the main mining regions, this has a direct impact on rural populations, whose daily survival and health depend largely on access to natural resources. For this reason, in some regions, environmental degradation resulting from mining activities, both in the past and the present, are already mortgaging the mining sector's participation in the "fight against poverty".

The 2002 Mining Law can certainly be considered to be an innovative instrument. In fact, new legal and institutional frameworks have been elaborated to ensure systematic accounting of environmental dimensions of all phases of mining projects. Yet, as we have shown, the Congolese State does not currently possess the means to adequately apply the law, and there is a large gap between the legal framework and actual practice. Moreover, the State is not in a position of strength to negotiate with investors regarding the responsibilities for dealing with environmental liabilities.

In this context, the companies are being called upon to take responsibility for, and include in their budgets the costs of, environmental protection. Some companies operating in the DRC do conform to the environmental requirements of the Mining Law and international codes of conduct, but this study shows that this is not yet the norm. Moreover, the artisanal sector, which has emerged informally, lies outside government planning and framework policies. This sector is currently a major source of pollution and is causing significant degradation of ecosystems. The informal, even illegal, presence of these *creuseurs* appears to be a major obstacle in addressing social and environmental responsibilities related to this type of activity.

In this context, the legal and institutional instruments adopted in the framework of reform seem to be poorly adapted to the situation in the DRC. The environmental and social impacts of mining are still largely considered as externalities,[29] one-third of the country's land area has been allocated to mining companies through concessions, and artisanal activity continues to proliferate. Given the state's dysfunctionality, there is therefore the question of responsibility of donors and companies to pursue the country's sustainable development objectives (Palan 1999).

---

[29] An externality exists when one agent's action influences the well-being of another agent, without mediation by the logic of the market-place.

Because it is at the centre of the economic, social, and environmental concerns that are the key to development of the country, the future of the Congolese artisanal mining sector needs to be seriously examined. Even in the absence of epidemiological data, one can assume that the *creuseurs* and their families are being exposed to numerous and unacceptably high degrees of pollution. Given that in the next few years many large mining companies plan to bring new mines into production, it seems all the more urgent to pose questions about the environmental consequencies of these activities.

In October 2007, a ministerial commission was mandated to examine one-sided joint venture agreements in the mining sector that were signed during the wars and the period of political transition. The Commission's findings concluded that none of the 60 contracts under review were valid. The aim was to modify these contracts to guarantee that they were in the economic interests of the State, but this could also have been a good opportunity for the government to improve transparency in the mining sector governance and insist on full adherence to the law by investors.

In April 2008, the government signed a collaboration agreement with Chinese investors for a total of US$9–14 billion. In exchange for being allowed to operate the mines, the Chinese investors committed themselves to constructing infrastructure (roads, hospitals, railways, school, airports). In several ways, these two types of construction, mines and infrastructure, constitute a real test of governance, both for the Congolese government and its economic partners. For instance, it will be essential to closely monitor if and how social development and environmental protection measures are incorporated into the agreements, and subsequently acted upon. Both the political will of the government and the responsibility of investors and donors will be critical if a positive evolution of environmental practices in the Congolese mining sector is to be achieved.

Acknowledgments This study was funded by CIRAD, the French Agricultural Research Centre working for International Development. It forms part of a PhD in sociology, supervised by Dr. B. Campbell at the Université du Québec à Montréal. Also, the author would like to thank Christian Lukusa Kantumunda, freelance attorney, for his pertinent review of the legal aspects of this chapter, and Chedly Boussetta for help with the layout.

Annex: Abbreviations Used in this Chapter

CAMI: *Cadastre minier*: Mining Registry

CPEM: *Comité permanent d'évaluation environnemental*: Standing Committee for Environmental Evaluation

CTCPM: *Cellule Technique de coordination et de planification minière*: Technical Unit for Coordinating Mining Planning

DPEM: *Direction de protection de l'environnement minier*: Directorate for the Protection of the Mining Environment

DSRP: *Documents de la Stratégie de la Réduction de la Pauvreté:* Strategy for Poverty Reduction Documents

EC: European Commission

EIS: Environmental Impact Study: *Etudes d'impacts environnementaux*

EMPP: Environmental Management Plan of the Project: *Plan de Gestion Environnementale de Projet (PGEP)*
EPE: *Engagement de protection environnemental*: Commitment to Environmental Protection
FDIs: Foreign Direct Investments
FEL: *Fonderie Electrique de Lubumbashi*: Lubumbashi Electrical Smelter
Gécamines: *Générale des Carrières et des Mines*
HDI: Human Development Indicator
IBRD: International Bank for Reconstruction and Development
IFIs: International Financial Institutions
LBG: *Ligue pour la Bonne Gouvernance*: Good Governance League
MIBA: Société Minière de Bakwanga
MRP: Mitigation and Rehabilitation Plan: *Plan d'Atténuation et de Réhabilitation* (PAR)
NDS: *Nouvelle Dynamique Sociale*: Union for a New Social Dynamic
OECD: Organisation for Economic Cooperation and Development
PAE: *Plan d'ajustement environnemental*: Environmental Adjustment Plan
PAR: *Plan d'atténuation et de rehabilitation*: Attenuation and Rehabilitation Plan
PNKB: *Parc national de Kahuzi-Biega*: Kahuzi-Biega National Park
PMURR: *Programme Multisectoriel d'Urgence pour la Reconstruction et la Réhabilitation*: Emergency Multisectorial Programme for Reconstruction and Rehabilitation
PNAE: *Plan National d'Action Environnemental*: National Plan for Environmental Action
SAESSCAM: *Service d'assistance et d'encadrement du Small Scale Mining*: Service for Assistance and Framework for Small-Scale Mining
SGE: *Système de gestion de l'environnement*: Environmental Management System
UNDP: United Nations Development Programme
UNEP: United Nations Environment Programme
WBG-IFC: World Bank Group–International Finance Corporation

References

Africa Intelligence Mining (2007) Qui veut acheter "Yellow cake"? l'AIEA enquête sur l'exploitation et la sortie d'uranium sous couverture d'exportations de minerais de cuivre, de cobalt et de nickel, La Lettre du Continent, p. 2

Annan K (2002) Allocution du Secrétaire général au Sommet pour le développement durable. Nations-Unies, Johannesburg

Armstrong W, D'Souza KPCJ, Chiruza H (2003) Etude d'échelle sur l'exploitation minière artisanale du coltan dans le Parc National de Kahuzi-Biega. Dian Fossey Gorilla Fund, Wardell Armstrong, Newcastle-Under-Lyme, 43p.

Banque Africaine de Développement (2007) RDC, Projet d'exploitation de cuivre et de cobalt de Tenke-Fungurume, Résumé analytique de l'étude d'impact environnemental et social. BAFD, 33p.

Banque mondiale (2008) République Démocratique du Congo, la bonne gouvernance dans le secteur minier comme facteur de croissance. Banque mondiale, Washington, DC, 146p.

Berke C, Pulkowski J, Martin N, Vasters J, Wagner M (2007) Les ressources naturelles en RDC, un potentiel de développement? BGR- KMW, Francfort, 108p.

CTCPM (2006) Termes de référence du plan minier. Ministère des Mines, République démocratique du Congo, Kinshasa

Campbell B (1997) Quelques enjeux conceptuels, idéologiques et politiques autour de la notion de gouvernance. In: Institut africain pour la Démocratie (ed) Bonne Gouvernance et Développement: Actes du Symposium International, Dakar, pp. 65–94

Campbell B (2001) La bonne gouvernance, une notion éminemment politique. In: Haut conseil de la coopération internationale (ed) Les non-dits de la bonne gouvernance. Karthala, Paris, pp. 119–149

Campbell B (dir.) (2005) Stratégies de lutte contre la pauvreté et espaces politiques: quelques inter-rogations. In: Campbell B (ed) Qu'allons-nous faire des pauvres ? Réformes institutionnelles et espaces politiques ou les pièges de la gouvernance pour les pauvres. L'Harmattan, Paris, pp. 13–45

CEI (2002) Etat de la diversité biologique en République du Congo, Niveau de connaissance, utilisation, gestion et menaces. PNUD, Kinshasa

De Failly D (2001) Le coltan: pour comprendre. In: De Marysse S, Reyntjens F (eds) L'Afrique des grands Lacs. Annuaire 2000–2001. L'Harmattan, Paris, pp. 279–307

De Failly D (2007) Production d'or, d'étain et de coltan au Kivu (RD Congo): qu'en est-il aujourd'hui? Bureau d'Etudes Scientifiques et Techniques (BEST), Bukavu

De Jonghe A, Berck A-S (2007) Des conflits liés aux ressources naturelles? Le pourquoi et le comment illustrés par le cas du Pérou et du Congo, Quel impact sur la souveraineté alimentaire? Justice et Paix, Montréal, 31p.

De Rivero O (2006) Letter dated 18 July 2006 from the Chairman of the Security Council Commit-tee established pursuant to resolution 1533 (2004) concerning the Democratic Republic of the Congo addressed to the President of the Security Council. Security Council, United Nations, New York, 51p.

Droits et Démocratie (2007) Etude d'impact des investissements étrangers sur les droits humains, tirer les leçons de l'expérience des communautés aux Philippines, au Tibet, en République démocratique du Congo, en Argentine et au Pérou. Droits et Démocratie, Montréal, 155p.

G8 (2007) G8 summit déclaration: growth and responsibility in the world economy. In: G8 (ed) G8 Heiligendamm Summet. G8, Heiligendamm, Point 80

Garret N (2008) Walikale, Artisanal Cassiterite Mining and Trade in North Kivu Implications for Poverty Reduction and Security. Communities and Small Mining, London, 88p.

Global Witness (2005) La paix sous tension: Dangereux et illicite commerce de la cassérite dans l'Est de la R.D.C. Global Witness, Kinshasa

Global Witness (2006) Une corruption profonde: fraude, abus et exploitation dans les mines de cuivre et de cobalt du Katanga. Global Witness, Washington, DC, 56p.

Habermas J (1978) Raison et légitimité: problèmes de légitimation dans le capitalisme avancé. Payot, Paris

Hilson G (2002) The future of small-scale mining: environmental and socioeconomic perspectives. Futures 34:863–872

Hocquard C (2007) La crise du tantale de 2000, ses répercussions sur la mine artisanale et les conflits de la région des Grands lacs africains. BRGM, p. 6.

International Finance Corporation (2007) IFC's Policy and Performance Standards on Social and Environmental Sustainability and Disclosure Policy: Progress Report on the First 18 Months of Application. IFC, Washington, DC, 54p.

Laforce M (2006) La gouvernance des activités minières en Afrique, Une responsabilité partagée. Perspectives internationales, Centre d'études internationales sur la mondialisation (CEIM-UQAM), pp. 5–7

Lawrence S, Medeiros M, Reisch N (2006) La Banque mondiale en République démocratique du Congo. EURAC, 24p.

Le Potentiel (2008) Congo-Kinshasa: Lubumbashi- l'entreprise Chemaf accusée de polluer la cité de Tshamilemba. Le Potentiel, Kinshasa

Maninga S, Mbikayi F (2007) Etat de l'industrie extractive en république démocratique du Congo, le cas du diamant, Rapport provisoire. SARW, Kinshasa, 57p.

Mazalto M, Campbell B (2004) Vers un meilleur équilibre dans la politique étrangère canadienne, réflexions à partir de l'étude du secteur minier en République démocratique du Congo au regard des recommandations de la revue des industries extractives. GRAMA, Chaire C.A. Poissant, Institut d'Etudes Internationales de Montréal, 21p.

Muhindo Songe Luyeye P (2007) Rapport annuel d'activités, Exercice 2006. République démocra-tique du Congo, Ministère des Mines, Direction de protection de l'environnement minier, Secrétariat Général des Mines, 31p.

Mustapha AO, Mangala P, Mbuzukongira MJ (2007) Occupational radiation exposures of artisans
    mining columbite-tantalite in the eastern Democratic Republic of Congo. Journal of Radiolog-
    ical Protection 27:187–195
Mutabazi A, Sanganyi Y (2007) Etat de l'industrie minière extractive en R.D.C: le cas du coltan
    au Sud-Kivu. South African Research Watch, 37p.
Mutombo WK (1999) Ordonnance-Loi no 81/013 du 2 avril 1981 portant législation générale sur
    les mines et les hydrocarbures. In: Business Service (ed) Législation minière congolaise de
    1888 à nos jours (textes et commentaires), Kinshasa
OCDE, Ministres de l'Environnement (2004) Recommandations du Conseil sur le flux de matières
    et la productivité des ressources. Entériné par les Ministres de l'environnement le 20 avril 2004.
    OCDE, p. 4
Onorato WT, Fox P, Strongman JE (1998) World Bank Group Assistance for Minerals Sector
    Development and Reform in Member Countries. Technical Paper. World Bank, Washington,
    DC, 48p.
Palan R (1999) Susan strange, 1923–1998: a great international relations theorist. Review of Inter-
    national Political Economy 6:121–132
Pasche A (2004a) Mine Uranifère de Shinkolobwe, République démocratique du Congo, Mission
    d'évaluation de la situation environnementale. Joint UNEP/OCHA Environment Unit, Genève,
    18p.
Pasche A (2004b) Rapport complémentaire sur la situation environnementale à la mine uranifère de
    Shinkolobwe en République démocratique du Congo. Joint UNEP/OCHA Environment Unit,
    Genève, 31p.
Radio Okapi (2007) Chemaf accusé de pollution. Radio Okapi, United Nations in DRC.
Ravallion M (1995) Living Standards Measurement Study. World Bank, Washington, DC.
RD Congo (2002) Loi portant code minier. RD Congo, Gouvernement Kinshasa, 137p.
RD Congo (2003a) Décret no 038/2003 du 26 mars 2003 portant règlement minier Gouvernement.
    RD Congo, Kinshasa, 179p.
RD Congo (2003b) Décret par le Président de la république: portant création d'un service public
    "Service d'Assistance et d'Encadrement du Small Scale Mining (SAESSCAM)" Président de
    la République, Kinshasa
Redmond I (2001) Coltan Boom, Gorilla Bust, the Impact of Coltan Mining on Gorillas and
    other Wildlife in Eastern DR Congo. Dian Fossey Gorilla Fund Europe, Born Free Founda-
    tion, Kigali, 26p.
Réseau Ressources Naturelles (2007) L'impact de l'exploitation minière sur l'environnement du
    Katanga. Table ronde. Réseau Ressources Naturelles, Lubumbashi, 43p.
Sakata G, Trefon T, Ngoy B (2007) Rapport d'étude sur le Service d'Assistance et d'Encadrement
    du Smal Scale Mining (SAESSCAM)/Katanga en RD Congo. Project Geology for an Economic
    sustainable development, (GECO 2007). Musée Royal de l'Afrique centrale, Tervuren, 34p.
Sarrazin B (2005) La construction des problèmes environnementaux en Afrique subsaharienne: la
    mise en place d'un diagnostic de Washington sur les ressources naturelles. Revue canadienne
    d'études du développement 26:799–815
Secrétaire Général des Mines (2005) Projet de création de laboratoire d'analyses minéralogiques et
    études environnementales pour l'administration du Ministère des mines. République démocra-
    tique du Congo, Mines Md, 12p.
Secrétaire Général des Mines (2008) Avis et considérations du secrétariat général des Mines sur
    le projet de loi-cadre sur la gestion et la protection de l'environnement. Ministère des Mines.
    République démocratique du Congo, Kinshasa, 9p.
SNC-Lavalin International (2004) Rapport préliminaire, Etude sur la restauration des mines de
    cuivre et de cobalt, République démocratique du Congo. Division internationale, environ-
    nement. Banque mondiale, Montréal, 204p.
Szablowski D (2007) Transnational Law and Local Struggles, Mining, Communities, and the
    World Bank. Hart Publishing, Oxford & Portland
UNDP (2007) UNDP and UNEP Cement Their Partnership with New Poverty and
    Environment Facility, United Nations Development Programme http://content.undp.org/
    newsroom/2007/february

Vande Weghe J-P, Franssen J, Kalambay G, Kramkimel J-D, Dieudonné M (2005) Etude Profil Environnemental (PEP) de la République Démocratique du Congo, rapport provisoire. Délégation de la commission européenne, EURATA, Kinshasa, 227p.

Waggitt P (2004) Radiological Report on an Inter-Agency mission to the Shinkolobwe mine site Democratic Republic of Congo. International Atomic Energy Agency, Joint UNEP/OCHA Environment Unit, Geneva, 22p.

Weber-Fahr M, Andrews C, Maraboli L, Strongman J (2002) Mining and development: an asset for competitiveness sound environmental management in mining countries. World Bank Group, International Finance Corporation, Washington, DC, 28p. www.natural-ressources.org/minerals/CD/docs/twb/asset-competitiveness.pdf

WHO, EHA/AFRO team, 2008, Weekly Emergency Situation Update, Vol 1, N°32, 20th October, regional Office for Africa, 1 p.

World Bank (2007) Democratic Republic of Congo, Growth with Governance in the Mining Sector. World Bank, Washington, DC, 76p.

Part IV
Mining and Societal Issues

Applications of Stakeholder Engagement and Eco-Efficiency as Enablers of Corporate Responsibility in the Australian Mining and Minerals Processing Industry

Turlough F. Guerin

Abstract This chapter describes case studies of community engagement and eco-efficiency across the Australian mining and minerals industry. Eco-efficiency includes new processes, systems, and initiatives for improving air quality, energy, and materials efficiency, waste minimisation, improved waste water management, and increased water use efficiencies across several sectors of the mining and minerals processing industry. Specific findings were as follows: (1) environmental and social improvements at operations and within communities in which they operate can realise economic benefits and will not always incur a major financial cost; (2) local communities provide the means by which a mining or minerals processing operation can realise its full potential in contributing to a region's economic and social well-being; (3) improvements to waste management practises and waste prevention can lead to cost reductions or even increased revenues; (4) water efficiency improvements will be needed by any mining company planning to remain viable in the future, particularly in Australia; (5) at the operations level, there needs to be clear commitment from senior management to make the case for change to a more sustainable mining or minerals processing operation; and (6) mining companies need to work closely with businesses and suppliers to identify new processes that enhance the sustainability of their businesses.

1 Introduction

As the minerals industry attempts to transition to sustainable development, leading companies are demonstrating their ability to make investments that are profitable

T. Guerin (✉)
Telstra Corporation Limited, Melbourne 3000, Australia
e-mail: turlough.guerin@hotmail.com

The views presented in this chapter are those of the author and do not necessarily reflect those of his employer, Telstra Corporation Limited.

while at the same time meeting social and environmental objectives. Fully integrating these ecological, social, and economic objectives is a step towards sustainable development, and represents a commitment to corporate responsibility (Cragg 1998; McAllister et al. 1999; Cooney 2000; Blignaut and Hassan 2002; Hilson and Basu 2003; Newbold 2003; Azapagic 2004; Bridge 2004; Altham and Guerin 2005; Guerin 2006a, b, c; Anonymous 2008).

The Australian mining industry, through the Minerals Council of Australia (a member of the International Council of Mining and Metals, ICMM), introduced a code of practice in 1996 to facilitate greater environmental stewardship across the industry. In 2004, the Minerals Council of Australia then released a framework for sustainable development called "Enduring Value",[1] which further underpins the Australian industry's commitment to sustainable development (Guerin 2006a). Enduring Value is based upon the ICMM's 10 principles of sustainable development published in May 2003.[2] The Australian mining industry has demonstrated that they are able to respond to stakeholder concerns by implementing these principles of sustainable development. This has become evident by the ratings of several mining companies on international indices such as the Dow Jones Sustainability Index[3] and the Australian Corporate Responsibility Index.[4] Eco-efficiency (or cleaner production) and stakeholder engagement are two elements, or approaches to business, that enable mining and minerals processing companies to transition towards sustainable development. This chapter discusses the application of these elements through several case studies.

1.1 Cleaner Production (Eco-Efficiency)

Resource- and eco-efficiency is a preventive strategy and, to improve its success in contributing to a business's sustainability, it should be linked to the core activities of the business (Schnitzer 1995; Haile 1998). It includes working efficiently, because efficiency is an important contributor to a successful business. Inefficiencies generate waste but continued technological development can reduce this waste and potentially convert it into commercially valuable resources. There is always scope to improve the efficiency of industrial processes, and this principle, also known as continuous improvement, underpins the ISO standards for environmental, quality, and safety management systems. Such an evolutionary or continuous improvement approach is generally better than a revolutionary approach, because people and organisations adapt better to gradual change.

---

[1] http://www.minerals.org.au/enduringvalue

[2] http://www.icmm.com/our-work/sustainable-development-framework/10-principles

[3] http://www.sustainability-index.com

[4] http://www.corporate-responsibility.com.au

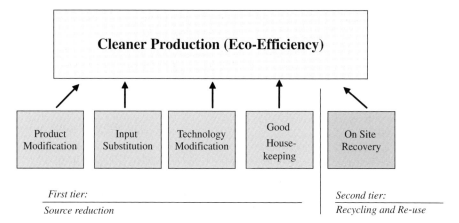

Fig. 1 Five eco-efficiency approaches. These approaches aim at making more efficient use of natural resources (raw materials, energy, and water) and reducing the generation of wastes and emissions at the source. This is generally achieved through a combination of product modification, input substitution, technology modification, good housekeeping, and (on-site) recycling and reuse

The main elements of resource- and eco-efficiency, as applied to the mining and the minerals industry (Watson 2002; van Berkel and Narayanaswamy 2004), can be summarised as follows (Fig. 1 and Table 1):

- Resource Use Optimisation: This includes the comprehensive utilisation of the mined resource through sequential mineral recovery, production of useful by-products, and conversion into geochemically-stable residues for safe storage.
- Input Substitution: Such as the use of less polluting process reagents and equipment auxiliaries (e.g., lubricants, coolants, and reagents that are available from reputable suppliers).
- Technology Modifications: These include improved process automation, process optimisation, equipment redesign, and process substitution.
- Good Housekeeping: This includes improvements in operational procedures and management in order to eliminate waste and emissions.
- On-site Recycling (Recovery or Reuse): This includes, for example, the useful application of process wastes (including emissions and process heat).

Resource- and eco-efficiency is an environmental improvement strategy, which leads to specific solutions applicable to any given business. However, resource- and eco-efficiency also draws upon, and is linked to, social and economic drivers. It is one of a number of ways in which an organisation can move toward sustainable development, and its success is linked to both internal and external stakeholders of a mining or minerals processing organisation. Resource- and eco-efficiency

Table 1 Broad applications of resource and eco-efficiency in the mining and minerals industry

Cleaner production element	Application	
	Mining	Minerals processing
Resource use optimisation	Improved separation of overburden and other wastes/materials to produce higher purity ore Enhanced modeling of ore body to optimise ore recovery Finding customers to match range of grades of ores produced	Sequential leaching to recover multiple minerals/metals from ore Conversion of process wastes and emissions into useful by-products Residue processing into geochemically stable forms for storage with minimised impact
Input substitution	Fluids selection for operating plant in process and non-process areas	Use of environmentally-friendly (e.g., biodegradable) reagents and process auxiliaries
Technology modification	Efficient mine design to minimise minerals movement during operation and for closure In-pit milling and separation Design of mine refuelling and lubricant dispensing facilities to enable lowest cost and safe supply to mobile plant	Alternative metallurgical processes (e.g., biotechnological) Use of energy-efficient fixed/mobile plant Application of fuel-efficient furnaces and boilers. Enhanced monitoring and control of leaching and recovery processes
Good housekeeping[a]	Monitoring and benchmarking of haulage fleet fuel efficiency.	Use of inventories to manage and control all inputs into process.
On site recycling	Composting of green wastes to produce heat/steam generation Reuse of overburden/waste rock in progressive rehabilitation of mine site Reuse of water collected in empty mine voids and pits On-site collecting and laundering of used oil	Recovery and reprocessing of un-reacted ore from processing waste Counter-current use of water for washing operations Reuse of ore process wash water Cleaning and reuse of metal-cutting fluids

[a] Specific examples are: staff training and enhancing their awareness of eco-efficiency gains across an operation; spill and leak prevention (e.g., of hydraulic oil, compressed air, water, chemicals); implementation of a maintenance strategy to enhance reliability of plant; lubricant cleanliness assessments and management programs that can reduce plant failures; and non-process waste segregation (i.e., from workshops).

complements life cycle analyses and product stewardship programs and initiatives, as well as stakeholder engagement.

1.2 Stakeholder Engagement

Stakeholder engagement has become increasingly important in the mining industry, and is now being recognised by mining managers as critical to being able to operate their mines (Togolo et al. 2001; Boutilier and Thomson 2003; Guerin 2006a, b, c;

Mtegha et al. 2007). The types of stakeholders relevant to mining operations include local, state, and federal governments, communities, landowners, suppliers, and customers. The way in which these relationships with an operation are managed is critical because this will impact upon all initiatives to enhance the operation's sustainability (Steele 1996; Low and Gleeson 1998; Humphreys 2000; Glauser et al. 2005).

Traditionally, Australian mining companies have had only minor involvement with local or remote communities, except in the case of purpose-built mining towns, where companies typically played the role of a paternalistic service provider. Furthermore, the way in which traditional landowners have been engaged with in the past has contributed to fragmentation and community disruption. Over the past decade however, companies within the mining sector have changed the way they interact with their communities. Most of the larger companies have made public commitments to engage with affected communities and other stakeholders on matters of mutual concern, a variety of formal and informal consultative processes have been established at the local level, and a growing number of operations are adopting community relations plans and programs.

Although the terms community and stakeholder are frequently used interchangeably in public commentary, they are not synonymous. In the mining industry the term "community" is generally applied to the inhabitants and landowners of immediate and surrounding areas who are affected in some way by a company's activities; these effects may be economic and social as well as environmental in nature. "Stakeholders", on the other hand, are those who have an interest in a particular decision, either as individuals or representatives of a group. This can include people who influence a decision, or can influence it, as well as those affected by it. Stakeholders include non-government organisations, governments, shareholders, and employees, as well as local community members.

Historically, mining companies tended to function as closed systems, largely insulated from the influence of public opinion. This was epitomised by the purpose-built mining town, where the company was the dominant employer, owned and provided most of the services (including housing), and managed the town as an essential element of the mine's production system. Today, by contrast, mines must increasingly operate as open systems. This shift in focus is attributable to the intersection of a number of factors, including heightened stakeholder and community expectations, global scrutiny (from the myriad of virtual social networks), the demise of the traditional mining town, and the growing influence of corporate social responsibility (i.e., corporate commitment to environmental, social, and governance issues: Habirono 2001; Jenkins and Yakovleva 2006; Hopkins 2007) across all sectors of business. Whereas governments previously regulated the mining industry with little direct community involvement, communities have now become active participants in the process.

The rights of people in affected communities need to be respected not just because this is good business, but because it is ethical and intrinsic to the role of a responsible corporation. Within the minerals industry, a major driver for companies to improve their community engagement practices has been the desire to reduce the community risks associated with current and planned operations, and for obtaining access to new resources.

Table 2 summarises the key social, community, and stakeholder engagement issues that have been identified in the Australian mining industry from a recent literature review by Solomon et al. (2007).

Table 2 Key social and community engagement issues in the Australian mining industry[a]

Theme	Description
Communities, community engagement, and community development	There is increasing emphasis on engagement integrated with business planning through the life of a mining project, from exploration through to closure; and community relations has emerged as an increasingly important strategic consideration.
	The most common community relations activities are consultation and engagement, public relations, sponsorship and donations, community programs, and responding to community complaints. Tools such as social impact assessment are the most commonly used method for assessing and mitigating the social issues associated with mine development and are commonly carried out by consultants.
	Companies are increasingly forming relationships directly with communities, unmediated by government; however, there can be a lack of clarity concerning the roles and power of participants in these relationships.
Sustainable development, social licence-to-operate interpretations	Sustainable development definitions tend to converge on Brundtland (WCED 1987), but interpretations of implementation remain largely contested.
	The most commonly employed mechanisms for implementing sustainable development in the sector are resource- and eco-efficiency, technology modification and onsite recycling (or reuse) of wastes, as well as stakeholder engagement.
	Governments commonly believe that sustainable development in the minerals sector calls for an effective legal, regulatory, institutional, and policy framework.
	A 'social licence to operate', an unwritten social contract as a complement to a government regulatory licence, has become a key principle of sustainable development for the industry.
Labour relations, management, and internal governance	Long work hours, shifts, and rosters in the industry are gaining increasing attention because of potential implications for health and safety, families, and gender equality.
	Fly-in fly-out (FIFO) practices are also gaining increased attention. A survey of industry professionals found that two main issues impacting on profitability were workforce turnover and skills shortages, and that lifestyle factors (e.g., worklife balance) were important when considering job options.
	Acute labour shortages are projected between 2006 and 2010, particularly in Western Australia and in non-professional occupations.

[a] Compiled from a recent review of the literature (Solomon et al. 2007).

2 Scope, Purpose, and Review Methodology

This chapter presents case studies from the Australian mining and minerals industries. Each of these case studies demonstrates the commitment of the individual organisations to improving their social, environmental, and financial performance. These case studies cover a cross section of the Australian mining and minerals sector, and were undertaken over a 17 year period from 1990. The case studies specifically address the following areas at a mine or mineral processing operation:

- Stakeholder engagement;
- Energy and materials efficiency;
- Waste minimisation;
- Wastewater management;
- Water efficiency.

For each case study, I provide a brief background to the site or company, a description of the existing processes prior to implementing the initiatives, a description of the sustainable development initiatives, and the drivers, barriers, and conclusions drawn from each of the initiatives. Table 3 provides a summary of the main outcomes from the case studies in terms of their contribution to sustainable development. It is the purpose of this chapter to highlight how corporate social responsibility (or corporate responsibility) can be successfully implemented at the operational level of a mining or minerals processing operation, using eco-efficiency tools and stakeholder engagement.

The material used in this chapter has been drawn from published papers (van Berkel and Narayanaswamy 2004; Guerin 2006a, b, c), the Australian Federal Government's Department of Environment and Water website (Anonymous 2008), company web pages, and from studies that have been previously reported at Minerals Council of Australia conferences in recent years (Guerin 2006a, b, c). The remainder of this chapter describes in more detail each of these case studies.

3 Stakeholder Engagement

3.1 Ravensthorpe Nickel Operations, Western Australia

3.1.1 Background

Nickel West's Ravensthorpe Nickel Operation (RNO) is a US$2.2 billion new mine development project located in the south coastal region of Western Australia.[5] The operation includes an open-cut mine and hydrometallurgical process plant, with a

---

[5]In January 2009, BHP Billiton announced that it would immediately commence the indefinite suspension of the Ravensthorpe Nickel Operation due to reduced prices for nickel and increasing capital costs.

Table 3 Case studies of corporate social responsibility, and resource and eco-efficiency initiatives in the Australian minerals industry

Case study focus	Minerals sector	Company and location	Description of case study (activity and outcome)
Stakeholder engagement	Nickel	Ravensthorpe Nickel Operations, Western Australia	Relationship building is key to managing socio-economic impacts; in particular, working with community and government during the study phase
	Energy	Wesfarmers Premier Coal, Western Australia	Continuing to build the relationship with local communities has been a driver for the initiatives described
	Aluminium	Alcoa World Alumina—various locations in Western Australia	Implementation of a range of resource- and eco-efficiency initiatives at bauxite mines and alumina refineries led to saving of approximately A$0.5 M annually
	Gold	JV between Delta Gold and Placer Dome Granny Smith Mine, Laverton, Western Australia	Developed range of sustainability practices, taking holistic approach to mining. Created opportunities for indigenous communities and reduced landfill waste
	Gold	Oxiana Golden Grove Operations, Western Australia	Initiatives reduced pollution and landfill waste, improved energy efficiency, reduced greenhouse gas emissions, and improved rehabilitation processes
	Diamonds	Argyle Diamond Mine (Rio Tinto), Kununurra, Western Australia	Mine was threatened with closure in 2001 but is still operating in 2008 as a result of efforts by management to implement a range of initiatives
Air emission management	Aluminium	Comalco Aluminium Limited (Rio Tinto), Bell Bay Smelter, Tasmania	Use of dry-scrubbing technology led to A$11 M in savings and contributed significantly to the local community
Energy and materials efficiency	Industrial minerals	Tiwest Joint Venture Pigment Plant, Kwinana, Western Australia	Modified process to recover synthetic rutile uses waste acid from neighbouring company to produce ammonium chloride for use in pigment production

Table 3 (continued)

Case study focus	Minerals sector	Company and location	Description of case study (activity and outcome)
	Coal	BHP Billiton Coal, Illawarra region, NSW	Capture of coal seam methane and piping it to surface where it generates 94 MW of energy through electricity generation (energy for 60,000 homes)
	Copper, lead and zinc	Xstrata's Mount Isa Mines, Mt Isa, Queensland	Program of innovations enabled opening of new mine, increased capacity while cutting electricity use and CO_2 emissions, and delayed demand for new power station
Waste minimisation	Aluminium	Alcoa Portland Aluminium, Portland, Victoria	Reduced waste going to landfill by evaluating processes, gaining the commitment of its workforce, and combining with waste minimisation concepts
Wastewater management	Steel	OneSteel Whyalla Steelworks, South Australia	Reed beds were introduced for treatment of industrial waste water to reduce water consumption and increase quality of discharged water
Water efficiency	Copper and uranium	BHP Billiton Olympic Dam Mine, South Australia	Production processes were modified so less water is used in mineral flotation/separation; recycle acid and saline water from mine tailings for drilling and dust control

projected life of approximately 25 years. The process plant was scheduled to be progressively commissioned during calendar year 2007, and planned to recover nickel and cobalt from laterite ores to produce a mixed nickel and cobalt hydroxide intermediate product. The process plant will have the capacity to produce approximately 50,000 t of nickel contained in mixed hydroxide product, and 1,400 t of cobalt per annum.

The project is owned by Nickel West (100% owned by BHP Billiton) and managed by RNO. Nickel West is the world's third largest producer of nickel-in-concentrate, providing 16% of global production. All its operations are based in Western Australia, and include the Mt Keith and Leinster Nickel Operations, the Kalgoorlie Nickel Smelter, the Kambalda Nickel Concentrator, the Kwinana Nickel Refinery, and the Ravensthorpe Nickel Operation. The project involves open-pit mining from three adjacent orebodies, and the process plant will treat both limonite and saprolite ores. Mining by open cut is the most appropriate method for the ore, which is shallow and flat lying. Rehabilitation objectives are achieved through a planned mining sequence that enables progressive backfilling of mined pits.

The Ravensthorpe Nickel Project is close to the towns of Ravensthorpe, Hopetoun, and Esperance. The Shire has a population of about 1,500 people within an area of 13,500 km^2. The project is located in the farming district of Jerdacuttup of Western Australia, which is characterized by farms growing wheat, oats, canola, and lupins, and carrying sheep and cattle. Resident famers number fewer than 100, many from families who cleared the land from virgin bush to establish their farms, so their ties to the land are very strong. The district has a rich biodiversity of flora and fauna, most of which is endemic to the region, together with a pristine coastline. Aboriginal landowners are also present in the region.

3.1.2 Pre-Existing Process

The Ravensthorpe mine is a greenfield operation. Since pre-feasibility commenced in 2002, the company has been addressing the local socio-economic issues that will arise from establishing such an operation within a small regional community. The project received investment approval in March 2004, with on-site development commencing the following month.

3.1.3 Description of Case Study Initiatives

Establishment of a Locally Based Workforce

With the operations phase of the project having a life of 25 years and the project being located on the coast, a rare opportunity has arisen to establish a locally based workforce rather than a fly-in/fly-out operation. An estimated 300 employees and hundreds more people indirectly employed by the project, plus their families, will become part of the local community. Considerable effort has been focused on establishing a close and productive relationship with the Ravensthorpe and Esperance Shire Councils that will host the incoming employees and their families. The company has also been collaborating with the Western Australian and Federal Governments in supporting the community by providing multi-user infrastructure, including residential land and upgraded water, power, roads, community services buildings, and educational facilities within the townships. The state government is also upgrading the regional Port of Esperance, which will be used for the import of raw materials and export of product. The company's project team has encouraged local and regional businesses to participate in the construction phase, and has established online registration of business capability and contact details so that local goods and services are visible to larger contractors who are new to the region. They have also actively supported initiatives by three business chambers within the region that focus on helping members adapt their businesses to the longer-term service and maintenance requirements of a large mining and processing operation. This will facilitate increased capacity building within local businesses, broader skills capability, and increased retention of younger people within the region, which will in turn enhance the sustainability of the local communities.

   The estimated investment of approximately A$950 M in the Ravensthorpe project will provide both direct and indirect benefits to Western Australia. During operations, a large number of directly and indirectly employed workers and their families will live, work, and shop in, and become an important part of, the Shire of Ravensthorpe and surrounding communities. The economy of the region will also receive a significant boost, with approximately A$33 M in annual salaries being paid locally. A construction workforce of up to 1,200 people (peak) will be employed during the 3 year construction period, and a further 1,800 people are expected to be indirectly employed through other industries, including subcontractors and suppliers to the operation. The project is linked to the proposed extension of the Yabulu Nickel Refinery near Townsville in north Queensland. Because nickel is primarily used in producing stainless steel, the combined Ravensthorpe-Yabulu Extension Project forms an integral part of BHP Billiton's strategy to add value to minerals mined in Australia, and produce high value products for sale internationally.

Community Engagement

Historically, community concerns for large resource projects have been addressed within the scope of an Environmental Impact Study. However, the project team recognised early on that locals were not confident that regulatory authorities would adequately address their concerns. In response, the company facilitated the establishment of two local committees to assist community participation in the decision-making process. The Community Liaison Committee (CLC) has a regional focus, whereas the Jerdacuttup RNO Working Group (JRWG) represents the project's near neighbours, including land owners. The JRWG is a member of the CLC, which aids the effectiveness of the two committees by enabling the JRWG, whose focus is narrowly project/neighbour-related, to be represented in discussion of broader issues, while avoiding the danger of it being bogged down by having to deal with such matters.

   The CLC was formed to help the wider regional community adapt to the arrival of a major mining and processing operation. The committee assists in reducing potential tensions that could be associated with the difference in disposable incomes between the mining and agricultural sectors, and with large numbers of new families becoming permanent residents in what has been a very small community that has changed little for decades. The CLC also administers and evaluates sponsorship applications from the community, and recommends to the RNO how to allocate its annual sponsorship funds. Community Liaison Committee membership comprises representatives of seven non-government organisations, two local governments, and an independent academic from the University of Notre Dame, Western Australia.

   The Jerdacuttup RNO Working Group was formed to ensure that no social harm occurs, and to minimise long term environmental damage arising from the project's operations. The group was involved in establishing environmental and community baselines prior to the commencement of the project. Jerdacuttup RNO Working Group membership comprises seven farmers, a Jerdacuttup Primary School representative, two RNO representatives, and an independent community advisor appointed by the JRWG. Additionally, there are two alternate farm business representatives and an alternate local school representative.

The independent community advisor is funded by RNO for the JRWG. The advisor's role has been to review the social and environmental impact assessment work completed by RNO as part of the government approval process and its planning for the mining and processing operation. Advice is provided on the quality and the scope of the work in addressing the issues of the near neighbours to the project. Additional studies and baseline work completed through the JRWG include trial blasting, air quality, farmland values, and groundwater, soil, and vegetation programs. Several programs are still in progress, including a community health self-assessment, surface water flow predictions, and an independent review of the proposed designs for the tailings storage facility and evaporation ponds.

The feasibility study phase for this project was between 1998 and early 2004, and public meetings were first held in 1998. In response to the RNO indicating its consideration of a locally based workforce, the State Government established a statutory committee for the purpose of defining the community infrastructure that would be essential to attract and retain such people. All of this was started 5 years before the project was given the go-ahead by BHP Billiton in March 2004.

Other Aspects of Corporate Responsibility at Ravensthorpe

The operation is located in the Bandalup Corridor, a band of remnant vegetation adjacent to the Fitzgerald River National Park, and falls within the buffer zone of the Fitzgerald River Biosphere, a world-renowned biodiversity area. The Western Australian Department of Conservation and Land Management (CALM) manages both the national park and the biosphere. One of the allowable activities within the buffer zone of a biosphere is mining, subject to responsible environmental management. At Ravensthorpe, the clearing of remnant vegetation for project development has two main impacts on biodiversity, including loss of habitat for fauna and, to a lesser extent, direct fauna impact from road traffic. The loss of fauna habitat has been compensated through the purchase of an adjacent 650-ha "bush block" as a conservation offset, together with the revegetation of approximately 600 ha of existing cleared farmland to allow its incorporation back into the Bandalup Corridor. At the completion of these revegetation activities and subsequent mine rehabilitation, the width of the Bandalup Corridor will actually be increased.

During the feasibility study, detailed ecological survey work has identified over 700 individual flora species within the project leases, a number of which are endemic to the project leases, and in some cases have been identified for the first time. The project team has focused on reducing clearing of remnant vegetation by locating as much infrastructure as practicable on adjacent historically cleared land. Where clearing is unavoidable, progressive rehabilitation including backfilling of mined areas has been included in the mine development schedule. Additionally, four mining exclusion zones have been established to preserve endangered species. Results from large-scale rehabilitation trials, translocation trials for priority species, genetic studies, and seed propagation studies led to the development of rehabilitation and priority species management plans. The project also has the additional advantages described below:

- The hydrogen sulphide plant and hydrogen sulphide gas are no longer part of the process, resulting in reduced atmospheric emissions. In addition, electrowinning of nickel metal will not occur at the Ravensthorpe plant site, resulting in significantly reduced capital costs of the project and reduced power requirements compared with other nickel process flows.
- The power requirements of the project during steady state operations will be fully met by utilising waste heat from the acid plant, thus reducing the diesel fuel usage and resulting in the plant being fully energy independent.
- Key infrastructure components have been relocated to cleared areas to avoid unnecessary impacts to flora and fauna communities within the Bandalup Corridor.
- Mining plans for the Shoemaker-Levy and Hale-Bopp deposits (near to the main depot) are now included in the overall mine plan, providing greater flexibility in mine planning, and creating opportunities to selectively backfill mine voids.

Incentives for corporate responsibility at the operation are as follows:

- A commitment to building positive relationships with the community as an essential factor in this process;
- Establishing a good reputation for the operation and its parent company, characterised by trust and openness;
- Conveying to the community that the resources sector generally, and the company in particular, is a responsible manager of the environment;
- Securing partnering assistance from the Australian Government to deliver community infrastructure to rural and regional Western Australia.

3.1.4 Drivers, Barriers, and Conclusions

The driver for the project has been to establish the mine and return value to shareholders. Barriers and challenges in relation to managing the community integration have been:

- Working with more than one community committee;
- Responding to personal agendas pursued by members of community committees;
- Facilitating liaison between large state government agencies and small local governments in regional Western Australia;
- Developing a mining operation within a community of farmers, retirees, and summer tourists, who are naturally protective of their rural and regional lifestyle.

In conclusion, relationship building is essential to managing the socio-economic impacts of the Ravensthorpe Nickel Project. The key message from the success of the new project has been that, when developing large projects, a proponent should engage as early as possible with community and government during the study phase. A major future challenge for the operation will be to ensure the sustainability of the communities after mine closure.

3.2 Wesfarmers Premier Coal, Western Australia

3.2.1 Background

Wesfarmers Premier Coal is a part of the Wesfarmers Energy Group of Wesfarmers Limited, and is the largest coal producer in Western Australia where it operates a major open cut coal mine at Collie. This mine provides nearly 50% of the energy for the state's southwest integrated electricity grid. Collie is a town of over 9,000 people, and Premier Coal is an important contributor to the local economy, providing employment for nearly 300 people and supporting a wide range of regional industries and community-based activities. The operation is now one consolidated open cut mine with reserves to operate beyond 2100.

3.2.2 Pre-Existing Process

Since the start of operations in 1950, 2,642 ha of land have been disturbed by mining and associated activities. Conventional revegetation commenced in 1975, and 1,253 ha (or 47% of the disturbed land) have been rehabilitated as of 2006.

3.2.3 Description of Case Study Initiatives

Value-Added Land Rehabilitation Initiatives

Corporate responsibility is incorporated at the company strategic planning level, and personal responsibility, practices, and programmes are driven through the certified company ISO 14001-2004 environmental management system. The system provides a framework for enactment of objectives and targets set during the strategic planning cycle. Research and development in this area forms a significant part of this programme.

The mine consults with the broader community on sustainability issues through a variety of avenues, including a Stakeholder Consultation Group and formalised meetings with immediate residents and landowners. On specific projects, consultation is done through the development of individual community working group committees to drive projects forward. Its focus on sustainable mining is reflected in challenging yet imaginative rehabilitation programmes to leave positive legacies by converting three former mining operations into valuable community assets for future generations:

- A recreational lake with a surface area of 103 ha;
- A motor sports and driver training complex; and
- An aquaculture precinct for research and economic development.

Successful rehabilitation work requires careful planning and diligent machinery operators. A key to this success is correct waste rock management, with materials likely to generate acidic conditions buried deep in the dump or backfill profile. Final dump surfaces are covered with a 1–2 m blanket of inert material, and spread with topsoil ready for revegetation. In 2006, the site commissioned a materials study, including rainfall simulations at various mine dumps around the coal basin. This was to improve dump designs and the rehabilitation program so as to minimise erosion, which has become more prevalent as a result of increased rainfall intensity at the operation. In-fill planting was undertaken with the local Aboriginal Ngalang Boodja Nursery, using seedlings raised by them. Children from the local Amaroo School assisted as part of National Tree Day. Due to evaporation over summer, the WO-5B void (former mine void, now Lake Kepwari) water level dropped 0.5 m. This was replaced by diversion of 439 ML of water from the Collie South Branch river. Keeping the void filled with water has assisted acidity control considerably, and reduced the natural fill time of 100 years to 5 years. The current pH of around 5 is considered an acceptable level for active water sports, and the area is now available for other valuable purposes such as tourism and recreation. It is also acceptable for aquaculture.

Research into aquaculture as a relinquishment option is occurring at this site. Construction of a hatchery commenced at the aquafarm under the guidance of the Centre of Excellence for Sustainable Mine Lakes. This is a group combining the research capacities of Curtin University, the University of Western Australia, Murdoch University, and Edith Cowan University, all in Western Australia. The mine is a principal industry sponsor of this research group, which is focused on mine lake rehabilitation, quality prediction, remediation techniques, and end use options.

The high winter rains saw the fill programme at WO-5H void completed, as the level rose 2.7 m causing overflow. Verve Energy (a local energy supplier/distributor) continued to use the lake as a temporary water supply.

The Collie Motorplex is based at the former Western 2 mine site, and has successfully built upon existing mine infrastructure including workshops, other buildings, and an extensive road network suitable for motor racing and training.

These "value-added" rehabilitation projects go beyond compliance requirements (with respect to the government) and provide new industry standards for reuse of rehabilitated land. Apart from the creation of economic benefits and regional employment, these projects have shown that industry can work with communities to establish assets of lasting social, economic, and environmental value after completion of mining.

Community and Government

The mine's stakeholders include its employees, customers, suppliers and other contractors, relevant government agencies, the local communities in which it operates, and shareholders in the parent company, Wesfarmers Limited. The mine has a senior manager whose role is to liaise with all authorities to ensure that relevant government agencies are kept informed of progress on existing issues and any new issues that arise. The mine is a member of the Local Emergency Management Action Committee (LEMAC), which is part of the Police and State Emergency Services plans. Each year, LEMAC reviews the emergency services plan for the Collie region. The mine is represented on various committees of the Chamber of Minerals and Energy, from the executive to subcommittees, allowing the mine to participate in discussion of issues concerning the mining industry in Western Australia.

The mine reports annually to a consortium of government departments through the Collie Coal Mines Environmental Committee. Information on issues and achievements is distributed in the mine's quarterly publication "Premier Post" which is sent to all employees and contractors, and is widely distributed throughout the community. The mine also distributes a summary of operations called "Fortnightly Focus" to all employees and on-site contractors by mail and email. This aims to keep employees and contractors up to date with key performance areas such as safety and the environment. During the year, the mine hosted several mine visits including school groups, universities, and international delegations. The mine also holds a Community Open Day in conjunction with the Chamber of Minerals and Energy Mine Open Day programme. The day includes a mine and workshop tour, followed by a visit to Lake Kepwari and the motorplex. Skiing and wake-boarding demonstrations at the lake provide an insight for visitors into the future recreational uses of the area. All of these activities are examples of engaging with local community stakeholders.

The mine's intranet provides information on environmental and safety standards and procedures to employees. The Stakeholder Consultation Group also provides a strong community interface and improves community awareness and involvement. This group provides assessments of the annual report, and in particular the scope, clarity, design, and ease of information access and feedback on any community concerns with current and planned operations. During the 2006 year, the Stakeholder Consultation Group met twice and participants were provided with data on the mine's business status, environmental performance, and community contributions. One meeting was also held with the mine's immediate neighbours at Buckingham. The Buckingham community is particularly concerned with heritage preservation of their precinct. These concerns are being addressed in an ongoing manner by the mine.

3.2.4 Drivers, Barriers, and Conclusions

The mine aims to continue to operate with the least amount of negative impact on the environment while at the same time giving back to the community. Continuing to build the relationship with local communities has been important in enabling the mine to operate profitably, which is the underlying driver for the initiatives described. Although the mine has only limited impact in terms of its geographic influence, the case study shows how such an approach to stakeholder engagement can work for other mines.

4 Energy and Materials Efficiency

4.1 Alcoa World Alumina, Western Australia

4.1.1 Background

Alcoa World Alumina Australia is a trading name of the unlisted public company Alcoa of Australia Limited. The company owns and operates alumina refineries at Kwinana, Pinjarra, and Wagerup in Western Australia, with a combined capacity of 7.3 Mt/year, equivalent to 15% of world demand. Alumina is exported worldwide from shipping terminals at Kwinana and Bunbury. The company also operates two bauxite mines at Huntly and Willowdale in the Darling Range south of Perth, which supply the three refineries. This three-refinery system evolved historically as a result of the large bauxite reserves which required extensive capital investment for refining. Alcoa now employs approximately 3,700 people in Western Australia.

4.1.2 The Pre-Existing Process

Access to bauxite is the keystone of Alcoa's activities in Australia. Darling Range bauxite is a low-grade resource to which value is added through refining and smelting. Bauxite is defined by the company as any ore in the lease which has a content of more than 27.5% aluminium oxide. Typically, it takes 7 t of Western Australian bauxite to yield 1 t of aluminium. Currently, 26 Mt of ore are mined each year from the Huntly and Willowdale mines. Darling Range bauxite is the lowest grade ore mined on a commercial scale anywhere in the world. At each new mining area, approximately 0.5 m of topsoil and overburden is removed and conserved for later rehabilitation, and the top 1–2 m of cemented caprock bauxite is drilled and blasted so that it can be extracted along with the more friable bauxite below. Alcoa has developed a sophisticated blast acoustic model to ensure that blasting noise is kept below acceptable levels. Once the ore has been broken, it is loaded onto haul trucks by excavators or front-end loaders, and transported to primary crushers at the mines. Ore mined at Huntly is transported by conveyor to supply the Pinjarra refinery and the Kwinana railhead stockpile. From the stockpile, bauxite is railed to the Kwinana Refinery. Bauxite from the Willowdale mine is conveyed to the Wagerup refinery.

Alcoa operates a three-refinery system in Western Australia between the capital city, Perth, and the port of Bunbury 200 km to the south. The bauxite is fed via conveyors to the alumina refineries where it is treated by the Bayer process to produce alumina. Alcoa's Kwinana refinery, which began operating in 1963, has a current rated capacity of 1.9 Mt/year. The Pinjarra refinery is one of the world's biggest with an annual capacity of 3.2 Mt, and Wagerup has a capacity of 2.2 Mt.

4.1.3 Description of the Case Study Initiatives

The key elements of the initiatives taken at the operation, were:

- Eco-efficient design, including environment and safety risk assessment and minimisation, reduction or substitution of process inputs, elimination of waste, and pollution prevention;
- Integration of environmental considerations into all business decision-making processes;
- Implementation of standardised environmental management systems (to ISO 14001:2004) emphasising continuous improvement.

Detailed examples of these initiatives are provided in the following sections.

Improved Vessel Descaling Practices

In the past, explosives, jackhammers, and high-pressure water have been used to remove scale from inside process tanks. These techniques are expensive, affect equipment availability, and can cause serious injury and health risks. They produce large volumes of waste product, with associated soil and groundwater contamination and waste disposal risks. Caustic and acid wash methods associated with changed operating and maintenance practices have now superseded these older techniques and they require less use of energy. This has resulted in reduced costs with capital costs of A\$0.8 M leading to benefits >A\$5 M/year.

Improved Heat Exchanger Maintenance

Process heat exchangers become scaled-up (contaminated) and require regular maintenance. In the past, prior to de-scaling, the contents of the exchangers were emptied onto the concrete floor resulting in temporary loss of process solutions (75,000 L for each event), degradation of the concrete, safety issues, and risks of soil and groundwater contamination. Now, cool, lower concentration water is used to push the process solution through the heat exchangers prior to drain down. This saves costs and avoids the previous problems.

Improved Bauxite Quality

The contamination of the refinery liquor stream by organic compounds is a major constraint on production. These organic compounds originate with the bauxite. During mining, the topsoil and overburden is removed using heavy equipment. Previously, pockets of overburden were mined along with the bauxite. Research indicated that this was a significant source of organic compounds. Mining practices were modified to carefully remove all overburden, thereby reducing the input of these organic compounds and subsequent waste. Net savings from waste reduction are estimated at >A\$14 M/year.

Reduction in Fine Alumina Waste

Fine alumina is produced by uncontrolled precipitation and product breakdown during materials handling. It is captured in electrostatic precipitators during calcination. In the past, product quality specifications resulted in the recycling of over 200,000 t/year of fine alumina in the three refineries. More than half of this was lost to residue. Technology and product modifications led to waste reduction refinements in precipitation control and calciner design, as well as reuse of some of the superfine materials as liquor-burning feed medium. After discussions with customers, some relaxation in the product specification further reduced the need for superfines to be reprocessed.

Oxalate Management

Sodium oxalate, an organic impurity, is removed from the liquor stream as a crystalline material. In the past, it was either treated with lime and disposed of in residue areas, or incinerated in a rotary kiln. While recovering some of the sodium value with the oxalate, these processes imposed a range of health, safety, and environmental risks. A new use for sodium oxalate has been found as a reagent for vanadium processing, and it is now being transported from Kwinana and Wagerup to Windamurra where a new vanadium mine and processing plant is being commissioned (in 2008). Oxalate kilns at Kwinana and Wagerup have now been shut down, resulting in cost avoidance, energy savings, and reduced emissions.

4.1.4 Drivers, Barriers, and Conclusions

The primary driver has been to maintain and improve site profitability. The principle incentives for these initiatives were reduced costs. There were, however, significant barriers to the project. These were primarily that it was "too hard" and that "everything had already been tried." In both cases the impetus of the team, with backing by management, led to successful implementation of the initiatives described, which has stimulated further improvements. This case study demonstrates how major economic benefits can accrue to a business by making changes to existing methods for minerals processing, focusing on improving resource utilisation, and enhancing waste management.

4.2 Granny Smith Gold Mine, Western Australia

4.2.1 Background

The Granny Smith gold mine is a joint venture between Delta Gold and Placer Dome (now part of Barrick). It is located approximately 25 km south-southwest of the township of Laverton in the northeastern goldfields region of Western Australia.

4.2.2 The Pre-Existing Process

Granny Smith was a typical gold mining operation that had minimal interaction with the local communities in which it operated. Employment of local indigenous people in particular was relatively low, and there were low levels of recognition of the local community.

4.2.3 Description of the Case Study Initiatives

The approach taken by Granny Smith ensures that significant steps continue to be taken to encourage benefits in terms of environmental, economic, and social outcomes. This approach applies to relations in both the immediate vicinity of the mine site and with the local community of Laverton. The result has been substantially improved dialogue and cooperation with this town of 500 people, a substantial proportion of whom are Wongutha, the traditional custodians of the surrounding country. By recognising the importance of sustainability, Granny Smith's gold operations have introduced a philosophy that recognises economic potential as only one of many values, such as social justice and conservation, which can be nurtured in concert with traditional business goals.

Community Engagement

Efforts to facilitate an increase in local employment opportunities for the indigenous people, in both the town itself and on the mine site, have become a significant gauge of social progress. Cultural initiatives that seek to encourage and support opportunities for local artists to display and sell their work have also become a normal part of the mine's development strategy. It was therefore determined that local arts and crafts, such as weaving, painting, pottery, wooden artefacts, and carvings in the form of traditional "tools of the trade" (e.g., shields and boomerangs) would benefit from the construction of a small tourist outlet to facilitate greater sales. In this way, the evolving process of consultation and implementation continues, with the result that the mine's sustainability strategy is becoming increasingly multi-dimensional.

A new satellite operation, named Wallaby, has been under review since its discovery in the 1990s. In conjunction with stakeholder consultation, exploration processes have continued, and mine development is currently underway. Of particular importance, due to the Wallaby site's proximity to Laverton, has been the series of pro-active steps taken to liaise with the local community. This community consultation conduit will remain open throughout the life of the mine.

The Laverton Leonora Cross Cultural Association (LLCCA) is working in conjunction with other key supporters to achieve its commitment to improve employment, retention, and training for Aboriginal people, and to assist in any community initiatives that seek to address their socio-economic disadvantages. A central theme is maintaining the flexibility necessary to accommodate cultural differences, and initiating mentoring courses. Selected employees with superior people skills are trained in communication and mentoring to help guide new trainees. Since 1997, the LLCCA has placed approximately 60 people per year, both Aboriginal and non-Aboriginal, into meaningful full-time employment. Pre-employment training has been provided for many others, as well as assistance with résumé writing and preparation.

Economic sustainability of a mining community is a difficult issue. Ultimately, mines do close and this can leave communities that have not found other means of generating income with few options. Presently, employment opportunities and income filter through to the community. Efforts to increase the penetration of this capital into the local economy are helpful to both short-term economic and longer-term education and other social prospects. Innovation is another means of harnessing previously undeveloped local potential, and is essential to providing a truly sustainable future for an area. With this in mind, the potential for olive farming, tourism, and crafts sales are being investigated as part of the overall project to provide real diversification of the local community. Another project is the current experimentation with various aquaculture techniques. It is envisioned that experiments currently underway will see exhausted open pits utilised for either commercial or recreational aquaculture activities. Species including trout, silver perch, black bream, barramundi, yabbies, and marron are currently being studied in order to determine their suitability to local climatic factors.

Land Rehabilitation

This program incorporates the tried and tested seed, save, and sow method, which is used extensively on mining sites. The revegetation program involves planning and design for both operations and closure at the outset. As progressive decommissioning of sites occurs over the life of the operation, revegetation follows in phases. The revegetation strategy includes final forming of disturbed land, planting schemes for tailings areas, and general rehabilitation of the Granny Smith location as extraction operations shift over to the Wallaby site. An emphasis has been placed on ensuring that a wide range of local species are seeded onto rehabilitation areas after earthworks. The original plant species at the dig site are de-seeded for propagation and eventual reseeding. Finally, when mining operations are complete, the pits can be backfilled or graded depending on cost.

Non-Process Waste Management

Granny Smith initiated the "Ruggies" charity recycling program in 1997 to recycle waste materials previously disposed of in landfills. Several mines have since joined the Ruggies program and thousands of tonnes of waste have been recycled. The program has also succeeded in cleaning up mine sites. Steel from mill balls, copper from cables, and aluminium from drink cans are just some of the waste items that are now being recycled. Transport contractors that once travelled to the mine sites with full trucks and returned back to Perth empty are now taking saleable cargoes back with them. The money raised from the Ruggies program benefits Western Australia's only children's hospital and other charities in the state. All the people working in the Ruggies Recycling initiative are doing so on a voluntary basis, reflecting the community spirit of the program.

4.2.4 Drivers, Barriers, and Conclusions

A key driver for the Granny Smith mine to improve its commitment to corporate responsibility was a need to engage more effectively with the community in which it operated. For most of the past century there have been few significant attempts to cultivate positive relations with local indigenous people in Western Australia. The gold mining industry has been particularly weak with respect to the employment of aboriginal people. This has been a costly and also morally wrong oversight and has only recently begun to be addressed. This case study highlights an example of this change in industry perspective to one that views gold mining as a potential major contributor to remote rural areas. Through considering the well-being of local communities in addition to the accepted needs for proper environmental and economic stewardship, isolated operations such as Granny Smith have made significant gains towards this goal in a relatively short period of time.

4.3 Argyle Diamond Mine, Western Australia

4.3.1 Background

The Argyle Diamond Mine is located in the remote East Kimberley region of Western Australia, upstream from Lake Argyle. The mine employs approximately 750 people, of whom three quarters work at the mine site. There are approximately 150 people employed in the Perth office, some of whom sort and polish the most valuable diamonds. The remaining diamonds are shipped to Antwerp for sale, 90% of which are then sent to India for cutting and polishing. Argyle helped to develop this Indian industry, which currently employs approximately 750,000 people. From India the diamonds are sold around the world to various markets. Some of these markets, such as the market for champagne and cognac diamonds, were created by Argyle through marketing displays of attractively set diamonds of different colours, created by world-class master jewellers. The result is that over 90% of Argyle diamonds are now sold as various grades of gemstones. The local Kimberley component of this is projected to increase as local employment, education, and contracting schemes contribute more to growth over time.

4.3.2 The Pre-Existing Process

Diamonds are usually found in kimberlite pipes that have been intruded and cooled rapidly during a volcanic event. The diamond pipe at Argyle was becoming harder to access as mining went deeper, meaning that the mining operation was becoming less viable. Management began to prepare the mine for closure in 2001, and in that year, 88 Mt of rock was mined yielding 10 Mt of ore. The ore is transported to the primary crusher, and waste rock is hauled to dumps. Within the processing plant the ore is crushed, scrubbed, and screened before gravity separation. The final step separates the diamonds from the ore by X-ray fluorescence, which picks out the diamonds from their light flashes and fires a pocket of air knocking the diamond off the conveyor belt and into a collector.

4.3.3 Description of the Case Study Initiatives

In 1998 a new management team was given an open brief entitled "Creating a Future" to investigate the possibilities of extending the mine life beyond 2001. Rapid action was needed to ensure that the window of opportunity was not lost. Once the technical feasibility of future mining was proven through to 2007, a positive attitude within the new management's ranks was required to guarantee that project delivery dates were met. Mobilising the workforce to implement this vision achieved a streamlined and successful transition process.

An important innovation emerged as a secondary benefit once the future of the project was ensured. A huge amount of earth around the orebody was stripped to create the necessary access for the new phase of mining. Previous to the extension of the life of the mine, the focus had been on final rehabilitation plans of the mine site. It had been determined that the waste dumps would need shoring up at an estimated cost of approximately A\$50 M. With the new stripping creating additional waste rock, this material itself could be sorted and used to achieve this end at no greater cost than ordinary operations. This kind of synergy is a hallmark of the new thinking that ensured the continuation of mining at Argyle. However, it also required substantial restructuring across the entire mining operation because the next phase of mining was going to produce less diamonds from the same amount of rock.

To make its future operations viable, Argyle needed to find productivity improvements and efficiencies across the business. Among these was the installation of new crushers, which enabled greater profitability by capturing a much higher percentage of Argyle's smaller diamonds through more precise separation of materials. Many of these smaller diamonds had previously slipped through with the tailings.

An enhanced water management regime has also created the potential for greater efficiencies and increased reuse. Greenhouse-friendly hydropower from Lake Argyle currently provides 98% of Argyle's electricity needs, with diesel generators providing back-up power during peak periods of use in Kununurra and other regional applications. Some efficiency has been gained in diesel usage by earthmoving equipment through the introduction of newer models; however, this remains a major environmental and financial cost for the mine.

The ecological and economic efficiency gains were matched with human efficiency gains in the workforce. This has produced significantly more cost effective diamond extraction. Thus, plans have changed to enable diamonds to be mined above ground up to 2008, while creating the time needed to assess the added possibility of extending the life of the mine further through underground operations. In 2005, mine owner Rio Tinto took the decision to go underground, and the mine's life has now been extended to approximately 2018. All of these managerial and efficiency achievements are significant advancements, but they are only part of the set of innovations that followed the decision to "Create a Future".

Waste Recycling

Programs are in place to manage, use, and recycle waste resulting from the mining operation, as well as the mining village. Used oil from the lubrication of machinery and vehicles is recycled. Markets for the oil are found in nearby communities and as far away as Darwin (in the Northern Territory), allowing the operation of the plant, after initial costs are factored out, to continue at no cost. Waste from the Argyle settlement is separated into colour-coded bins as a normal part of life in the isolated mining camp, and back-loaded to Perth at minimal cost. A culture of greater care among employees has begun to take hold at the same time as ecological gains are made through recycling and greater transport efficiency.

Land Rehabilitation

An innovative approach that has been adopted in the mining operations is the reha-bilitation of land that has been stripped for alluvial mining. This process began in 1988. The goal is to rehabilitate the area such that the original biodiversity is re-established. Over 100 ha/year is rehabilitated across former mined areas. The process involves:

- Immediate soil replacement leading to self-seeding;
- Contouring and deep ripping to ensure water retention on the areas;
- Gathering seed stocks through local harvesting;
- Developing nursery stocks of important species;
- Planting those areas where these particular species are in low abundance; and
- Monitoring to ensure the areas are restored.

A breakthrough in this rehabilitation at the mine came when it was discovered that many of the important and rare species in the rehabilitation sites were "bush tucker" (i.e., edible plants), species that the local Aboriginal community was keen to see included in the rehabilitation program. Thus they were invited to play a role in seed collection to help in the development of nurseries and to ensure that a full representation of these important species was grown throughout the sites. There are more than 50 plants of ethno-botanical significance in the area and these have been documented with the assistance of Aboriginal traditional owners, so that the value of the plants for nutrition, medicine, and other household applications can be passed on to the community. The result of this approach has been an important development of the relationship with the local Gija (Aboriginal) people.

Engagement with Aboriginal Community

The level of trust within local communities surrounding the mine has grown cau-tiously, through Argyle's involvement in many community development projects across the Kimberley region of Western Australia. This is a direct result of a deliberate policy that attempts to improve the proportion of aboriginal people employed at the mine. This change has not come easily. It meant that the management team had to become aware of many subtle changes in the culture of how people were hired in the company. The approach of using written résumés and direct interviews is now not used solely, because it was evident that this selected out many suitable applicants in the past. Instead, the process involves a recruiting workshop, to which people are brought for up to a week, and where they can slowly to adjust to living in the community. They are given tasks by an Aboriginal instructor including training in truck driving at a mini-mine site. This process gives managers and superintendents the opportunity to observe potential employees involved in a range of team and individual activities, and to assess their skills in this way. These same people have become the "champions" of Aboriginal employment, working to mentor and support new aboriginal employees in the workplace. Training is also provided continuously once people are employed. The culture gap is now being bridged rather than ignored and/or accommodated, resulting in greater trust and employment retention rates.

Health and Safety

The development of a safe and healthy work environment is an essential part of a sustainable operation, because it is a basic necessity for retaining the workforce. Argyle has instituted a range of programs including:

- Fit for work (including monitoring for substance and alcohol abuse);
- Healthy Lifestyle Program (promoting healthy eating habits and alternative fitness programs);
- Employee Assistance Program (psychological and emotional wellbeing services, monitoring, and promotion);
- Plant safety (detailed courses and training for management in recognising patterns of unsafe work practices);
- Emergency response (fully trained teams are based at Argyle, who also assist with local emergencies outside the mine site);
- Health information (e.g., on skin care for cancer prevention).

The results for the mining operation have been reductions of total injuries, severity of injuries, and time lost to injury since 1998. Employee retention rates remain high relative to the industry average.

Integration of Common Goals Between Departments

Health and safety, environment, economic efficiency, and community relations become increasingly sustainable when integrated. This is one of the key elements discovered by Argyle in the process of "Creating a Future". This means that there is a structure to ensure that these factors are all part of the one operation. Each is considered with respect to the others before major steps are taken, to maximise the benefits and avoid any potential problems. It also means that the process of decision-making must be able to recognise innovation at all levels of involvement in the company.

4.3.4 Drivers, Barriers, and Conclusion

As Argyle mine approached closure, the driver for action was the fact that drastic changes were required across the mine to enable it continue to operate. The response was the need for staff to be innovative and forward-thinking in their ideas, their technology, and their management systems. The mine has reaped multiple benefits from integrating environmental and social objectives into its economic performance. Sustainability improvements will outlast the life of the mine and will be used by Argyle people in other mining operations.

4.4 Golden Grove Mine, Western Australia

4.4.1 Background

Oz Minerals (formerly Oxiana) owns the Golden Grove base and precious metals mine located in Western Australia, approximately 350 km northeast of Perth. Oxiana acquired Golden Grove from Newmont Mining Corporation in July 2005. The mine produces concentrates of zinc, copper, lead, and precious metals. These are exported through the nearby Port of Geraldton to smelters in Asia and Europe. Golden Grove includes the Gossan Hill and Scuddles underground mines. Between 2001 and 2004, Golden Grove produced (metal in concentrate) an annual average of 55,420 t of zinc, 25,070 t of copper, 1.3 Moz of silver, and 15,476 oz of gold. Approximately 55% of the operation's revenue is derived from the sale of zinc concentrates, about 25% from copper concentrates, and about 20% from gold, silver, and lead.

4.4.2 The Pre-Existing Process

The following aspects of the mine were limiting its move to sustainability:

- An estimated 1,000 t of concentrates with a value exceeding A$1 M were lost each year as airborne emissions;
- Poor practices, such as depositing oily wash pad sediment in areas receiving groundwater inflow;
- Pumping oily wash-pad water directly into the mine water discharge dam;
- Solvent degreasers (i.e., petroleum-based hydrocarbons) were used on site for cleaning equipment;
- Two billion litres of water were being discharged each year, which represents a large volume of water for a waste stream.

4.4.3 Description of the Case Study Initiatives

Containment of Spilt Product

Construction of a storage shed for concentrates in 1998 gave a financial pay back period of 18 months for the A$1.5 M spent. It also reduced the contamination of surrounding bush-land and soils with metal sulphides. This contamination had caused vegetation die-back, and the acidified soils would not support plant growth. The environmental and financial benefits of the storage shed helped to justify further expenditure on spillage containment, including a A$1 M upgrade to spillage containment structures in the process plant completed in 2001. It also enabled operations staff to spend more time on operating the plant efficiently without the distraction of constant spillage clean up.

Underground Petroleum Hydrocarbon Management

Golden Grove discharges over 2 billion litres of water to a salt lake each year under rigorous license conditions, including a requirement that oil and grease contamination be less than 10 ppm. The maintenance of clean discharge water is, therefore, fundamental to the company's licence to operate. Mine discharge water is also used as process water in the plant, and oil contamination, if present, dramatically inhibits mineral flotation. Persistent oil contamination can therefore result in millions of dollars worth of potentially recoverable product being lost to the tailings dam. The main source of underground petroleum-based hydrocarbon contamination is hydraulic hose failure on loaders, which is common and can result in the immediate loss of 500 L of oil. In most cases, oil will spill directly into water being pumped out of the mine.

When monitoring of oil and grease in discharge water commenced in 2000, contamination of up to 30 ppm was commonly observed. This contamination is thought to be at least partially responsible for the intermittent poor performance of the plant at that time. An improvement programme for underground petroleum hydrocarbon management commenced in 2000. The program included routine maintenance, checking and change-out of hydraulic hoses at the first sign of cracking to minimise failures, equipping underground machinery and workshops with peat-based oil absorbent to immediately clean-up spills, and delivering hydrocarbon awareness training to all operators. The reporting of underground oil spills as environmental incidents was introduced. A hydrocarbon accounting system was also introduced to track oil usage and recovery.

The impact of the above measures was remarkably successful, with the average level of oil and grease contamination in the discharge water decreasing from 10 ppm in 2000 to 2 ppm in 2002, and with 7 of the 12 months having undetectable levels of contamination (<1 ppm). Oil spills are reported by operators, and there is a general high awareness and diligence in managing this issue in the work-place. In early 2002, an ecological risk assessment on the impact of oils and greases in mine discharge water on Lake Wownaminya was completed. This assessment was based largely on indicator organisms such as dragon and damsel flies. It indicated that the current level of oil and grease contamination is having no detectable impact on the ecological health of the lake, and that the ecology could accommodate intermittent contamination levels of 10 ppm which would occur from time to time following hydraulic hose failures on loaders. The Western Australian Department of Environmental Protection subsequently approved a licence amendment to increase the upper allowable level of oil and grease from 5 to 10 ppm.

Land Contamination

Golden Grove currently has spillage and emission controls to prevent land contamination from acidic soil contaminated with metal sulphide. A large volume of material (>5,000 t) from severely contaminated areas has been removed to the tailings dam, and replaced with clean soil. Areas of lesser contamination are being remediated in-situ with significant financial, energy, and land disturbance savings. In the past 2 years, Golden Grove has applied over 1,000 t of lime to sulphide-contaminated soil and the response has been significant. Bare trees, thought to be dead, have sprouted foliage, and seedlings are emerging in areas that were recently bare acidic deserts. The process of liming has been assisted by the strategic placement of irrigated topsoil windrows directly planted out with trees. In the past 7 years, Golden Grove has planted >10,000 seedlings. The in-situ liming of contaminated soils has already reduced the final closure liability of the mine by >A$0.5 M.

Solid Waste Management

Golden Grove has effective waste management procedures based on the tiered approach of minimising waste at the source, maximising recycling, and appropriately disposing of the remainder. The success of this approach is reliant on work-force commitment, which in turn is developed through a combination of training, rewards, and recognition, and making staff and contractors accountable for non-compliance with standards. Monthly work area inspections have been a particularly effective tool in this process. Each operational area is jointly inspected each month by a representative from the environmental department and the area supervisor, and is scored for performance on a range of aspects including waste management. The best performer on site receives an excellence certificate. Poor performance over the course of the year results in a poor performance appraisal, which is linked to witholding of bonus payments, or, in the case of contractors, to non-renewal of contracts. Competition for the excellence certificates is intense.

Energy Management

Golden Grove has been participating in the Australian Greenhouse Challenge program since 1999. The success of the mine's energy efficiency programs is a good example of site performance enabling the achievement of corporate targets. The eco-efficiency target for energy set by management for the operation in 2001 was for a 10.9% improvement by 2004. Golden Grove's reduction in kilograms of CO_2 released per tonne of ore milled dropped from 78.49 kg/t in 2001 to 68.25 kg/t in 2002, a reduction of 13%. The main reason for this change was improvement in mining efficiency. Waste rock that was previously brought to the surface is now being deposited directly into available stopes underground, and this has had the added advantage of limiting the expansion of waste rock piles.

4.4.4 Drivers, Barriers, and Conclusion

Drivers at the operation for improving the mine's corporate responsibility were the need to reduce emissions, reduce water contamination and wastage (through the use of incentives), and to improve the processing of ore at the operation. One of the greatest barriers to implementing these initiatives was the resistance of employees to work differently. Overcoming these barriers had significant financial benefits for the company, which has facilitated management support for environmental programs on site because of the realisation that environmental expenditure can give a high financial return on investment.

4.5 Comalco's Bell Bay Aluminium Smelter, Tasmania

4.5.1 Background

Comalco, a subsidiary of Rio Tinto Plc, is a major bauxite mining and aluminium smelting company with operations in Australia and New Zealand. Comlaco owns and operates an aluminium smelter at Bell Bay, located on the Tamar River in northern Tasmania (Australia).

4.5.2 The Pre-Existing Process

The process of smelting aluminium is a continuous operation. Fumes produced need to be "scrubbed" to remove fluoride. Prior to the commissioning of dry scrubbing technology, the smelter relied on wet scrubbing for the treatment of potroom emissions, whereby alkaline liquor was brought into contact with hot gases and the fluoride chemically removed. The wet scrubbing process had two stages. The first stage included a cyclone for the removal of coarse particles, which were recycled back into the process by blending with the primary alumina. Removal of particles was inefficient. The fumes passed through the cyclones then passed through the second stage. This essentially consisted of a series of sprays, whereby water with alkaline chemicals added contacted the fumes and absorbed the hydrogen fluoride. The hydrogen fluoride-laden water was then piped into a treatment plant, where fluoride materials were removed by precipitating cryolite, using chemical reagents. The residual water was later discharged into the Tamar River. The precipitated cryolite was dried to a solid using a rotary kiln that burnt fuel oil, before being recycled back into the smelting process. Cryolite was used as an electrolyte in the chemical bath of the reduction cell or pot. With wet scrubbing, water consumption for the site was 90 ML/month. Up to 60 h/month of downtime was incurred for maintaining each of the six scrubbing systems, and large amounts of chemical reagents were required to neutralise the hydrogen fluoride.

4.5.3 Description of the Case Study Initiatives

Dry scrubbing has now replaced wet scrubbing for treating fumes from the smelter, and is the most technologically advanced fume scrubbing system available for the aluminium industry. Hydrogen fluoride is captured in a gaseous form to be combined with alumina in a reactor. The alumina absorbs the fluoride and returns fluoride-rich alumina to a silo ready for recycling into the smelting process. In the early 1990s, existing dry scrubbing technology was complex, and required the same alumina to be recycled many times in order to remove sufficient quantities of the hydrogen fluoride. This resulted in higher operating costs and scaling, whereby alumina with small amounts of hydrogen fluoride and water would stick to steel surfaces, causing blockages and flow problems. Getting good contact between the hydrogen fluoride and the alumina, moving the fumes from the potlines to the scrubber, transporting the alumina around the site, all proved challenging. In 1995, Comalco finished the first installation of its own dry scrubbing technology at its New Zealand aluminium smelter. The pilot plant and the first full-scale commercial installation at the New Zealand smelter had proven excellent contact between the alumina and the hydrogen fluoride, and the alumina only needed to pass through the system once.

Dry scrubbing was installed at the Bell Bay smelter in 1997. In 1999 Comalco implemented plans to construct two more industrial fume scrubbers in its drive to continually improve environmental performance. The two projects at the Bell Bay smelter, involving green carbon fume scrubbing and carbon baking furnace fume scrubbing, have an estimated capital cost of A$18–20 M. With the installation of dry scrubbing, Comalco has achieved world's best practice for potroom fume scrubbing technology, and reduced fluoride emission by 33%. The A$44 M dry scrubbing project has not only delivered a significantly improved environmental performance, but is an inherently resource- and eco-efficient process. The specific benefits are described in the following paragraphs.

Financial benefits have been:

- Improved business viability through efficiency gains and lower cost performance;
- Elimination of the use of chemicals required for wet scrubbing;
- A$5.0 M saved from reduced chemical usage;
- A$4.5 M saved from recycling fluoride-rich alumina, reducing aluminium fluoride costs;
- A$0.5 M saved from reduced maintenance;

- S$1 M in miscellaneous savings as a result of dry scrubbing, including A$0.25 M savings in water consumption;
- Total savings are estimated at A$11 M/year.

Environmental benefits have been:

- 95% reduction in pot room-ducted fluoride emissions;
- 70% overall reduction in the operation's fluoride emissions;
- 70% reduction in the operation's water consumption;
- Improved fluoride regulation requirements;
- Negligible particulate emissions;
- Reduced discharge of water into the Tamar River;
- Cleaner working environment;
- Substantial reduction in chemical usage;
- Increased potroom fume removal rates have improved the capture of process fumes and significantly reduced fugitive emissions;
- Significant increase in recycling of materials, with the operation having more than halved the use of aluminium fluoride;
- 95% reduction of fluoride produced in potlines.

The benefits to the community have included:

- Improved occupational health and safety performance;
- Capital expenditure for the project has meant A$18 M invested directly in Tasmania, of which A$4 M was invested in the Tamar Valley region;
- Aesthetic improvements, because plumes from multiple stacks have been reduced;
- Peak construction workforce of 200 during project commissioning;
- Confirmation that the company is committed to continual improvement, particularly in environmental performance, and is prepared to invest significant capital in this area of its operations.

4.5.4 Drivers, Barriers, and Conclusions

A key driver for the process changes at the smelter was to reduce costs and to improve environmental performance. In the past, uncertainty surrounding the operation's future imposed constraints for site improvements. However, with a power supply agreement in place to extend the life of the Bell Bay operation to 2014, the investment of A$44 M to provide world's best practice fume scrubbing technology was made possible. All of the improvements to environmental performance and the benefits to the community ensure the continued viable operation of the smelter. This means continued employment, and the economic benefits that flow from the smelter's operation are likely to benefit Tasmania for many years to come.

5 Energy and Materials Efficiency

5.1 Tiwest's Kwinana Titanium Dioxide Pigment Plant, Western Australia

5.1.1 Background

Tiwest Joint Venture (Tiwest) is an equal joint venture between Ticor Resources Pty Ltd. and KMCC Western Australia Pty Ltd. The operations include: a titanium mine and wet processing plant at Cooljarloo (in Western Australia) producing titanium minerals concentrate; road transport to a dry separation plant at Chandala 60 km north of Perth, which produces ilmenite, rutile, zircon, and leucoxene; a plant at Chandala to upgrade ilmenite to synthetic rutile; a plant at Kwinana, south of Perth, converting synthetic rutile to titanium dioxide pigment; warehouses at Henderson, south of Perth, for storing the pigment; and exporting and other facilities at the Kwinana port. The principal end-product, titanium dioxide pigment, has broad application in, for example, paint, plastics, ink, and pharmaceutical products.

5.1.2 The Pre-Existing Process

The original process was based on chloride technology adopted during the 1980s and 1990s to improve pigment quality and reduce effluent rates. Using feedstock from the adjacent Chandala complex, the operation reacts synthetic rutile with petroleum coke and chlorine in fluidised bed reactors or chlorinators. The reaction process produces titanium tetrachloride, which is purified by condensation and fractional distillation. The remaining gases are systematically treated by scrubbing and incineration. Liquid effluent goes to the wastewater treatment plant, and treated effluent goes to ponds where further settlement takes place. The treated water from the ponds is discharged to the ocean under strict environmental controls.

The next stage of the process uses a special process for reacting titanium tetrachloride with superheated oxygen and support fuel to produce base titanium dioxide pigment. The base pigment goes through a finishing process which involves milling, classification, surface treatment, filtering, drying, micronising, and bagging. Various grades of pigment are produced to meet market requirements. Residue from the operation is separated and returned to Cooljarloo where it is encased in specially constructed clay-lined pits and used as part of the mine rehabilitation programme.

5.1.3 Description of the Case Study Initiatives

Various initiatives have been implemented under the broad headings of energy, materials, and water efficiency. These are discussed in the following sub-sections.

Energy Efficiency

A major initiative has been the installation of a cogeneration plant, commissioned in 1998, and owned by Western Power (an electricity supply and distribution company in Western Australia). A gas turbine generates electricity, and the exhaust gases that would have otherwise been vented into the atmosphere are used to generate superheated steam for the microniser, the final part of the production process. The plant generates all of Tiwest's power requirements, plus surplus electricity for the southwest Western Australia interconnecting grid. It also reduces steam demand from the package boilers, and reduces greenhouse gas emissions. Some of the other energy efficiency initiatives implemented include:

- A procedure to prevent excessive temperature of superheated steam, which is wasted by subsequent cooling of the excess heat;
- Tuning to prevent excessive burning of natural gas on the waste gas incinerators when on minimum fire;
- Tunnel driers tuned to prevent unnecessary over-heating of pigment;
- Commissioning of a second waste gas incinerator, producing steam and thereby reducing demand from the central plant boilers;
- Replacing the effluent pond transfer pump with an overflow weir;
- Installing non-return valves on all sump pumps to eliminate back flow and reduce the frequency of operation;
- Reduced air puffer frequency on non-critical bag filters in the pigment finishing section;
- Improvements in process monitoring and control, reducing the requirement to re-process off-specification material;
- Initiatives to reduce petroleum coke consumed by the chlorination reaction included tightening feed control and improving reaction efficiency;
- Initiatives to reduce consumption of gas as support fuel for the oxidation reactors, including improving gas flow metering, monitoring oxidation reaction stability, and improving combustion efficiency in the oxidisers.

Materials Efficiency (Production of Hydrochloric Acid)

Dilute hydrochloric acid (HCl), generated from scrubbing the gas stream from chlorination, was previously neutralised in the waste treatment plant. Two initiatives were realised to recover the HCl: first, to recover acid for direct sale, and second, for use in making ammonium chloride at the Chandala operation. For this purpose, a second scrubber was installed to produce HCl at a higher concentration, which enabled its reuse as a low quality acid. In the conversion process to ammonium chloride, waste HCl is transferred to neighbouring Coogee Chemicals, which converts it to ammonium chloride and drives it by tanker to Chandala for use in the production of synthetic rutile. The supply of the ammonium chloride to Tiwest is achieved at a significantly lower cost than previous imports.

Rutile Recovery Plant

The aim of this initiative has been to recover synthetic rutile from process effluent. After chlorination, all metallic chlorides go to a sump and to the wastewater treatment plant. Overflow from the fluidised-bed reactor is rich in unreacted synthetic rutile and petroleum coke, but this previously went with the metal chlorides to effluent treatment. A new plant was installed and commissioned in 2000 to recover synthetic rutile from the effluent using hydrocyclones, which separate by particle size. The titanium-rich fraction is filtered on a belt filter, washed, dried in a fluidised bed drier, and returned to the chlorinator with the normal input material. The titanium-poor fraction then continues to the wastewater treatment plant.

Use of Supplementary Fuel

In 2001, a switch was made to using an alternative fuel that produces less water to react with the chlorine and form HCl. This has improved overall chlorine efficiency.

Water Efficiency

In 1995, Tiwest conducted a water audit, which identified opportunities for reducing water consumption and identified areas for savings and reuse. Successful projects have included the commissioning of counter-current washing in pigment filtration, the reuse of microniser condensate in pigment filtration, and the installation of a recovery tank for water reuse. The use of groundwater and reprocessed water are currently being pursued as part of the Kwinana Waste Water Recycling Plant. Water consumption has been cut by 50% since plant start-up.

Production of Hydrochloric Acid

Hydrochloric acid production provides a mutual benefit for Tiwest and Coogee Chemicals. Tiwest benefits by having a local means for re-using much of its waste HCl in its operations, and saving the costs of neutralization. Coogee Chemicals benefits from being able to produce ammonium chloride cheaply, and having an ongoing local supplier and customer. Benefits that have been achieved include:

- Cost savings in the use of ammonium chloride;
- Reduced quantity of lime for neutralisation;
- Reduced waste for disposal.

Rutile Recovery Plant

The rutile recovery plant is designed to recover up to 21,000 t/year of unreacted synthetic rutile and coke (based on 180,000 t/year) which equates to net savings of approximately A$31,000 per day on an investment of A$6 M.

5.1.4 Drivers, Barriers, and Conclusions

Besides Tiwest's commitment to environmental improvement, these initiatives have been driven by cost, productivity, and environmental compliance considerations, as well as environmental benefits such as reduced greenhouse gas emissions. There were initial barriers to implementing this type of industrial ecology at the Kwinana industrial area. From 1990 to 2000, this type of synergy between companies in Kwinana increased by more than five times, indicating that the barriers were overcome through increasing engagement among companies in the area.

5.2 BHP Billiton Coal Operations, Illawara, New South Wales

5.2.1 Background

BHP Billiton Coal Illawarra operates four underground coal mines in and around the Illawarra region of New South Wales, which is situated 75 km south of Sydney. Three of these mines, the Appin, Tower, and West Cliff mines, produce approximately 3.5 Mt of coal per year. The coal is primarily used for domestic steel making, although some coking and energy coal is also exported.

5.2.2 The Pre-Existing Process

Gaseous methane is contained within subterranean coal seams, and is a potential explosion hazard. Methane is a greenhouse gas with high global warming potential, but is also a sought after fuel, which by not being recovered was a wasted resource.

5.2.3 Description of Case Study Initiatives

In 1995, BHP Billiton, in conjunction with Energy Developments Limited and Lend Lease Infrastructure, developed a power generation plant that uses waste methane to generate up to 94 MW of electricity. This is sufficient to provide energy to 60,000 homes. Supply of the fuel for electricity generation is achieved by capturing methane from within and below the coal seam (approximately 250 Mm^3/year). It is piped to the surface where it is distributed to a series of modular gas engines that drive electrical generators. Natural gas supplied by pipeline is used as supplementary fuel in the event of a shortfall in methane supply from the mines. BHP Billiton pays a fee to Energy Developments Limited to operate the generation plant; however, the energy that is generated is sold by BHP Billiton to the electricity grid. Some of the gas collection costs incurred by BHP Billiton are recovered in this way. Methane drainage of the mines is required to allow mining to continue safely. Utilisation of the methane provides an important energy resource, while reducing greenhouse gas emissions by approximately 50%. This represents a reduction in greenhouse gas output of the equivalent of approximately 3 Mt of CO_2 per year. In addition to providing an independent source of electricity for the community and the mines, the utilisation of this otherwise wasted resource reduces the amount of coal consumed by New South Wales power stations.

5.2.4 Drivers, Barriers, and Conclusions

The capture and utilisation of methane from coal seams provides, in conjunction with the legal requirements to do this, environmental benefit through reduced release of greenhouse gases. In addition, the economics of this process was also a significant driver in establishing the project. The difficulty of estimating future electricity prices due to deregulation of the power industry was a key consideration in determining the economic viability of the project. This case study demonstrates how the use of resources can be optimised, providing a new stream of income where there was previously a waste.

5.3 Xstrata's Mount Isa Pb-Zn Mine, Queensland

5.3.1 Background

Local discoveries of many new ore-bodies and expansion of current operations meant the inevitable increase in power demand for Xstrata's Pb-Zn operations in the Mount Isa Region of Queensland. It also meant that Xstrata had either to invest in new generating capacity or utilise existing generating plant more effectively. More effective use meant that Xstrata was required to undertake rigorous energy management reviews within its own operations.

5.3.2 Pre-Existing Process

There were several processes on site that resulted in high levels of electrical energy use. These were:

- Operating a generator above ground to deliver electricity at the mining operations;
- A low efficiency underground cooling system;
- No direct linkage made between what mine operators did and energy usage.

5.3.3 Description of the Case Study Initiatives

Major reductions in energy consumption, peak demands, and greenhouse emissions were achieved by implementing initiatives in a number of areas.

Underground Cooling

A 1,000 kW impulse turbine (a turbine which uses high pressure water) and generating set was installed 1,000 m underground. Chilled water at 1°C is discharged at around 100 L/s down a vertical pipe from the surface to underground. Prior to the installation of the set, the water gained around 2.5°C between the surface inlet and the underground outlet, resulting in an outflow temperature of 3.5°C. The installation of the set recooled the water by 2°C down to 1.5°C, reducing the chilled water requirement by 11%. Running the set during times of peak demand could also lower required generating capacity. Many of the mine cribrooms were subsequently fitted with dedicated refrigerated air conditioners, reducing pumping costs and the need for chilled water.

Ventilation Efficiency

Twelve 2 MW axial ventilation fans on the surface are mounted over vertical shafts (typically 1,000 m deep) are used to either extract or supply air to the underground workings. The pitches of the fan blades are automatically changed at regular intervals during the day by a process controller installed on the surface. Fan blades are driven at minimum pitch at times when ventilation of the whole mine is not required, such as during shift changes.

Dispersed throughout the mine are approximately 1,000 smaller ventilation fans, each fan having an average connected load of 11 kW. These fans increase general air movement underground and direct ventilation to priority areas. The ventilation system is controlled in such a way that energy use is minimised.

Energy Accountability

It was realised that even greater energy efficiency and reduced emissions would result if operators were made more accountable for energy use. In early 1997, a lease-wide personal computer-based energy and emission management system (PC:EMS) was installed. The system was designed to allow easy data input together with meaningful displays. Plant operators are the key to PC:EMS: they set daily operational forecasts at half hour increments via their terminals. In the central database, the forecast demand and energy requirements are calculated by multiplying the proportion of plant estimated to be operating by the full load rating (MW) of the plant.

Each plant operator, therefore, is able at any time to see the energy and environmental costs of running their plant, and any other plant, in dollars, energy consumed in MWh, peak demand in MW, and tonnes of CO_2 emitted to generate the necessary power. At present, the target is for forecasts to be within a band between 110 and 90% of actual demand. Because local operators may not be immediately aware of influences outside their control, their forecasts are subject to adjustment by the PC:EMS Administrator. These adjusted or final forecasts are then issued to the power station. Final forecasts are usually within 105 and 95% of the actual value, and are typically 104% of the actual or measured value. As a result of these initiatives, a number of conclusions can be drawn:

- Operators must "own" reductions;
- Improved operating practices usually result not only in lower operating costs but also improved safety awareness;

- Lower maximum demand means better utilisation of generators, ending in lower capacity charges:
- Energy management ensures most efficient and cost effective participation in a future emissions trading scheme;
- Any mine or minerals processing facility lacking systematic energy management will have higher production costs than its more efficient competitors, will attract lower public opinion, and will potentially suffer penalties for exceeding greenhouse emission limits;
- Enhanced energy management leads to a more profitable, better run operation.

5.3.4 Drivers, Barriers, and Conclusions

Xstrata was subjected to increased market competition with lower prices for the sale of metals. In this context, a reduction in operating costs was essential. The company was also looking to meet growing environmental concerns internationally by reducing CO_2 emissions from its coal and oil-fired generators, and postponing or removing the need for additional energy generation capacity. One of the main barriers was that suitable "off-the-shelf" software for the demand management initiative was not available. Therefore, the management team had to write, install, debug, and commission software whilst training operators in its use. The operators were scattered over a large geographical area, and during operator training and equipment commissioning production had to continue.

In conclusion, Xstrata has implemented a program of innovations, which has resulted in reduced energy consumption and the postponement of further capital outlay for generating plant. As a direct result, CO_2 emissions have been substantially reduced. The company has been able to open the deepest mine in Australia, with all the additional power requirements that this entailed, while still reducing total energy use. Carbon dioxide emissions related to metal produced have fallen by more than 10% since 1995/1996.

6 Waste Minimisation

6.1 Alcoa's Portland Aluminium Smelter, Victoria

6.1.1 Background

The Portland Smelter in south-western Victoria, Australia, produced its first aluminium in 1986 upon commissioning the first of two smelting potlines. The second potline, commissioned in 1988, raised the intended production capacity of the plant's 408 smelting pots to approximately 320,000 t of primary aluminium per year. Portland Aluminium is a joint venture between Alcoa of Australia, Eastern Aluminium, the Chinese Government, and the Japanese trading company Marubeni. On commissioning, the smelter was placed on a full production setting.

6.1.2 The Pre-Existing Process

Portland Aluminium uses the Hall/Heroult process to convert alumina to aluminium. Alumina ore (containing aluminium oxide) is subjected to high electric current in smelting pots. The electrical current passes from an anode through a molten bath of cryolite to a cathode which removes the oxide, leaving a cake of molten aluminium. Substantial infrastructure surrounds the complex chemical and physical aluminium production process. Alumina, the major component in the process, is produced at the Alcoa refineries in Western Australia and delivered to Portland by sea. Carbon products that are processed to produce 170,000 carbon anodes per year are also delivered by sea. Electricity is transmitted 700 km from brown coal-fired power stations in the Latrobe Valley via a 500 kV tower system. Natural gas is supplied by underground pipeline, primarily to fuel the anode baking furnaces. Raw materials are vacuum unloaded at the wharf and conveyed to the plant 4.2 km away, via an underground and overland enclosed belt conveyor. The final product from the process is near-pure aluminium in the form of 22.5 kg ingots.

6.1.3 Description of the Case Study Initiatives

In 1990, Portland set itself two basic, and at the time, unique goals. The first objective was to have no process materials going to landfill; the second was to have zero general waste going to landfill by the end of 1995. At the time that these goals were set, the overall financial loss to landfill was estimated at around A\$1.3 M/year. Resource- and eco-efficiency initiatives were subsequently undertaken in the three key stages of the aluminium production process: electrodes, smelting, and casting.

Electrodes

The anode baking furnaces require major maintenance after approximately 100 baking cycles. Maintenance was traditionally done in-situ. A method has been developed at Portland to allow full units of wall sections to be pre-built off-site, transported, and installed with minimal need for people to enter the hot baking furnace. Bricks from carbon bake furnace re-builds are returned to the manufacturer for re-processing. New products are manufactured from the recycled bricks and sold back to Portland Aluminium or on the open market. The aim of the furnace maintenance initiative is to totally recycle all furnace components to the specification standards for new materials.

Off-site maintenance of the anode baking furnaces minimises health and safety risks, increases furnace life, and reduces the generation of refactory waste. Furthermore, by improving the quality of the anodes, smelting pot processing performance is enhanced. The initiative cost A\$9 M and has an estimated payback period of 6–8 years. Emissions have been substantially reduced as a result of the initiative.

Smelting

After smelting pots have been relined, placed in the potline, and made ready for commissioning, they are pre-baked using a natural gas-fired baking system. This reduces start-up stress, extends pot life, and subsequently reduces the amount of spent potlining material generated. A specially designed dustproof and soundproof facility has been constructed where smelting pot shell demolition and maintenance is conducted. Specialised equipment has been developed to allow processing of large aluminium metal pads left in the pots, and to recover for recycling steel cathode bars, steel vapour barrier, and pot repair steel.

The smelting pot reline facility allows several pot shells to be re-built at once. Part of this operation is to prepare new cathodes for pot installation. The cathodes have a steel electrode glued into them, and during initial heating of the pots the glue gives off solvent gases into the working area of the pot rooms. Therefore, prior to pot start-up, the cathodes are baked in a special purpose, state EPA-licensed furnace; volatile gases are collected and scrubbed within the furnace prior to atmospheric discharge.

A process has been developed in conjunction with Ausmelt, an Australian furnace development company, which is unique to the world aluminium smelting industry. It is projected that the process will allow the old linings from smelting pots to be heat treated to remove residual cyanide and to recover the valuable fluorides.

Major up-grades of the bath handling facility have reduced both occupational health and environmental risks. Bath products are recovered during the recycling of spent anodes from the smelting pots. Traditionally, this caused an extremely dusty work environment, but dust has now been greatly reduced and contained.

The cost of refurbishing each smelting pot is A\$80,000–100,000. In the process, approximately 100 t of spent potlining is generated, which requires treatment at a cost of >A\$250 per tonne. Considerable savings are derived by extending pot life beyond the normal production period of 5,000 days. The smelting pot maintenance processes developed by Portland Aluminium were designed to minimise downtime and increase productivity.

Casting

The quality and presentation of aluminium ingots is important to market acceptance, and a continual effort is being made to reduce the level of impurities in metal during smelting. The installation of robots to skim ingots during casting has complemented these efforts. Skimming improves surface quality, reduces the risk of moisture ingress during storage and transport, and the subsequent dangers during ingot re-melting. This task was previously performed manually with operators exposed to risk of both hot metal splashes and fumes.

6.1.4 Drivers, Barriers, and Conclusions

With so much raw material being wasted and going to landfill in the early years of smelter operations, there was a strong incentive to recover as much raw material as possible to reduce costs. Setting such demanding goals of zero waste from the process and zero waste to landfill provided a cultural challenge to the operation, and waste minimisation has now become an integral part of Portland Aluminium's operations. The company actively embraces resource- and eco-efficiency practices as a means of achieving improved process performance in all aspects of its operations. Process changes reflect the company's commitment to its environmental, health, and safety policies. Waste management and resource- and eco-efficiency tools have provided the means for change, impacting strongly on the economic viability of the plant while reducing environmental risk and creating a healthier and safer working environment. Portland Aluminium's methods are now used by other smelters to obtain improved performance, reduce emissions, and lessen health and safety risks.

7 Wastewater Management

7.1 OneSteel Whyalla Steelworks, South Australia

7.1.1 Background

The scarcity of water is an intrinsic part of life in most of Australia. Better use and conservation of this vital resource will improve the quality of life for many Australians, especially those who live and work in the semi-arid regions. One such location is Whyalla in South Australia, where OneSteel's Whyalla Steelworks has been operating since 1964. In the year 2000, the Whyalla Steelworks (formerly owned by BHP Steel) became a separate entity called OneSteel. The operation produces approximately 1.2 Mt of raw steel per annum, principally billet steel feed for OneSteel's operations in the Newcastle region of New South Wales. Approximately 35% of the operation's raw steel production is also converted to finished products for the construction and rail industries. Iron ore for the Steelworks is sourced from OneSteel's mines 80 km away in the South Middleback Ranges (South Australia).

7.1.2 The Pre-Existing Process

Water is essential to the operation of a steelworks, being used for cooling, cleaning, lubrication, and numerous other purposes. In Whyalla, it is a scarce and expensive resource. Studies over a number of years were reviewed by the steelworks, and it was evident that there were unacceptable and unsustainable wastewater discharges into the Spencer Gulf.

7.1.3 Description of the Case Study Initiatives

In Australia, reed beds have been used for stormwater run-off and sewage treatment, but little was known about applications to steelworks wastewater treatment. The Llanwern plant at Whyalla represented the first reed bed technology trial on coke oven effluent. In soil-based reed bed systems, the effluent to be treated percolates through the biologically active soil and roots of a large bed of reeds, and then drains through a pipe at the base of the bed. The function of the reeds is to pump oxygen into the soil through the roots. Near the roots, there is an aerobic (oxygen-containing) zone, and further away there is an anaerobic (oxygen-free) zone. Thus, within the soil, a range of processes exist that allow the transformation of environmentally undesirable components of wastewater. Construction of the trial beds commenced in February 1993. Surveys of reeds in the surrounding areas were undertaken and information on reeds best suited to waste water treatment was reviewed. Accordingly, five reed varieties, all Australian native species, were selected to make up the trial. Once the system was established, the process of adapting the reeds to the effluent was started. The trial lasted 18 months and paved the way for the planning of the full-scale system. After the trials, a large scale (2 ha) trial system was constructed and commissioned in 1997. This involved the adaptation of the plants and biological life within the system to pollutants in the wastewater. Ongoing work is occurring to increase the effluent load removed by the reed beds. Currently, in excess of 70% of the ammonia is removed from the treated coke ovens effluent, with removal of other organic and inorganic materials running at or above 90%.

The benefits of the process were:

- Future recovery of a valuable resource of fresh water for recycling to the plant;
- Improving the quality of OneSteel's waste water discharges into the Spencer Gulf;
- Improving the quality of reclaimed land that previously had no value in the coke ovens area, while improving the visual appearance of that part of the plant;
- Reducing the impact of wind blown dust in an area with no vegetation; and
- Providing a physical shield (for the steelworks) against the hot north winds during the summer months.

7.1.4 Drivers, Barriers, and Conclusions

OneSteel needed to reduce water consumption across the steelworks, and also required a more efficient effluent treatment process for coke oven discharges. Specifically, previous studies identified the effluent from coke ovens as a significant source of organic matter and ammonia. OneSteel needed to reduce and/or eliminate these materials from its wastewater prior to discharging it. Strong winds, dusty conditions, hot summers with high evaporation rates, and salts within the soil and water have all posed problems at various stages of implementing the wastewater treatment initiative. These setbacks have been overcome through modifications of the design to suit Whyalla's environment. The reed bed waste water treatment system provides a low cost solution (particularly in terms of maintenance) to the coke oven discharge problems, but has also enhanced the environmental value of the land at the steelworks.

8 Water Efficiency

8.1 BHP Billiton's Olympic Dam Mine, South Australia

8.1.1 Background

BHP Billiton produces refined copper, gold, and silver, and uranium oxide concentrate at its Olympic Dam mine, near the township of Roxby Downs, South Australia. Western Mining Company (WMC) owned and operated the mine until April 2005. The mine's current capacity is 10 Mtpa, and in 2008 the mine is proposing an expansion to 70 Mtpa. Approximately A$2 billion has been invested in this integrated underground mine and processing plant to extract these minerals. The mine is close to the rim of the Great Artesian Basin, which covers 1.7 M km^2 in Central Australia (one of the largest sedimentary basins in the world). It is estimated that 425 ML of water flow into the South Australian section of the Basin each day. The overall flow is so large that water under pressure seeps to the surface in mound springs throughout several regions of arid South Australia; these springs are of high conservation significance.

8.1.2 The Pre-Existing Process

The mining operations use approximately 29 ML of water a day, while the nearby town of Roxby Downs, which was built to support the operations, uses another 2.8 ML. Roxby Downs has grown to become an important regional centre. In the 1980s, water was extracted from a single borefield located 110 km from the mine and town. The water requires desalinisation before it is suitable for human consumption or can be used for plant processes at the mine. An extensive monitoring program ensures that impacts on mound springs on the southern rim of the Great Artesian Basin are minimised. In 1996, the company was granted permission to begin drawing water from a second borefield a further 90 km into the basin. The goal was to reduce withdrawal pressures on the original borefield, and to ensure an additional water supply for expansion of mine operations. One of the conditions for approval was that the operation should monitor carefully the water flows to nearby water mounds.

8.1.3 Description of the Case Study Initiatives

Since 1997, the overall approach to reducing water consumption at Olympic Dam has been to:

- Develop more efficient work practices;
- Substitute lower quality recycled water where practicable; and
- Modify metallurgical processes to reduce water consumption or increase water recovery.

Various processes have been modified so that less water is used in flotation and separation of the minerals from the ore. These have included:

- Use of high density thickeners to reduce water passing to the tailings system;
- Recycling the acidic liquids from mine tailings that historically had been evaporated;
- Using highly saline water which seeps from the mine for drilling and dust control (water quality being of lower concern in such instances); and
- Implementing various other minor water conservation programs, including reuse of wash waters.

In the financial year ending 2007, the mine delivered water efficiency savings of 4 ML/day.

Since 1998, BHP Billiton has also worked closely with the 3,000 residents of Roxby Downs in water conservation activities, such as:

- Providing trees, plants, and drip irrigation systems to households;
- Encouraging use of and providing advice on arid zone gardens;
- Fostering mulching to retain garden moisture;
- Introducing low water consumption trees and shrubs;
- Recycling treated town effluent to the local golf course and sports ground;
- Introducing synthetic turf for recreational use; and
- Capturing and reusing storm water.

The cost of water conservation initiatives within the plant are many times higher than the cost of equivalent potential water savings in the pastoral industry. The operation has therefore offered to assist pastoralists in the borefields region to more efficiently utilise their water by providing assistance for the closure of boredrains, and their replacement with piping, tank, and trough systems. This initiative has been well received, and the potential water savings are between 14.6 and 23.8 ML/day, which is significantly greater than any potential water savings available at the mine and town. The operation will continue to research and develop methods to reduce water consumption and to encourage water conservation through educational programs for employees and other Roxby Downs residents.

8.1.4 Drivers, Barriers, and Conclusions

For individuals and companies living and operating in an arid climate such as South Australia, water conservation and management is an important consideration. BHP Billiton recognises its responsibilities to assist with management of the Great Artesian Basin, and is strongly committed to minimising its withdrawals, to conserving water, and to recycling water whenever possible. Because of the relatively high cost of water, the operation has considerable incentive to minimise its own use and to encourage residents to do the same. As of 1998, the unit cost of water for the project was A$1.61 per kL for general process water, and A$2.40 per kL for potable water.

This compares with the cost of A$0.88 per kL charged by the South Australian Government to other users. It was a company decision that water be sold to Roxby Downs residents at the same rate as paid elsewhere in the state of South Australia.

The main barriers to further water conservation are process constraints, and the capital and operating costs of recycling equipment. Owing to the high cost of supply, the processing facilities at Olympic Dam were designed and built, and are being operated, to be efficient in the use of water. However, the continuous reuse of process water results in a build-up of salts (particularly chlorides) and other contaminants, which originate either from the initial water source or from the ore or process chemicals used. Eventually, the concentrations of some salts and other contaminants become so high that they have detrimental effects on process efficiency.

This case study provides an example of where modifications were made to the minerals separation processes to substantially reduce water use. This type of innovation is critical for mining companies to remain in business, because water is commonly a limiting factor in the viability of a mining operation and the communities in which they operate.

9 Conclusions

The mining and minerals industry in Australia is embracing the concept of corporate responsibility, and working in such a way that it is contributing to the global transition of the industry to the more efficient use of resources and wealth creation. The case studies presented above cover a wide range of the environmental, social, community, and business issues facing modern mining operations in Australia. They demonstrate:

- A need for an integrated response by mining companies to the environmental, social, and economic impacts at their operations.
- That different operations face different challenges, depending on the location of their operations, and indicate the need for mining operations to work with the natural environment in their area. For example, the BHP Billiton Olympic Dam case study illustrates measures undertaken to improve water efficiency in remote, arid regions.
- The importance of mining operations working with and providing employment for indigenous communities. This is becoming an increasingly important expectation for many mining companies, and the efforts of Argyle Diamond Mine are an example of this.
- That there were relatively few examples where a mining operation had engaged explicitly with suppliers to enhance its own move towards resource- and eco-efficiency, a notable exception being the Tiwest Joint Venture. This suggests that there are barriers to the current way in which suppliers are engaged by mining operations, and that these barriers need to be overcome before the full benefits will be realised.

Based on an analysis of these case studies, the following observations can be made regarding the use of resource- and eco-efficiency, life cycle analysis (LCA), product stewardship, and stakeholder engagement in the Australian minerals industry:

- Technology modifications, including process equipment redesigns and on-site recycling (or recovery), are the most common applications of resource and eco-efficiency tools.
- Stakeholder engagement was identified as an important approach being used, with more than 50% of case studies demonstrating its important role.
- Resource use optimisation was also an important resource- and eco-efficiency tool, with approximately 50% of the reported case studies deploying this approach. This included optimisation of how ore resources and other process raw materials are used, as well as improvements in how other inputs were used, including energy sources and chemical reagent inputs into downstream processes.
- Input substitution, good house keeping, LCA, and product stewardship were less prominent, with less than 50% of case studies demonstrating that they were used to reduce consumption at mining and minerals processing operations.
- Pigment production at the Tiwest Joint Venture in Western Australia is an example where numerous tools for ecological efficiency were effectively used in an integrated manner.
- Coal seam methane capture at BHP Billiton's operations in the Illawarra region of New South Wales, and efficient water use at its Olympic Dam operations in South Australia, demonstrate where resource- and eco-efficiency technologies and stakeholder engagement have been integrated effectively.

When the financial performance of five of the case studies (two aluminium smelters, an alumina refinery, a gold mine, and a pigment production plant, all of which provided cost, benefit and payback data estimates) were aggregated, capital investments of A\$62 M yielded annual returns of A\$32 M. This represents an estimated average payback of 1.9 years for these environmental improvement projects (Table 4).

In summary, the case studies highlight the following important messages regarding the implementation of resource and eco-efficiency in the Australian minerals industry:

- Minerals companies in Australia are putting corporate responsibility and resource- and eco-efficiency into operation at their sites. The extent to which this is being done varies from operation to operation.
- Environmental and social improvements at operations and communities in which they operate can realise economic benefits, and will not always incur a major financial cost for a mining operation.
- Local communities provide the means by which a mining or minerals processing operation can realise its full potential in contributing to a region's economic and social well-being.

Table 4 Energy efficiency investments and payback periods in selected case studies

| Case study | Industry sector | Case study focus | Description of case study (activity and outcome) | Description of investment | Value (A$M) | | Payback period (years) |
					Initial investment	Annual saving	
Alcoa World Alumina— various locations in Western Australia	Aluminium	Energy and materials efficiency	Implemented range of eco-efficiency initiatives at bauxite mines and alumina refineries	Vessel descaling equipment	0.8	5	0.2
Comalco (Rio Tinto) Bell Bay Smelter, Tasmania	Aluminium	Air emission management	Installation of dry-scrubbing technology led to A$11 M in savings and contributed significantly to the local community	Dry scrubbing technology for treating flue gases	44	11	4
Alcoa Portland Aluminium, Portland, Victoria	Aluminium	Waste minimisation	Reduction in waste going to landfill by evaluating processes, through gaining the commitment of its workforce and combining with waste minimisation concepts	Improved raw materials handling, recycling of anodes, and improved smelting pot refurbishment	10	2.4	4.2

Table 4 (continued)

Case study	Industry sector	Case study focus	Description of case study (activity and outcome)	Description of investment	Value (A$M)		Payback period (years)
					Initial investment	Annual saving	
Oxiana Golden Grove Operations, Western Australia	Precious metals	Energy and materials efficiency	Initiatives reduced pollution and landfill waste, improved energy efficiency and reduced greenhouse gas emissions, and improved rehabilitation processes	Building of storage shed for concentrates, construction of spillage containment structures	1.5	2.25	0.7
Tiwest Joint Venture Pigment Plant, Kwinana, Western Australia	Industrial minerals	Energy and materials efficiency	Modified process to recover synthetic rutile using waste acid from neighbouring company to produce ammonium chloride for use in pigment production	Enhancement of rutile recovery plant	6	11.3	0.5
				Totals/average	62	32	1.9

- Improvements to waste management practises and waste prevention can lead to cost reductions and even increased revenues.
- Water efficiency improvements will be needed by any mining company planning to remain viable in the future, particularly in Australia where water is a major limiting resource. Similarly, energy efficiency improvements will also be important.
- At the operations level, there needs to be clear commitment from senior management to make the case for change to a more sustainable mining or minerals processing operation.
- Mining companies need to work closely with other businesses (e.g., neighbours) and suppliers to identify new processes, products, and knowledge that will increase the contribution of their businesses to broader sustainability objectives.

References

Altham J, Guerin TF (2005) Cleaner production. In: Rajaram V, Dutta S, Parameswaran K, editors. Sustainable Mining Practices. London: A.A. Balkema Publishers (Francis & Taylor Group Plc), pp. 93–120

Anonymous (2008) Australian Government Department of Land & Water: Eco-Efficiency & Cleaner Production Case Studies by Industry Type. From www.environment.gov.au, accessed on 1 August 2008

Azapagic A (2004) Developing a framework for sustainable development indicators for the mining and minerals industry. Journal of Cleaner Production 12(6): 639–662

Blignaut JN, Hassan RM (2002) Assessment of the performance and sustainability of mining sub-soil assets for economic development in South Africa. Ecological Economics 40(1): 89–101

Boutilier R, Thomson I (2003) Assessing the state of stakeholder relationships. Mining Environmental Management March: 12–15

Bridge G (2004) Contested terrain: Mining and the environment. Annual Review of Environment and Resources 29: 205–259

Cooney JP (2000) People, participation and partnership. CIM Bulletin 93(1037): 46–48

Cragg AW (1998) Sustainable development and mining: Opportunity or threat to the industry? CIM Bulletin 91(1023): 45–50

Glauser S, McAllister ML, Milioli G (2005) The challenges of sustainability in mining regions: The coal mining region of Santa Catarina, Brazil. Natural Resources Forum 29(1): 1–11

Guerin TF (2006a) A survey of sustainable development initiatives in the Australian mining and minerals industry. Minerals & Energy – Raw Materials Report 20(3): 11–44

Guerin TF (2006b) Realising minerals theories "Down Under". Mining Environmental Management March: 12–16

Guerin TF (2006c) Cleaning up the act [sustainable development in mining industry]. Materials World 14(8): 28

Habirono H (2001) Community based sustainable development: A corporate social responsibility. Mining Environmental Management July: 13–17

Haile S (1998) Environmental management systems and cleaner production. Journal of Environmental Planning & Management 41(2): 268–269

Hilson G, Basu AJ (2003) Devising indicators of sustainable development for the mining and minerals industry: An analysis of critical background issues. International Journal of Sustainable Development and World Ecology 10(4): 319–331

Hopkins M (2007) Corporate Social Responsibility. London, England: Earthscan

Humphreys D (2000) A business perspective on community relations in mining. Resources Policy 26(3): 127–131

Jenkins H, Yakovleva N (2006) Corporate social responsibility in the mining industry: Exploring trends in social and environmental disclosure. Journal of Cleaner Production 14(3–4): 271–284

Low N, Gleeson B (1998) Situating justice in the environment: The case of BHP at the Ok Tedi copper mine. Antipode 30(3): 201–226

McAllister ML, Scoble M, Veiga M (1999) Sustainability and the Canadian mining industry at home and abroad. CIM Bulletin 92(1033): 85–92

Mtegha HD, Cawood FT, Minnitt RCA (2007) National minerals policies and stakeholder participation for broad-based development in the southern African development community (SADC). Resources Policy 31(4): 231–238

Newbold J (2003) Social consequences of mining and present day solutions – Region II in Chile highlighted. Sustainable Development 11(2): 84–90

Schnitzer H (1995) Environment and innovation: Introducing cleaner production. Innovation 8(3): 309–317

Solomon F, Katz E, Lovel R (2007) Social Dimensions of Mining in Australia – Understanding the Minerals Industry as a Social Landscape. Melbourne, Australia: CSIRO

Steele BW (1996) Product stewardship at Dofasco: A discussion paper. Canadian Mining and Metallurgical Bulletin 89(999): 71–76

Togolo M, Rae M, Omundsen T (2001) Meeting society's expectations through stakeholder engagement. Mining Environmental Management March: 10–11

van Berkel R, Narayanaswamy V (2004) Sustainability as a framework for innovation in minerals processing. Journal of the Australian Institute of Mining & Metallurgy July/August: 80–86

Watson J (2002) Cleaner coal technology transfer to China: A 'win-win' opportunity for sustainable development? International Journal of Technology Transfer & Commercialisation 1(4): 347

World Commission on Environment and Development (WCED) (1987) Our Common Future, Report of the World Commission on Environment and Development (the Brundtland Commission). Oxford: Oxford University Press

Socio-Economic Impacts of the Nanisivik and Polaris Mines, Nunavut, Canada

Léa-Marie Bowes-Lyon, Jeremy P. Richards, and Tara M. McGee

Abstract Nunavut has gained importance in the last few years as an area of high mineral potential with exploration leading to discoveries of several mineral deposits that have, or will in the coming years, become mines. As a territory with an economy based in large part on government employment, new mining operations have the potential to provide Nunavut with an alternative way to develop its economy through job creation, local business opportunities, royalties, and taxes. Mining could provide the people of Nunavut with training opportunities for various jobs that can later be used for employment with community-based businesses. Understanding the socio-economic impacts of past Arctic mining operations, such as the Polaris and Nanisivik Mines, is important if these opportunities are to become reality and if Nunavut is to benefit as much as possible from future mining operations.

The Polaris and Nanisivik lead-zinc mines closed in 2002 after over 20 years of operation. Each mine was located near an Inuit community: Polaris, situated 100 km northwest of Resolute, was a fly-in/fly-out operation that used Resolute as a staging point; and Nanisivik was a community-based operation connected to Arctic Bay by a 21 km-long all-weather road. The differences between the physical connections of the mines to Inuit communities provide a good opportunity to compare and contrast their socio-economic impacts on each community.

This qualitative study uses company and government reports and the data from 51 interviews, conducted over a 4-week period in January–February 2005 with residents of Resolute and Arctic Bay, to learn how current and future mining in Nunavut might better establish long-lasting, positive socio-economic benefits according to current sustainable development practices (Bowes-Lyon 2006).

The results reveal that community members believe the mines had some positive short-term impacts, but few lasting benefits. Positive economic impacts included the creation of new businesses and jobs at the mines, with resulting increased incomes; however, many of these benefits disappeared after mine closure. Positive social

J.P. Richards (✉)
Department of Earth and Atmospheric Sciences, University of Alberta, Edmonton, AB,
T6G 2E3, Canada
e-mail: Jeremy.Richards@ualberta.ca

J.P. Richards (ed.), *Mining, Society, and a Sustainable World*,
DOI 10.1007/978-3-642-01103-0_13, © Springer-Verlag Berlin Heidelberg 2009

impacts were restricted to better education opportunities at the Nanisivik school, whereas negative impacts included increased alcohol consumption. The benefits of employment were mixed because, although employment was available to local Inuit, training was limited and did not in general lead to any certification that could be transfered to other jobs after mine closure.

Overall, it is concluded that because the benefits were not numerous and mostly did not persist after mine closure these mines did not contribute to the long-term sustainable development of the region. To increase socio-economic benefits and assist communities with fulfilling their sustainable development objectives, mining companies in Nunavut should emphasize education and training for locals and encourage local business development and partnerships, through strong relationships and close communication with involved stakeholders.

1 Introduction

Nunavut, Canada's newest territory, has seen a recent upsurge in mineral exploration activity, and development of several promising mineral deposits including the Hope Bay, Meliadine, and Meadowbank gold deposits, the Mary River iron deposit, the Izok Lake zinc resources, the High Lake copper resources, and the Jericho diamond deposit, Nunavut's first diamond mine.

As a result of the signing of the Nunavut Land Claims Agreement in 1993 and the creation of Nunavut in 1999, the Inuit beneficiaries of the agreement now control a portion of these mineral resources through surface and subsurface land ownership. They are currently owners of the largest block of privately owned mineral lands in Canada, totaling 37,870 km^2 (NTI 2007).

The Inuit of Nunavut will benefit from this land ownership through royalties and taxes collected from mines on Inuit-owned subsurface land, and through transfer payments from the Government of Canada for royalties and taxes collected from mines on Inuit-owned surface land. In addition, they will benefit from the Inuit impact benefit agreements signed between mining companies and regional Inuit associations when a mine is built on Inuit-owned land.

Nunavut's economy is considered mixed, with both the traditional land-based economy and wage economy playing very important roles. The traditional economy is considered to be worth C$40 to C$60 million annually (Vail and Clinton 2001), while the real gross domestic product is C$1,020 million (Statistics Canada 2006). Nunavut's economy relies heavily on both federal and territorial governments. In 1999, government spending accounted for over 55% of the total domestic demand in Nunavut, compared to 22% nationally (Vail and Clinton 2001). Furthermore, the government employed almost half the total workforce of Nunavut in 1999 (Vail and Clinton 2001). The territory faces other development challenges because it has the youngest population in Canada, an overall low level of education, and a high level of unemployment.

Nunavut's leaders want to develop the territory, and realize that mining is one way to bring economic prosperity to, and improve the social well-being of, its people (SEDSG 2003). One way to ensure that mining brings socio-economic benefits to the territory is to follow a sustainable development approach (Sherlock et al. 2003a,b; Richards 2005, 2006a) whereby mining companies, communities, Inuit organizations, and governments work together to maximize the overall economic and social benefits of mining while minimizing the environmental impacts (Togolo et al. 2001; Veiga et al. 2001). These benefits include employment, business development, education, capacity building, and training (e.g., O'Faircheallaigh 1999; IFC 2000; McAllister and Milioli 2000; Habirono 2001).

In order to determine the approaches best suited for creating an environment in Nunavut where sustainable socio-economic development can take place, it is necessary to look critically at past mining operations and their impacts on the communities closest to them. Examining these impacts provides insights into how individual communities and mines interact in remote regions, and how these interactions might be better managed to achieve sustainable development outcomes.

Two communities in Nunavut that have been closely exposed to mining in the last 25 years are Arctic Bay, near the Nanisivik mine, and Resolute, near the Polaris mine. Using a qualitative research approach, this study examines and compares the economic, social, and employment impacts of these two mines, both during mining operation and after closure, to determine their effects on the development of the Arctic Bay and Resolute communities. This study did not attempt to address the environmental impacts of the mines. Since closure, both the Polaris and Nanisivk mine sites have been extensively remediated.

2 The Nunavut Land Claims Agreement

The territory of Nunavut, inaugurated on April 1, 1999, is the culmination of many years of negotiations between the Inuit of Canada's Arctic, who wanted to settle their land claims, and the Government of Canada. In 1982, the Tungavik Federation of Nunavut (TFN) was created to negotiate an Inuit land claims agreement with the federal government. Negotiations over the subsequent 10 years culminated in the signing of the Nunavut Land Claims Agreement in 1993.

In exchange for loss of aboriginal title to most of their lands, the Inuit negotiated for a quantum of land where they would have legal title (McPherson 2003). Selecting land with high mineral potential was seen as a means to allow Inuit to benefit from any mineral resource development, and to help them achieve their goals of economic self-sufficiency (NLCA 1993). The Inuit subsequently obtained title to 355,842 km^2 of land, of which 37,870 km^2, or approximately 2% of Nunavut, includes mineral rights. Companies undertaking mineral exploration and development on Inuit-owned subsurface land pay royalties to Nunavut Tunngavik Incorporated (NTI), a not-for-profit corporation that administers all claim settlement benefits for the Inuit in Nunavut. On Crown Lands managed by the federal government, NTI also receives 50% of the first C$2 million paid in royalties to the Crown, and 5% thereafter on an annual basis (NLCA 1993; Richards 2006a). An additional way for the Inuit to potentially benefit from mineral development on Inuit-owned land is through Inuit impact benefit agreements signed between the Regional Inuit Association of the Nunavut region where development activities would be located, and the company responsible for the development activities. Such agreements permit the Inuit to negotiate training, hiring, housing, and business opportunities with individual companies operating in their region (NLCA 1993).

In addition to land ownership, the Inuit were given partial control of all Nunavut lands, waters, wildlife, and development through five management boards: the Nunavut Impact Review Board, the Nunavut Planning Commission, the Nunavut Water Board, the Nunavut Wildlife Management Board, and the Surface Rights Tribunal.

3 Mining and Sustainable Development in Nunavut

Mining in Nunavut has been limited in the past, and continues to be limited, due to lack of infrastructure, remoteness of resources from markets, high transportation costs, lack of a skilled labour force, sparse geological information on mineral potential and other resources, and strict and potentially overlapping regulatory requirements (Vail and Clinton 2001). Sustainable development investigations and planning for the north must therefore take these limitations into account.

The potential importance of mining for the development for Nunavut's economy has been recognized by numerous organizations, such as the Conference Board of Canada (Vail and Clinton 2001) and the Sivummut Economic Development Strategy Group (2003). They have also emphasized the importance of developing social programs while building the economy to gain the maximum socio-economic benefit from mines and other projects, and to maximize sustainable development in Nunavut.

For mining and other non-renewable resource development projects, the application of sustainable development principles implies resources will not be wasted, and that the development of the resource will be done in a socially, environmentally, and economically responsible manner (Crowson 2002; Richards 2002, 2005). Most importantly, the wealth created by extraction of non-renewable resources needs to be invested in other forms of human and social capital that can survive beyond the life of a mine (Richards 2006a, b). In this regard, the mining industry can apply sustainable development principles by focusing on the mineral resources (natural capital), and transforming that natural capital into human capital (through education and training) and social and physical capital (through wealth creation and infrastructure development), both in communities affected by mining operations and in wider society (Jackson 2005).

Job creation, transfer of skills and technology, and local infrastructure and services development are potential benefits of mining operations, but will only be enjoyed beyond a mine's production years if there are the incentives and knowledge to turn the investments into longer term *sustainable* development (MMSD 2002).

Mining industry attempts to address the sustainability of mining operations have occurred mainly due to pressure from NGOs, governments, and investors. Efforts, such as the Mining Minerals and Sustainable Development Project (MMSD 2002), the Global Reporting Initiative (www.globalreporting.org/), the Global Mining Initiative (www.icmm.com/gmi.php), the Mineral Policy Research Initiative (www.idrc.ca/en/ev-70315-201-1-DO_TOPIC.html), the International Council on Mining & Metals Principles (www.icmm.com/icmm_principles.php), the Environmental Excellence in Exploration program (www.e3mining.com), and others have created a plethora of guidelines and codes of conduct specifically for, or that apply to, mining companies. The goal of these guidelines is to help industry become more socially, environmentally, and economically responsible.

The MMSD Project was important to those working towards applying sustainable development principles to the mining industry. A principal recommendation to the MMSD project from MMSD North America was to "Enhance effort to address the legacy of past mining and mineral activities" (MMSD North America 2002, p. 55). This reflected the poor environmental and social record of past mining operations, which has created a certain level of public distrust of the mining industry. The current study addresses this recommendation by examining the socio-economic impacts of two now-closed mines on the communities closest to them, and offers suggestions for future mine development in Nunavut that could lead to sustainable development outcomes.

4 Methodology

A qualitative approach (Bryman 2001) was used for this study, in order to understand the different experiences and opinions of people affected by the mines through first-hand observations and interviews.

Fieldwork was conducted over a 4-week period in January and February 2005, with 2 weeks spent in each of Arctic Bay and Resolute. Informants were chosen by using purposive and directed sampling, which allows the researcher to select "information-rich" cases that help with identifying the themes important to the study (Erlandson et al. 1993; Patton 1990). This sampling method requires as many interviews as are necessary to get the best and most in-depth information to help answer the study's questions (Erlandson et al. 1993). Each informant was drawn from one of eight categories: former mine employees, mine employee family members, businesses, elders, community members/leaders, nurses, Royal Canadian Mounted Police (RCMP) officers, and school officials. Mine employees were interviewed in the greatest numbers because they were knowledgeable about mine employment and the mines' impacts on them and their families. Business people and elders were interviewed in large numbers as well because the businesses had important information about the economic impacts of the mines, and the elders had lived in the

Table 1 Distribution of
participants for each
interview category for Arctic
Bay and Resolute

	Resolute	Arctic Bay
Mine employees	7	9
Businesses	5	4
Elders	4	4
Mine employee family members	0	4
Community members/leaders	2	4
Nurses	1	2
RCMP	1	1
School Officials	2	1
Total	22	29
Total population in 2001	215	645
Total population in 2006	229	690

communities for the duration of the mine projects, and so were in a good position to discuss changes in the communities resulting from the mines. Interviews in the other categories reflected the smaller numbers of total potential informants within the two communities. Table 1 shows the distribution of participants for each interview category for Arctic Bay and Resolute.

The interviews included two main parts. The first part, which was similar for all interviews, gathered demographic information from each informant to be used to create a profile of the people interviewed. The second part sought information about each informant's perceptions of the impacts of the mines on topics such as education, mine employee family circumstances, community–mine relationships, communities' economic conditions, and future mining scenarios. A semi-structured interview format was used. Interviews were conducted in the language chosen by the informants. When Inuktitut was chosen, an interpreter was present to assist with the interview.

Interviews were transcribed and then analyzed using NVivo, a qualitative research analysis software. Analysis was performed by identifying recurring themes pertinent to the research from the interviews. These themes focused the interpretation of results and allowed for direct comparison between comments made by individual participants.

5 Description of Study Sites

5.1 Arctic Bay (Nanisivik Mine)

Arctic Bay (Figs. 1 and 2) is located on the northwestern tip of Baffin Island and has a population of 646 (2004 data; Statistics Canada 2005a). The current location of the community was first permanently settled in 1926 when a Hudson's Bay Company Post was built (Bissett 1968a). About 92% of the inhabitants claim Inuktitut as their first language (Statistics Canada 2005a), and hunting is still a large part

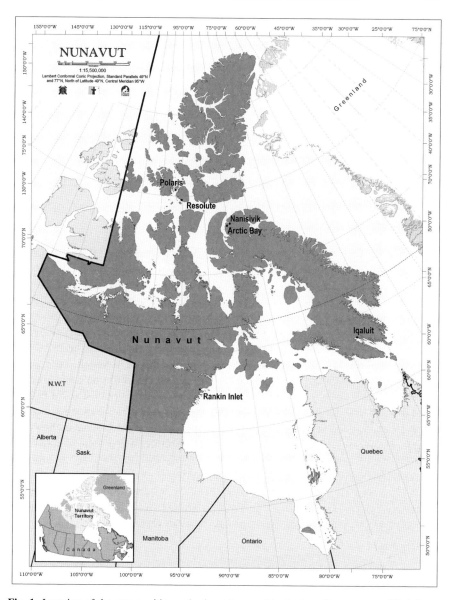

Fig. 1 Location of the communities and mines discussed in the text. Base map modified from The Atlas of Canada (atlas.nrcan.gc.ca; reproduced with permission of Natural Resources Canada 2008, courtesy of the Atlas of Canada)

of local activities. Since the mid-1960s, the community has gained basic services, some business, and scheduled air service from Iqaluit on a gravel runway suitable for jets, built specifically for the Nanisivik mine in the late 1970s. Local businesses offer some employment, but many jobs are related to community services offered by the hamlet, or are government-related.

Fig. 2 Community of Arctic Bay in July 2004

5.2 Nanisivik Mine, 1976–2002

The Nanisivik lead-zinc mine (Figs. 1 and 3) was Canada's first high Arctic mine, and is located 30 km by road from Arctic Bay. The Strathcona Agreement, signed on June 18, 1974, between the Department of Indian Affairs and Northern Development (DIAND) and Mineral Resources International (MRI; Government of Canada and MRI 1974), gave MRI permission to develop Nanisivik. For DIAND, Nanisivik would "provide employment and other socio-economic opportunities for Canadians, particularly those resident in the said Territory, and to obtain information on resource development possibilities in Arctic areas of the Northwest Territories"

Fig. 3 Nanisivik town site in July 2004. The town site is in the foreground, while the hill in the background represents the location of the deposit

(Government of Canada and MRI 1974, p. 1). The agreement also included C$16.7 million in grants and loans from the federal government for the construction of a town site at Nanisivik, an airport, a 30 km road linking Arctic Bay and Nanisivik, and a deep-sea port. In return for its investment, it was agreed that the federal government would receive an 18% equity interest in Nanisivik Mines Ltd., the mine would operate for 12 full production years, and the Nanisivik mine workforce would be 60% Inuit by the third year of production.

Approval for the project was rushed through, with only 3 months elapsing between Cabinet's approval of the project on March 28, 1974, and signing of the Master Agreement on June 18 of the same year. Significantly, approval was given before major environmental and socio-economic studies were performed (Hickling-Partners 1981). The first socio-economic impact study of the mine was conducted in 1979 by the Baffin Region Inuit Association, 3 years after mine production had started (BRIA 1979). The results of the study were intended to serve as guidelines for the government and the Nanisivik mine to address problems that had arisen in the early years of operation.

Production at Nanisivik began in 1976 and ended in 2002, 14 years later than originally planned, due to extension of ore reserves. The ownership of the mine changed eight times during its 26-year life (Burns and Doggett 2003), with Breakwater Resources being the sole owner since 1996. In anticipation of the mine's closure, a socio-economic impact study was carried out in 2002 using interviews conducted with community members to outline the mine's legacy and the impacts of closure (Brubacher & Associates 2002). This study, conducted before mine closure, provided a detailed representation of residents' feelings about the mine, and described the predicted impacts of mine closure. Reclamation at the mine site is ongoing as of 2006.

The Nanisivik town site was created jointly by the Government of the Northwest Territories and Nanisivik Mines Ltd. Employees from southern Canada and surrounding Inuit communities moved to the mine site to enable them to live close to the mine and with their families. In the community of about 300 were a health centre, school, community centre with gymnasium and pool, skating rink, post office, RCMP detachment, and a small general store. Inuit workers living in Arctic Bay traveled between Arctic Bay and Nanisivik every day. Most employees worked 6 days a week for 91 days and then had 21 days off (Storey and Shrimpton 1989).

Inuit training and employment at Nanisivik was only partially successful because the 60% Inuit employment rate specified in the Strathcona Agreement was never achieved. The highest rate of Inuit employment reached 30% during mine construction, but during operation, the highest level achieved was 28%, which slowly declined over the years (McPherson2003). The mine offered Inuit workers on-the-job training, but often found that there were not enough local people interested in mine employment to fill the positions available to them (McPherson 2003).

5.3 Resolute (Polaris Mine)

Resolute (Figs. 1 and 4), located on Cornwallis Island, has a population of 215 (2004 data; Statistics Canada 2005b). It was established in 1953 when four Inuit families from Inukjuak, Quebec, and one from Pond Inlet on Baffin Island, were resettled there by the Department of Resources and Development and the RCMP (Bissett 1968b). Resolute's location close to areas of oil, mineral, and scientific interests, with scheduled air service from major centres such as Montreal and Edmonton since the mid-1960s, caused an influx of researchers and businesses into the community.

Fig. 4 Resolute in January 2005

5.4 Polaris Mine, 1982–2002

The Polaris lead-zinc mine, the most northerly in the world, was located 100 km northwest of Resolute on Little Cornwallis Island (Fig. 1). The mine was owned by Cominco Ltd. (which became Teck Cominco in 2001) throughout its history. It started production in 1982, closed in 2002, and reclamation was completed in 2004.

Although opened only 6 years later than Nanisivik, the Polaris mine was different in most aspects of its operation, from planning through to reclamation. Almost 10 years before production started, Cominco began visiting local communities that would be northern points of hire for Polaris (Yellowknife, Resolute, Grise Fjord, Cambridge Bay, Pelly Bay, Spence Bay, and Gjoa Haven). These visits, led by consultant J.E. Barrett, were organized to obtain a better understanding of Inuit culture and to record Inuit impressions about employment at the future mine (Barrett 1973). Barrett returned four times in 1976 to continue developing a relationship between the mine and these communities (Barrett 1976). Reports prepared following these visits detailed reaction of the Inuit to the project, recommendations about how best to include the Inuit in the mine's workforce, and examined whether people would prefer working at a community-based mine or a rotation mine (Barrett 1973, 1976).

In 1980, Outcrop, a consultant group retained by Cominco, predicted the potential socio-economic impacts of the Polaris mine on Resolute, other surrounding communities, and elsewhere in the Northwest Territories (Outcrop 1980). The report included recommendations to Cominco about Inuit employment, community involvement, and local business partnerships. By the time mining operations began, communities contacted by Cominco supported its approach to community consultation (Graham 1982). Cominco had conducted many environmental studies before starting construction, in contrast to Nanisivik where no environmental or socio-economic impact studies had been prepared before government approval for the mine was granted.

A Letter of Understanding between Cominco and the Government of Canada, detailing the basis upon which the project was approved, was signed on February 4, 1980 (Graham 1982). A Memorandum of Understanding with the Government of the Northwest Territories, signed on August 12, 1981, outlined the socio-economic requirements of the project (Cominco and GNWT 1981).

By 1981, the Government of the Northwest Territories' policy was to discourage the construction of new communities dependent on one resource (Hickling-Partners 1981). Though Cominco contemplated the possibility of developing a community at the mine site (Barrett 1973), it eventually developed Polaris as a fly-in/fly-out operation. All personnel and equipment arrived in Resolute via jet, and were transferred to a Twin Otter airplane for the short trip to the mine site. Southern Canadian employees worked a 63-day shift with 28 days off (Storey and Shrimpton 1989). Inuit workers had a choice to work the same rotation or a shorter rotation of 42 days with 28 days off (Storey and Shrimpton 1989) which gave them more time for traditional pursuits. Due to frequent flights between Polaris and Resolute, Inuit working the longer rotation also had an opportunity to go home for periods ranging from a few hours to overnight, 4 weeks into their rotation.

Unlike the Strathcona Agreement for the Nanisivik mine, the Memorandum of Understanding between the Government of the Northwest Territories and Cominco did not specify the Inuit employment rate to be achieved during the mine's life (Cominco and GNWT 1981). The Outcrop report (1980) claimed that a workforce with 20–40% Inuit from surrounding communities could be attained by 1985. However, at the peak of mine operations it employed less than 30 Inuit, or less than 10% of the total Polaris workforce (Di Menna 2004).

6 Results

Mining operations in themselves are not sustainable, but the wealth generated and commodities produced have the potential to contribute to sustainable development. In order to maximize the sustainable impacts of mining, it is essential to ensure that the benefits are long-term, lasting well beyond mine closure, rather than short-term benefits that disappear at mine closure or shortly thereafter. Negative impacts need to be minimized throughout the life of a mine and after closure.

The economic, social, and training impacts that the Nanisivik and Polaris mines had on the communities of Resolute and Arctic Bay are described below, divided into short- and long-term impacts. Short-term impacts are defined as those that occurred only during the life of the mine, and long-term impacts are those that persisted for an extended period after mine closure.

6.1 Short-Term Economic Impacts

6.1.1 Employment Income

Increased employment income was the most significant short-term economic benefit of the mines on the communities, due to substantial contributions to the total
personal incomes in each community. In Arctic Bay, for example, this amounted
to C$1 million annually (Brubacher & Associates 2002). The benefits, such as
increased disposable income and a higher standard of living, were most significant
for mine employees and their immediate families. The effects of increased community income reached others in the community as well through sharing between
mine employees and their extended families, and increased sales at the local general
stores.

However, when the mines closed, mine employees lost their main source of
income. Former workers with families have had a difficult time adjusting to this
loss of income, which effectively reduced their ability to purchase groceries and
household items:

> "It's a big adjustment too for people. (. . .) You're used to bringing home 3000 dollars a
> month, now all of a sudden you get 300 a month. What a kick in the head." *(Bowes-Lyon
> 2006 , p. 25.)*

The loss of employment income from mine closure also had a negative effect on
the general stores. With reduced incomes, buying power in the communities was
substantially reduced, creating a decline in sales.

6.1.2 Standard of Living

The presence of the mines had an important short term impact on the standard of
living in both communities, stemming mainly from the improved economy of the
communities.

During mine operation, mine salaries combined with low freight and airfare costs
(especially for Nanisivik mine employees, who received a monthly ship-bourne
freight allowance) facilitated the purchase of household items and hunting equipment needed to pursue traditional activities (e.g., snowmobiles and boats), which
would otherwise have been out of the financial reach of most families. Furthermore,
through extended family sharing, traditional activities were more easily enjoyed by
more members of the communities. A community member whose nephew worked
at Nanisivik reported:

> "(. . .) just to give you an example, for a boat, my nephew, he bought (. . .) a 24 footer boat
> and for him to bring it (. . .) up here he paid 250 dollars and for me to (. . .) ship it up here
> would cost me 8000 to 13,000 dollars." *(Bowes-Lyon 2006, p. 29.)*

Other factors that improved Arctic Bay residents' standard of living over the
short-term were the facilities at the Nanisivik community centre, which included a
swimming pool, gymnasium, and restaurant, giving them the opportunity for alternative family activities not possible in Arctic Bay. In Resolute, residents benefited

from the donation of surplus mine materials such as building supplies, and from help the mine provided to hunters passing through the area who were running low on fuel, needed repairs, or needed medical assistance.

6.1.3 Jet Service

Jet service to the communities was not provided by the mines, but was a direct consequence of the mines' existence. This was a short-term benefit because jet service to the communities terminated in November 2005, 11 months after interviews were conducted and 3 years after mine closure. Jet service provided residents with access to cheaper airfares and airfreight costs, direct flights to southern cities, and cheaper and more abundant fresh produce deliveries. The benefits extended to local businesses because of the lower cost for airfreight, and to local artist and tourism industries because of the cheaper airfares that promoted tourism to those areas.

The loss of jet service to the communities, replaced by lower capacity turbo prop aircraft, has affected all aspects of daily life by increasing airfares and airfreight charges, grocery bills, and the cost of purchasing supplies needed by local businesses. In Resolute, for example, residents had already noticed an increase in airfare and airfreight costs after Canadian North, one of two airlines servicing the community, stopped flying there in 1999. Prices again increased after the Polaris mine closed in 2002, and are now much higher. Similarly, the tourism industry had already seen a decline in visitors passing through the community since 2002, and the airport gift shop experienced a 50% reduction in sales. A decline in visitors will mean a loss of community employment and the eventual closing of some of the hotels.

6.2 Long-Term Economic Benefits: Business Creation

The impact of the mines on Arctic Bay's and Resolute's local business communities was dependent on the relationship between the mines and these businesses, and the nature of the business environment in the communities before the opening of the mines. This is a long-term benefit because, although some of the businesses have disappeared since mine closure or have suffered loss of revenue, all surviving businesses gained valuable experience during mine operation that has persisted after mine closure.

In Arctic Bay, where few businesses existed before the construction of the Nanisivik mine, some local entrepreneurs took advantage of the presence of the mine to open businesses. For example, a local Inuk man negotiated a contract to run an employee shuttle service between Arctic Bay and Nanisivik, as well as a janitorial contract with the Government of the Northwest Territories to clean government buildings in Nanisivik. Another entrepreneur started a courier service to haul freight from the airport to Arctic Bay. These businesses provided employment for approximately one to four Arctic Bay residents per business. In Resolute, where several businesses were established before the construction of the Polaris mine, the businesses contracted with the mine to provide flight, cargo handling, and accommodation services.

The closing of the mines affected each business differently, but mostly through loss of revenue. All the surviving businesses have used their experience of providing various services to continue without the mines' custom. The long-term success of the businesses is exemplified by the Inuk entrepreneur in Arctic Bay. He was able to start a business and diversify it over time into a heavy equipment operation plus a hotel. Although his employee shuttle and the janitorial services are no longer needed, he developed other services to continue his business, having learned from his experiences when the mine was operating.

6.3 Short-Term Social Impacts

6.3.1 Alcohol Use

A significant short-term negative impact of both mines was the increased availability of alcohol in both communities. Although a permit granted by the local alcohol committee was required for every order of alcohol in Arctic Bay and Resolute, purchasing alcohol at Polaris and Nanisivik was unrestricted and unmonitored.

Problems with alcohol stemming from the presence of the mines were most prevalent in Arctic Bay and Nanisivik. The road linking the two communities made it easy for Arctic Bay residents to obtain alcohol in Nanisivik. It was also possible for Nanisivik residents to order alcohol at a reduced cost because of the cheaper shipping freight costs available to Nanisivik employees, as discussed above. This situation led to drunk-driving incidents between Arctic Bay and Nanisivik, and difficulties for Inuit employees who were late or missed work shifts because of alcohol, and subsequently lost their jobs:

> "I lost my job due to alcohol and drinking too much and not showing up (...) for work 'cause of the night before. That was a big factor up there for a lot of (...) Inuit employees, because alcohol abuse was pretty rampant among Inuit workers up there." *(Bowes-Lyon 2006 , p. 31.)*

Many Arctic Bay residents were disappointed that alcohol was freely available in Nanisivik, and suggested that more Inuit might have worked at the mine, and for longer, had alcohol not been so easily available.

Consultation prior to mine development at Nanisivik might have alerted officials to the problems that could arise from an increased availability of alcohol in the area. Also, consultations with Arctic Bay residents throughout Nanisivik's life may have had an effect in altering the mine's stance on alcohol, and action could have been taken to reverse the effects that alcohol had had thus far. Because no action was taken, Arctic Bay residents developed some animosity toward the mine.

In Resolute, alcohol issues related to the mine did not appear to have been as prevalent as in Arctic Bay, mainly because few mine employees came into contact with community residents, and so the possibility of residents obtaining alcohol from workers was minimal. There were, nonetheless, problems with alcohol for

Inuit employees at Polaris, causing some to be late or miss shifts. Being away from home for extended periods of time and making substantial amounts of money meant that many also consumed a large quantity of alcohol once home after the end of their rotation.

6.3.2 Education

Apart from the school at the Nanisivik mine site, both mines had only a small positive impact on education in the communities in the short-term through scholarships for students. According to school officials in Arctic Bay and Resolute, the high school graduation rates in those communities were not affected by the presence of the mines, and no students pursued an education in mine-related fields (an interesting finding in itself, and deserving of further study). Also, although mine officials visited the schools on a sporadic basis and some children had the opportunity to visit the mines each year, there was apparently little interaction between the mines and the schools, and few children knew what the mines brought to the communities.

6.4 Long-Term Social Impacts

6.4.1 Education

The only significant long-term positive social impact of the mines was education-related. Both communities benefited from the construction of a gymnasium and the donation of money to their school by their respective mines.

The mines, however, had little impact on the actual education of local students except in the case of the Nanisivik school. The differences between the school in Nanisivik, built for mine employees' children, and the school in Arctic Bay, reveal that Nanisivik had an indirect impact on education in the area. The Nanisivik school had a mix of children with southern Canadian parents and Inuit parents, and apparently had stronger discipline, more stringent homework requirements, and a higher level of English instruction than the school in Arctic Bay. Inuit children who attended the Nanisivik school and transferred to the Arctic Bay school once their parents moved back to the community, found the discipline to be laxer in Arctic Bay, and generally did better in school, especially in English, than their peers who had only attended the Arctic Bay school. Though the Nanisivik students did poorly in Inuktitut classes, their overall grades and motivation about school were generally higher than the Arctic Bay students. This information suggests that education in Nanisivik was higher quality than in Arctic Bay, and that after the Nanisivik school closed, the overall quality of education in the area decreased. The consequence of this difference between schools has been that former Inuit students of the Nanisivik schools continue to have better jobs in Arctic Bay, due most likely to their higher level of English and the fact that they are more accustomed to working with southern Canadians. As one participant said:

> "We seem to get better jobs compared to them and that's what I don't like 'cause it's only half an hour away and it was such a big difference (. . .)." *(Bowes-Lyon* 2006 *, p. 32.)*

6.4.2 Infrastructure

Both Arctic Bay and Resolute benefited from infrastructure from the mine. Arctic Bay received ten houses that were relocated from Nanisivik, whereas in Resolute, the school and the hamlet received office furniture, and the Arctic College campus received books from Polaris. The local RCMP detachment in Resolute also obtained some of the mine's gym equipment after its closure, and has since set up a small workout facility and training program for the local youth.

6.4.3 Community Engagement

Although the Nanisivik and Polaris mines had an impact on the communities of Arctic Bay and Resolute, community members felt that they had few forums in which to voice their concerns or discuss mine plans with mine management. Both mines met with each community's hamlet council, were available for meetings when the councils wanted them to attend, and informed the hamlet councils of mine management decisions with potential impact on communities. Despite these efforts, residents felt left out of decision-making. This is true of both communities, but more so of Arctic Bay where residents felt that Nanisivik had been more prominent in their daily lives. Arctic Bay residents wanted to take an active part in decision-making about the Nanisivik mine because it was so close to their community:

> "I just wish that they consulted with the community and the elders especially during the operation of the mine. Things I think would have gone a lot better if they worked closely together with the elders and the community." *(Bowes-Lyon* 2006 *, p. 34.)*

More extensive community engagement could have contributed to creating positive long-term social impacts by allowing residents to discuss their concerns about the mines' impacts on the communities and to feel that these concerns were being taken seriously, and would have enabled issues to have been resolved more quickly (e.g., Laurence 2006).

The reclamation of the Nanisivik townsite was the responsibility of Breakwater Resources, the owner of the mine. Infrastructure such as houses and a community centre were available in Nanisivik after the mine closed. As part of the process to determine what to do with this infrastructure, public hearings were organized by the Government of Nunavut, in partnership with the federal government, the mine, Arctic Bay, and regulatory bodies (GN 2002). During interviews for this study, it became apparent that ten houses had already been moved from Nanisivik to Arctic Bay before the hearings, had been retrofitted, and made available to community members. As a result, during the first public hearing held in June 2004, two years after mine closure, residents expressed a desire for more homes to be moved to Arctic Bay to alleviate their housing shortage, while others suggested that the facilities of the community centre should be converted into a training school, or left for the use of Arctic Bay residents. Because this first public hearing was perceived to be a consultative meeting to determine the fate of the houses in Nanisivik, residents were confident that by the second hearing, held in February 2005, discussion would continue. However, the government instead announced that no more houses would be moved to Arctic Bay, citing unspecified problems of contamination, and the entire town site at Nanisivik would be torn down. Residents had no opportunity to learn any more about the decision, leaving them disappointed that their suggestions were ignored. Had discussions and plans been undertaken earlier in the reclamation process, or earlier during the operation of the mine, many felt that the reclamation of the houses and community centre could have been completed causing less uncertainty for residents, and in a more timely manner.

Eventually, the reason given to residents for not moving more houses to Arctic Bay was the elevated level of contamination of the buildings. However, from conversations with Arctic Bay residents, it was clear that many of them, including at least one official on the socio-economic sub-committee (a committee created to assess the best way to minimize the socio-economic impacts of mine closure on Arctic Bay), did not know what form this contamination took. This illustrates the lack of communication between territorial government officials (who had the report explaining the contamination) and community residents, and is the cause for some distrust of the territorial government amongst community members.

6.5 Short-Term Employment Impacts

The Nanisivik and Polaris mines offered mining employment to Arctic Bay and Resolute residents for the first time. Once the mines closed, however, employment opportunities at the mines no longer existed and local mine employees had to look for work elsewhere within the communities. The majority of Inuit employees at Nanisivik and Polaris worked in jobs requiring little training and minimal education, with the majority of male Inuit employed as seasonal labourers during the summer shipping season, and the majority of female Inuit employed as housekeepers or cleaning staff. Male employees working year-round generally worked as heavy equipment operators, after receiving on-the-job training. A few Inuit employees at Nanisivik, who showed special interest and worked well at the mine, were trained for other skilled jobs, such as heavy equipment mechanic, electrician, carpenter, and plumber. However, only a handful of Inuit employees worked underground regularly, principally at Nanisivik. The main reasons for this were claustrophobia, and that being underground was linked to thoughts of being buried after death.

There was an expectation from the communities of Arctic Bay and Resolute that community members would be hired to work at the mines. Many were disappointed when the percentage of Inuit employed at each mine remained quite low (a maximum of 30% at Nanisivik and 10% at Polaris). This was especially disappointing to residents of Arctic Bay where the Strathcona Agreement stated that there would be a 60% Inuit work force at Nanisivik after the third year of production. The government of the Northwest Territories, the federal government, and Mineral Resources International set up a Training and Employment Advisory Committee before construction began at Nanisivik, which was to meet monthly to review the progress in the employment objectives listed in the Strathcona Agreement. These objectives principally consisted of finding ways to ensure that as many Inuit as possible could be trained and hired for employment at Nanisivik. The committee met for the first few years after Nanisivik was approved, but then stopped meeting. Many in Arctic Bay felt that the training and employment committee did not do enough to improve the employment situation of Arctic Bay residents, and felt that more Inuit would have been employed at the mine if more attention had been paid to the Strathcona Agreement.

6.6 Unrealized Long-Term Training Opportunities

Both the Nanisivik and Polaris mines trained new employees and provided apprenticeships for those interested in obtaining trade certificates. Nevertheless, in the end, the number of local Inuit that completed apprenticeships at the mines was quite low. The majority of these people were trained as heavy equipment operators, but it appears that several of them did not receive the official territorial certification that would have allowed them to work as heavy equipment operators elsewhere in Nunavut and the Northwest Territories once the mines closed.

7 Discussion

The *Nunavut Economic Outlook* (Vail and Clinton 2001) predicted that a minimum of four new mines will open in Nunavut before 2012 and that the mining industry will be the territory's largest wealth-creating industry. Using Arctic Bay's and Resolute's experiences with mining as a starting point, and keeping the limitations of working in the North in mind, it is possible to determine what steps need to be taken by communities, governments, and industry to ensure that today's mining operations in Nunavut contribute to the sustainable development of the communities closest to them.

7.1 Economic Impacts

Mining can have positive economic impacts through GDP growth, tax revenues, direct employment, indirect economic spin-offs, and supplier partnership programs (Ericsson and Noras 2005). The *Nunavut Economic Outlook* forecast estimates that mining will contribute more than C$110 million in real GDP per year to Nunavut's economy, but cautions that little of this money will directly benefit residents of Nunavut unless more are hired to work at the mines, and mines purchase more goods and services from Inuit-owned businesses (Vail and Clinton 2001). Therefore, in order for communities to benefit from mining, emphasis must be placed on community-specific economic benefits such as direct employment, economic spin-offs, and supplier partnership programs.

For residents of both Arctic Bay and Resolute, increased employment income from working at the mines was their most significant short-term economic benefit, but was unfortunately not a longer term sustainable benefit. As was experienced in the two communities studied, mine closure rapidly reduced total employment income through loss of employment. This can lead to a return to pre-mining economic conditions if no longer-term economic benefits are developed in the community, although it is too early to tell if this will occur at the sites studied (this survey was conducted only 2 years after mine closure).

Both the Nanisivik and Polaris mines contributed to Arctic Bay's and Resolute's business development. At Arctic Bay, several small businesses were created as economic spin-offs from Nanisivik and created additional employment within the hamlet. At Resolute, Polaris was able to directly partner with existing local businesses, increasing their workload and their employee base, while giving them experience with the mining industry. However, analysis of businesses in Arctic Bay and Resolute shows that, generally, they have had difficulties since the mine closures. In Arctic Bay, for example, most of the businesses created to service the mine, or as spin-offs from the presence of the mine, have not persisted after mine closure. In contrast, Resolute's businesses have continued to operate due to the diversified services offered and their adaptability, but revenues have declined.

Business training for community members in Arctic Bay and Resolute could have increased both the number of businesses in each place and the success rate of these new businesses, thereby directly contributing to sustainable development in each community. Micro-enterprise and supplier development programs have been successful in other mining areas, by helping to create businesses that can supply goods or services on a larger scale (IFC 2000). The shuttle service discussed above is a good example of a business that would fall under a micro-enterprise development program. In this case, companies work with communities to help them develop businesses that answer the specific needs of a mining company, such as catering, cleaning, and construction services. These programs work towards ensuring that purchasing policies for mines exist, that upcoming contracts are available early so that local businesses have time to prepare a bid, that local businesses have the resources available to make their bids competitive, and that they can learn from the process (IFC 2000). Despite the possibilities suggested by the above examples, it is not clear whether these types of programs would succeed in the long term in Nunavut, where communities are remote, the population remains small, and business opportunities are very limited.

Thus, in general, little *sustainable* economic development occurred in Arctic Bay and Resolute as a result of the mines. Key steps that were missed were not hiring more Inuit for work at the mines, and the mining companies not participating more actively in the development of the communities' economies.

7.2 Social Development

The most direct community involvement that the Nanisivik and Polaris mines had with the communities was through educational scholarships, mine site visits for students, and occasional information meetings for each community's council. In addition, a community centre with a pool was built in Nanisivik that Arctic Bay residents could use, but it was demolished during the reclamation process. However, for residents of both Arctic Bay and Resolute, the most important social benefit arising from the presence of the mines was the short-term improvement in standard of living.

This short list of the mines' involvement with their respective communities reveals minimal interaction, and missed opportunities. There are numerous contributions that a mine can make to a community's social development. One approach is to directly contribute to community services such as health, education, transportation, water supply services, energy, information and telecommunications, and the patronage of small businesses and vendors (Ericsson and Noras 2005). However, neither Arctic Bay nor Resolute saw any major contribution to their community services in these ways.

The agreements signed between the mining companies and the Government did not sufficiently address the relevant social issues that could have potentially affected community members during operation and after closure. The issues surrounding the presence of alcohol in Nanisivik is a good example of this. The minimal contributions of the mines to the social sustainability of their communities point to areas of weakness in the social planning of mining companies, and highlights the importance for social development to be addressed in the socio-economic impact assessments required for project approvals in Nunavut today.

7.3 Training and Employment

Job creation at new mining projects was identified as one of the best contributions mining can make towards sustainable development (MMSD 2002).

Both the Nanisivik and Polaris mines had relatively few Inuit employees throughout their years of operation. The low levels of education, training, and work experience of the local residents made it difficult for the Nanisivik and Polaris mines to hire more local residents, resulting in lost opportunities for economic growth and future local employment. Additionally, both mines started production during a transitional period for the Inuit when many were still making a move to permanent communities and beginning to participate in the wage economy. Low local employment levels at mines continues to be a problem today for mining as well as other industries in Nunavut, and will limit Nunavut's opportunities to develop the territory if not addressed (Vail and Clinton 2001). Unfortunately, as seen in this case study, employment at mines in Nunavut is currently a short-term impact for communities located near a mine. Once reclamation is finished, mining jobs within a community are no longer available, unless, of course, another mine exists in the area. Hence, this type of short-term impact is not generally conducive to the longer-term sustainable development of a community. The short-term nature of mining employment can be offset by the long-term benefits of training received by employees while working at a mine, as long as alternative work opportunities exist within the community to employ those leaving mine work. Employment opportunities within communities would exist with the various government agencies present in each community, local businesses, and local construction businesses. This could possibly allow Inuit to be employed in positions that southern Canadians currently hold, effectively making communities more self-sufficient.

Mine employees at both Nanisivik and Polaris were trained for work at the mines, which enabled some to obtain work within their communities after mine closure. This shows that the experience gained while working at a mine can be useful for work in other parts of the local economy (MMSD 2002). However, the contribution of training to the sustainable development of a community depends on the amount and depth of training given. In Arctic Bay and Resolute, mine employees received at least basic training, but few were trained in trades, partly due to the fact that each apprentice needed to complete courses in the southern Northwest Territories or northern Alberta. Having a trade school in Nunavut would go a long way towards helping with the training of a workforce capable of performing mine (and other) work. In February 2006, the Government of Nunavut announced funding for a trade school in Rankin Inlet with an opening date of 2009 (GN 2006). This trade school should help to address some of the employment issues facing Inuit.

Nevertheless, mining companies will still have to be proactive in training Nunavummiut. Current mines in the North have had more success with hiring and training larger numbers of aboriginals, most with little or no previous experience in mine work, than Nanisivik or Polaris did. The Red Dog mine, located in northwestern Alaska, for example, is working towards achieving a 100% aboriginal employment rate and, as of 2001, had a 62% aboriginal employment rate (Werniuk 2001). Similarly, the Diavik mine, in the Northwest Territories, had a 38% aboriginal employment rate in 2004, only 2% less than their target (DDMI 2004). The success of these mines appears to come from their flexibility in hiring local employees, providing significant training at the mines and in surrounding communities, and providing aboriginal and Inuit community residents with the opportunity to work in a wide range of jobs, including management positions, something not done at Nanisivik and Polaris. Furthermore, these new mines have signed agreements with Inuit organizations stating training responsibilities and hiring targets.

To increase the reach of mine training and ensure that more local residents are hired to work at mines, a greater emphasis needs to be placed on community engagement and education (e.g., Laurence 2006). This did not occur in Arctic Bay and Resolute, where community residents had a poor understanding of mine processes due to the limited interaction between the mines and residents. By teaching residents what happens at mines and what mine training can lead to, more residents may participate in and benefit from mine work.

8 Conclusions and Recommendations

This study has investigated the economic, social, and employment impacts of the Nanisivik and Polaris mines on their neighbouring communities, Arctic Bay and Resolute. The economies of both towns were temporarily improved by the presence of the mines, but few of these benefits have lasted beyond mine closure (cf. Laurence 2006).

The mines had few direct social impacts because they had little interaction with the communities. The major negative social impact of the mines was increased alcohol usage in the communities, whereas the most important benefit was the increased level of education available at the Nanisivik school. The lack of interaction between mines and communities, however, disappointed many residents because it meant they were rarely consulted and were left out of decision-making processes.

The mines provided employment for some Inuit from Arctic Bay and Resolute, but the Inuit employment rates at the mines fell far below expectations. The Inuit who did work at the mines gained new skills, but commonly did not receive proper certification for their training. Nevertheless, some have been able to use these skills in other work since mine closure, and have the experience to obtain work in Nunavut's new mines if they so choose.

The Nanisivik and Polaris mines did not provide an environment that promoted sustainable development within the communities closest to them because the benefits that the mines brought to the communities did not persist after mine closure, and minimal efforts were made to create, capitalize on, and maximize the potential benefits that could have arisen from the relationship between the mines and their respective communities.

At the time when Nanisivik and Polaris began production, the Inuit had no control over the approval process for mining projects. Their only involvement was through community consultation organized by the companies or the government. The suggestions made by residents during these consultations were, however, not always taken into consideration in company or government decisions about the development of the mines. Had the companies and the government worked more closely with each other, and with Arctic Bay and Resolute residents, the development of the mines could have happened in a way that would have increased positive social and economic impacts, and would have had a more lasting positive impact on the development of the communities. As it stands, it is almost as if the mines had never existed, so few are the lasting positive (or, it has to be said, negative) impacts of the mines.

For the communities to have benefited more fully from the operations and for the short-term developments that arose from mining to have been more sustainable, there needed to have been better cooperation between all stakeholders. Furthermore, the mining companies and the government should have focused on transforming some of the wealth generated by extraction of the mineral resources into opportunities for developing human skills and financial and infrastructural benefits in the communities (Jackson 2005).

Fortunately, with the signing of the Nunavut Land Claims Agreement (NLCA), the approach to mining in Nunavut has become more aligned with the principles of sustainable development. Through the NLCA, the Inuit have more control over decisions made about each mining project, they stand to benefit from royalties collected from mining on Inuit owned land, and they will be able to create more lasting benefits for communities through environmental impact assessments and Inuit impact benefit agreements.

What is interesting about the Nanisivik and Polaris cases is that, despite recommendations made in a socio-economic impact assessment in Polaris's case, and despite legal requirements in Nanisivik's case, neither Nanisivik nor Polaris were held accountable for their failures to follow their own guidelines. With the signing of the NLCA, Inuit impact benefit agreements are required before any major development project can go ahead (NLCA 1993). Inuit impact benefit agreements are signed between the mining company and the communities to be affected, and specifically outline how negative impacts will be mitigated and positive impacts maximized. This process should help communities with their socio-economic development. Unfortunately, there are still no clear guidelines for tracking or enforcing accountability in these agreements in Nunavut today aside from what each involved party agrees to within individual impact benefit agreements (NLCA 1993 §26.9.1). This could lead to a wide range of standards for the impact benefit agreements across Nunavut, as well as a varying degree of accountability for upholding the principles within those agreements. Enforcement is required to prevent companies from making statements in socio-economic impact assessments and then ignoring them once their operation phase has begun. Due diligence investigations should also be performed after project approval to ensure that socio-economic impact assessments are accurate (Joyce and MacFarlane 2001).

The renewed presence of mining companies in Nunavut, such as Tahera Diamond Corporation, Hope Bay Mining, Newmont, and Cumberland Resources will provide Nunavut with its first opportunities to apply the NLCA guidelines to mining projects. The Jericho diamond mine, owned by Tahera, was the first mine to operate under these socio-economic and environmental guidelines and will provide an important test of the NLCA[1]. The lessons learned from Nanisivik and Polaris about sustainable development in the North will also be invaluable to communities and mining companies in their work towards applying sustainable development principles in Nunavut.

Three recommendations are listed below for ensuring that mining has as many positive benefits as possible in Nunavut in the future:

1. Teach business skills to Nunavummiut to increase the number of Inuit-owned and Inuit-operated businesses. Mining companies can join with the Government of Nunavut to create courses geared towards teaching business skills. Alternatively, mining companies can create their own courses and participate more directly in community activities, perhaps as a negotiated aspect of impact benefit agreements.

---

[1] Unfortunately, the Jericho diamond mine closed in 2008.

2. Achieve an effective community consultation/participation mechanism so that communities affected by mining have the opportunity to actively participate in mine-related decisions that have the potential to affect the community. Mining companies need to become more open to the idea of including community input into mine decision-making in order to create trust between communities and mining companies. Additionally, information about the mining process should be made available to the community through information sessions, pamphlets, radio interviews, and mine visits.

Community consultation is one of the most important steps for creating socio-economic development programs during mine planning, because it raises community awareness of the project's impacts and allows for agreement on management and technical approaches in order to maximize the benefits and reduce the negative consequences of a project (IFC 1998). Accordingly, it is necessary to engage in community consultation in order to create assessments that integrate the values and expectations of the community both with government policies and with the goals of the mining project. Such assessments would better address impacts and make project direction more adaptable to unforeseen circumstances throughout a mine's life.

3. Ensure that training opportunities are numerous and varied to allow employees to improve their skills, get promoted, and receive appropriate certification. Mining companies should also participate in educating people outside the work environment through the sponsoring of community-based educational activities, such as mentoring programs for young community members (MMSD 2002), and schools dedicated to workforce-specific training. This would increase the number of Nunavummiut employable from the start of mine construction and production, and open up the possibility for on-the-job training to other Nunavummiut, as well as potentially generate interest in mine work within the community at large.

Acknowledgments This project was made possible by the financial and logistical assistance of numerous organizations, and we are grateful for their support: the Canada-Nunavut Geoscience Office, Natural Resources Canada, Indian and Northern Affairs Canada, Nunavut Tunngavik Incorporated, the Government of Nunavut, and the Polar Continental Shelf Project. We especially thank Ross Sherlock for initiating this project while at the Canada-Nunavut Geoscience Office, securing the initial funding, and providing feedback about sustainable development in Nunavut. We thank two anonymous reviewers for their helpful and constructive comments on an early draft of this chapter, and Simon Handelsman for careful and thorough editing.

References

Baffin Region Inuit Association (1979) Socio-economic impacts of the Nanisivik mine on North Baffin region communities
Barrett JE (1973) Employment of the Inuit at Polaris, Little Cornwallis Island, Nunavut
Barrett JE (1976) The Polaris project and the Inuit
Bissett D (1968a) Northern Baffin Island: An area economic survey. Industrial Division, Department of Indian Affairs and Northern Development, Government of Canada, Ottawa

Bissett D (1968b) Resolute: An area economic survey. Industrial Division, Department of Indian Affairs and Northern Development, Government of Canada, Ottawa

Bowes-Lyon L-M (2006) Comparison of the socio-economic impacts of the Nanisivik and Polaris mines: A sustainable development case study. Unpublished Master of Science Thesis, University of Alberta

Brubacher & Associates (2002) The Nanisivik legacy in Arctic Bay: A socio-economic impact study. Brubacher & Associates, Ottawa

Bryman A (2001) Social research methods. Oxford University Press, New York

Burns NR, Doggett M (2003) Nanisivik mine – A profitability comparison of actual mining to the expectations of the feasibility study. Exploration and Mining Geology 13:1–10

Cominco and Government of the Northwest Territories (1981) Memorandum of understanding. Unpublished Internal Report, Cominco

Crowson P (2002) Sustainability and the economics of mining – what future? Minerals & Energy 17:15–19

Diavik Diamond Mines Inc (2004) Sustainable development report 2004. Retrieved May 9, 2006, from http://www.diavik.ca/News/2005/Diavik%202004%20SD%20Report.pdf

Di Menna J (2004) Mining in Nunavut. Canadian Geographic 124:52

Ericsson M, Noras P (2005) A viable alternative. Mining Environmental Management, December 2005:7–9

Erlandson DA, Harris EL, Skipper BL, Allen SD (1993) Doing naturalistic inquiry: A guide to methods. Sage Publications Inc, Newbury Park, CA

Government of Canada and Mineral Resources International Limited (1974) The Strathcona Agreement

Government of Nunavut (2002) Government will continue seeking positive legacy from Nanisivik mine closure, minister says. Retrieved May 10, 2006 from http://www.gov.nu.ca/Nunavut/English/news/2002/oct/oct1.shtml

Government of Nunavut (2006) Trade school to bring training opportunities. Retrieved May 15, 2006 from http://www.gov.nu.ca/Nunavut/English/news/2006/feb/feb18.pdf

Graham KA (1982) Eastern Arctic study case study series: The development of the Polaris mine. Centre for Resource Studies, Kingston, ON

Habirono H (2001) Community based sustainable development: A corporate social responsibility. Mining Environmental Management, July 2001:13–15, 17

Hickling-Partners Inc. (1981) Evaluation of the Nanisivik project

International Finance Corporation (1998) Doing better business through effective public consultation and disclosure. International Finance Corporation, Washington, DC

International Finance Corporation (2000) Investing in people: Sustaining communities through improved business practice. International Finance Corporation, Washington, DC

Jackson R (2005) The challenges of sustaining mining benefits. Mining Environmental Management, May:6–9

Joyce SA, MacFarlane M (2001) Social impact assessment in the mining industry: Current situation and future directions, No. 46. International Institute for Environment and Development and World Business Council for Sustainable Development

Laurence D (2006) Optimisation of the mine closure process. Journal of Cleaner Production 14:285–298

McAllister ML, Milioli G (2000) Mining sustainability: Opportunities for Canada and Brazil. Minerals & Energy 15:3–14

McPherson R (2003) New owners in their own land: Minerals and Inuit land claims. University of Calgary Press, Calgary, AB

Mining, Minerals and Sustainable Development North America (2002) Towards change: The work and results of MMSD-North America. Final Report. International Institute for Sustainable Development, Winnipeg, Manitoba

Mining, Minerals, and Sustainable Development Project (2002) Breaking new ground. Earthscan Publications Ltd, London

NLCA (1993) Agreement between the Inuit of the Nunavut settlement area and her Majesty the Queen in right of Canada. Indian and Northern Affairs Canada and Tungavik Federation of Nunavut, Ottawa

Nunavut Tunngavik Incorporated (2007) Mining policy. Retrieved 15/10/2007 from http://www.tunngavik.com/publications/information-brochures/mining-policy-in-en-inu.pdf

O'Faircheallaigh C (1999) Making social impact assessment count: A negotiation-based approach for indigenous peoples. Society & Natural Resources 12:63–80

Outcrop Ltd (1980) Potential socio-economic impacts of the Polaris mine project

Patton MQ (1990) Qualitative evaluation and research methods. Sage Publications Inc, Newbury Park, CA

Richards JP (2002) Sustainable development and the minerals industry. Society of Economic Geologists Newsletter 48:1, 8–12

Richards JP (2005) The role of minerals in sustainable human development. In: Marker BR, Petterson MG, McEvoy F, Stephenson MH (eds) Sustainable minerals operations in the developing world. Geological Society of London Special Publication 250, London, pp. 25–34

Richards JP (2006a) The application of sustainable development principles in the minerals industry: Examples of best-practice and challenges for the future. International Journal of Interdisciplinary Social Sciences 1:57–67

Richards JP (2006b) "Precious" metals: the case for treating metals as irreplaceable. Journal of Cleaner Production 14:324–333

Sherlock RL, Scott DJ, MacKay G (2003a) Bringing sustainability to the people of Nunavut. Geotimes, December 2003:18–20

Sherlock RL, Scott DJ, MacKay G, Johnson W (2003b) Sustainable development in Nunavut: The role of geoscience. Exploration and Mining Geology 12:21–30

Sivummut Economic Development Strategy Group (2003) Nunavut economic development strategy: Building a foundation for the future: http://www.nunavuteconomicforum.ca/public/files/strategy/NUNAVUTE.PDF

Statistics Canada (2006) Real gross domestic product, expenditure-based, by province and territory. Retrieved 13/03/08 from http://www40.statcan.ca/cgi-bin/getcans/sorth.cgi?lan=eng&dtype=fina&filename=econ50.htm&sortact=2&sortf=6

Statistics Canada (2005a) Community highlights for Arctic Bay. Retrieved 02/01/2006 from http://www12.statcan.ca/english/Profil01/CP01/Details/Page.cfm?Lang=E&Geo1=CSD&Code1=6204018&Geo2=PR&Code2=62&Data=Count&SearchText=arctic%20bay&SearchType=Begins&SearchPR=01&B1=All&Custom

Statistics Canada (2005b) Community highlights for Resolute. Retrieved 02/07/2006 from http://www12.statcan.ca/english/profil01/CP01/Details/Page.cfm?Lang=E&Geo1=CSD&Code1=6204022&Geo2=PR&Code2=62&Data=Count&SearchText=resolute&SearchType=Begins&SearchPR=01&B1=All&Custom =

Storey K, Shrimpton M (1989) Long distance labour commuting in the Canadian mining industry. Working Paper No. 43. Centre for Resource Studies, Kingston, Ontario

Togolo M, Rae M, Omundsen T (2001) Meeting society's expectations through stakeholder engagement. Mining Environmental Management, March 2001:10–11

Vail S, Clinton G (2001) Nunavut economic outlook: An examination of the Nunavut economy. The Conference Board of Canada, Ottawa

Veiga MM, Scoble M, McAllister ML (2001) Mining with communities. Natural Resources Forum 25:191–202

Werniuk J (2001) Cominco's Alaskan triumph. Canadian Mining Journal 122:16

Assessing the Socio-Economic Impacts of Mining: Case Study of the Landau Colliery, South Africa

Anthony Dane

Abstract This chapter analyses the socio-economic impacts of the Landau Colliery in South Africa, an operation of Anglo Coal. The aim is to contribute to an understanding of how large-scale mining activities in low and middle income countries can enhance socio-economic development, and what the components of "success" are. The project was undertaken by Anglo American and follows largely the methodology of the International Council on Mining and Metals' Resource Endowment Initiative. The study firstly assesses the extent to which a South African mine contributes towards achieving the Millennium Development Goals, and secondly investigates the nature and components of "success", including the key elements that help to bring it about, and the steps needed to advance it further. The study shows that within a context of transformation and required social and infrastructural development, the mine has achieved positive contributions in terms of procurement, employment, human capital development, social, business and community development, and economic contributions (contributions to government revenue, GDP and exports, and the host economy). Positive contributions have particularly benefitted local communities and previously disadvantaged South Africans. Furthermore, efforts have been made to contribute towards creating sustainable communities post closure in 2025. Additionally, the study highlights weaknesses that have restricted achievement of the full potential benefit of the mine. These are largely the result of a lack of collaboration between the communities, the mine, government, NGOs, and other players in the private sector. The study concludes with recommendations for improving the mine's socio-economic contribution, most of which require the urgent need for more coordinated efforts.

A. Dane (✉)
Anglo American, Johannesburg, South Africa
e-mail: imbovane@gmail.com

J.P. Richards (ed.), *Mining, Society, and a Sustainable World*,
DOI 10.1007/978-3-642-01103-0_14, © Springer-Verlag Berlin Heidelberg 2009

1 Introduction

Sustainable development is increasingly impacting the relationship between firms and their various stakeholders: investors, customers, suppliers, employees, communities, landowners, and governments. Firms are increasingly expected to deliver positive sustainable development outcomes. This is particularly the case in the mining industry where individual operations can have enormous impacts on local environments and communities, as well as having substantial economic impacts on the national economies in which they operate.

Schools of thought on mining are divided between those who argue that mineral resources are a "curse" resulting in social and economic decline instead of growth (Sachs and Warner 1995; ICMM 2006a), and those who consider that mineral resources are an endowment that can promote growth and development (Pedro 2006). Mineral resource-rich countries have achieved mixed success with regard to translating their mineral wealth into economic growth and development. All have benefited from the mineral revenues generated, but not all countries have witnessed an improvement in basic living standards and socio-economic conditions of its citizens. Pedro (2006) argues that mineral extraction can spur growth and reduce poverty if deployed under appropriate conditions. But what conditions are necessary to ensure that mining contributes as much as possible to growth and poverty reduction?

With this as a backdrop, the International Council on Mining and Metals (ICMM), in May 2004, initiated the Resource Endowment Initiative to better understand how large scale mining activity in low and middle income countries can enhance socio-economic development, and what the components of "success" are: Pedro's (2006) "appropriate conditions". This chapter aims to contribute to this understanding through a case study of the socio-economic impacts of the Landau Colliery in South Africa, an operation of Anglo Coal. The study firstly assesses the extent to which a South African mine contributes towards achieving the Millennium Development Goals[1] (MDGs; IIED 2007), and secondly investigates the nature and components of "success", including the key elements that help to bring it about, and the steps needed to enhance success further. This understanding throws light on appropriate policies for companies to achieve successful socio-economic outcomes.

A number of key questions are addressed:

- What has been the overall contribution of Landau Colliery to socio-economic development at national, regional and especially local levels (particularly looking at the MDGs)?
- What has been the impact of the South African context on the ability of the mine to deliver positive socio-economic outcomes?

---

[1]The Millennium Development Goals (http://www.un.org/millenniumgoals) are an ambitious agenda for reducing poverty and improving lives that world leaders agreed on at the Millennium Summit in September 2000. For each goal, one or more targets have been set, most for 2015, using 1990 as a benchmark (International Institute for Environment and development 2007).

– What strategies have been effective in managing revenues generated from the mine for sustainable development and poverty reduction?
– What can be done to improve the socio-economic development contribution of Landau Colliery?
– What are the responsibilities of development partners in terms of these recommendations?

This report takes the form of a case study, and examines the socio-economic development performance of Landau Colliery in South Africa using the ICMM's Resource Endowment Toolkit[2] (ICMM 2006c) as a framework. This study was conducted for Anglo American, and the Resource Endowment Toolkit was chosen as a framework for the methodology as part of the company's commitment to the initiative.

This chapter starts by looking at the context in which Landau Colliery operates, in order to better understand the socio-economic impact of the mine. It seeks to gain an understanding of the role that mining has played, and is currently playing, in the South African economy, and what impact this has had on the past and present ability of Landau Colliery to achieve positive impacts. The second section deals with results obtained for Landau Colliery's impact and the methodology used to obtain those results. Finally, Landau's contribution to socio-economic growth and development is examined by looking at a summary of the mine's impact using a SWOT analysis. The chapter concludes with recommendations for improving the socio-economic impact of the mine.

2 Context

Mining does not take place in isolation: the overall context in which a mine operates therefore plays an important role in the socio-economic impact of that mine. This chapter begins by examining the impact of mining in South Africa (from the perspective of the resource curse debate), followed by an analysis of the current context in which Landau Colliery operates.

2.1 Minerals in the South African Context

The mining industry has shaped the current context in which Landau operates. There is general agreement that a rich endowment of natural resources, particularly in minerals, has played a critical role in the evolution of the South African economy, and a significant legacy has been the emergence of prominent industrial centres and

---

[2]The Resource Endowment Toolkit (http://www.icmm.com/document/183) provides a systematic and consistent approach to documenting the impacts (both good and bad) of individual mining projects on a local, regional, and national level (ICMM 2006c).

towns such as Emalahleni (previously Witbank) (Dales et al. 2004). Not only has
the minerals sector contributed significantly to the country's GNP (Gross National
Product), providing capital for reinvestment and new developments, but it has also
provided the impetus for the development of a diverse secondary and tertiary sector,
as well as an extensive and efficient physical infrastructure (Dales et al. 2004). But
does South Africa really benefit from its resources, or is the country a victim of the
so-called resource curse?

When engaging in the resource curse debate, it is important to define "successful" and relatively "less successful" mineral resource-dependent countries. Previous
studies have used a single criterion such as Gross Domestic Product (GDP) growth
to make this distinction. The ICMM (2006b), however, suggests that development
is more complex, and that a broader range of economic and social indicators are
needed to assess the full impact of mining. Two broad areas of performance have
been investigated: economic growth and poverty alleviation.

Data on the rate of growth of GDP and non-mineral GDP for South Africa were
obtained from the beginning of the life of the Landau mine until 2006, and are
presented in Fig. 1. Mineral GDP has grown at an average of 0.4% from 1993 to
2006. However, the relative contribution of mineral GDP growth to overall GDP
growth has decreased over the same time. Figure 1 indicates that the rate of growth
in non-mineral GDP has increased slightly more than the overall rate of growth in
GDP, indicating that the minerals industry has played a decreasing role in the growth
of the economy since 1993. While mining as a whole is a declining percentage of the
national economy, the economy is diversifying, and it is important to remember that
this was driven largely by the mining industry. A diversifying economy indicates
that mineral wealth is not coming at the expense of other growth, and does not
necessarily indicate a "less successful" mineral resource-dependent country (ICMM
2006b).

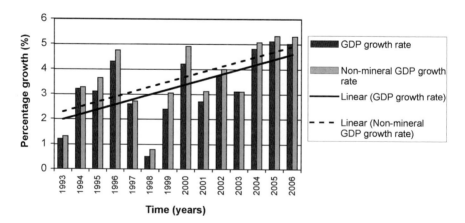

Fig. 1 South Africa's GDP and non-mineral GDP growth rate over time

Based on a global assessment made by the ICMM's Resource Endowment Initiative[3] (ICMM 2006b) and the United Nations Development Programme[4] (UNDP 2004, 2008), South Africa is ranked quite low in key development indicators, suggesting that the contributions to a growing and diversifying economy (as indicated above) have not been fully realised, as measured by many socio-economic development indicators.

2.2 The Context in Which Landau Colliery Operates

Landau Colliery is situated on the outskirts of Emalahleni town, in the Emalahleni Local Municipality of the Nkangala District, Mpumalanga Province, South Africa (Fig. 2). The total population of Emalahleni is 276,412 persons, which constitutes 27% of the total Nkangala District population and 9% of Mpumalanga's population. Emalahleni town offers a full spectrum of business and social activities, and has a large industrial area.

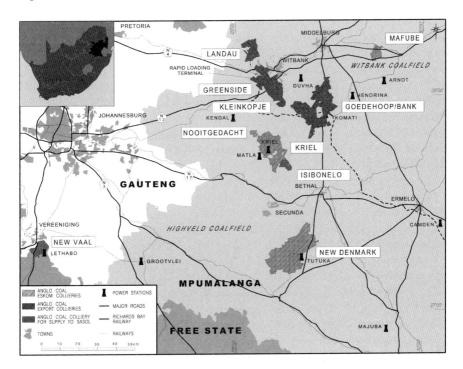

Fig. 2 Location of Landau Colliery

[3]Compared to 33 other countries, South Africa was found to be a weaker performer overall, but with relatively better performance in a few economic indicators.

[4]South Africa ranked 121st out of 177 countries in terms of the Human Development Index (HDI). The HDI combines measures of life expectancy, school enrolment, literacy, and income to allow a broader view of a country's development than does income alone.

The Emalahleni area forms part of the precinct referred to as the Energy Mecca of South Africa, due to its rich deposits of coal reserves and power stations. Coal mining in the area commenced in 1889, and as a result the town of Emalahleni was established around 1905. The Emalahleni Integrated Development Plan (IDP) suggests that mining has had a significant negative impact on the environment, resulting in sinkhole formation, subsidence, underground fires, and seepage of water from underground workings. It has also had a significant economic impact, with some of the mining towns closing down and people being retrenched (Emalahleni Local Municipality 2007). However, the mining sector has, second to the electricity sector, contributed significantly to the local economy. Twenty-three percent of the workforce are in the mining and quarrying industry. This represents a considerable percentage of the population, considering that unemployment in the Emalahleni area (38.4%) is higher than the national average (25.5%). The average monthly household income in the area was approximately R3,721 (R2,531.27 in Nkangala District Municipality). Mining and quarrying, in 2006, contributed 19.4% (R23,086 million) to Mpumalanga's GDP at constant prices (Statistics South Africa 2007).

However, there is evidence to suggest that mining, historically, has not translated significantly into improved living standards for the local population. This has largely been the result of poor planning and administration. The development pattern in the area is fragmented due to the previous dispensation of separate development based on race. As such, Landau's neighbouring communities of Clewer, Ackerville, and Schoongezicht were separated from Emalahleni Town (as well as being separated from many of the economic opportunities available in the town). This has prevented local communities from capturing many of the spillover benefits that normally arise from the presence of a large mine in the area. The Emalahleni IDP (Emalahleni Local Municipality 2007) further suggests that the situation is exaggerated in areas such as these due to large areas being undermined or having mining rights, further restricting the access of these communities to Emalahleni town. Natural features like floodplains and marshlands restrict the opportunities for the physical integration of these communities even further. There is therefore a need for a coordinated sustainable development programme in the area, involving community representatives, mining companies, non-governmental organizations (NGOs), public sector officials, and other interested parties.

In addition to the local socio-economic conditions and government development plans, Landau Colliery's approach to socio-economic development has been influenced by national legislation. The South African government has recently introduced several interventions aimed at overcoming some of the negative socio-economic impacts caused by previous government policies.

Historically, black South Africans were prohibited from being employed in most of the administrative, skilled, or supervisory posts and trade unions, which were developed along racially separate lines. The new government has, however, tried to create an enabling environment for further economic growth that allows the broader participation of previously disadvantaged South Africans. In creating such an environment, a number of barriers to entry for Historically Disadvantaged South

Africans[5] (HDSAs) in mining needed to be overcome. These included a lack of: access to mineral rights; access to reasonably priced finance; business, technical, and managerial skills and experience; capacity to handle the legislative burden and the compliance costs of being in business; and a lack of access to markets (Department of Minerals and Energy 2007).

With the aim of restructuring ownership and participation patterns within the local minerals industry, the government, in consultation with stakeholders, developed the Broad-Based Socio-Economic Empowerment Charter for the Mining Industry (The Mining Charter; Department of Minerals and Energy 2007). Emphasis is afforded to the incorporation of previously disadvantaged South Africans into all levels of management and production, socio-economic and environmental accountability, and further beneficiation. The goal of the Charter is to "create an [mining] industry that will proudly reflect the promise of a non-racial South Africa."[6] (Department of Minerals and Energy 2007).

The government then produced measures for assessing the progress of mining companies with respect to a number of key areas as they relate to socio-economic goals. This document is known as the "Mining Scorecard" (Department of Minerals and Energy 2007) and consists of a number of elements that have substantially influenced Landau Colliery. These include the imposition of targets in a number of areas such as human resources development, employment equity, mine community and rural development, procurement, ownership, and joint ventures (Department of Minerals and Energy 2007).

The new Minerals and Petroleum Resources Development Act (MPRDA 2002) legislates the official policy concerning the exploitation of the country's minerals. The restructuring of the South African economy and changing local and international circumstances were taken into consideration by the Department of Minerals and Energy (DME) while drafting the new Act (Department of Minerals and Energy 2007). The Act recognises the need to expand opportunities for historically disadvantaged persons to enter the mining and minerals industry, or to otherwise benefit from the exploitation of the nation's mineral resources.

In terms of the conversion process of the MPRDA, Landau Colliery is required to convert its mining rights to new order rights within 5 years from May 2004. A Social and Labour Plan must be submitted as part of the conversion application, with the objectives of:

- Promoting employment and advancing the social and economic welfare of all employees at Landau Colliery, as well as all relevant stakeholders of the communities within which the mine operates;
- Contributing to the transformation of the mining industry; and

---

[5]HDSA refers to any person, category of persons, or community disadvantaged by unfair discrimination before the Constitution of the Republic of South Africa, 1993 (Act No. 200 of 1993) came into operation.

[6]http://www.dme.gov.za/minerals/mining_charter.stm

- Ensuring that the holders of mining rights contribute towards the socio economic development of the area in which they operate, including the major
  labour sending areas.

These objectives will guide the mine in, among other things, its future socio-
economic development endeavours. It is suggested that the MPRDA, the Mining
Charter, and other legislation will continually be amended in consultation with
stakeholders so as to improve their effectiveness in contributing to growth and socio-
economic development objectives within the country (Sonjica 2008).

Secondly, the Emalahleni IDP and Local Economic Development Plan (LED) set
out development priorities, and it is these priorities that have an important influence
on the social and infrastructure development initiatives that the mine may choose to
carry out. The IDP provides an important insight into the current socio-economic
state of the area, as well as giving an indication of the socio-economic priorities,
and how government plans to spend in order to address those priorities. The LED
outlines the economic objectives of government. The goal of the Nkangala District Municipality LED is to transform the district into a hive of economic activity
characterized by high levels of investment, sustainable job creation, and improved
income levels, in a way that builds on the distinctive potential of each municipality
and preserves the integrity of the environment (Anglo Coal 2007a).

These development strategies are guided by a Provincial Growth and Development Strategy (PGDS) that addresses the most fundamental issues of development
arising from international community targets such as the Millennium Development
Goals (IIED 2007), national policies and strategies (such as Vision 2014: Department of Trade and Industry 2007), provincial strategies, and local government plans
and strategies (e.g., IDPs). The PGDS therefore ensures the alignment of provincial
and local level strategies with national and international policies. The mine contributes towards local socio-economic development planning through its involvement in the Emalahleni Local Municipality LED and IDP forums (South African
Coal Estates Collieries 2005), as well as the Nkangala District Municipality's IDP
(Anglo Coal 2007a).

The LED and IDP, together with legislation such as the Mining Charter and the
MPRDA, govern the mine's approach to transformation, the social and economic
welfare of its employees, and the socio-economic development of the local area.
Future developments within the industry will continue to be driven by legislative
pressures pertaining to Black Economic Empowerment (BEE) and equality in the
industry, environmental sustainability, socio-economic responsibility, and corporate
best-practice (Dales et al. 2004).

2.3 Profile of the Operation

Mining in the Landau Colliery area began in 1895. The mine currently consists of
two separate operational sites, the Kromdraai Opencast and Navigation Plant. The

Kromdraai operation will close in 2015, but mining will continue in the Navigation West section (started December 2007) until 2023. The coal is mined using opencast methods, processed at the Navigation Plant and then sent to Richards Bay by rail. Approximately 4 million tons of coal is produced each year with an annual turnover of approximately R1 billion in 2007. Ninety percent of the coal is exported and 10% is sold to domestic markets.

3 Methods

The methodology used in this assessment follows largely that proposed in ICMM's Resource Endowment Toolkit (ICMM 2006c). The assessment approaches have been kept simple, and use a range of techniques to estimate economic and socio-economic impacts. The research focused on a body of qualitative and quantitative data arranged around six areas: a comparison of local versus national socio-economic indicators; social and infrastructure development; employment; development of human capital; procurement; and economic contributions. Data were collected from both primary (interviews) and secondary (reports, websites, etc.) sources. Primary data were obtained from interviews and meetings with various stakeholders. Information obtained from interviewees was neither personal nor commercially sensitive.

3.1 Local Versus National Socio-Economic Trends

For the purposes of this study, "local" was defined as the Nkangala District Municipality. Where reference is made to the Emalahleni Local Municipality, this is explicitly stated.

Landau Colliery is a member of the South African Coal Estates (SACE) community of mines. For operational reasons, especially optimal utilisation of the remaining resources, the Landau, Greenside, and Kleinkopje collieries form a combined SACE operation. These collieries are located within close proximity to each other and commonly embark on collaborative efforts regarding socio-economic development in the local area. This is evidenced by the fact that a combined Socio-Economic Assessment Toolbox (SEAT) report was undertaken by these SACE collieries. The SEAT approach has been undertaken by Anglo Coal to improve the management of each operation's impact on their socio-economic environment (SACE 2005). This combined approach to socio-economic development of communities in the area made it difficult to isolate the individual impact of Landau Colliery. In addition, Shared Services, which provides support services to all Anglo Coal South Africa mines, captures data at a SACE level (rather than at the level of individual mines). At times, certain assumptions had to be made to obtain data specific to Landau Colliery. This was necessary in order to conform to the ICMM's (2006c) objective of analysing the socio-economic contribution of a single operation. These assumptions included splitting certain SACE and Anglo Coal data based on Landau Colliery as a percentage of revenue, volume produced, or number of employees, depending on which was deemed to be the most appropriate.

3.2 Social and Infrastructure Development

A list of recent social and infrastructure development projects was obtained from Landau Colliery's Human Resources and Projects Departments. Primary data collection consisted of structured interviews with stakeholders (project representatives/beneficiaries, mine management, government representatives, and community representatives). The intent of these interviews was to gain an understanding of completed and anticipated initiatives, what were felt to be priority development areas, to what extent the mine's approach was aligned with that of the local government and communities, and the relationship between the mine, government, and local communities. In addition, in order to gain an understanding of the relationships between the mine, local communities, and government, interviews were conducted with representatives from the communities of Ackerville and Schoongezicht (during a monthly community engagement forum), with a representative of the Clewer community, and with a local government official. Additional meetings with other government officials were requested but these did not materialise.

3.3 Employment

Employment generated by the mine includes the following elements:

- Direct employment, which consists of staff on the payroll, offsite management and support staff (including Shared Services staff based in Emalahleni town, and Head Office staff based in Johannesburg), and core contractors that are based on site permanently;
- Indirect employment, consisting of non-core contractor employees working on the operation in fulfilment of contracts, employment created through the supply chain, and employment created through social and infrastructure provision; and
- Induced employment, which is employment that is generated by the spending of direct and indirect employees.

3.3.1 Direct Employment

Information on the number of direct employees was obtained from the Human Resources departments of the mine, Shared Services, and Head Office. Information was obtained on the characteristics of those employees: the percentages of

Historically Disadvantaged South Africans (HDSA), the percentage of women, and the percentage of employees regarded as living in the local area (Nkangala District municipality).

3.3.2 Indirect Employment

The social investment activity at the mine has in certain cases created meaningful employment (this is over and above the temporary employment created during construction phases). Representatives from projects where permanent employment had been created were interviewed to determine the number of employees/beneficiaries of those projects, as well as their characteristics. In accordance with the Resource Endowment Toolkit (ICMM 2006c), full time equivalent jobs were estimated based on Landau Colliery's contribution to the project, the number of beneficiaries, and the proportion of employment derived from the project.

Indirect employees created in the supply chain are those who work for the operation's suppliers and contractors, and whose employment is dependent, at least in part, upon custom from the operation. These data were obtained through a sampling of suppliers to identify the suppliers' annual turnover, total employment, and the characteristics of those employees, using a formula provided in the Resource Endowment Toolkit (ICMM 2006c).

A list of all Landau Colliery's suppliers, the value of the goods or services procured from those suppliers (procurement expenditure), and their characteristics was obtained. A stratified random sampling approach was used based on BEE status, total procurement expenditure by the Landau Colliery, and whether or not the company was local. Due to time constraints, the sample was based on procurement expenditure over 3 months (August–October 2006). Shared Services in Emalahleni town confirmed that the months chosen would provide a representative sample. Six local and seven non-local suppliers were chosen. The local suppliers were interviewed individually during the data collection period in Emalahleni town. Non-local suppliers, in the interests of time, were interviewed using an email questionnaire. Suppliers were asked to indicate their total employee complement, the characteristics of those employees, and their total annual turnover.

Total procurement expenditure on each supplier interviewed was obtained from the procurement department; this was used to estimate Landau Colliery's percentage contribution to each supplier's annual turnover. Using this percentage, the total number of employees attributable to business with Landau Colliery could be estimated. Because the sample was assumed to be representative, the average number of indirect employees per supplier was multiplied by the total number of suppliers to obtain an estimate of Landau Colliery's total indirect employment creation through its supply chain. Care was taken to communicate the purpose of the project and to dispel any confidentiality concerns. In addition to providing a verbal outline of the purpose of the study, a letter was given to each supplier stating that company names and information would not be published in any form; only aggregated data would be published. In addition, any information supplied would not have any direct impact on the current relationship between the mine and the supplier.

As a final example of indirect employment, the Richards Bay Coal Terminal (RBCT) exports coal for Landau Colliery, as well as many other collieries in South Africa. The volume of Landau coal exported from the terminal on an annual basis was used to determine the relative contribution of Landau Colliery to the total annual turnover of coal exported from Richards Bay. From this, a percentage estimate was obtained to determine the number of employees at RBCT that can be attributed to Landau Colliery's exports.

3.3.3 Induced Employment

Induced employment was calculated using a multiplier based on an average number of 12.18 induced jobs created per R1 million of GDP (Dales et al. 2004).

3.3.4 Dependants

In order to determine the average number of employee dependants, a stratified random sampling was conducted, based on different salary brackets, HDSA, and gender. Employees were asked to enumerate all those living in the same household who were financially dependent on the Landau Colliery employee. An average number of dependants was calculated, and this value was multiplied by the total number of employees to obtain the total number of household dependants supported directly by the mine.

3.4 Procurement

The supply chain department at Head Office provided Landau Colliery's procurement expenditure per supplier, and also included characteristics such as whether the supplier was a BEE company and where the supplier was located. Data for the most recent period (September 2006–August 2007) were summarised to produce a list of all the suppliers to the mine, the percentage that were BEE companies and their associated procurement expenditure, and the percentage that were local and their associated procurement expenditure. In addition, the same sample used in determining indirect employment created in the supply chain (see above) was used to gain a greater understanding of the mine's relationship with suppliers, and whether the mine had assisted the individual businesses in any way.

3.5 Economic Contributions

3.5.1 Contribution to Government Revenue

Based on the ICMM's (2006c) methodology, all payments made to the public sector were identified for 2003 and 2007, and an estimate was made for 2011. It was also attempted to determine to which level of government those payments were made.

3.5.2 Contribution to the Host Economy

Total value added was calculated as total revenue less the value of payments to suppliers. Landau Colliery provided data from 2003 to the present, and data were then forecast until end of mine life (2025). The local element of value added was determined by subtracting flows of money that leave the economy (following procedures outlined in ICMM 2006c). These flows included repatriated profits, interest payments to foreign banks, management fees to overseas headquarters, and the savings of expatriate workers that are transferred overseas. In order to produce more accurate estimates of value added, the concept of opportunity cost was factored into the value-added calculation (ICMM 2006c).

3.5.3 The Opportunity Cost of Economic Resources

The opportunity cost of economic resources represents the returns or income generated by those resources if they were to be used for an alternative purpose. The cost represents the opportunity foregone by using those resources for their current use when other alternatives are available. The total annual opportunity cost (for 2007) was determined by estimating the opportunity costs of the three factors of production: land, capital, and labour.

The fixed asset value of Landau Colliery in 2007 was used as an estimate of the mine's capital resources. In order to determine the income stream (or return to capital) from an alternative use of the capital, it was assumed that the capital could be invested, and the interest earned would be regarded as the opportunity cost of that capital. Using the government bond rate as the risk free rate (it is assumed that government will always be able to repay loans), the return to capital for 2007 was estimated.

The opportunity cost of the land was estimated by multiplying the total mining rights area by the value of the land. In the same way as was done to determine the opportunity cost of capital, the government bond rate was applied to determine the income stream should the land, as capital, be invested and therefore earn interest.

In order to determine the opportunity cost of labour, the number of employees that would be expected to gain alternative employment (in the case of the mine not existing) was calculated. In addition, workers with different skills would earn different incomes and therefore represent a different opportunity cost. An attempt was made to match the skill levels/professions of the Landau employees to average incomes in the area for those professions. Average incomes in Mpumalanga for three different employee categories (professional, technical, administration and managerial; clerical and sales; and production, transport, service, etc.) were obtained from Statistics South Africa (2000). Estimates for 2007 were calculated by increasing the 2000 figures by CPIX (consumer price index excluding mortgage costs) plus 2% each year. Landau employees were then grouped according to the following categories: management and officials: senior skilled: skilled: and learners. It was assumed that a certain percentage (38.4%) of the lower skilled workforce would, if the mine was not operational, be unemployed and therefore represent no opportunity cost. This estimate was based on unemployment figures for the Emalahleni Local municipality (Census 2001). The numbers of employees in the two categories that would be expected to gain alternative employment were multiplied by the relevant average income for each category, to determine a total opportunity cost of labour.

3.5.4 Contribution to GDP and to Exports

Data on national level GDP and exports for 2007 were obtained. The retained value, or the local element of value added (as calculated for Sect. 3.5.2), is regarded as a representation of the mine's contribution to GDP (following ICMM 2006c). The value of coal that is anticipated to be exported in 2007 was obtained from Landau Colliery's finance department, and was used to determine the mine's percentage contribution to national exports.

4 Results

4.1 Comparison of Local Versus National Socio-Economic Trends

Development indicators at the local scale were difficult to obtain. Data were more readily accessible at the national and provincial level, but where district and municipal data were available they were included in the analysis.

In terms of health indicators, Mpumalanga province experiences very similar conditions to the country as a whole. The province does have a higher under-five mortality rate for children, but a significantly lower maternal mortality ratio. Human immunodeficiency virus (HIV) prevalence among antenatal clinic attendees in 2005 (Department of Health 2006) shows Mpumalanga to have an occurrence of 34.8% (second highest behind KwaZulu Natal), compared with 30.2% in South Africa as a whole. In 2003, 12% of Landau Colliery's workforce received Voluntary Counselling and Testing (VCT).

In 2003, six Landau Colliery employees were on HIV Disease Management Programmes (HIVDMP), of whom four were receiving anti-retroviral therapy (ART). By 2007, VCT participation was 82% of the workforce, with 68 on HIVDMP, of whom 14 were receiving anti-retroviral therapy. This can be regarded as a direct benefit for local communities (considering that 99% of the permanent employees on the mine are local), and it is assumed that there would have been positive (spillover) effects, such as increased education and awareness, as a result of the communities' interaction with these employees. This positive impact is likely to increase in the future because the company has plans to extend HIVDMPs, including ART, to dependants. This finding is in contrast to the commonly held view that mining results in an increase in diseases and other negative health-related effects (Groom 2008; Stephens and Ahern 2001). Provincial data are available, but adequate health-related data for the local area could not be obtained, so it is unclear as to whether the mine has resulted in an increased incidence of diseases or not.

Poverty-related indicators are again very similar at a provincial and national level. A poverty study analysing household consumption welfare indicators (Alderman et al. 2000) estimated that the Magisterial District of Witbank (presently the Emalahleni Local Municipality) had an imputed mean monthly expenditure of R3,525 per household (at 1996 prices), and fell into the category "Least poor" where less than 20% of households were below the poverty line of R800 or less per month. This value is well above the national average of R2,789 per month (Alderman et al. 2000). Unemployment at the local level, particularly at the municipality level (38.4%), is significantly higher than the national level of unemployment (25.5%). Thus, although there were fewer jobs available, those jobs were higher paying, and households were therefore receiving a higher income than on average in South Africa.

Although adequate data were not available, case studies analysed by the ICMM (2006b) indicate that income gains contribute to broadly generalized improvements in poverty levels, and to social welfare more generally. It is likely that Landau Colliery has indirectly had a positive impact on HIV/AIDS treatment in the local communities through the various initiatives offered to its permanent workforce.

4.2 Social and Community Development

The mine has contributed in excess of R49 million and 3,300 management hours[7] towards over 30 social and community development objectives since 2002, creating an estimated 136 temporary and permanent jobs over that time. Projects include building physical infrastructure, building and maintaining schools, educating and training local community members, assisting community HIV/AIDS initiatives, local enterprise and skills development, and other general community development. The most notable contribution to infrastructure is the mine's 10% (R30 million) contribution to the Emalahleni Water Treatment Plant, which treats and restores to potable standards the mine water from several local mines. The water is sold back to the local authority. This will contribute to long term water security in the area, as well as reducing the mine's environmental impact.

In addition, as a business unit of Anglo Coal, Landau Colliery has had an indirect impact on several other programmes, but accurate estimates of financial and management time contributions, as well as the full extent of the benefits provided, were not available. As such, the estimates provided above can be regarded as conservative. Furthermore, significant goodwill is created through a well run programme, but the impact of this goodwill is difficult to quantify.

Landau Colliery has committed, as part of the requirements of Regulation 46 of the MPRDA, to support a number of future projects as discussed with the IDP forum, and which are part of the Nkangala District IDP (Anglo Coal 2007a). Landau Colliery anticipates contributing R14.04 million towards infrastructure development

---

[7]Management hours are defined as the estimated time contributed by managers and technical staff.

between 2008 and 2011, and in excess of R4.8 million towards future social projects (Anglo Coal 2007a).

The social and economic impacts of these future projects will be guided by the SEAT study. It will be particularly important to establish partnerships with organisations that have the required skills. In addition, this will involve closer links with both Anglo Zimele (a business development branch of Anglo Coal) and the Chairman's Fund to make better use of Anglo's resources.

4.2.1 Local Business Development

Landau Colliery supports business development at an individual project level, and also indirectly contributes to this objective through Anglo Coal's business development branch, Anglo Zimele. Anglo Zimele aspires to create sustainable, commercially viable businesses, in geographical proximity to company mines or based in adjacent communities. Their lines of business, however, need not be directly linked to mining. The idea is to promote diversification in local communities away from direct employment in mining. The focus lies in investment in black-empowered Small/Medium Enterprises (SMEs) that demonstrate a commercially viable and sustainable business or business plan. Typically, Anglo Zimele provides financial support and expertise in a hands-on approach, through strategic guidance and skills transfer (Anglo Zimele 2007). Landau currently supports 17 Anglo Zimele-assisted businesses through the procurement of goods and services. In 2006, Landau Colliery spent a total of R9.4 million on these companies. Anglo Zimele-supported businesses have a survival rate after 5 years of 3 times the South African average (Anglo American 2007a).

Anglo Zimele has begun a process of rolling out "business hubs" to provide these opportunities to a greater proportion of the surrounding communities. These business hubs, it is suggested, will provide a mechanism to devote more resources to local enterprise development. In addition, they aim to provide skills and opportunities in a variety of industries outside of mining. It is hoped that this will create sustainable communities that last beyond closure of Landau Colliery and the general decline of mining in the area. A business hub, which is accessible to Landau Colliery's surrounding communities, was opened in Emalahleni town in 2007.

4.2.2 Community Development

Social and infrastructure development projects undertaken by the mine are guided, supported, and communicated through collaborative relationships between government, the mine, and surrounding communities. Fundamental to the success of these projects is the relationship with local communities and government. The establishment of productive dialogue with local and regional communities and governments can result in a greater chance that social investments will be tailored to meet government and community needs and expectations, and thus have a greater chance of being regarded as a positive commitment to local community development (ICMM 2006b; Pedro 2006).

Stakeholder interviews indicated that the mine has increasingly improved its relationships with surrounding communities. Traditionally, there was little community engagement and relationships were poor. Currently there are monthly community forums with the three surrounding communities (Clewer, Ackerville, and Schoongezicht). The intention of these meetings is to open the channels of communication, allowing the mine to provide information, and the communities to raise issues and concerns (the mine provides feedback on the issues/concerns raised by the communities). In addition, the mine hosts an annual open day to provide information to local farmers and address their concerns.

The current community engagement forums are expected to continue. In addition there will be a SACE stakeholder forum to engage all communities affected by the SACE group of mines, and there are plans to recruit a dedicated community development officer at Landau Colliery.

In the same way, the increased number of meetings between mine management and government, as well as evidence from interviews with representatives from both parties, suggest that the relationship between the mine and local government is improving. The mine has demonstrated commitments to align its socio-economic development goals with those of the government (see Sect. 2.2). Currently, the mine, in choosing which social and infrastructure projects to invest in, chooses from the government's LED "wish list" (communities are involved in determining the "wish list"). However, it is suggested that such efforts could be more effective if government played a greater role in recommending which projects mines should invest in. Comments from interviewees indicate that there is a need for improved communication between the mine and the government, as well as more partnerships.

4.2.3 Future Approach to Community Development

There are a number of general weaknesses that have been identified in many of the projects undertaken by Landau Colliery. These include a lack of evaluation of the impact of initiatives, a lack of provision of long-term employment, a lack of collaboration with other units, poor engagement with communities, and projects being undermined due to some of the negative effects of poor socio-economic conditions in local communities. In the latter case, for example, materials such as wood, fencing, and scrap metal have been stolen or used for fuel from some agricultural and playgrounds projects.

Anglo Coal has identified a number of focus areas to address these weaknesses, many of which are common to several operations. These include:

- Small, Medium, and Micro Enterprise (SMME) development: in collaboration with Anglo Zimele and linked to procurement policies;
- Agriculture: to provide the opportunity for HDSA community members to farm on rehabilitated mine lands;
- Skills development/capacity building;
- Infrastructure such as the Emalahleni Water Treatment Plant;

- Health, with a particular focus on HIV prevention and treatment. The company has plans to provide HIV disease management programmes (HIVDMPs), including anti-retroviral therapy(ART) to dependants;
- Education (holistic approach);
- Environmental conservation: for example, cleaning and waste management campaigns, as well as trying to set up businesses around environmental conservation. Projects such as Basa Njengo Magogo (make fire the traditional way) will be implemented that aim to educate communities in a new way to stack and light coal fires for cooking and heating, that releases less pollution and therefore has environmental as well as health benefits. This will be driven by Anglo Coal, but is likely to at some point impact the communities adjacent to Landau Colliery.
- Sports, culture, and heritage.

The new drive within Anglo Coal is to focus on projects that benefit external communities more, as well as emphasising the sustainability of projects, poverty alleviation, job creation, and uplifting the lives of communities. These new focus areas were endorsed by government and will be incorporated into future projects.

4.3 Employment

Mining remains an important source of employment in South Africa. According to the Department of Minerals and Energy (2005), 2.6% of the economically active population in South Africa is employed in the mining industry. If the multiplier and induced effects of the industry are used, the contribution to employment in 2006 as a result of mining rises to about 20% of total non-agricultural formal sector employment in South Africa (Chamber of Mines 2006). Job creation is a key contribution of Landau Colliery to development, and is in line with national objectives. Jobs provide an opportunity for people to enter the cash economy, to generate income for purchasing basic goods (which stimulates induced employment), and to support their families (World Business Council for Sustainable Development 2007).

Landau Colliery has displayed a gradual increase in permanent employees over time (Fig. 3). This may be the result of the implementation of new shift structures (allowing for the hiring of an additional 60 employees since 2006), and increasing the ratio of permanent employees to contractors. The implementation of these strategies coincided with an increase in production from roughly three million tons to over four million tons of coal per year. 2016 marks a reduction in anticipated employment numbers from 482 to 380 (estimate). This is due to the closure of the Kromdraai section of the mine at the end of 2015. The Navigation West section will be mined from the end of 2007 to 2025. In 2007 the mine provided employment to 910 direct employees and 4,615 indirect employees. Table 1 provides a summary of Landau Colliery's current total employment levels, including contractors and other indirect employees.

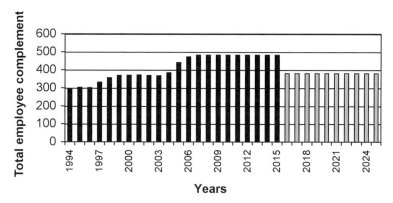

Fig. 3 Total permanent employees at Landau Colliery over time (employment level after 2015 falls due to closure of the Kromdraai pit)

In addition, while not included in the snapshot of employment shown in Table 1, Landau Colliery has created periodic employment over the last 5 years through social and infrastructure projects, particularly in construction-related activities. Wherever possible, these part-time/project-orientated jobs have been awarded to local candidates and HDSAs.

Landau Colliery's induced employment is 3,244 (Table 1). The ICMM Toolkit (2006c) suggests that induced employment can be assumed to be 165–250% of the sum of direct and indirect employment. This equates to a range of between 2,234 and 5,660 induced employees. The value of 3,244 (using the method suggested by

Table 1 Landau Colliery employment (2007)

Employment category	Number of employees
Direct employment	
Permanent employees (on the mine)	493
Core contractors (on site)	331
Head office employees (proportion attributable to Landau)	9
Shared services employees (proportion attributable to Landau)	77
Total	910
Indirect employment	
Other contractors	166
Created in the supply chain	217
Created through social provision	50
Other (e.g., Richards Bay Coal Terminal)	28
Total	461
Induced employment	3,244
Total employment	4,615
Dependants (of direct employees only)	1,923

Dales et al. 2004) fits within this range, and is therefore regarded as an acceptable estimate of the employment induced by the colliery. This very significant number indicates that the largest impact on employment lies outside the mining industry altogether, and gives an indication of the overall significance of the operation as a job creator in the domestic economy (Dales et al. 2004). In comparison, direct employment at the mine represents only 0.64% of locally employed persons.

It is important when trying to gain an understanding of how many people the mine is supporting, to analyse the number of dependants. The data obtained from the questionnaire indicated that the average number of dependants was 3.9 people per direct employee (493), leading to an estimate of the total number of dependants of 1,923 (Table 1).

The characteristics of employees in terms of the proportion of HDSA and women were not recorded in 2002, because the Mining Charter had not yet come into effect. Since then, the mine has begun to report on these characteristics. Landau Colliery's employment policy in terms of preferential hiring of local people is constrained by the Labour Relations Act, but the mine attempts to hire local people where possible. In attempting to achieve targets set out in the Mining Charter, the mine has experienced a significant growth in the percentage of HDSAs, the percentage of HDSAs in management, and also the percentage of women on the workforce. Landau Colliery is currently achieving the target of 10% women in technical mining roles, and is expected to be ahead of its plan to achieve 40% HDSA in management by 2011. Table 2 provides a summary of the current characteristics of direct and indirect employees (where these data could be obtained).

Table 2 Characteristics of employees at Landau Colliery

		% HDSA	% Local	% Women	% Women in technical mining roles
Direct employment	Permanent employees	77.96	98.8	12.1	10.1
	All direct employees	77.3	97.9	14.3	–
Indirect employees	Includes employment created through the supply chain and through social provision only	74.4	55.9	29.6	–
Induced	Could not be determined				

Note: Percentages have been weighted depending on the relative proportion of employees that each component represents.

Women currently comprise 12.1% of Landau Colliery's permanent workforce (HR Month end report: October 2007). There has been a concerted effort to hire more women, for example by changing the shift structure to accommodate an additional shift of women. A study was conducted to identify the areas of under representation within each occupational category and level, and appropriate strategies and targets are currently under consideration. This includes the continued review of policies and procedures aimed at attracting and retaining women (Anglo Coal 2007a). Landau Colliery has provided the necessary infrastructure and facilities to support more women in mining, and has set up an employment equity committee (consisting of unions, workforce representatives, and mine management) which identifies jobs that would be more suitable for women.

It is important to analyse the quality of the jobs provided by the mine in order to gain a greater understanding of the true socio-economic impact of employment creation. Landau Colliery is affiliated with the Chamber of Mines, and all employment packages are determined through collective bargaining (between the company and unions). The colliery provides medical aid, production- and safety-related bonuses, a housing allowance (including assistance in various ways to promote house ownership and the conversion of single accommodation to family units), and competitive salaries. The total payroll was R53,045,000 in 2003, R107,352,000 in 2007, and is expected to be in excess of R150,000,000 by 2012 (Anglo Coal 2007a).

Despite the highly mechanized nature of modern industrial mining, the absolute numbers of people employed both directly and indirectly (1,371; Table 1) are substantial and economically and socially important. The high percentage of local South African permanent employees (98.8%) and the low numbers of expatriate (non-South African) staff (1.2%) is in contrast to common perceptions of job allocations within the mining sector (ICMM 2006b). All local South African employees live in the Nkangala District municipality. Furthermore, the direct employment created by the mine can be regarded as high quality employment due to the additional benefits, such as medical coverage, a housing allowance, bonuses, and salaries that are competitive within the industry.

Successful efforts have been made to hire more women at the mine through Anglo Coal's Women in Mining (WIM) programme. Management and unions agreed with the WIM concept and this resulted in Anglo Coal WIM becoming one of the benchmark initiatives in the mining industry (Anglo American 2007b). 12.1% of Landau Colliery's direct employees are women (currently above the Mining Charter target). This is striking when compared to the case studies analysed by the ICMM, which found that Chile, with 3.7% women, had the largest percentage of women in mining in the countries studied (ICMM 2006b). Employment of women, through efforts such as WIM and women employed through social and infrastructure initiatives, has substantial positive socio-economic impacts. Pronyk et al. (2006) suggested that the empowerment of women in South Africa not only contributes towards improved economic wellbeing, but also results in an increase in communication concerning HIV, increased testing, and reduced HIV infection rate and occurrence of unprotected sex.

A possible negative impact that the mine could have caused is a high level of induced unemployment due to inward migration resulting from the perceived employment opportunities in the area. In 2001, 76,668 people in Emalaheni Local Municipality were employed; however, 47,703 were unemployed, and 66,514 were not economically active (Census 2001). Although Landau Colliery contributes towards an absolute increase in employment, the high level of unemployment (38.4%) could be partly attributed to inward migration.

4.4 Human Capital Development

The Mining Charter recognises that scarcity of relevant skills is one of the key barriers to HDSAs entering the mining industry. Human resource development therefore remains a key challenge facing the future success of the industry (Dales et al. 2004). Developing skills in a particular field where those skills are scarce ensures that the pool of skills in that field is increased. In the context in which Landau Colliery operates, this is particularly necessary with regard to the training of artisans, and has been an area of focus for the mine.

Since 1995, Landau Colliery has provided literacy and numeracy classes to all employees who were interested, as well as to their spouses and dependants. Currently Adult Basic Education and Training (ABET) classes are also extended to members of the community, but, because the workforce is largely literate, enrolment in ABET classes has been low. As an alternative, the mine has offered its employees life skills programmes, providing training in carpentry, bricklaying, welding, computer skills, and financial skills.

The importance of these programmes is recognised in the fact that the value of employees increases with additional training, as reflected in the salaries of different skilled workers. Landau Colliery's minimum skills requirement for a learner miner is a grade 10 (normally age 16) level of education, a driver's license, and to be able to speak English. A learner miner will earn roughly R5,272 per month (2007 rates). Through on-the-job training and external training, a learner miner can advance to become a skilled miner, who earns between R6,934 and R8,706 per month, and receives a housing allowance of R2,636 per month (2007 rates). On-the-job training represents a large portion of the increased value of an employee; however, it was not possible to obtain a quantitative estimate of the value of this training. A summary of the training budget is provided in Table 3.

The mine has demonstrated a commitment to training, by developing the skills of its workforce as well as those of the community where possible. It has done so through external and on-the-job training, as well as by providing opportunities such as career progression and mentorship plans. Cognisance is given to HDSAs with emphasis on development of women, and skills training in identified fields. The programmes are formalised and include internal and external programmes, complemented by a structured mentorship programme. The outcomes of investments

Table 3 Landau Colliery training analysis

	2001	2003/2004	2007/2008
Annual training budget (Rands)	1,889,765	3,561,000	5,562,134
Percentage of payroll	3.6	5.4	5.9

made in training are reflected in the current value of the mines' employees (reflected through salaries). The overall skill level at Landau Colliery relative to other Anglo Coal mines in the area in high, which is perhaps a reflection of the quality and commitment of Landau Colliery training initiatives.

4.5 Procurement

Procurement creates business for a company's suppliers. This leads to further economic activity, including jobs in the supply chain, induced employment, and business development through the spending of the suppliers' employees.

The Mining Charter requires demonstrable BEE procurement as a pre-requisite for mining rights conversion. Anglo Coal South Africa's actions predate this requirement, and demonstrate a desire to increase procurement from BEE suppliers motivated by a business case rather than one of simply complying with requirements (Anglo Coal 2007b). Landau Colliery does not afford any formal preferential treatment based on whether a supplier is local or has a BEE status. However, Anglo Zimele, by encouraging existing suppliers to form partnerships and helping to develop HDSA procurement capacity, has been able to increase the number of HDSA suppliers. This has resulted in the company and the mine being able to choose suppliers based on cost and quality of the good or service. In this way, the mine has been able to comply with the Mining Charter's requirement to, among other things, commit to a progression of procurement from HDSA suppliers, and to achieve those targets.

An informal attempt has also been made to encourage purchasing from local suppliers, where possible. This has been driven from the mine level, whereby the mine nominates local suppliers, through a tender committee, to be considered for certain supply requirements. In areas where local suppliers have the necessary capacity, the mine encourages the purchasing of goods and services from those suppliers. HDSA companies are encouraged to tender for contracts, and attempts are made to continuously increase the number of companies in the HDSA database (Anglo Coal 2007a). This approach, in addition to helping to develop a more reliable and racially representative local supply base, also strengthens the mine's local "social license to operate" (World Business Council for Sustainable Development 2007).

The extent to which procurement has an impact on a particular location depends on the amount of procurement in the supply chain that is sourced from that location (WBCSD 2007). As can be seen in Table 4, 48% of all suppliers to Landau Colliery (353 suppliers) are local; however, 52% of procurement expenditures (R165.8 million in 2006/2007) goes to local suppliers. Although only 23% of all suppliers are BEE companies, they receive 65% of the total procurement expenditures (R205.8 million). International suppliers represent only 4.1% of the total number of suppliers, but they receive 24.5% of total procurement expenditures (R78.2 million). All of these suppliers do have offices within South Africa, so although some of the economic contributions (such as value added) will leave the economy, the country

Table 4 Characteristics of procurement expenditures (September 2006–August 2007)

	Expenditure (Rand millions)	Expenditure as % of total	Number of suppliers	Number of suppliers as % of total
Non-BEE	113	35.5	564	76.6
BEE	205.8	64.6	172	23.4
Non-local	153	48	383	52
Local	165.8	52	353	48
Domestic suppliers	240.6	75.5	706	95.9
International suppliers	78.2	24.5	30	4.1
Capital	10.7	3.3	18	2.5
Consumable	234.5	73.5	401	54.5
Service	73.7	23.1	317	43
Total	318.8	100	736	100

will nevertheless experience some socio-economic benefits through employment creation, and contributions to government revenue through taxes.

The high local procurement expenditure would likely make a large contribution towards induced employment and poverty alleviation in the area in which the mine operates. This is the result of consumption linkages such as the spending by the colliery and its local employees on local goods and services. Some of the success of the mine in this regard can be attributed to the efforts of Anglo Zimele in the areas surrounding the mine. Growth of small business, according to the WBCSD (2007), is an important contributor to overall development. Enterprises, which are locally owned and locally managed, provide employment opportunities to local citizens. Additionally, more enterprises in the formal sector means increased contribution to taxes and value added to the economy, as well as greater compliance with regulations such as labour conditions.

In the context of transformation in South Africa, pressure has been placed on Anglo Coal in terms of its power to influence suppliers in this regard. In terms of the Broad-Based Black Economic Empowerment (BBBEE) codes,[8] Landau Colliery (through Anglo Coal) must assist government to enforce those codes. The company, however, cannot force suppliers to comply with the codes but, can and does encourage them to do so, and monitors their compliance.

4.6 Economic Contributions

Mining continues to be the single most important earner of foreign exchange in the South African economy (Dales et al. 2004), and currently accounts for 6.2%

---

[8]http://www.info.gov.za/gazette/acts/2003/a53-03.pdf

of the country's GDP (Chamber of Mines 2006). On a local level, mining contributes 25% of the Emalahleni Local Municipality's Gross Geographic Product (GGP) (Emalahleni Local Municipality 2007). During the life of the mine, revenue contributed to the host economy has changed significantly over time. Landau Colliery contributes 0.25% of all national exports, and 0.015% to the national GDP (Table 5). Significantly, however, only 0.004% of the total number of employed persons in South Africa work towards making these contributions, implying that the mining industry has a relatively efficient and productive workforce.

Table 5 Summary of economic contributions made by Landau Colliery in 2007

Economic contributions	2007
Contribution to national revenue	
Taxes borne	R195.32 million
Taxes collected	R19.10 million
Total	R214.42 million
The opportunity cost of economic resources	
Capital	R107.70 million
Land	R22.60 million
Labour	R30.20 million
Total	R160.50 million
Contribution to the host economy (retained value 2007) less the opportunity cost of economic resources	R105.50 million
Percentage contributions	
Percentage contribution to GDP	0.015%
Percentage contribution to GDP (Mpumalanga)	0.224%
Percentage contribution to exports	0.248%
Landau employees as a percentage of the total number of employed persons in South Africa	0.004%
Total direct and indirect employment as a percentage of local employment	1.5%

Tax contributions from Landau Colliery have increased from R48.2 million in 2003 to R214.4 million in 2007 (Table 5), and there will likely be further changes in the contribution to government revenue in the future. The closing down of the Kromdraai pit in 2015 would result in a significant decrease in the profit taxes paid to government (the largest contributor to government revenue). There would, however, likely be increases in tax contributions from the mine due to a number of charges. Contributions in 2011 are estimated at R396.0 million. Firstly, royalty payments are mooted to begin in 2009, and will be based on the ash content of the coal (Department of National Treasury 2006a). Landau anticipates paying R21.6 million in royalties in 2009. Secondly, if fuel prices continue to rise, it is assumed that fuel taxes will do the same; however, it is difficult to estimate the value of these future fuel costs and associated taxes. And thirdly, the price of water is likely to increases due to increasing scarcity and new legislative charges.

Specific legislative measures include:

1. Air quality permits: It is possible that permits will be used as economic instruments (as in the United States) to limit air pollution. If this does come into effect, then it would not likely be regarded as a contribution to government revenue (Department of Environmental Affairs and Tourism 2007).
2. Waste Discharge Charge System: A charge is to be levied on waste water discharge based on a number of criteria. Pricing arrangements are still under review and the current estimate is that this charge will affect the mine in 2009 (Department of Water Affiars and Forestry 2003).
3. Raw Water Pricing Strategy: This is an administration (a charge for water resource management) and abstraction charge. This charge, which fluctuates depending on the availability of water, has been established, but is in its early implementation stages. It will likely influence the mine more in the future as water becomes increasingly scarce (Department of Water Affairs and Forestry 2007).

Currently the mine does not pay any property taxes. Fiscal contributions are paid directly to the national government, and local and regional governments must rely on central allocation of funds in a "trickle down" effect. However, there is discussion on imposing rates and taxes in the future that would be payable to the local government authority, but no agreement has yet been reached.

The total value added from Landau Colliery from 2003 until the end of life of mine is presented in Fig. 4. 2016 signals a sudden decrease in the value added, again due to closure of the Kromdraai pit in 2015, with subsequent production coming only from the Navigation West section of the mine. Retained value (value added less flows of money that leave the economy) increased from R150 million in 2003 to R266 million in 2007 (Table 6).

The mine is able to directly influence the relative impact of benefits relating to social and infrastructure development, employment, human capital development, and procurement. However, there are substantial contributions to government revenues over which the mine has only very limited influence, mainly through stakeholder participation in socio-economic development planning. While this revenue contribution is substantial, it cannot be assumed that it will follow a simple and lin-

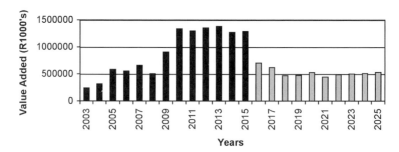

Fig. 4 Landau Colliery value added

Table 6 Financial contributions to national revenue from Landau Colliery since 2003

Contribution to national revenue	2003 (Rand millions)	2007 (Rand millions)	2011 (anticipated) (Rand millions)
Taxes borne	38.87	195.32	370.87
Taxes collected	9.36	19.10	25.13
Total	48.23	214.42	396.00
Contribution to the host economy (value added); see Sect. 3.5.2	242.53	659.90	1,296.70
Retained value (value added less flows of money that leave the economy)	150.10	266.00	659.74

ear path, and translate into growth. Whether the full benefits of these contributions are realised depends mostly on government policies and decisions. It is the role of government to ensure the allocation of financial resources towards priority sectors or pro-poor growth in order to achieve poverty reduction (ICMM 2006b; Pedro 2006). Indeed, there is debate regarding, firstly, the long term effect of these taxes on the mining industry in general and, secondly, regarding the manner in which the revenues should be spent. In the context in which Landau Colliery operates, there have been particular criticisms regarding the local government's ability to effectively and efficiently spend those revenues in ways that reflect local priorities.

5 Discussion

5.1 How Effective Is Mineral Taxation at Improving Socio-Economic Outcomes? A Brief Look into the Debate

Theoretically, a socially-related tax introduced at the correct rate can achieve a social objective and raise revenue in the process (Department of National Treasury 2006b). However, ICMM (2007) notes that royalty payments and other taxes imposed on the mineral industry can have a dramatic impact on the profitability of a project and the economic cut-off point, which can lead to reduced investment and premature mine closure. There is indeed contention surrounding the taxing of minerals in South Africa, and the effect that the various legislations will have on the future size of the industry and the benefits it will bring. Legislation such as the MPRDA is relatively new and has been subject to a much criticism (e.g., Rostoll 2007).

Additionally there is debate around whether local communities benefit more from tax regimes where a portion of the revenues are distributed to the local/regional levels, as opposed to accruing at the national level (ICMM 2007), or whether or not funds should be ring-fenced or earmarked. Requests to earmark mineral revenues, either for a specific cause or to be spent in the specific geographic area from which the revenues were generated, stem from the hope that this will guarantee and possibly increase the source of funding for that cause, or improve the benefits that the communities receive for the exploitation of minerals in their areas (Department of National Treasury 2006b).

In the case of South Africa, the majority of taxes are raised at national government level, the argument being that national taxes are easier to administer, and avoid the problems of duplication associated with a decentralised tax system. However, certain expenditure responsibilities have been devolved to the provincial and local government sphere. Provinces, therefore, have a limited degree of fiscal and political autonomy (Department of National Treasury 2006b). Revenue for this autonomous expenditure could, as allowed for in the Constitution, be generated through the imposition of taxes, levies, and duties (other than income tax, value-added tax, or customs duties), and also by raising loans. Local governments are allowed by the Constitution to impose rates on property and surcharges on fees for the services they are responsible for, as well as to raise loans subject to restrictions (Department of National Treasury 2006b). Additionally there is also a loose association between the source of the revenue and the spending programme, but no direct earmarking of funds (Department of National Treasury 2006b).

Some argue that more effective governance and a more efficient utilization of public resources can both be achieved if expenditure decisions are entrusted more to regional and local authorities rather than at the national level (ICMM 2007). Furthermore, specific funds or earmarking can promote transparency, greater funding certainty, and a more direct relationship between payments and benefits, which may enhance both equity and efficiency (Department of National Treasury 2006b). The negative impacts of mining (e.g., new inward migration, potentially leading to induced unemployment) and the potential positive effects associated with the new opportunities for broader economic development, require regional and local governments to have the capacity to plan and invest at much higher levels than is the case in non-mining areas (ICMM 2007). There is a broad global trend for greater decentralization driven by the argument that the negative dimensions and potential positive development benefits could be more effectively managed at regional and local level (Pedro 2006; ICMM 2007). However, the lack of regional and local level government capacity in South Africa provides a strong argument against such a system in that country.

5.1.1 The Case of Landau Colliery

In the case of Landau Colliery, the imposition of the MPRDA and mineral taxation has not resulted in a change in the life of mine. Of greater importance in the context of the mine's socio-economic impact, therefore, is how the revenues from Landau Colliery have been spent.

An important area for future study would be to analyse the capacity of the local government to effectively spend revenue from an earmarked fund. It is possible that the local government in which Landau operates does not have the capacity to effectively spend revenues, as suggested by poor local socio-economic development indicators (where they could be obtained) and certain criticisms made by the media regarding lack of transparency (Business Day 1998a), allegations of unauthorised expenditure and the widespread absence of important documents (Business Day 1998b), corruption (Business Day 2004), and allegations of fraud by government officials (Business Day 1997). Furthermore, the Institute for Study of Democracy in South Africa (INDASA) ranked Mpumalanga as ninth out of ten provinces for measures of parliamentary democracy (Judson 2001).

As indicated in Sect. 4.6 above, currently the mine does not contribute any revenues that are autonomously spent by the local government. However, with the imposition of rates and taxes expected in the future, some revenue would be generated by the mine for this purpose. If local government was found to have adequate capacity, this could result in improved outcomes if one was to accept the arguments for greater decentralization. Large capacity-building efforts would most likely be needed, but this could represent an extremely important source of improvement in the mechanisms whereby mining investments such as Landau Colliery deliver greater benefits to both the local and broader society (ICMM 2007).

Furthermore, South Africa's poor HDI ranking and the relatively poor performance of Mpumalanga with regard to certain indicators suggest that revenue is possibly not being spent efficiently and in a manner that reflects local priorities. The full benefits of the taxes paid by Landau Colliery may not be being realised.

5.2 Impacts of Landau Colliery After Closure

Pedro (2006) highlights the importance of developing diversified and alternative activities that would be sustainable after the life of the mine. A concerted effort is being made to reduce direct dependence on the mine, and the mining industry in general, after Landau Colliery closes. This is being achieved largely through Anglo Zimele and the development of business hubs, which are aimed at developing non-mining-related businesses and local community capacity-building, through skills and training programmes as well as informal knowledge and technology exchanges (Anglo Coal 2007a). However, mining has been taking place in the area for a long time, and the industry is less transient than is often perceived. Furthermore, when analysing the specific effect of the closure of Landau Colliery, it is important to note that the negative socio-economic impact will be diluted by the fact that there will likely continue to be substantial mining activity in the area. Landau Colliery is a part of the Witbank cluster (Dales et al. 2004), which has engendered multiplier effects and spillovers at local, community, and rural levels. The economic base is broadened to develop not only direct linkages between mining and other sectors, but also various indirect (supply and support services) and induced contributions to positive, sustainable outcomes (Pedro 2006).

5.3 Landau's Contribution to Socio-Economic Growth and Development

The overall effect of Landau Colliery has been positive, particularly at a local level (Table 7). However, it is difficult to determine the exact extent of the impacts of the mine, and indeed of mining as a whole, on the local area. While there are measurable positive impacts, these would ideally need to be compared to similar case studies, and there are also some negative impacts (Table 8). The lack of availability of other local socio-economic development indicators restricted the degree to which the socio-economic impacts of mining in the local area can be assessed. Improved measurement of indicators, such as the MDG indicators, is therefore needed to better assess the impacts of mining on the local economy, as well as the specific impact of one operation such as Landau Colliery.

5.4 SWOT Analysis of Landau Colliery

A SWOT (strengths, weaknesses, opportunities, and threats) analysis was carried out to identify the key areas influencing the mine's ability to achieve positive socio-economic outcomes. A summary of the analysis is provided in Table 9.

The aim of this study was to determine how Landau Colliery could enhance socio-economic development in the area in which it operates, and what would be the components of "success". This SWOT analysis has been used to better understand two of the study's main objectives:

- What strategies have been effective in managing revenues generated from the mine for sustainable development and poverty reduction?
- What can be done to improve the socio-economic development contribution of Landau Colliery?

Key strengths in effectively managing revenues from the mine for sustainable development include the ongoing efforts of Anglo Zimele, and a commitment from the mine with regard to the local community. Mine management has demonstrated a desire to hire more local community members and more women, to engage with surrounding communities, and to help develop local businesses in collaboration with Anglo Zimele. While legislation pertaining to socio-economic development governs the mine's actions, management has displayed commitment and achieved results that exceed that which is required to meet minimum requirements. Moving into the future, there is an opportunity to utilise the growing resources available for socio-economic development. The mine, along with government and other development partners (both private sector and NGOs), can potentially have a greater positive impact if they work together more effectively.

High community expectations will, if unmanaged, remain a threat to the mine's ability to improve its socio-economic performance in the future. Community

Table 7 Summary of positive impacts of the Landau Colliery

Impact	Description/details
Economic impact	
Physical infrastructure	Emalahleni Water Treatment Plant (EWTP) (contributing to the long term water availability in the area and mitigating an environmental risk) and other financial contributions totaling R38.4 million, and over 2,000 management hours contributed since 2002
Employment	Direct: 910 (78% HDSA, 99% local, 12.1% women, 10.2% women in technical mining roles)
	Employee benefits:
	• Medical
	• Housing allowance
	• Bonuses
	• Salary
	Indirect: 461 (74% HDSA, 56% local, 30% women)
	Total direct and indirect employment as a percentage of local employment: 1.5%
	Induced: 3,244
	Total: 4,615
	Total cost (payments to employees): R107.35 million (2007)
	Payments to employees (2003) as a percentage of total payments to employees in Nkangala (2003): 0.5%
	Dependants supported (by direct employees): 1,923
Procurement	Local procurement expenditure (% of total): 52%
	BEE procurement expenditure (% of total): 64.6%
	Total procurement expenditure: R318.90 million
Contribution to human capital development	Training expenditure: R5.56 million
	On-the-job training: could not be quantified
Economic contributions	Contribution to national revenue: R214.42 million
	Contribution to the host economy (value added): R659.90 million
	Retained value (value added less flows of money that leave the economy): R266.00 million
	Contribution to the host economy less the opportunity cost of economic resources: R105.50 million
	Percentage contribution to GDP (National): 0.015%
	Percentage contribution to GDP (Mpumalanga): 0.2%
	Percentage contribution to exports: 0.25%
Enterprise development	Anglo Zimele: Landau Colliery supports 17 BEE companies that have been assisted by Anglo Zimele
	Local community capacity building (specific to Landau Colliery) through:
	• Projects/initiatives
	• Empowerment through training (red ticket and induction) and provision of PPE
Education	Landau ABET centre provides training beyond the basic ABET level. The opportunity is extended to general community members

Table 7 (continued)

Impact	Description/details
	Internship and bursary plans offered by Anglo Coal provide the opportunity for people, especially HDSAs and women (and people from communities surrounding operations), to develop their skills
Social impact	
Social development projects	Physical infrastructure: improvements in community safety (through the construction of a fence along the railway line)
	Education: R10 million in financial and over 1,136 management hours contributed to local schools
Employment and procurement	Income gains contribute to broadly generalized improvements in poverty levels and to social welfare more generally
Health	Empowerment of women (increasing the % of women employees) results in an increase in education and communication about HIV, uptake of HIV testing, and reductions in the levels of unprotected sex
	HIV/AIDS awareness in collaboration with Yebo Youth (previously with Love Life)
	Indirect contribution to local communities through HIV/AIDS campaigns, testing and treatment of permanent employees who are local
	Basa Njengo Magogo: (Anglo Coal project): alternative stacking and fire-lighting method that reduces air pollution in the local area in which Landau Colliery operates. The benefits of the method have been demonstrated to over 31,000 households since 2006

Table 8 Summary of negative impacts of the Landau Colliery

Impact	Description/details
Employment	Possible induced unemployment due to inward migration (perhaps reflected in the high local unemployment rate relative to the national rate)
Environment	Environmental damage such as dust, acid mine drainage, spontaneous combustion, etc. can contribute to negative health and other social impacts. Interviews with communities indicated that environmental damage was not a key concern. A lack of adequate local health data made it difficult to determine the extent of this impact

engagement needs to be continually improved, promises need to be kept (the mine must deliver on any commitments made to the local communities), and results need to be achieved. The mine, in collaboration with other members of the private sector and government, needs to achieve meaningful and visible socio-economic benefits for these communities. In addition, better measurement of socio-economic variables in conjunction with setting clear project success criteria is needed to better measure, prioritise, and monitor impacts.

Table 9 SWOT analysis

Strengths	Weaknesses
Community engagement	Social projects
• Community engagement is continually improving	• Success criteria of projects are not identified and measured
Improved ability to enhance community business development and skills transfers, particularly through Anglo Zimele	
• The facilitation of knowledge and competencies creation for more local processing and value addition (Pedro 2006) and the promotion of diversification in local communities away from direct employment in mining	
Commitment to socio-economic development	
• A management team that aims to increase employment, particularly of local community members and women	
• A commitment to achieving, and at times exceeding, socio-economic-related legal requirements	
Health related programmes	
• HIV/AIDS programmes that have positive benefits for local communities both directly and indirectly	
Improved opportunities to hire more women and HDSAs	
A willingness to devote substantial financial and human resources to achieving positive socio-economic outcomes	
Opportunities	**Threats**
Social situation	*Social situation*
Abundant supply of labour for lower-skilled positions	High community expectations
	Poor socio-economic conditions:
• Many community members fit the current minimum job entry requirements (grade 10, a South African driver's license, and a minimum level of English proficiency)	• Most notably unemployment and a lack of access to adequate services
	• Crime and theft can threaten the success of projects
Current societal-orientated development paradigms	Lack of supply of artisans and other more skilled labour
	Economic situation
• Opportunities for a competitive advantage in the marketplace by achieving positive socio-economic outcomes (Pedro 2006)	Expected volatility of global markets
	• Uncertain impacts on the coal industry and Landau Colliery
	Infrastructure
Resources for socio-economic development	Unreliable supply of energy
• There are many development partners (both private sector and NGOs) who are either required or willing to contribute substantially towards socio-economic development	• Evidenced by South Africa's energy provider Eskom's supply crisis in 2007 and 2008

Table 9 (continued)

Opportunity for greater involvement of local communities in socio-economic development-related strategies and initiatives	*Environment* Climate change
	• Uncertainty regarding the future impacts on the mine and its ability to produce positive socio-economic outcomes
Economic situation	
High demand for coal expected throughout the life of mine	Existence of burdens from previous mining
Infrastructure	• This may have contributed to perceptions that current mining activity will continue to have a negative environmental impact
Good rail infrastructure to Richards Bay Coal Terminal	
Good supply of water	• The responsibility for remediation of such burdens is often unclear, and expectations of Landau are high
• Assisted largely by the development of the EWTP	
Environment	*Other*
Growing environmental awareness in this area	Lack of local level government capacity
Other	• Landau Colliery has, at times, had to step in and fulfill the role of development planners in the absence of an adequate guiding force, and in certain instances has had to take on the role of government
Technological advances	
• These may yield benefits particularly in health related impacts	• Lack of capacity to effectively spend resources generated by the mine, prioritise areas for intervention, co-ordinate efforts, and monitor impacts
	A discretionary regulatory environment
	• This may contribute to the full benefits of the taxes paid by Landau Colliery not being realised

The mine does possess many strengths; however, most of these are relatively new and therefore efforts to continually improve these strengths are necessary. To date, the mine has not been able to deliver positive socio-economic outcomes to its full potential due to external threats. Certain actions and interventions are needed to overcome these threats.

5.5 *Improving Landau Colliery's Socio-Economic Impact*

It is evident that, while local communities are benefitting from Landau Colliery, there is opportunity to extend that benefit more widely. Improving the benefit of the mine is the shared responsibility of Landau Colliery, all spheres of government (particularly local government), and other development organisations (through partnerships). Steele (2004) suggests that development outcomes can be enhanced by forging tri-sector partnerships and creating "coalitions of change" between public, private (the mine and NGOs), and local stakeholders to maximise positive socio-economic development outcomes. Although some of the following recommendations are targeted at specific stakeholders, most require a coalition for change characterized by increased collaboration and shared decision-making.

Landau Colliery needs to implement a new approach to socio-economic development in order to create projects that are sustainable, and community engagement needs to be monitored and continually improved. Government needs to improve the coordination of social and infrastructure developments at different mines in order to channel resources towards priority areas, and to avoid situations whereby too many similar initiatives are undertaken at the expense of others.

Pedro (2006) suggests that successful outcomes regarding poverty impacts require a focus on setting clear objectives and identifying priorities for intervention in a consultative process. However, poor monitoring of the MDGs and other indicators at the local level makes it difficult to identify, prioritise, and measure socio-economic impacts. This is an area where government, the mine, and NGOs need to collaborate more effectively. There is a need for continuous efforts to share resources and improve coordination of socio-economic development efforts, and to develop appropriate plans to address the MDGs and measure the impacts of strategies. All actors have to be committed to collectively chosen, well-understood goals and objectives. In order to translate those goals into practical actions leadership and capacity building is necessary at mine, government, community and other stakeholder levels (Pedro 2006).

In terms of the effectiveness of social development projects, the most successful have been those in which Landau employees have possessed appropriate skills. An example is the coal bagging project where a women from the local community was given the opportunity to develop a coal bagging business with access to Kromdraai's weathered coal and coal left on spoil tips. The coal is handpicked, bagged, and then sold to communities. Skills necessary to run the business were transferred from various employees of the mine. Currently the project is self sustaining. However, it is also important to establish projects that are not mining-related to ensure the sustainability of communities when the colliery closes. In such cases, it is essential that Landau outsource the necessary skills to develop these projects. This would provide an opportunity for NGOs and other development partners to collaborate to ensure that social projects are being driven by appropriately skilled individuals, thereby increasing the chances of success.

A second key requirement to improving the effectiveness of projects is to identify their success criteria. In this way project effectiveness can be monitored and resources deployed in the most appropriate manner in which to fulfill these criteria.

Improved communication and collaboration is also needed between the mine and government in order to increase business development initiatives in the area. Anglo Zimele has indicated that they have received great interest and commitment from members of the communities, but they have limited resources. Some interviewee responses indicate that there is a huge demand for business training and assistance.

In the absence of local government capacity, the ICMM (2006b) has argued that the outcomes of recommendations such as those mentioned above, are unpredictable. Improved governance, through capacity building, is therefore critical to the realization of the full economic and social benefits of mining investments (ICMM 2006b). A key to many governments' poor performance is lack of capacity, particularly at the local level. It is not the role of mines such as Landau Colliery to take up the role of government with regard to socio-economic development. However, by aligning the mine's goals with those of the local and provincial governments, resources can be shared, and more effective socio-economic outcomes should be realised. Interviews with local government representatives and mine and head office management indicate that there is an increasing level of collaboration in this regard. Socio-economic development plans, as outlined in IDPs and LEDs, are aligned across government levels, and have received contributions from the private sector (including Landau colliery and Anglo Coal) and other stakeholders in their design. Anglo Coal collieries have increasingly tried to align their socio-economic development goals with those of the local government, by contributing to policy design and working towards improving the relationship between the mine and local government. Although both parties acknowledge the need for greater collaboration, there are indications that the relationship is improving, and that their socio-economic development goals are becoming more closely aligned.

6 Conclusions

Mining in the South African context does not support the resource curse hypothesis. The industry has driven the diversification of the economy, and continues to contribute directly and indirectly through employment, foreign investment, foreign earnings, and value-added. Landau Colliery, within this context, has contributed to positive socio-economic development.

The social and community impacts of Landau Colliery are in part a function of the direct and indirect impacts of mining on incomes and employment, training and skills development, direct social and infrastructure development, and broader economic impacts. These impacts are fundamentally influenced by regulations such as the Mining Charter and MPRDA, national policies for revenue redistribution, and the effects these have on local government investments in social and infrastructure development. The overall contribution of Landau Colliery to socio-economic development at national, regional, and especially local levels has been predominantly positive and has improved over time. The mine has, through its value-added, contributed to the local, regional, and national economies. At a national level, contribution to GDP and to exports are significant, especially when taking into consideration the proportion of Landau employees relative to the total number of employed persons in South Africa.

The contribution to local and non-local employment in absolute terms is significant, and economically and socially important. The mine has demonstrated efforts to increase employment, with particular focus on transformation priorities over and above legislative requirements. The indirect and induced employment impacts, and the indirect impacts through employee dependants are substantial, especially within the local context where poverty and unemployment are high relative to the rest of the country. Although Landau Colliery contributes towards an absolute increase in employment, the mine could also have contributed to a high level of induced unemployment due to inward migration resulting from perceived employment opportunities in the area. The high levels of domestic procurement confirm that proactive company policies and local enterprise development has had an important impact. Local enterprise development has been particularly successful, driven largely by the approach of Anglo Zimele.

The current South African context has, through various requirements of the MPRDA and the Mining Charter, encouraged Landau Colliery to transform and improve its socio-economic development performance. These improved outcomes have, however, not been solely driven by these instruments. This is evidenced by the fact that many initiatives and aims superseded current legislation or exceed current targets. Examples include the efforts of Anglo Zimele, the Women in Mining initiative, the Socio-Economic Assessment Toolbox, and a mine management that has displayed genuine commitment to improved socio-economic outcomes of the local area. Successful strategies include developing local HDSA suppliers, developing enterprises that are not dependent on the mining industry, initiating projects that utilise the available skills within the mine's workforce and concentrate on creating self-sustaining projects, commitments to developing current and potential employees, and a focus on benefitting local communities.

The current context in terms of the contribution to government revenue and how those revenues are being spent is surrounded by much debate. It is generally accepted that stable and internationally competitive, if not necessarily low taxes, are critical to attracting and retaining foreign direct investment (FDI). It has also been suggested that the level of taxation (and other legislative requirements) in South Africa, as well as the unpredictable discretionary environment, will possibly have a negative impact on the future size of the industry. There is no evidence that alleged issues with the legislation in South Africa have had a detrimental impact at Landau in terms of the economic cut-off point, which might lead to the premature closure of the mine. However, the full benefits of the taxes paid by Landau Colliery may not be being realised.

Achieving better redistribution of the benefits of mineral wealth through improvements in governance and management of revenue flows is the responsibility of a coalition of change. Improved socio-economic outcomes would result from continued commitments towards socio-economic development of the local area, and the adoption of some of the recommendations listed below:

- Identify criteria for success of social and infrastructure projects, and continuously monitor them;
- Continuously monitor and improve community engagement;
- Government must co-ordinate social and infrastructure developments at different mines in the area;

- Improve communication and collaboration between government, the mine, and NGOs;
- Monitor the Millennium Development Goals: Government, the mine, and NGOs together need to identify, prioritize, and measure socio-economic impacts;
- Focus on projects where the mine possess relevant skills or resources; and
- Increase local government capacity: the mine should assist where possible.

These improved outcomes would result in a significant pay-off in both business and political senses. It is evident from this study that it is the shared responsibility of all development partners and communities to collaborate more effectively in order to achieve improved, shared, socio-economic benefits.

Mines have a social obligation to ensure that their operations generate the maximum positive impact possible, and have the wealth and position to do so. At the same time, however, provincial departments and municipalities have a mandate to promote industrial development and growth in their jurisdictions. Overcoming socio-economic realities in a sustainable manner requires the collaboration of all participants; it should not be the responsibility of one stakeholder.

Acknowledgments I would like to thank the members of the Ackerville, Schoongezicht, and Clewer communities, the Emalahleni Local Municipality, management at Landau Colliery, and all other Anglo employees who assisted in this study.

In addition I would like to thank the team of project supervisors: Ian Emsley, Jonathan Samuel, Karin Ireton, Stan Pillay, Stephen Bullock, and Yvonne Mfolo.

References

Alderman H, Babita M, Lanjouw J, Makhatha N, Mohamed A, Ozler B, Qaba O (2000) Chapter 2: Combining Census and survey data to construct a poverty map of South Africa. In Statistics South Africa: Measuring poverty in South Africa. Staistics South Africa. http://www.statssa.gov.za/publications/publicationsearch.asp. Accessed 18 September 2007

Anglo American (2007a) Responsible mining. Women in Mining: Promoting Human Rights and Economic Development. http://www.angloamerican.co.uk/aa/development/performance/miningbriefs/mb_women/mb_women.pdf. Accessed 28 August 2008

Anglo American (2007b) Responsible mining. Enterprise Development: Helping Small Businesses to Stand on Their Own Feet. http://www.angloamerican.co.uk/aa/development/performance/miningbriefs/mb_entdev/mb_entdev.pdf. Accessed 28 August 2008

Anglo Coal (2007a) Social and Labour Plan. Unpublished Report

Anglo Coal (2007b) South Africa BEE Procurement Strategy. Unpublished Report

Anglo Zimele (2007) Independence Through Enterprise. www.anglozimele.co.za. Accessed 21 November 2007

Business Day (1997) R16m Mpumalanga Fraud Bid Exposed. http://www.businessday.co.za/Articles/TarkArticle.aspx?ID=264308. Accessed 14 October 2007

Business Day (1998a) Mpumalanga Government Keeps Mum on Spending Figures. http://www.businessday.co.za/Articles/TarkArticle.aspx?ID=310635. Accessed 14 October 2007

Business Day (1998b) Mpumalanga to act at highest level on damning report. http://www.businessday.co.za/Articles/TarkArticle.aspx?ID=322697. Accessed 14 October 2007

Business Day (2004) Mpumalanga is ANC's Corruption Capital. http://www.businessday.co.za/Articles/TarkArticle.aspx?ID=969124. Accessed 14 October 2007

Census (2001) http://www.statssa.gov.za/census01/html/default.asp. Accessed August-December 2007

Chamber of Mines (2006) Facts and Figures 2006. http://www.bullion.org.za/Publications/Facts& Figures2006/F&F2006.pdf. Accessed 26 August 2008

Dales D, Walker M, Black P, Botha A, and Mtegha H (2004) Minerals Cluster Policy Study in Africa: Pilot Studies of South Africa and Mozambique. Economic Commission for Africa, Addis Ababa. http://www.uneca.org/eca_programmes/sdd/documents/mineral_cluster_study_sa_mozambique.pdf. Accessed 20 September 2008

Department of Environmental Affairs and Tourism (2007) The 2007 National Framework for Air Quality Management in the Republic of South Africa. Draft for public comment, July 2007. http://nfaq.csir.co.za/cocoon215/nfaqg/docs/files/NationalFramework(Revision15a)2July07.pdf. Accessed 5 September 2008

Department of Health (2006) National HIV and Syphilis Antenatal Sero-prevalence Survey in South Africa 2005. http://www.doh.gov.za/docs/reports/2005/hiv.pdf. Accessed 23 September 2007

Department of Minerals and Energy (2005) The State of Mining in South Africa. http://www.dme.gov.za/minerals/about_minerals.stm. Accessed 14 September 2007

Department of Minerals and Energy (2007) Broad-Based Socio-Economic Empowerment Charter for the South African Mining Industry. http://www.dme.gov.za/minerals/mining_charter.stm. Accessed 20 August 2007

Department of National Treasury (2006a) Mineral and Petroleum Resources Royalty Bill. http://www.treasury.gov.za/legislation/bills/2006/Royalty%20Bill%202006.pdf. Accessed 20 August 2007

Department of National Treasury (2006b) Draft policy paper. A Framework for Considering Market-Based Instruments to Support Environmental Fiscal Reform in South Africa. www.treasury.gov.za/search.aspx?cx=018115738860957273853%3Aj5zowsrmpli&cof= FORID%3A11&q=blignaut#922. Accessed 19 November 2007

Department of Trade and Industry (2007) Vision 2014. http://www.thedti.gov.za/thedti/vision.htm. Accessed 21 August 2007

Department of Water Affairs and Forestry (2003) Towards a Strategy for the Waste Discharge Charge System. http://www.dwaf.gov.za/Dir_WQM/docs/Waste/Towards% 20a%20Strategy%20for%20a%20WDCS%20Sept03.pdf. Accessed 12 December 2007

Department of Water Affairs and Forestry (2007) Government Notice 29697,3. A pricing strategy for Raw Water Use Charges. http://www.dwaf.gov.za/Documents/Notices/29697a.pdf. Accessed 15 August 2007

Emalahleni Local Municipality (2007) Integrated Development Plan. http://www.emalahleni.gov. za/idp_sdbip.html. Accessed 21 August 2007

Groom J (2008) TB, HIV/AIDS, Malaria in mining – ICMM good practice guidance released. Mineweb. http://www.mineweb.co.za/mineweb/view/mineweb/en/page68?oid=61307&sn= Detail. Accessed 20 September 2008

ICMM (International Council on Mining and Metals) (2006a) Analytical framework. The Challenge of Mineral Wealth: Using Resource Endowments to Foster Sustainable Development. www.icmm.com/document/184. Accessed 10 September 2007

ICMM (International Council on Mining and Metals) (2006b) Synthesis of four Country Case Studies. www.icmm.com/document/197. Accessed 10 September 2007

ICMM (International Council on Mining and Metals) (2006c) The Resource Endowment Toolkit. The Challenge of Mineral Wealth: Using Resource Endowments to Foster Sustainable Development. www.icmm.com/document/184. Accessed 25 August 2007

ICMM (International Council on Mining and Metals) (2007) Taxing minerals: a briefing note based on the ICMM review of the topic. Unpublished report

IIED (International Institute for Environment and Development) (2007) About the MDGs. http://www.iied.org/Gov/mdgs/about.html. Accessed 18 November 2007

Judson F (2001) The dynamics of transition governance in South Africa: Voices from Mpumalanga. Africa Today 48: 2. Academic Research Library p. 58.

MPRDA (Mineral and Petroleum Resources Development Act) (2002) Act No.28 of 2002 http://www.info.gov.za/gazette/acts/2002/a28-02.pdf. Accessed 15 July 2007

Pedro AMA (2006) Mainstreaming mineral wealth in growth and poverty reduction strategies. Minerals and Energy 21:2–16

Pronyk P, Hargreaves JR, Kim J, Morison L, Phetla G, Watts C, Busza J, Porter JD (2006) Effect of a structural intervention for the prevention of intimate-partner violence and HIV in rural South Africa: a cluster randomised trial. The Lancet 368:1973–1983

Rostoll L (2007) MPRDA causing SA's mining ranking to tumble, says Leon. Creamer Media's Mining Weekly. www.miningweekly.co.za/print_version.php?a_id=119463. Accessed 18 November 2007

Sachs JD, Warner A (1995) Natural resource abundance and economic growth: (December 1995). Cambridge, MA, National Bureau of Economic Research, Working Paper W5398, 54p

Sonjica B (2008) SA: Sonjica: Mining Indaba Conference. Keynote speech by the Minister of Minerals and Energy, Ms Buyelwa Sonjica, at the Mining Indaba Conference in Cape Town. http://www.polity.org.za/article.php?a_id=126359 Accessed 19 March 2008

South African Coal Estate Collieries (SACE) (2005) Socio-Economic Assessment Toolbox (SEAT) Report. Unpublished Report

Statistics South Africa (2000) Income expenditure by households: Mpumalanga. http://www.statssa.gov.za/publications/publicationsearch.asp. Accessed 14 September 2007

Statistics South Africa (2007) Gross Domestic Product. Annual estimates: 1993–2006. Annual estimates per region: 1993–2006. Third quarter 2007. www.statssa.gov.za/Publications/P0441/P04413rd Quarter2007.pgf. Accessed 14 September 2007

Steele P (2004) Pro-poor Growth or Boom and Bust? Coalitions for change to increase sustainability: The contribution of natural resources and the environment to pro-poor growth (A draft framework), Institute of Policy Studies, UK

Stephens C, Ahern M (2001) Worker and Community Health Impacts Related to Mining Operations Internationally a Rapid Review of the Literature http://www.iied.org/mmsd/mmsd_pdfs/worker_community_health_impacts_literature_review.pdf. Accessed 13 November 2007

UNDP (United Nations Development Programme) (2004) Human Development Report 2004: Cultural liberty in today's diverse world. http://hdr.undp.org/en/reports/global/hdr2004. Accessed 13 November 2007

UNDP (United Nations Development Programme) (2008) South Africa: The Human Development Index — Going beyond income. http://hdrstats.undp.org/countries/country_fact_sheets/cty_fs_ZAF.html. Accessed 20 August 2008

WBCSD (World Business Council for Sustainable Development) (2007) Measuring Impact Framework. Draft Methodology. www.wbcsd.org/web/projects/sl/MIMethodology.pdf. Accessed 13 November 2007

Public Policy Processes and Sustainability in the Minerals and Energy Industries

Ciaran O'Faircheallaigh

Abstract This chapter considers the extent to which public policy processes are geared towards sustainability in the minerals and energy industries, and how they may need to change to take the requirements for sustainability into account. It begins with a general discussion of sustainability in the minerals and energy industries, drawing on examples from three countries. As these examples show, mineral extraction is neither inherently sustainable nor unsustainable, but rather is *made* sustainable or unsustainable, in large part by public policies applied to the industry. The nature and content of public policies reflect, in turn, the policy-making processes utilised in formulating and implementing policy, and so it is critical to consider the links between policy processes and sustainability. Nine specific variables are identified that shape these processes, including: the identity and number of interests or groups involved in making policy and how much influence they exert; the range of policy alternatives considered; time frames and information sources for policy making; the time periods over which the impacts of policy alternatives are considered; and the criteria applied by policy makers. The chapter argues that current policy-making processes in relation to the minerals and energy industries are often incompatible with the pursuit of sustainability, and that radical changes are required if public policies are to support sustainabilty in these industries.

1 Introduction

This chapter focuses on the processes used in making public policies, both in relation to specific mining and energy projects and to policy issues that affect mineral development. This is an issue of central importance to sustainability in the minerals and energy industries, given that in most countries minerals are largely or entirely

C. O'Faircheallaigh (✉)
Griffith University, Nathan, QLD 4111, Australia
e-mail:Ciaran.Ofaircheallaigh@griffith.edu.au

J.P. Richards (ed.), *Mining, Society, and a Sustainable World*,
DOI 10.1007/978-3-642-01103-0_15, © Springer-Verlag Berlin Heidelberg 2009

in public ownership, and that governments play a critical role in resource development through policy and legislation dealing, for instance, with planning and land use, environmental impact assessment, environmental management, taxation, corporations, foreign investment and trade, and indigenous land rights. If the content and impact of public policies are not conducive to sustainability in the minerals and energy industries, it is inconceivable that sustainability can be achieved. To put the argument somewhat differently, mining as an industry (as opposed to mining of individual ore bodies) is not either inherently sustainable or inherently unsustainable. It is *made* sustainable or unsustainable in large part by public policies, a point illustrated in Sect. 2. And mining cannot be made sustainable if public policies are incompatible with the pursuit of sustainability.

The content of public policies is very much influenced by policy-making *processes*, by *the way in which* policies are conceived, developed, endorsed, and implemented. It follows that the nature of those processes has profound implications for the prospects of achieving sustainability. The implications of public policies for sustainability in mineral industries are increasingly recognised (MMSD 2002, pp. xiv, xviii, xxii, xxvi, xxix, 336, 356–357; Richards et al. 2004, pp. 62–64; MacDonald and Gibson 2006, pp. 4,13–15; ICMM 2008, pp. 4–5, 9, 17). Much less attention has been devoted to the relationship between public policy processes and sustainability, despite the recognition by the Mining, Minerals and Sustainable Development (MMSD) Project, for instance, that "Decision making-processes are as vital as the end results" (2002, p. xvii). To the extent that policy processes do attract attention, this tends to focus heavily the issue of public participation and stakeholder engagement (Day and Affum 1995; Cheney et al. 2002; Gao et al. 2002; MMSD 2002, pp. xx, xxviii-xxix, 346, 354, 401; Weitzner 2002; Environmental Law Institute 2004; ICCM 2008, pp. 17, 21). This is also a feature of the general literature on sustainable development (see for instance Lyons et al. 2001; Stratford and Jaskolski 2004). But public participation is just one of many variables shaping policy processes (see below). This chapter develops a more broadly-based approach in analysing the way in which public policy processes affect, or can affect, sustainability in mineral development.

Although most democratic governments, including Australia's, have committed themselves in principle to the pursuit of sustainability (COAG 1992), it is quite a different matter to devise policy-making processes that reflect and promote those principles. Existing processes have developed incrementally over long periods of time when sustainability was not an explicit goal of government policy, and they are not simple or easy to change (Howlett 2002, p. 241). In addition, there is the very important issue of what sorts of processes are likely to facilitate sustainability in the development of natural resources (or in policy making generally). Even where governments have both the will and capacity to bring about major changes to public policy processes, what sorts of processes should they change *to*?

This chapter begins by examining the link between public policy processes and sustainability in the minerals and energy industries. It briefly discusses the link between policy processes and policy content and outcomes, and then identifies nine variables that, in combination, are critical in defining the nature of public decision making in the context of the minerals and energy industries. It illustrates the operation of these variables through two short case studies from Australia: one of state government policy on natural gas development in Western Australia, and the other of federal government policy towards the uranium industry. The discussion highlights the dynamic nature of current processes in Australia, the ways in which they diverge from an approach that would promote sustainability, and the issues involved in pursuing such an approach in relation to resource development in a "real world" context. Although this chapter considers the impacts of the nine variables in an Australian context, they have been identified through a review of international literature on policy making and on sustainability in mineral industries. Thus, the approach developed here is of wide (international) applicability.

2 Public Policies and Sustainability in the Minerals and Energy Industries

Drawing on the standard definition of sustainability provided by the Brundtland Commission, sustainability in the minerals industries requires that development of mineral resources should meet "the needs of the present without compromising the ability of future generations to meet their own needs" (Brundtland 1987). It is possible to argue that given the non-renewable and ultimately finite nature of mineral resources, sustainability in the minerals industries is impossible to achieve, because use of resources by the current generation must inevitably deplete the availability of resources at some time in the future and so compromise the needs of future generations (Ali and O'Faircheallaigh 2007, p. 7). This argument may be valid if the analysis is restricted to the "physical" or "geological" sustainability of mining and is considered over an infinite future, because at *some* point all mineral resources are potentially subject to exhaustion and this exhaustion may compromise the needs of the generations then living. However, sustainability can also be considered in environmental, economic, and social terms. Once it is so considered, quite different and more complex conclusions may be reached regarding the "sustainability" of the minerals industries.

On the one hand, mining activities that appear sustainable over long periods of time in physical or economic terms may prove unsustainable in environmental or social terms. To provide an obvious example, copper mining on Bougainville Island in Papua New Guinea did not cease in 1989 because the Panguna ore body was depleted in a physical sense or was no longer profitable to mine, but because social conflict surrounding the project and its environmental impact led to an armed rebellion which permanently closed it (May and Spriggs 1990). Similarly, a failure to achieve social sustainability led Rio Tinto to cease developing one of the largest, low-cost uranium deposits in the world at Jabiluka in Australia's Northern Territory. The Aboriginal traditional owners for the area, the Mirrar, had determined on the basis of their experience with the nearby Ranger uranium mine that, because of its social impact, uranium mining was incompatible with their survival as a people. As a result, they undertook a major and ultimately successful legal and political campaign to stop Jabiluka (Katona 2002).

On the other hand, exploitation of a physically non-renewable resource "may be deemed 'sustainable' if there is an effective conversion of the natural capital, represented by the resource, to social capital that would allow for long-term livelihoods" (Ali and O'Faircheallaigh 2007, p. 6; see also MacDonald and Gibson 2006, p. 3).

I am not suggesting that different dimensions of sustainability can exist independently of each other in the sense that mineral development could continue if some of these elements were "present" and others "absent". In the real world, the various dimensions of sustainability are inextricably linked. However, for analytical purposes it is critical to understand that sustainability does have these different dimensions, and that a rigorous analysis of sustainability and the minerals and energy industries requires a focus on all of them. And once environmental, economic, and social dimensions of sustainability are incorporated into the analysis, public policies, and therefore public policy processes, become critical. This point, and the need to avoid any a priori assumptions about mining and sustainable development, can be illustrated by briefly examining specific examples of mineral development in three different countries.

The first involves phosphate mining on the Pacific island of Nauru. Mining of Nauru's rich phosphate deposits began in 1906 when the island was still part of the German Empire. Production expanded after World War I when Nauru was entrusted by the League of Nations to the United Kingdom, Australia, and New Zealand, and further growth occurred after World War II when Nauru was made a Trust Territory of the United Nations under the trusteeship of Australia. Mining continued after Nauru achieved independence in 1968. The resource is now close to exhaustion, little agricultural land remains, most of the island has been reduced to what has been called a lunar wasteland, and Nauru's economy is in a parlous state (Howard 1991; Weeramantry 1992; Australian Broadcasting Commission 2005). These outcomes appear certain to compromise the needs of future generations of Nauruans in fundamental ways.

Given the limited size of Nauru and its phosphate deposits, commercial mining was not sustainable in a physical sense over the long term. However, it does not necessarily follow that mineral development was *inevitably* unsustainable in economic, environmental, or social terms. A critical factor in the eventual outcome was the policies that the United Kingdom, Australia, and New Zealand developed towards the island. The colonial powers allocated only a tiny proportion of the value of phosphate exports to the Nauruans, and failed to take any meaningful measures to rehabilitate the island. A Rehabilitation Fund set up for this latter purpose "stood at the meagre level of $599,325" when the UN trusteeship ended in 1968 (Howard 1991, pp. 166–194, Weeramantry 1992, pp. 231–235, 283).

At independence, Nauruans were well aware of the problems they faced. The Nauruan government decided to allocate a substantial proportion of the revenues generated by phosphate mining to the Nauru Phosphate Royalties Trust, which undertook a major program of overseas investment designed to create an asset base that would generate ongoing revenue for Nauru after mining ceased. The Trust was used to establish a national airline and shipping company, to build up an international investment portfolio, and to purchase office blocks, hotels, and land, especially in Australia and the United States. Many of these investments ran into financial difficulties, with Air Nauru in particular incurring heavy losses and requiring a subsidy equivalent to a third of the government's budget in 1980–1981. Other ventures also ran into financial difficulties and profits on investments were well below expectations (Howard 1991, pp. 175–194; Kearney 2008). By 2002, Nauru was in a dire financial situation and had to seek emergency financial aid of US$4.8 million from Australia. By 2004, the country was on the brink of bankruptcy, with receivers taking control of Nauru's property investments in Australia due to the country's outstanding $US165 million debt to America's General Electric Capital Corporation (Australian Broadcasting Corporation 2005).

Nauru was also experiencing significant social problems. The growth of royalty revenue, part of which was paid to individual landowners, led to dramatic changes in the lifestyle of Nauruans. All citizens were guaranteed a basic level of income by government, and certain individuals and families whose land was mined after royalty levels increased received substantial incomes. Lack of agricultural land due to the impact of mining combined with substantial cash incomes and a general tendency to expend these incomes on consumer goods, led to a dramatic increase in consumption of imported packaged foods, and alcohol consumption also rose rapidly. Over a number of decades, serious health problems resulted, with high rates of obesity and of "lifestyle" illnesses such as diabetes and heart disease. Significant inequalities in income also emerged. In the 1980s, the Nauruan government made what Howard (1991, p. 179) described as a "half hearted program" to address health problems, with limited results. Thus, by the turn of the 21st century Nauru's social sustainability was also under threat (Australian Broadcasting Commission 2005; Kearney 2008).

Phosphate mining on Nauru has proved not to be sustainable in physical, economic, environmental, or social terms. However, different outcomes could have eventuated if: (1) the colonial authorities had pursued different policies, and allocated a significant portion of phosphate revenue to environmental and economic sustainability funds over the decades prior to independence; (2) Nauru's investment strategy had been more successful; and (3) the Nauruan government had pursued different and more effective health and social policies.

The second example involves the Ok Tedi copper mine in the Western province of Papua New Guinea. The Ok Tedi mine is still operating, but it could be argued that it has long since passed the point at which its operations could be said to be sustainable in a broader sense, and that even if it operated for hundreds of years into the future it could never operate sustainably. This is because its presence has destroyed the resource base of hundreds of Papua New Guinea villagers, as a result of the devastating environmental impacts of the release of its tailings and waste rock into the Ok Tedi and Fly rivers. This outcome emerged in large measure from the decision of the Papua New Guinea government to allow the project to proceed following its developers' refusal to continue with construction of a tailings dam after the original dam site was inundated by a huge land slip. There have also been serious failings in the environmental management and regulatory systems at the the mine, which resulted, for instance, in the release of cyanide into the Ok Tedi river in 1984, killing thousands of fish, crocodiles, and turtles. The project has certainly imposed huge costs on future (as well as on current) generations, because it will take decades for the river systems to return to health, and because large areas of agricultural land have been permanently lost (Kirsch 1997, pp. 121–126, 2004; MMSD 2002, p. 348).

Some villagers of the current generation have been compensated for their losses, others have been only partially compensated, and some have received little or no compensation (Banks and Ballard 1997a, p. 6). It is not at all clear that future generations will receive any compensation (Banks and Ballard 1997b). Even where people have been compensated for economic losses, they have also incurred social and cultural losses that are impossible to quantify or to address in monetary terms. As Kirsch (2004) notes, the landscape on which personal and social histories were inscribed, and through which social relations were defined and maintained, has changed irrevocably, and and the birds, animals, and fish with whom, in Yonggom cosmology, they maintained intimate and vital relationships have disappeared. In his words, for Yonggom landowners along the Ok Tedi river, "environmental damage implies a fundamental change in how [they] view their relationship with the world around them. With reference to past events becoming increasingly structured by abstract chronologies, their remembrance of things past are no longer linked to their surroundings.... Their environment is no longer a site of productivity, but a scene of loss. It no longer provides them with security, but confronts them with new, indecipherable risks" (Kirsch 2004, pp. 191, 197).

In this case, lack of sustainability has largely reflected corporate and government decisions, rather than any inherent characteristics of a large copper mining project. It is of course possible that the project was not, as its developers claimed, economically viable if it had to bear the costs of tailings storage. If this is the case, sustainability could only have been advanced by a decision not to mine.

Let us consider a third scenario, bauxite mining on Cape York in northern Australia, by Rio Tinto Alumina (formerly Comalco Ltd). In this instance mining has already lasted for 40 years, is based on a very extensive and low-cost resource base, and is likely to continue for decades or even hundreds of years into the future. For the first four decades of its life, few economic benefits accrued to the traditional Aboriginal owners of the land on which mining occurred, while at the same time they incurred significant social, cultural, and environmental costs (O'Faircheallaigh 2005). In 1992, Australia's High Court recognized the existence of Aboriginal native title in Australia in the *Mabo* case, and in the following year the Federal Government provided legislative recognition of native title through the introduction of the *Native Title Act 1993* (NTA). A number of traditional owner groups subsequently lodged native title claims over sections of Comalco's mining leases. Partly in response to these changes, Rio Tinto altered its policies towards Aboriginal communities affected by its operations, and as a result invited traditional owners to negotiate a legally binding agreement covering a range of issues and impacts associated with Comalco's operations (Cape York Land Council/Comalco 2001).

The Western Cape Communities Co-existence Agreement (WCCCA), signed in 2001, includes a system for minimising or avoiding damage to Aboriginal cultural heritage, a review and upgrading of environmental management systems and Aboriginal participation in environmental management, and extensive programs to increase Aboriginal employment and business development opportunities. In terms of economic and social sustainability, a critical part of the agreement involves payment of substantial royalties to the traditional owners, both by Queensland's state government and by the project operator, and investment of over half of this revenue flow in a long-term capital fund, which operates on prudential rules designed to maximise returns while minimising risk. Income is reinvested for 20 years, after which it becomes available for current expenditure, but the capital base is to be preserved in perpetuity.

It can be argued that mineral development can contribute to sustainable development in this case even if mining ends long before it is currently expected to do so, because the agreement's cultural heritage and environmental management provisions will minimise any costs borne by future generations, and because they will enjoy a substantial and ongoing revenue flow along with the economic and social opportunities associated with it, which would not be available in the absence of mining. This final example highlights again the critical impact of public policies. Indeed, it could be argued that the sustainability of mining in this case increased dramatically over a short period of time because of the High Court's recognition of indigenous rights in land, the federal government's action in giving legislative expression to this recognition, and the Queensland government's decision to divert a share of its own royalty income into the WCCCA.

3 Policy Processes, Policy Outcomes, and Sustainability

Public policies play a central role in determining whether or not mining will be sustainable. The content and outcomes of public policies are, in turn, intimately linked to the nature of public policy processes, a link that is well documented across a wide variety of political systems and policy contexts. For example, Alston et al. (2006) found that the driving force behind policies in contemporary Brazil was the specific nature of interactions between key political actors in different policy arenas, and that "the dynamics of the policymaking game" in these arenas yielded predictable outcomes in terms, for example, of the stability and adaptability of policy. The specific features of policy-making processes were in their view "key determinants of the characteristics of public policies", characteristics they labelled as "the dependent variable" (Alston et al. 2006, pp. 6–11). Howlett (2007) found that the nature of policy-making subsystems in Canada influenced both the pace and nature of policy change, and that changes in policy-making processes were linked in a predictable manner to changes in the nature and rate of policy change (for other examples see

Connick and Innes 2001; Gains 2003; Aninat et al. 2006; Stein et al. 2008). At a more specific level, Howard (2007) showed how one aspect of the policy process, the availability to participants of information on the probable impact of policy alternatives being considered, influenced the likelihood of departures from the *status quo* in relation to water management policy in Australia. Stratford and Jaskolski (2004, pp. 321–332) argued that another specific aspect of the policy process, absence of appropriate leadership, acted as a barrier to the development of policies conducive to sustainability in local government in Tasmania.

As noted earlier, in the context of sustainability in the mineral and energy industries, considerable attention has focused on the link between public participation in policy-making processes and the nature of policy outcomes, generally based on an assumption that greater public participation increases the likelihood that public policies will promote sustainability, and, according to some analysts, that it is indispensable to the pursuit of sustainability (Clark and Clark 1999, p. 189; MMSD 2002, p. 346). However, the nature of any causal relationship between the extent of public participation and public policy outcomes requires investigation, and, in addition, the extent of public participation is only one of many variables that shape the nature of policy-making processes. The following sections discuss nine such variables, and the ways, both individually and in combination, in which they relate to sustainability. The variables and some key questions in relation to each are presented in Table 1. I have identified these variables based on a review of general literature on public policy making (e.g., Hogwood and Gunn 1989; Lindblom 1993; Parsons 1995; Pal 2006), and the literature on sustainability in the minerals and energy industries (e.g., Labonne 1999; Warhurst and Mitchell 2000; MMSD 2002; Weitzner 2002; Environmental Law Institute 2004; MacDonald and Gibson 2006).

I do not claim that this analysis is comprehensive, and accept that other variables may be relevant in particular cases, that individual variables considered here require more extensive analysis, and that further research is also needed to explore the ways in which they inter-relate and the manner in which they link to sustainability. The objective is to demonstrate the need for a systematic analysis of the variables involved in public policy making, and of their role in relation to the pursuit of sustainability in the minerals and energy industries, and to offer an initial contribution to such an analysis.

3.1 Inclusion: Who Is Involved in Policy Making?

The first issue involves the identity and interests of those who participate in policy making. Sustainability will not be prioritised if those participating in decision making do not value it, and if those who do value sustainability are excluded. The importance of this issue is highlighted by the examples discussed earlier. For instance, over some 60 years, almost no provision was made for investment of Nauru's phosphate revenues in economic, environmental, or social sustainability, because policy making involved only the colonial powers and they were not concerned about

Nauru's sustainability, but rather about securing cheap and reliable supplies of phosphate for farmers in their home countries (Weeramantry 1992).

Historically, policy making in Australia and other free market economies has been dominated by individual mining companies, mining industry associations, and government agencies (in particular treasuries and mines and energy departments). Companies play a critical role because mineral development is driven fundamentally by their investment decisions (where to explore, what to explore for, which projects to develop, which projects to close, and when to close them), and because larger companies in particular have substantial capacity to lobby public policy makers. At a broader level, industry associations also have a significant capacity to influence government, both directly by lobbying, and indirectly by helping to mobilise public opinion around specific issues such as land access and indigenous rights (Libby 1989; Lindblom 1993).

Table 1 Nine variables that shape the character of public policy processes

Variable	Some key questions
1. Participation in decision making	Who participates and who is excluded? What mechanisms foster, or militate against, inclusion? How much priority do those included/excluded attach to sustainability?
2. Authority over outcomes	What sources of power are available to participants? Who has the capacity to determine outcomes? How "open" or "closed" is decision-making in terms of participation and influence over outcomes?
3. Decision alternatives	What range of alternatives is considered? How do these relate to the policy "status quo"? How is the range of alternatives affected by the degree of openness in decision-making?
4. "One-off" or adaptive decision making	Do policy decisions tend to be "one-off", or can they be revisited regularly? What are the implications of this for sustainability?
5. Time allocated for decision making	How much time is available for policy making? How and by whom is this determined? Does time for decision making accommodate cultural differences?
6. Resources available to support decision making	What resources are available to support decision making, and to whom are they available? Are there major disparities in the resources available to different participants?
7. Time period over which effects of policy decisions are considered	What time frames are utilised? Who determines these time frames?
8. Information available to support policy making	What sorts of information are available to inform policy making? Are specific types of information privileged, or discounted?
9. Criteria used in policy making	What impacts and issues are considered in policy making, and which are excluded? What weight is attached to various impacts? What methodologies are used in assessing impacts, and are some privileged over others?

Mining companies may be very concerned about sustainability in corporate terms: ensuring that they achieve the levels of profitability and growth required to secure their own survival in a competitive environment. However, they do not necessarily attach priority to the economic, social, or environmental sustainability of individual communities, regions, or countries (Trebeck 2007). Although in theory democratic governments should reflect the diversity of interests that comprise their electorates, in reality they have often been driven by the need to demonstrate their ability to foster economic growth, particularly by supporting large projects that offer highly visible additions to employment and exports. As a result, they have in the past prioritised rapid development of mineral resources (Harman and Head 1981; Howlett 2007).

More recently, the introduction of environmental impact assessment legislation has provided an opportunity for other interests to participate in public policy processes, for instance through public submission and objection processes that allow interested groups to argue that projects should not proceed, should be modified substantially before proceeding, or should be subject to environmental and other conditions (Cheney et al. 2002, p. 2). Legal recognition of indigenous rights has provided an avenue for indigenous participation in decision making. Under Australia's NTA, registered native title claimants have a "Right to Negotiate" with applicants for mining leases, and with relevant state authorities regarding the terms on which development can occur. The opportunity to negotiate with developers and governments allows native title groups to articulate their views on mineral development, and some have concluded legally-binding agreements that impose conditions on mining projects related to economic, social, and environmental sustainability (O'Faircheallaigh and Corbett 2005), a point illustrated by the earlier discussion of the WCCCA.

Finally, the adoption of corporate social responsibility policies by many of the world's leading mining companies (MMSD 2002; ICCM 2008) creates another opportunity to expand the range of interests involved in policy making. Many major companies have established policies that require systematic identification of and engagement with their stakeholders, including community, environmental, and indigenous groups. BHP Billiton, for instance, requires all of its sites "to identify their key stakeholders and consider their expectations and concerns for all operational activities, across the life cycle of operations. Sites are also required to specifically consider any minority groups (such as Indigenous groups) and any social and cultural factors that may be critical to stakeholder engagement" (BHP Billiton 2008).

3.2 Authority: Who Determines Policy Outcomes?

Participation in policy decision making is one matter, but the capacity to determine outcomes is another (Howard 2007). Companies and governments have sources of power and influence that are unequivocal and substantive, and particularly in relation to individual projects each enjoys a "yes/no" power of decision over whether or not development will proceed. Companies have an investment veto, because in a market economy government cannot force companies to invest. Governments have a regulatory veto, because projects cannot be developed without government approvals. The power to decide whether or not a project will proceed provides both parties, in turn, with substantial influence over a range of other decisions, because companies and governments can insist that they will only invest in a project, or approve a project, under the "right" conditions. For companies, these conditions may relate to royalty rates, public funding of mine infrastructure, project scale, or environmental conditions. For governments, they may relate to company funding of infrastructure that will have multiple users, domestic processing of minerals, or allocation of a proportion of energy output to domestic markets.

The capacity of environmental, community, or other civil society groups to shape policy outcomes is less clear cut. It can be substantial in some cases. For example, environmental groups successfully mobilised public opinion and lobbied Australia's federal government to maintain a ban on development of new uranium projects for more than a decade after 1983. Environmental groups prevented development of the proposed Windy Craggy copper mine in British Columbia in 1993. However, in relation to mining projects that do not attract widespread public attention, either because of the nature of the mineral being mined, or of their location in environmentally or culturally sensitive areas, the power of environmental or community groups in public policy making is still limited. For example, over the last 30 years, political actions have resulted in only a handful of mining projects in Australia and Canada being refused development approval or being substantially modified.

In some cases, indigenous people have a legal right to determine whether or not development can proceed. For instance, in Australia's Northern Territory, the consent of Aboriginal people granted freehold title under the *Aboriginal Land Rights (Northern Territory) Act 1976* is required for the granting of exploration and mining leases, allowing them to exclude mining altogether from areas they regard as highly sensitive. Their veto also allows them to apply stringent environmental and cultural heritage protection to, and obtain substantial economic benefits from, projects they do allow to proceed. In combination, such actions have the potential to greatly enhance the contribution of individual mining projects to economic, social, and environmental sustainability, especially where revenues extracted through negotiated agreements are invested in social and economic development in areas where mining is prohibited (O'Faircheallaigh 2002).

Indigenous people may be able to use political pressure to win for themselves a de facto right of veto even where such a right does not exist in law. For example, after Innu and Inuit traditional owners halted development of Inco's Voisey's Bay nickel project through direct action and litigation, the Government of Newfoundland and Labrador informed Inco that it would not allow the project to proceed without the consent of the Innu and Inuit (Newfoundland and the Innu Nation 2002). Voisey's Bay provides a clear example of the way in which control over decision making affects the sustainability of mining. One of the conditions on which the Innu and

Inuit eventually allowed the project to proceed was that annual nickel output would be less than half that originally planned. This reduced environmental impacts and provided local communities much more time to capitalise on opportunities created by Voisey's Bay (for instance skills and enterprise development), allowing them to build an economic base that could survive the end of mining.

Although the NTA does not confer a veto on Aboriginal landowners in Australia, it does give them some capacity to influence outcomes through their right to negotiate with governments and developers, a capacity bolstered by their ability to use their procedural rights to delay projects if agreement is not reached. The evidence suggests that Aboriginal groups that are cohesive, well organised, and reasonably well resourced have been able to use the NTA to enhance environmental and cultural heritage protection and extract significant economic benefits. However, to date there have been few if any cases where it has been used to prevent mining in specific areas, or to change project configurations in the way achieved at Voisey's Bay (O'Faircheallaigh and Corbett 2005; O'Faircheallaigh 2008).

The extent to which policy processes incorporate and confer power upon a wide and diverse range of interests determines the degree of openness that characterises policy making. A process that incorporates or confers authority only on mining companies and government economic agencies, for example, could be characterised as highly closed. A process that also involves and confers influence on a range of other actors, including government agencies operating in environmental and social fields, and environmental, indigenous, and community groups, could be described as highly open. It is useful to think of various configurations as aligned along a spectrum that includes many different approaches, including, for instance, ones where the canvassing of policy options is very open, but where the making of final decisions is closed. There can be tension between the degree of openness of a policy process and its capacity to generate binding and timely decisions, with the possibility that excessive openness can result in "policy paralysis" arising from the fact that multiple contending interests exercise a capacity to veto decisions they do not approve of (Jones 2001). What is required to promote sustainability in the minerals and energy industries is a balance that confers a degree of power on groups that have most to gain from sustainable outcomes, but does not create a policy system so open and fluid that it destroys the certainty needed to facilitate large capital investments in resource projects. Achieving this balance in practice is a substantial challenge. Its pursuit must begin with recognition of the need to reform the traditionally closed policy systems that have operated around the minerals and energy industries.

3.3 Policy Alternatives

Policy processes can vary according to the range of policy alternatives that are considered. In some cases an incremental approach is adopted, with only a narrow range of alternatives, representing minor adjustments to the *status quo*, being considered. At the other end of the spectrum, the full range of conceivable alternatives may be evaluated (Pal 2006, pp. 271–293). In the context of mineral development and sustainability, an incremental approach might involve a continuation of traditional approaches based essentially on economic calculations, with gradual adjustments being made to project design and operations to enhance environmental and social sustainability. These might involve, for instance, reduction of water and energy consumption (which may of course improve project economics in any case), or establishment of corporate foundations to channel additional social benefits to affected communities. Alternatively a much wider and more radical range of alternatives might be considered. These could include fundamental changes to project scale and design to enhance sustainability, and a "no project" alternative where prospects for sustainability appear poor. The latter alternative is sometimes considered in conventional environmental impact assessments (EIA), but usually immediately dismissed on the basis of the economic benefits that would be foregone (e.g., Chevron Australia Pty Ltd. 2005, p. 14). It can be argued that given the magnitude of environmental and social problems associated with recent patterns of industrial development, an incremental approach is no longer adequate, because at least in the short to medium term it replicates approaches that have caused the problems in the first place.

The range of decision alternatives considered is linked to the openness of decision-making processes. A closed process dominated by interests that have benefited from the status quo is likely to be associated with an incremental approach to policy making, and unlikely to facilitate a consideration of radical alternatives, whereas "opening up" the decision-making process is likely to have the opposite outcome. This is illustrated by the example of Voisey's Bay. It is very unusual for a mining project to proceed at half the scale identified as optimum by the project proponent, and this outcome followed on from the success of the Innu and Inuit in "opening up" the decision process and establishing themselves as powerful actors within it.

3.4 "One-Off" Versus Adaptive Policy Making

Another dimension of policy making involves the issue of whether a process focuses on single, "once and for all" decisions in relation to mining projects, or creates systems that allow for ongoing, adaptive changes to project design and operation. Historically, the emphasis has been on the former, with a single critical set of decisions determining whether a project can proceed and, assuming it can, the conditions under which it will operate for the remainder of project life (or at least over long periods of time). Governments tend to impose or to negotiate provisions on key issues such as royalties, mineral processing, and infrastructure provision with potential investors in a new mine, and to leave these in place for the whole of the project life. In Australia, the agreed conditions are often enshrined in legislation, creating additional barriers to their amendment (Fitzgerald 2002). Such an approach has been deemed essential in order to provide the certainty demanded by potential investors (MacDonald and Gibson 2006, p. 14). Similarly,

EIA processes generally produce one decision, or one advice to government, as to whether a project should proceed, and establish a single set of environmental or other conditions that will be in place indefinitely (Joyce and MacFarlane 2001, pp. 3, 12; Howitt 2003, pp. 337–338; O'Faircheallaigh 2007, pp. 322–333). These matters are rarely revisited. I am not aware of any legislation, for example, that requires a further EIA to be undertaken after, say, 10 years, to establish whether project impacts are as expected and whether the decision to allow the project to operate, or the conditions imposed on it, should be revisited.

It can be argued that such an approach is inadequate if sustainability is to be achieved. Given the rapid changes that can occur in understanding of environmental and social impacts, and in mining, mineral processing, and waste treatment technologies, it may be essential to develop "adaptive" approaches that regularly revisit relevant issues and enhance responses to them. The requirements for adaptive management have started to attract substantial attention in the academic literature, and attempts are being made to apply them in the environmental management of some projects (Morrison-Saunders and Arts 2005; O'Faircheallaigh 2006). However, it can be argued that policy-making systems are still very much dominated by a "one off" approach, as indicated by the fact that even where adaptive approaches are required by legislation, they are rarely applied in practice (O'Faircheallaigh 2007).

3.5 Time Allocated for Policy Making

The time period allowed for policy making is an important variable because it influences the potential to involve a range of interests in decision-making and the way in which they can be involved. It also shapes the type and range of information considered in making decisions (discussed below), for instance because considerable time may be required if policy makers are to gain access to information dealing with long-term natural cycles, or information from oral sources such as interviews with indigenous elders. Generally speaking, the shorter the time frame applied, the narrower the opportunities that exist in terms of involving diverse interests and a wide range and depth of information. Particular problems may arise where significant cross-cultural differences exist in policy-making processes. For instance, in some indigenous societies, decision-making is highly consensual and time consuming, and it may prove impossible to establish and articulate indigenous positions within the time frame allowed by mainstream policy processes.

Two influences tend to be paramount in setting time frames for policy making in mineral industries. The first are statutory requirements, with mining, environmental impact assessment, and native land title legislation usually establishing requirements on government to take decisions within specific time periods (e.g., Western Australia 2008). The second are the financial and market pressures facing mining and energy companies, which often mean that they are working to tight time frames, especially if they are competing with other potential suppliers to meet limited market opportunities.

Decision-making time frames for major mining projects are not always unduly short. For example most legislation allows government ministers some discretion to extend statutory time frames. In some cases companies may be developing projects over extended periods of time as part of long-term development and production schedules, for example when they hold long-term contracts and/or are mining raw materials for use in vertically-integrated operations. However, in general, companies tend to press hard to minimise decision-making time frames, to some extent because they want to build in some "fat" in case of unforseen delays arising from other factors.

Governments tend to be responsive to what they see as commercial imperatives. The result can be that insufficient attention is paid to sustainability issues in policy making. This is not to argue for open-ended time frames. Market and financial imperatives are real, and a project that ends up being uneconomic because it has missed a market opportunity is of course incapable of contributing to sustainable development. However, sustainability in mineral industries can also not be achieved if time frames for decision-making make it impossible to incorporate those with a strong interest in sustainability, or to gather and analyse information that is critical in establishing requirements for sustainability. Thus, if mineral and energy industries are serious about pursuing sustainability, they cannot insist that time frames be driven solely by commercial factors.

3.6 Resources Available to Support Policy Making

Another important variable involves the availability of resources to support participation in policy making. Corporations fund their participation in decision-making on the assumption that they can recoup their costs from projects that proceed, and government agencies and actors have access to public budgets. For other potential participants, in particular indigenous and civil society groups, obtaining the resources required to participate in policy making is often difficult. In certain jurisdictions (for instance Canada) intervener funding is sometimes available to support participation in EIA processes. However, the resources involved are commonly very limited; for example, a total of only C$1.5 million was available for all non-government groups wishing to make submission to the EIA for the Mackenzie Valley Pipeline, one of Canada's largest proposed resource projects (Green 2008). In addition, an EIA constitutes only one stage of a decision-making process, which begins with an application for exploration licences and ends only with project decommissioning.

Indigenous groups face particular challenges because their members tend to be economically disadvantaged and to live in remote regions. The latter means that the cost of bringing people together to discuss proposed projects is high, especially because consensual decision making approaches can require multiple meetings. The cost of making representations to governments in distant capital cities is similarly high. Acquiring access to the technical expertise required to participate effectively in decision-making processes is also costly. While resource constraints may not stop indigenous, environmental, and community groups from participating effectively, they often do require them to focus their efforts on a small number of projects and to ignore others, an outcome that may have important implications for sustainability. For instance, many of the Native Title Representative Bodies that represent Aboriginal landowners in Australia regularly ignore a large proportion of notifications in relation to exploration permits and decline to take advantage of relevant procedural rights because they lack the resources to do more than focus on a small number of mining lease applications whose potential impacts are immediate and substantial (Corbett and O'Faircheallaigh 2006, p. 174). One important consequence of this situation is that cumulative impacts of multiple developments, many of which escape any scrutiny, cannot be addressed.

The resources available to support policy making and the time available for making decisions are inter-related, but in ways that are variable and contextual. For example, provision of additional resources can allow specific activities required for policy making (for instance collection of information) to be undertaken more rapidly. On the other hand, resources may be used wastefully if insufficient time is available to develop a coherent and logical plan to identify, collect, and analyse relevant information.

3.7 Time Periods Over Which the Effects of Policy Alternatives Are Considered

Central to the concept of sustainability is a consideration of the effects of current actions and choices on future generations. Thus, the time frame over which the policy process considers the impact of policy alternatives is critical. If future generations are to be considered, time frames must obviously be extended (for instance 60–100 years if the impact on two generations is to be considered).

In reality, the time frames over which the impacts of major mining and energy projects are analysed as part of public policy making are generally much shorter than this. Typically, the time frame on which attention is focused is the life of project (e.g., Blacktip Joint Venture 2004; Moolarben Coal Mines Pty 2006a). This can be as short as 10 years. The assumption generally is that economic impacts, for instance, only need to be considered during project life, and that environmental impacts will be remediated before a project is abandoned. In reality, while the direct economic impacts of a project may largely coincide with its productive life, its indirect effects may last longer (see Bowes-Lyon et al., this volume), and indeed *must* last longer if the project is to make a lasting contribution to economic sustainability. There are many cases of "legacy" mines whose negative environmental impacts persist long after mining has stopped (MMSD 2002, pp. 407–408) and, as the earlier discussion of Nauru illustrates, social impacts can also be long lived. Thus, a policy process that considered whether a project would generate net public benefits over an operational life of 20 years might yield a different outcome to one that assessed net benefits over 40 years.

3.8 Information for Policy Making

If decision-making is to attach value to and promote sustainability, then decision makers must have access to relevant information, and in particular to information regarding the likely impacts of proposed projects over extended periods of time, without which likely impacts on future generations cannot be assessed. Projects approved in the absence of such information will not necessarily be unsustainable, but in the absence of the requisite information, policy makers cannot attach appropriate weight to the requirements for sustainability. Information must also be available to consider the various dimensions of sustainability (economic, environmental, and social).

Given that the time frames over which project effects are considered are often short (see Sect. 3.7), information regarding impacts on future generations is frequently unavailable. In relation to environmental impacts, an additional and particular problem arises from the fact that baseline data is commonly available only for the limited time period over which a project developer has been undertaking environmental studies in a particular area. Given uncertainty regarding the nature of the "background" environment over longer periods of time, project proponents may argue that it is in fact impossible to accurately assess the implications of the impact represented by a mining project, and may use the absence of existing information as a rationale for not attempting to do so.

Quite apart from the issue of whether projects should be approved if their long term impacts cannot be evaluated, application of traditional ecological knowledge (TEK) held by indigenous groups may assist in providing long term baseline information, given that the observations of living generations, combined with knowledge inherited from their forebears, can cover extensive periods of time. Because of its overwhelmingly oral nature, access to that information requires the participation of the people who possess it. Some attempts have been made to incorporate TEK into decision-making processes, for instance in relation to management of diamond mines in Canada's Northwest Territories, with limited success to date (O'Faircheallaigh 2006). As noted above, issues relating to decision time frames, resources to support participation in decision-making, and the legal rights of indigenous people must be addressed if indigenous people are to participate more activity and so make their TEK available to policy makers.

More generally, there has been a tendency to focus on assembling information relating to economic and environmental aspects of project impacts, and to pay considerably less attention to social impacts, with the MMSD noting, for instance, that social factors have only recently started to "creep" into environmental impact assessment (MMSD 2002, p. xxi). In the absence of information on current social realities and the potential social impact of proposed mining projects, policy makers clearly cannot take into account the requirements for social sustainability. As noted earlier, failure to address these requirements can result in the abandonment of major ore bodies whose exploitation is viable in technical and economic terms.

3.9 Criteria for Policy Making

A final issue involves the decision criteria that are employed by policy makers. Which impacts are considered and which are excluded? What weight is attached to different impacts, for example as between economic, social and environmental effects? Are all potential impacts evaluated using similar methodologies? What weight is attached to short-term impacts as opposed to long-term impacts? To express this last point somewhat differently, what discount rate is applied by decision makers? Application of a high discount rate will mean that short term effects are weighted much more heavily than long term effects (MMSD 2002, p. 347), whereas use of a lower rate will moderate this effect. If appropriate decision criteria are not employed then there is little prospect that policy processes will help ensure that mining promotes sustainable development, regardless for instance of what information resources or time frames are applied to policy making.

As indicated above, historically, assessment of mining projects has tended to focus almost entirely on economic criteria. This is evident, for instance, from the matters addressed in agreements under which Australia's state governments provided approval for major resource projects (Fitzgerald 2002, p. 139). Introduction of environmental impact assessment and environmental protection legislation in the 1970s required a focus on the physical environment, but a great deal of latitude remained and remains today in determining *which* environmental impacts or issues are taken into account. For instance, in September 2007, Australia's Federal Court rejected an attempt by environmental groups to challenge the Federal Minister for the Environment's decision in relation to a proposed coal mine on the basis that the Minister had not properly assessed the adverse effects on greenhouse gas emissions that would be caused by burning coal produced by the mine.

There is also the issue of the relative weight attached by policy makers to economic and environmental impacts, however the latter are determined. As discussed above, a small number of projects in Australia have been substantially modified or halted because of their expected environmental impacts. But the vast majority are approved on the basis that their promised economic benefits exceed any potential environmental costs, and the significance of social and cultural impacts have continued to be downplayed, if not excluded, by policy makers. Where social or cultural effects are considered, the analysis often lacks the rigorous, quantitative methodology applied in assessing economic impacts (e.g., Dames and Moore 1994; Chevron Australia Pty Ltd. 2005; Moolarben Coal Mines Pty Ltd. 2006b). This is important because it may be difficult for a policy maker to insist on project changes to address potential social and cultural impacts expressed in general, tentative, or hypothetical terms, given anticipated economic benefits that are clearly quantified and substantial and so appear certain to eventuate.

Another key issue involves the discount rate applied to future expected costs and benefits. In calculating economic benefits of mining projects, the conventional approach for companies is to use a discount rate that is equivalent to their average cost of capital (Van Rensburg and Bambrick 1978, pp. 149, 153). Using a discount rate in the region of 10% (Van Rensburg and Bambrick 1978, pp. 151–152), this would obviously render negligible the current value of any benefits or costs accruing to "future generations", in comparison to short term benefits or costs. It should, however, be noted that companies undertake some exploration expenditures that are not expected to yield an income until a decade or more into the future, and in this case lower discount rates are presumably applied because such expenditures are indispensable to the long-term "physical" and "corporate" sustainability of mining.

Public policy makers may use discount rates somewhat lower than companies, though treasury bond rates, which are sometimes used as a basis for discount rates, are often in the range of 5–10%. If calculations are made over periods of 40 or 60 years, for instance, application of such discount rates would still have the effect of rendering negligible the current value of long-term benefits and costs. Other participants in policy making may prefer lower discount rates. For instance, indigenous people who hope that their traditional territories will provide sustenance for future generations may not discount long-term environmental impacts at anything like as high a rate as would non-indigenous decision makers. At a more fundamental level, it can be argued that affording quantitative cost–benefit techniques such as discounting a central role in decision making immediately devalues the importance of factors that are not quantifiable, or not easily quantifiable, such as indigenous attachment to ancestral land, unique cultural heritage values or ways of life, or biological diversity.

Application of alternative and non-conventional policy-making criteria may not, it should be stressed, always result in potential mining projects appearing unsustainable, and so lead to a decline in the overall level of mineral development. For example potentially positive social and cultural impacts may be undiscovered where the focus of decision making is exclusively economic and/or environmental. If lower discount rates are applied, projects that would otherwise not be developed because they generate substantial long-term benefits but few short-term ones may appear sufficiently attractive to ensure their development.

4 Case Studies

The relevance and applicability of the nine policy process variables identified above can be illustrated through two short case studies related to contemporary energy development in Australia.

4.1 Liquefied Natural Gas in Western Australia

During the last decade, large reserves of natural gas, amounting by some estimates to a third of Australia's total reserves, have been located about 300 km offshore from the coast of the Kimberley region in northwest Western Australia, an area little affected to date by large-scale industrial development. Additional and highly prospective parts of the offshore region will be made available for exploration in 2008. Natural gas will be piped ashore to gas processing and liquefaction plants, and converted into Liquefied Natural Gas (LNG), mainly for export, and LPG and condensates for sale in domestic markets. A number of major oil and gas companies are now involved in development activity, usually as part of joint ventures, including Shell, Woodside, Total, and Inpex, a Japanese company whose largest shareholder is the Japanese government. Inpex's project is the most advanced, with a site identified on the Maret Islands, some 70 km from the Kimberley Coast, preliminary environmental and engineering studies completed, and a decision to proceed to FEED (Front End Engineering and Design) anticipated in the near future. Woodside is also actively engaged in site selection, with other companies expected to follow.

Given the very large gas reserves discovered to date and the prospect of additional discoveries, the physical sustainability of gas production seems assured for decades into the future. Gas development raises major issues in relation to the environmental, economic, and social sustainability of coastal regions, and the Aboriginal peoples whose ancestors have occupied the area for thousands of years. Construction of pipelines, the building and operation of LNG processing plants, and shipping of LNG raise issues regarding the integrity of coastal environments. These environments support the wildlife and fish populations on which many Aboriginal people depend for their subsistence, some of which (for example turtles) are also of great cultural significance. There are particular concerns regarding the possibility that the various companies that have discovered gas will each establish separate processing facilities, affecting in total large areas of coastline and generating cumulative impacts that might threaten the viability of coastal ecosystems (Rothwell 2008).

Environmental groups have tended to oppose gas development outright. For example, the Wilderness Society argues that gas development in the Kimberley "should be ruled out straight away as incompatible [with environmental values], destructive and inherently unsustainable" (Wilderness Society 2008). The position of Aboriginal groups including the regional land organization, the Kimberley Land Council (KLC), reflects the fact that Kimberley Aboriginal communities already face serious issues in terms of the social, cultural, and economic sustainability of their communities. Less than 20% of working-age Aboriginal people are in formal employment, and there is a heavy reliance on welfare payments; life expectancy is some 20 years lower than for non-Aboriginal Australians; access to education and housing is poor; and communities face serious social issues, including substance abuse, family violence, and child abuse.

Aboriginal leaders and the KLC believe that against this background they cannot ignore the potential opportunities associated with gas development, such as the creation of educational and employment opportunities and provision of revenue streams that can create further economic opportunities and support health, housing, and other services. Thus, they favour what they call "responsible development", which involves the concept of one or two industrial hubs where all gas processing and related industrial facilities would be located, minimising potential environmental impacts and facilitating environmental monitoring. It also requires effective protection of the environment and of Aboriginal cultural heritage, and that Kimberley Aboriginal people share substantially in the benefits of resource development (KLC 2007a, 2008a).

Historically, policy-making processes relating to mineral development in Western Australia have been dominated by mining companies and state government agencies, whose central role is to facilitate and encourage resource exploitation. Policy has been driven by an "ideology of development", which assumed that the interests of Western Australia were served by fostering development of the state's mineral resources as fully and as rapidly as possible. Policy-making criteria reflected this assumption, and removal of any barriers to development has been a central policy objective. For example, the State Government used police on a number of occasions to suppress attempts by environmental and Aboriginal groups to prevent exploration or mining in ecologically or culturally sensitive areas. Time frames for decision-making reflected the needs of developers. The views of Aboriginal people or environmental or community groups were afforded scant regard, and no attempt was made to draw on TEK in policy making. Sustainability in any sense of the term was not a major focus of policy. One specific result of this policy approach has been the widespread loss of Aboriginal cultural heritage, including the complete destruction of a number of major Aboriginal sites by the Argyle diamond mine, and extensive damage to some of the world's oldest rock art on the Burrup Peninsula (Harman and Head 1981; Hawke and Gallagher 1989; Dixon and Dillon 1990; Howlett 2007).

There are indications that public policy processes in relation to gas development will depart from the past in important ways. The recognition of native title has given Aboriginal people a legal basis on which to be involved in policy making. This is reflected in the fact that Aboriginal traditional owners, supported by the KLC, have participated in negotiations with Inpex regarding its proposed Maret Islands processing plant, and with the Western Australian government regarding this project and gas development more generally. When traditional owners were concerned that negotiations with Inpex were not resulting in effective measures to protect cultural heritage and the environment, the KLC initiated legal proceedings to prevent the grant of licences that Inpex required for geophysical work on the Maret Islands (KLC 2007b). The company then quickly concluded protocols with traditional owners covering cultural heritage and its conduct of environmental studies.

As mentioned earlier, the Native Title Act does not give Aboriginal people a veto over decisions in relation to resource development. In 2006 and 2007, the Western Australian Prime Minister, Alan Carpenter, made a number of statements, including one before Parliament, stating that gas development in the Kimberley will not proceed unless is creates significant economic and social benefits for Aboriginal people and unless it has the support of Kimberley traditional owners (Carpenter 2006). The Deputy Premier and Minister for State Development, who has portfolio responsibility for gas development, reiterated this position in February 2008, stating that "LNG processing... will only go ahead with the fully informed consent of the traditional owners and their substantial economic participation" (Australian Broadcasting Commission 2008). Also in February 2008, the Western Australian and Federal governments released draft terms of reference for a strategic assessment of a plan for a common-user LNG precinct, which state that the "informed consent" and "support" of traditional owners will be a condition of any site approval (Department of Industry and Resources 2008).

This approach represents a radical departure in the context of Western Australia, which, if the State and Federal Governments adhere to their commitments, will afford Aboriginal people a central place in policy making. This in turn would mean that the criteria used in making decisions about resource projects will be considerably broader than in earlier decades. More generally, it would have major implications for the sustainability of gas development, because in other contexts Kimberley traditional owners have been willing to support resource development only where proposed environmental and cultural heritage protection measures have been enhanced, and when long-term investment funds and other initiatives to enhance economic and social sustainability have been established (Bergmann 2006; KLC 2008b).

Major issues continue to arise in relation to other policy process variables identified in the earlier discussion. Time frames for policy making continue to be influenced by developer schedules, and at times they are entirely inadequate to allow appropriate consultation and decision-making processes within Aboriginal communties. For example, the Northern Development Taskforce established by the Western Australia Government to identify potential sites for a common-user LNG precinct informed the KLC in early December 2007 that it would determine a short list of approved sites by March 2008. Large areas of the Kimberley are annually inaccessible for much of the period between December and March because of the wet season, so this time frame would make it impossible to effectively engage Aboriginal traditional owners in the process of identifying acceptable sites. This in turn could result in decisions that are not based on community consensus, thereby creating social discord and contributing to a loss of social capital over the long term.

In relation to resources to support Aboriginal participation in policy making, both Inpex and the Western Australian government have signed multi-million dollar agreements to help meet the cost of community consultations and to provide Aboriginal groups with access to technical expertise. On the other hand, no effective use is being made of Aboriginal TEK in policy making, with Inpex's environmental studies, for instance, being undertaken with virtually no active Aboriginal input, despite repeated attempts by the KLC to secure the involvement of traditional owners. This failure has potentially serious consequences for the environment, especially given that Inpex's studies have only been under way for a few years and that its baseline data are consequently very limited, severely curtailing the time frames over which potential environmental impacts of various decision alternatives can be reliably assessed. Traditional owners have already challenged the company's studies on turtle ecology on the basis that company counts of turtles yield numbers that bear little relationship to traditional owners' observations over decades of turtle harvesting (KLC 2007, personal communication). In addition, the fact that governments and developers are ignoring and so devaluing TEK can create social disquiet and undermine the authory of Aboriginal elders, thereby disrupting the social sustainability of Aboriginal communities (Nadasdy 2003).

While gas development off the Kimberley coast is in its early stages, this case study highlights the interrelationship between various dimensions of sustainability, and also the possibility that mineral development may in some circumstances *enhance* the economic and social sustainability of affected communities. This is certainly what Aboriginal leaders believe will occur if gas resources are developed "responsibly". It illustrates again the significance of public policy processes, with recent changes in policy making in Western Australia likely to significantly enhance the prospects of achieving development that is ecologically, economically, socially, and culturally sustainable. The case study also reveals the variable and dynamic nature of policy processes, with some aspects of policy making (for instance the failure to utilise TEK) remaining unchanged at the same time that others (for example Aboriginal participation and the provision of resources to support it) are changing substantially.

4.2 Uranium Policy in Australia

Australia accounts for some 38% of the world's uranium reserves, but over the last 30 years its share of world output has always been well below this level, and currently stands at about 20% (Commonwealth of Australia 2006a, p. 1). This situation reflects policy decisions by successive Australian federal and state governments to limit development of the uranium industry, primarily because of concerns regarding the potential impact of uranium mining on the environment and on Aboriginal people, and regarding the link between uranium production and nuclear proliferation (O'Faircheallaigh et al. 1989).

In 1977 the then Liberal/National Party federal government approved development of a number of major mines in the Northern Territory, subject to their compliance with an extensive regulatory regime dealing with environmental management and rehabilitation, nuclear non-proliferation, and monitoring of impacts on affected Aboriginal communities, and to the negotiation of agreements between project developers and affected Aboriginal communities (Commonwealth of Australia 1977). On coming to office in 1983, the Labor federal government imposed a prohibition on the development of any additional mines beyond the three mines then operating or under development (the "Three Mines Policy"), a position it maintained until its removal from office in 1996. Prohibitions on uranium mining were also introduced by Labor governments at the state level, including Queensland and Western Australia.

The Liberal/National party government led by John Howard, elected in 1996, was opposed to Labor's Three Mines Policy. Slow growth in the nuclear power industry and the associated low level of prices for uranium meant that little pressure existed to open new mines, while the earlier closure of the Nabarlek project meant that there was "room" under the existing policy for the establishment of an additional mine, which occurred in 2001 with the opening of the Beverley project in South Australia. After the turn of the century this situation began to change as concerns regarding global warming led to renewed interest in nuclear power, and as uranium prices began to recover. In 2006, the Howard government decided to reassess Australia's policy on uranium mining and Australia's role in the nuclear fuel cycle, and the Federal Opposition Leader, Kim Beazley, indicated Labor's support for such a move.

In order to provide a basis for this assessment, in June 2006, the Federal Government established the Uranium Mining, Processing and Nuclear Energy Review (UMPNER) ("the Review"). A Review Taskforce would undertake "an objective, scientific and comprehensive review of uranium mining, value added processing and the contribution of nuclear energy in Australia in the longer term" (Commonwealth of Australia 2006b). The UMPNER also included an Expert Panel to review scientific aspects of the Taskforce's findings. The Review's Terms of Reference required it to consider Australia's capacity to increase uranium mining and exports and the potential to establish other steps in the nuclear cycle in Australia; the extent to which nuclear energy could contribute to the reduction of greenhouse gas emissions; and health, safety, and security implications of nuclear energy (Commonwealth of Australia 2006b, pp. 137, 147).

The UMPNER released a Draft Report for public comment and for peer review by the Expert Panel on 21 November 2006 (Commonwealth of Australia 2006a), and submitted its Final Report to the Federal Government in December 2006 (Commonwealth of Australia 2006b). The Review supported expansion of uranium mining in Australia because of the additional economic opportunities this was expected to provide, and predicted that the development of a nuclear power industry in Australia could contribute significantly in achieving cuts in greenhouse gas emissions. It found that environmental and health and safety risks associated with uranium mining and disposal of nuclear waste are manageable, as are the risks of nuclear proliferation (Commonwealth of Australia 2006a, pp. 1–10; 2006b, pp. 2–9).

The Review made no findings regarding the potential social impacts of expanding the uranium industry, including its potential impact on Aboriginal people, and the issue is not even mentioned in the summary of either the Draft or Final Reports (Commonwealth of Australia 2006a, pp. 1–10; 2006b, pp. 2–9). The Review's Terms of Reference did not explicitly require the social sustainability of uranium mining to be addressed. The omission is nonetheless surprising given that, as mentioned above, opposition from Aboriginal people concerned about uranium mining's social impact had recently led Rio Tinto to abandon one of Australia's largest uranium deposits at Jabiluka, highlighting the link between social sustainability and "the capacity to increase uranium production".

Looking at the way in which decision-making was configured in this case and beginning with the issue of participation in and influence over decision-making, the first point to note is the composition of the Review Taskforce and Expert Panel. Of the six members of the Taskforce, four were nuclear physicists, one an energy economist, and the sixth the Chairman of a power generating company. Thus, the Taskforce was heavily dominated by people associated with the nuclear and power industries, not in itself likely to encourage consideration of the full range of sustainability issues raised by uranium mining. The Expert Panel was somewhat more broadly based, but again consisted exclusively of people with a background in science and business, and included no-one whose professional expertise related to the social impacts of uranium mining (Commonwealth of Australia 2006b, pp. 138–139, 147–150).

The Review received submissions from some 80 organizations and undertook "consultations" with a similar number. The large majority of organizations making submissions were mining or energy companies, industry associations, and environmental and anti-nuclear groups. None of the organizations that made submissions was Aboriginal, and none had a primary focus on social issues or delivery of social services (Commonwealth of Australia 2006a, Appendix C). The large majority of the 80 organizations consulted by the Taskforce was involved in the minerals and energy industries, whereas only three were environmental groups and one was Aboriginal (the Northern Land Council, which represents Aboriginal people in Australia's major uranium producing region in the Northern Territory). None of the Aboriginal organizations specifically affected by uranium mining (for instance the Gunjehmi Association, which represents the Mirrar owners of the Ranger and Jabiluka leases areas) were consulted.

The decision alternatives considered by the Review were limited, which is not surprising given that its Terms of Reference assumed a focus on expansion of the uranium industry, and given the composition of the Review Panel and Expert Group. For instance, there was no consideration of a "no expansion" option for Australia's uranium industry based on pursuit of alternative energy policies as a means of reducing Australia's greenhouse gas emissions. Neither was there any consideration of an option for expansion of the uranium industry that would include a strong focus on social sustainability, based for instance on incorporating Aboriginal traditional owners into site selection and development planning in the way that is being attempted in relation to LNG development in Western Australia.

The time frame allocated for policy making could be regarded as very truncated. Only 5 months were available for the Taskforce to consider the wide and complex array of issues involved in expanding uranium mining and Australia's role in the nuclear fuel cycle, and only 3 weeks were available for preparation of submissions on the Taskforce's Draft Report. This would certainly be entirely inadequate, for instance, for any Aboriginal groups that wished to discuss the Draft with their constituents and prepare a response. Given that submissions on the Draft Report were due on 12 December 2006 and the Final Report was completed by the end of 2006, the Taskforce would only have had a matter of days to consider submissions. The Final Report does not indicate how many submissions were made or how they were addressed. The fact that the summaries of the Draft and Final Reports are almost identical indicates that little opportunity existed for the Taskforce to pay any regard to the submissions.

As indicated earlier, a substantial number of the organizations that participated in the Review were individual companies or industry or professional associations, which would have access to substantial resources to support their participation, as would a number of major environmental organizations. However, smaller environmental groups, NGOs involved in delivery of social services, and Aboriginal groups have access to very limited resources. Indeed, given that most Aboriginal organizations struggle to fund their "core" organizational activities, some of which involve statutory obligations, their participation would have required provision of funding by the Taskforce or the Federal Government. In fact, no funding was provided by the Government to facilitate public participation in the UMPNER (Commonwealth of Australia 2006a). The fact that no public hearings were conducted as part of the Review also militated against more broadly based public participation.

Given the time available to conduct its inquiry, the information base available to the Taskforce largely consisted of existing research regarding economic, technical, and environmental aspects of the uranium and nuclear industries. The Taskforce did commission three studies, dealing with greenhouse gas emission of nuclear energy, electricity generation, and global developments in uranium markets. Given the time frames available, these studies also relied largely on collation of existing research (Commonwealth of Australia 2006b, pp. 151, 155). There is no indication from its Draft or Final reports that the Taskforce sought, or had access to, information on the social sustainability of uranium; or regarding the environments that would be affected by an expansion of uranium mining, including Aboriginal knowledge of those environments; or regarding Aboriginal understandings of the environmental impact of existing uranium mines in Australia.

The criteria used in policy making related essentially to the achievement of economic growth (relevant in particular to an expansion of uranium mining) and to the need to reduce Australia's greenhouse gas emissions (relevant to the establishment of nuclear power generation). Other criteria related to the avoidance of any health and safety problems potentially associated with the uranium and nuclear industry (Commonwealth of Australia 2006a, pp. 7, 15). The implications of uranium mining for social sustainability was not utilised as a decision-making criterion.

In important respects, the processes utilised in undertaking the UMPNER stand in marked contrast to those being employed in relation to LNG development in Western Australia. This highlights the dynamic and variable nature of decision-making processes even within a single jurisdiction, and the quite different implications for sustainability of different approaches to decision-making.

5 Conclusion

Public policies are central to the prospect of achieving environmental, economic, and social sustainability in the minerals and hydrocarbon industries. A failure to achieve these dimensions of sustainability can, in turn, destroy the viability of otherwise valuable ore bodies, as occurred with Panguna and Jabiluka, or mean that mining is ultimately unsustainable, as on Nauru and at Ok Tedi. On the other hand adoption of appropriate public policies can result in mineral development enhancing social, economic, cultural, and environmental sustainability, as is occurring with bauxite mining in Western Cape York, and as Aboriginal leaders hope will occur with gas development in the Kimberley.

The content and impact of public policies are shaped, in turn, by the nature of public policy processes. This fact is rarely recognised in the literature on sustainability in the minerals and energy industries, and there have been few if any attempts to analyse the variables that determine the nature of policy processes, or their implications for sustainability. This chapter identifies nine key variables that characterise public policy processes in relation to mineral development, and offers an initial analysis of how they relate to different dimensions of sustainability and to each other. More research is needed in this area, but it is already clear that the way in which each variable is dealt with in specific policy processes has important implications for the prospects of achieving sustainabilty.

This point is well illustrated by approaches to public policy making in relation to gas development in Western Australia, and to Australia's policy on uranium mining and nuclear energy. In the former case, certain aspects of the policy process (for instance, government recognition of the need for Aboriginal participation in and influence over policy making) are likely to enhance the prospects for sustainable development. The provision of government funding to Aboriginal groups to allow them to take advantage of opportunities for participation illustrates the way in which individual variables in the policy process are related. However, the way in which other variables are being addressed (e.g., the short time frames applied to policy decisions, and the failure to mobilise Aboriginal traditional knowledge for use by decision makers) is likely to militate against sustainability. The same applies more generally to policy making on uranium mining and nuclear energy. Here, the failure to provide the funding or time required to allow participation by Aboriginal or environmental interests, to consider social sustainability as a decision criterion, or to consider more than a narrow range of decision alternatives will reinforce each other and undermine the prospects for sustainable development. In a wider context, to the extent that such approaches characterise public policy making in relation to the mineral and hydrocarbon industries, their sustainability will be under threat.

There is considerable scope for further research in relation to the nine policy process variables identified in this chapter. One issue is whether there are other important variables that should be considered. There is also a need for more detailed analysis of the way in which individual variables interact in actual policy-making processes. Some insights into this area have been offered in the initial discussion of the nine variables and in the two case studies, but, for instance, more systematic attention could usefully be focused on the interaction between time, information, and values (i.e., the time frames applied to the policy process, the information available to policy makers, the criteria applied in policy making, and the time periods over which the effects of policies are considered). Important insights could be gained by extending the focus to social and political contexts other than those involving indigenous peoples, which have been the major focus here. Finally, additional work is required to consider how analysis of the nine variables might be applied in policy making, for instance through use of a matrix that systematically relates the variables to different "dimensions" of sustainability (economic, social, environmental), or through development of a weighting system that affords greater importance to certain variables depending on the prevailing policy context and policy goals.

References

Ali S, O'Faircheallaigh C (2007) Extractive industries, environmental performance and corporate social responsibility. Greener Management International: The Journal of Corporate Environmental Strategy and Practice 52: 5–16.

Alston LJ, Melo MA, Mueller B, Pereira C (2006) Political Institutions, Policymaking Processes and Policy Outcomes in Brazil. Inter-American Development Bank, Washington, DC.

Aninat C, Landregan J, Navia P, Vial J (2006) Political Institutions, Policymaking Processes and Policy Outcomes in Chile. Inter-American Development Bank, Washington, DC.

Australian Broadcasting Commission (2005) Country Profile – Nauru http://www.radioaustralia.net.au/news/countries/NAURU.htm, accessed 11 January 2008.

Australian Broadcasting Commission (2008) Minister accuses opposition of ignoring Kimberley concerns. ABC News Online 28 February.

Banks G, Ballard C (1997a). Introduction: Settling Ok Tedi. In Banks G, Ballard C (eds) The Ok Tedi Settlement: Issues, Outcomes and Implications. Australian National University, Canberra, pp. 1–11.

Banks G, Ballard C (eds) (1997b). The Ok Tedi Settlement: Issues, Outcomes and Implications. Australian National University, Canberra.

Bergmann W (2006) Negotiating Prosperity: Maximising Resource Benefits to Our People. Keynote Address, Conference on Achieving Objectives: A New Approach to Land Claim Agreements in Canada. Ottawa, June.

BHP Billiton (2008). Our Stakeholders. http://www.bhpbilliton.com/bb/sustainableDevelopment/ourStakeholders.jsp, accessed 12 March 2008.

Blacktip Joint Venture (2004) Social Impact Assessment Blacktip Project Wadeye, Northern Territory. Impaxisa Consulting, Brisbane.

Bruntland G (ed) (1987) Our Common Future: The World Commission on Environment and Development. Oxford University Press, Oxford.

Cape York Land Council/Comalco (2001). 'A Way Forward Together.' Press Release, Cairns, 11 March.

Carpenter A (2006) West Kimberley Onshore Liquefied Natural Gas Processing Facilities: Statement by Premier. Legislative Assembly, Western Australia, 21 November.

Cheney H, Lovel R, Solomon F (2002) People, Power and Participation: A Study of Mining-Community Relationships. MMSD Australia, Melbourne.

Chevron Australia Pty Ltd (2005) Draft Environmental Impact Statement/Environmental Review and Management Programme Executive Summary. Chevron Australia Pty Ltd, Perth.

Clark AL, Clark JC (1999) The new reality of mineral development: social and cultural issues in Asia and Pacific nations. Resources Policy: 189–196.

COAG (Council of Australian Governments) (1992) Inter-Governmental Agreement on the Environment. COAG, Canberra.

Commonwealth of Australia (1977) Uranium: Australia's Decision. Australian Government Publishing Service, Canberra.

Commonwealth of Australia (2006a) Uranium Mining, Processing and Nuclear Energy – Opportunities for Australia? Draft Report. Department of Prime Minister and Cabinet, Canberra.

Commonwealth of Australia (2006b) Uranium Mining, Processing and Nuclear Energy – Opportunities for Australia? Department of Prime Minister and Cabinet, Canberra.

Connick S, Innes J (2001) Outcomes of Collaborative Water Policy Making: Applying Complexity Thinking to Evaluation. Institute of Urban and Regional Development, University of California, Berkeley.

Corbett T, O'Faircheallaigh C (2006) Unmasking native title: The national native title tribunal's application of the NTA's arbitration provisions. University of Western Australia Law Review 33(1): 153–177.

Dames and Moore (1994) The Century Project: Draft Impact Assessment Study Report, Volume 1, Volume 2, Volume 3. Dames and Moore, Brisbane.

Day JC, Affum J (1995) Windy Craggy: Institutions and stakholders. Resources Policy 21: 21–26.

Department of Industry and Resources (Western Australia) (2008) Strategic Assessment Agreement, Terms of Reference, Draft Selection Criteria. www.doir.wa.gov.au, accessed 17 March 2008.

Dixon R, Dillon M (eds) (1990) Aborigines and Diamond Mining: The Politics of Resource Development in the East Kimberley. University of Western Australia Press, Nedlands.

Environmental Law Institute (2004) Prior Informed Consent and Mining: Promoting the Sustainable Development of Local Communities. Environmental Law Institute, Washington, DC.

Fitzgerald A (2002) Mining Agreements: Negotiated Frameworks in the Australian Minerals Sector. Prospect Media, Chatswood.

Gains F (2003) Executive Agencies in Government: The Impact of Bureaucratic Networks on Policy Outcomes. Journal of Public Policy 23(1): 55–79.

Gao Z, Akpan G, Vanjik J (2002) Public Participation in Mining and Petroleum in Asia and the Pacific: The Ok Tedi Case and Its Implications. In Zillman DN, Lucas AR, Pring G (eds) Human Rights in Natural Resource Management. Oxford University Press, New York, pp. 679–693.

Green J (2008) Information underload. Uphere Business March: 23–27.

Harman EJ, Head B (eds) (1981) State Capital and Resources in the North and West of Australia. University of Western Australia Press, Nedlands.

Hawke S, Gallagher M (1989). Noonkanbah, Whose Land, Whose Law. Fremantle Arts Centre, Fremantle.

Hogwood BW, Gunn LA (1989) Policy Analysis for the Real World. Oxford University Press, Oxford.

Howard J (2007) Do stakeholder committees produce fair policy outcomes? In Wilson AL et al. (eds) Proceedings of the 5th Australian Stream Management Conference. Charles Sturt University, Thurgoona, 157–162.

Howard MC (1991) Mining, Politics and Development in the South Pacific. Westview Press, Boulder, CO.

Howitt R (2003) Rethinking Resource Management: Justice, Sustainability and Indigenous Peoples. Routledge, London and New York.

Howlett C (2007) The Role of the State in Mining Negotiations: A Case Study of the Century Mine, North Queensland. Doctoral dissertation, Griffith University, Brisbane.

Howlett M (2002) Do networks matter? Linking policy network structure to policy outcomes. Canadian Journal of Political Science 35(2): 235–267.

ICCM (International Council on Mining & Metals) (2008) Annual Review 2007: Essential Materials, Produced Responsibly. ICMM, London.

Jones DR (2001) Political Parties and Policy Gridlock in American Government. Edwin Mellen Press, New York.

Joyce SA, MacFarlane M (2001) Social Impact Assessment in the Mining Industry. Mining Minerals and Sustainable Development, London.

Katona J (2002) Mining uranium and indigenous Australians: The fight for Jabiluka. In Evans G, Goodman J, Lansbury N (eds) Moving Mountains: Communities Conflict Mining & Globalisation. Zed Books, London, pp. 195–206.

Kearney S (2008) Aussie police call on kids to help quell violence in Nauru. The Australian, 10 March.

Kirsch S (1997). Is Ok Tedi a precedent? Implications of the law suit. In Banks G, Ballard C (eds) The Ok Tedi Settlement: Issues, Outcomes and Implications. Australian National University, Canberra, pp. 118–140.

Kirsch S (2004) Changing views of place and time along the Ok Tedi river. In Rumsey A, Weiner J (eds) Mining and Indigenous Lifeworlds in Australia and Papua New Guinea. Sean Kingston Publishing, Wantage, UK, pp. 182–207.

KLC (Kimberley Land Council) (2007a) Sharing the Resource Boom Key to Aboriginal Employment. Media Release, Broome, 30 November.

KLC (2007b) Kimberley Land Council moves to protect Maret Island from disturbance. Media Release, Broome, 28 April.

KLC (2008a) Informed Consent of Traditional Owners Key to Sustainable Gas Development in the Kimberely. Media Release, Broome, 11 January.

KLC (2008b) Completed Agreements. http://www.klc.org.au/agrees_complete.htm, accessed 18 March 2008.

Labonne B (1999) The mining industry and the community: Joining forces for sustainable social development. Natural Resources Forum: 315–322.

Libby R (1989) Hawke's Law: The Politics of Mining and Aboriginal Land Rights in Australia. University of Western Australia Press, Nedlands.

Lindblom CE (1993) The Policy-Making Process. Prentice Hall, Englewood Cliffs, NJ.

Lyons M, Smuts C, Stephens A (2001) Participation, empowerment and sustainability: (How) do the links work? Urban Studies 38(8): 1233–1251.

MacDonald A, Gibson G (2006) The Rise of Sustainability: Changing Public Concerns and Governance Approaches Towards Exploration. Society of Economic Geologists Special Publication 12: 1–22.

May R, Spriggs M (eds) (1990) The Bougainville Crisis. Crawford House, Bathurst.

MMSD (Mining Minerals and Sustainable Development) (2002) Breaking New Ground: Mining Minerals and Sustainable Development. International Institute for Environment and Development, London.

Moolarben Coal Mines Pty Ltd (2006a). Moolarben Coal Project: Environmental Assessment Report, Vol. 1. Wells Environmental Services, East Maitland.

Moolarben Coal Mines Pty Ltd (2006b). Moolarben Coal Project: Socio-economic Impact Assessment. Hunter Valley Research Foundation, Hamilton.

Morrison-Saunders A, Arts J (2005) Assessing Impact: Handbook of EIA and SIA Follow-Up. Earthscan, London.

Nadasdy P (2003) Hunters and Bureaucrats: Power, Knowledge and Aboriginal-State Relations in the Southwest Yukon. University of British Columbia Press, Vancouver.

Newfoundland, Innu Nation (2002) Memorandum of Agreement Concerning the Voisey's Bay Project. Innu Nation, St Johns.

O'Faircheallaigh C (2002) A New Model of Policy Evaluation: Mining and Indigenous People. Ashgate Press, Aldershot.

O'Faircheallaigh C (2005) Creating Opportunities for Positive Engagement: Aboriginal People, Government and Resource Development in Australia. Paper presented to the International Conference on Engaging Communities, Brisbane, 12–17 August.

O'Faircheallaigh C (2006) Environmental Agreements in Canada: Aboriginal Participation, EIA Follow-Up and Environmental Management of Major Projects. Canadian Institute of Resources Law, University of Calgary, Calgary.

O'Faircheallaigh C (2007) Environmental agreements, EIA follow-up and aboriginal participation in environmental management: The Canadian experience. Environmental Impact Assessment Review 27(4): 319–342.

O'Faircheallaigh C (2008) Negotiating protection of the sacred? Aboriginal-mining company agreements in Australia. Development and Change 39(1): 25–51.

O'Faircheallaigh C, Corbett T (2005) Indigenous participation in environmental management of mining projects: The role of negotiated agreements. Environmental Politics 14(5): 629–647.

O'Faircheallaigh C, Wade-Marshall D, Webb A (1989) Uranium in Australia: An Annotated Bibliography. Australian National University, Darwin.

Pal LA (2006) Beyond Policy Analysis. Thomson Nelson, Toronto.

Parsons W (1995) Public Policy: An Introduction to the Theory and Practice of Policy Analysis. Edward Elgar, Cheltenham.

Richards JP, Dang T, Dudka SF, Wong ML (2004) The Nui Phao tungsten-fluorite-copper-gold-bismuth deposit, northern Vietnam: An opportunity for sustainable development. Exploration and Mining Geology 12: 61–70.

Rothwell, N (2008) On the brink. Weekend Australian Magazine 15–16 March 2008.

Stein E, Tommasi M, Scartascini C, Spiller P (2008) Policymaking in Latin America. Harvard University Press, Harvard.

Stratford E, Jaskolski M (2004) In pursuit of sustainability? Challenges for deliberative democracy in Tasmanian local government. Environment and Planning B 31: 311–324.

Trebeck K (2007) Tools for the Disempowered? Indigenous Leverage Over Mining Companies. Australian Journal of Political Science 42(4): 541–562.

Van Rensburg WCJ, Bambrick S (1978) The Economics of the World's Mineral Industries. McGraw-Hill Book Company, Johannesburg.

Warhurst A, Mitchell P (2000) Corporate social responsibility and the case of the Summerville mine. Resources Policy 26: 91–102.

Weeramantry C (1992) Nauru: Environmental Damage Under International Trusteeship. Oxford University Press, Melbourne, Australia.

Weitzner V (2002) Through Indigenous Eyes: Towards Appropriate Decision-Making Processes Regarding Mining on or Near Ancestral Lands: Final Synthesis Report. The North-South Institute, Ottawa.

Western Australia (2008) Approval Flowchart – Environment Review and Management Program. www.doir.wa.gov.au/documents/investment/ERMP(i).pdf, accessed 12 March 2008.

Wilderness Society (2008) A perfect storm is gathering in the Kimberley region of northern WA. http://www.wilderness.org.au/articles/kimberley-in-crisis, accessed 2 July 2008.

The Extractive Industries Transparency Initiative: Panacea or White Elephant for Sub-Saharan Africa?

Gavin Hilson and Roy Maconachie

Abstract This chapter critically examines the challenges involved in implementing the Extractive Industries Transparency Initiative (EITI) in sub-Saharan Africa. The EITI is a policy mechanism being supported by donors and Western governments as a key to facilitating economic improvement in resource-rich developing countries. Proponents of the EITI argue that poor governance and a lack of transparency are the main reasons why resource-rich sub-Saharan Africa is underperforming economically, and that implementation of the EITI, with its foundation of "good governance", will help address these problems. However, as the chapter illustrates, this task is by no means straightforward: the EITI is not necessarily a blueprint for good governance in the region's resource-rich countries. Although it is acknowledged that the EITI is a policy mechanism that could ultimately prove effective in generating significant institutional change in host African countries, on its own it is incapable of facilitating reduced corruption, prudent management of mineral and/or petroleum revenues, or mobilizing citizens to hold corrupt government officials accountable for embezzling profits from extractive industry operations.

1 Introduction

In September 2002, former British Prime Minister Tony Blair launched the Extractive Industries Transparency Initiative (EITI) at the *World Summit on Sustainable Development* in Johannesburg. The EITI is portrayed in Western policy making circles as an intervention capable of resurrecting the stagnating economies of natural resource-rich Africa, Asia, and Latin America. It calls on signatories to verify and disclose details "of company payments and government revenues from oil, gas and

G. Hilson (✉)
School of Agriculture, Policy and Development, The University of Reading, Reading, RG6 6AR, UK
e-mail: g.m.hilson@reading.ac.uk

J.P. Richards (ed.), *Mining, Society, and a Sustainable World*,
DOI 10.1007/978-3-642-01103-0_16, © Springer-Verlag Berlin Heidelberg 2009

mining",[1] which, proponents argue, is a key to "to improv[ing] a country's credibility among foreign investors and the international banking community ... [and] its potential for future development" (EITI 2006, p. 23). The revenues derived from oil, gas, and mining operations are vital for economic growth in more than 60 developing countries, but of the combined 3.5 billion citizens residing in these countries, an estimated 1.5 billion subsist on less than US$2 per day.

The inability to unlock mineral wealth for the benefit of the citizenry of developing countries—a phenomenon that has become known as the "resource curse" (Auty 1994)—has spawned extensive debate. The focus of several recent analyses (e.g., Ross 2001; Duruigbo 2005; Pegg 2006a, b) is the World Bank, which, since 1990, has provided more than US$2.75 billion in loans and guarantees to support mineral and petroleum development in Africa alone (Pegg 2003). Growing condemnation of the involvement of the Bank in the extractive industries led former president James Wolfensohn to commission the Extractive Industries Review (EIR 2004)[2], a recently-concluded independent assessment conceived to solicit feedback on the organization's involvement in the extractive industries. Much to the surprise of Bank officials, the EIR generated a flood of critical assessments (e.g., Caruso et al. 2003; Campbell 2003a, b). Former Indonesian Minister for Development Supervision and the Environment, Dr. Emil Salim, was the "eminent person" selected to lead the exercise, and in March 2004, tabled his findings to Mr. Wolfensohn, which included, *inter alia*, a call for the Bank to phase out support for oil and coal projects in developing countries. Management, needless to say, rejected the recommendation outright, countering that the Bank "would continue investments in oil, gas, and mining projects, as these remain an essential part of the development of many poor nations."[3] Critics, however, see the emphasis on "development" here as Bank officials' unwillingness to abandon lucrative activities.

For the Bank, the launch of the EITI was timely, deflecting criticism from the EIR and the organization's involvement in the extractive industries in general, and shifting the focus of the resource curse debate toward developing world governments. Chief supporters of the EITI—principally, G-8 governments and the Bank itself—today argue that the host governments of many resource-rich developing countries are largely to blame for their development failures, and link their poor economic performance to revenue mismanagement and corruption. Proponents note that "in many countries, money from oil, gas and mining is associated with poverty, conflict and corruption ... [which are] often driven by a lack of transparency and accountability around the payments that companies are making to governments, and the revenues that governments are receiving from those companies" (EITI 2005, p. 2). The aim of the EITI is largely to increase transparency over payments by

---

[1] "EITI Factsheet" http://www.eitransparency.org/UserFiles/File/keydocuments/factsheetaugust05. pdf (accessed 11 November 2007).

[2] http://web.worldbank.org/WBSITE/EXTERNAL/TOPICS/EXTOGMC/0,,contentMDK:20306686~menuPK:336936~pagePK:148956~piPK:216618~theSitePK:336930,00.html

[3] Extractive Industries Review–Management response http://www.ifc.org/eir (accessed 13 November 2007).

companies to governments and government-linked entities, as well as transparency over revenues received by host country governments. It is believed that a regular provision of quality information will create a system of "checks and balances" that will assist in holding companies and governments accountable, resulting in improved economic importance, political stability and a better investment climate.

At the time of writing, 16 countries in sub-Saharan Africa had pledged, in principle, to fulfill the objectives of the EITI, an "achievement" that has heavily overshadowed the potential shortcomings of the initiative and the simplicity with which its authors have diagnosed the resource curse "epidemic" in the region. First, proponents espouse country ownership of the EITI, arguing that its implementation would enable citizens to hold their governments responsible for any mismanagement of revenues accrued from extractive industry projects. But in countries such as Chad and Equatorial Guinea, where the regimes rule oppressively and have long embezzled funds at the expense of peoples' needs, making this information available—that is, detailing where revenues originating from extractive industry projects are being channeled—will not ignite backlash from citizens, who are well-aware of this corruption; nor is it likely to trigger a radical change in these governments' approaches toward community development.

Second, is "good governance" necessary in order to attract investment in sub-Saharan Africa? Proponents of the EITI proclaim that "political instability caused by opaque governance is a clear threat to investments in extractive industries, where investments are capital-intensive and dependent on long-term stability to generate returns" (EITI 2005, p. 5). Mismanaged revenue flows and corruption, however, did little to discourage investment in petroleum projects in the past in countries such as Nigeria and Chad, which rank among the most corrupt countries in the world (see Moody-Stuart 2004; Pegg 2006b); nor have prolonged autocracy and widespread civil violence diminished Western interests in diamonds and other natural resources in the Democratic Republic of the Congo (DRC) and Sierra Leone (see Le Billon 2006; Ross 2006).

This raises a third, and final, key issue: that of good governance itself, and more specifically, the ambiguity of the term in this context. Among the 16 African candidate and potential candidate countries, four (Equatorial Guinea, Chad, Niger, and Guinea) are ruled by presidents who gained control via a *coup d'état*, and two others (Cameroon and Gabon) are ruled by presidents who have been in power for over 25 years; even Nigeria and Ghana, the supposed "trailblazers" of the EITI, are both less than 10 years removed from dictatorship rule. Within most candidate and targeted countries, human rights abuses are widespread. It appears that the EITI, which is a voluntary pact that imposes no penalties on regimes that violate its principles, is far from being a recipe capable of offsetting the resource curse in sub-Saharan Africa.

This chapter critically reflects upon the challenges of implementing the EITI and with facilitating good governance in resource-rich sub-Saharan Africa. While proponents argue that the EITI is a blueprint for good governance, the analysis that follows argues that systems of good governance must first be in place in order for the objectives of the EITI to be fulfilled. The EITI is a policy mechanism that could prove effective with significant institutional change in host countries but is likely incapable of facilitating reduced corruption on its own.

2 Beyond the Resource Curse: Is Good Governance the Magic Bullet?

Most countries in sub-Saharan Africa with economies now dependent upon exports of natural resources are characterized by poor economic growth, low living standards, corruption, and political authoritarianism (Le Billon 2005). Six of sub-Saharan Africa's most Highly Indebted Poor Countries are major fuel exporters (Weinthal and Luong 2006), as well as several of the region's major solid mineral producers, including Ghana (gold), Tanzania (gold), Guinea (bauxite), Zambia (copper) and the DRC (copper, cobalt, coltan). Over the last two decades, a large body of analysis on the resource curse has emerged that seeks to explain the empirical correlation between natural resource wealth, a poor political environment, and underdevelopment in sub-Saharan Africa. This debate is broadly rehearsed within the following three sub-literatures: (1) the relationship between resource wealth and economic performance; (2) links between resources and civil violence; and (3) resource abundance and the nature of political regimes.

At the heart of most resource curse critiques are the seminal studies of Sachs and Warner (1995 1997), who were among the first to argue that developing countries with high ratios of natural resource exports to GDP tend to experience low economic growth. Building on this foundational work, many analysts (e.g., Collier and Hoeffler 1998, 2005; Elbadawi and Sambanis 2002; Pegg 2003) have since observed that developing countries rich in mineral and oil resources are commonly characterized by: widespread poverty and/or civil violence, particularly in rural areas; poorly developed agricultural and manufacturing sectors, a phenomenon referred to as "Dutch Disease"[4]; and, because of their one-dimensional economies, high susceptibility to fluctuations in the market values of minerals. The logic behind the resource curse thesis is that economic rents generated from the export of minerals and petroleum "induce governments to rely on such flows instead of having to impose taxes on corporate and personal incomes, to allow exchange rates to appreciate so as to dampen manufacturing and agricultural exports, to overspend in periods of high resource prices, and to deplete natural resources without replacing the declining capital stock" (Heller 2006, p. 25).

Reflecting upon the abundance of material presented on the resource curse (e.g., Wheeler 1984; Gelb & Associates 1988; Auty 1994, 1995), Sachs and Warner

---

[4]The term Dutch Disease describes a condition where revenues accrued from a resource boom are typically spent on non-tradable goods, a move that stimulates an appreciation of the real exchange rate, and, in turn, draws resources (namely labour and capital) from other potentially-productive sectors of the economy. The term surfaced in the late-1970s to describe the decline of the manufacturing sector in the Netherlands following the discovery of natural gas in the 1960s.

(2001, p. 828) concluded that "empirical support for the curse of natural resources is not bulletproof, but it is quite strong." A wealth of case study material reinforces this claim. Ross (2001), for example, analyzed the performance of oil- and mineral-dependent countries, and argued that the majority rank low on the Human Development Index (HDI). In his more recent studies, Ross has examined the relationship between resource abundance and civil war, observing, *inter alia*, that: oil increases the likelihood of conflicts; "lootable" mineral resources (i.e., resources such as diamonds and gemstones which have a high value-to-weight ratio, and can be easily appropriated and transported by unskilled workers) tend to lengthen existing conflicts; and petroleum and diamond production indeed fuels civil violence or warlordism, albeit in different capacities (Ross 2004a, 2006).

Pegg (2003) drew on a wealth of empirical analyses which collectively illustrate that African economies reliant upon minerals and/or oil for sustenance have achieved minimal economic growth, have underdeveloped agricultural and manufacturing sectors, and have weak linkages to the global economy. For example, in Equatorial Guinea, one of the region's newest oil producers, cocoa and coffee declined from approximately 60% of GDP in 1991 to less than 9% in 2001 (Weinthal and Luong 2006). In Zambia, Africa's largest copper producer, the economy contracted at an annual rate of 4% between 1975 and 1996, and today, household income levels are almost one half that of 1960 levels (PREM 2007). These and related findings reinforce arguments presented by Weber-Fahr (2002), whose World Bank study showed that GDP per capita contracted by 2.3% annually in countries where mining contributes to more than 50% of exports; and that in sub-Saharan Africa in the 1990s, the economies of mining countries contracted by 1% per year, or 20% more than the region as a whole.

In explaining *why* countries in sub-Saharan Africa dependent upon mining and/or oil production are performing so poorly, donors have tended to shy away from placing blame on the foreign companies that generally control extractive operations, and from implicating Western parties in general. In this regard, the EITI is by no means unique, its conceivers espousing good governance to be the key to resuscitating the region's petroluem- and mineral-dependent economies. In other words, it is not a case that host countries receive *insufficient* amounts of revenues from oil and mining operations, but rather that host governments are mismanaging the royalties they receive. Proponents of this view argue that Dutch Disease or, more broadly, the resource curse, can be alleviated "by adopting appropriate government policies" (Mikesell 1997, p. 194), an outcome that has hitherto proved elusive in sub-Saharan Africa. This, as Weinthal and Luong (2006, p. 38) explain, is because "countries rich in minerals . . . fail to develop a robust central bureaucracy because their ability to rely on an external revenue source engenders rigid and myopic decision making." There is now a wealth of scholarly literature (e.g., Atkinson and Hamilton 2003; Shaxson 2005) that suggests that the paradox of a resource curse in mineral- and oil-rich regions of sub-Saharan Africa is largely due to corruption and revenue mismanagement within host countries.

The EITI is premised upon the notion of good governance, its authors arguing that the resource curse is, in fact, attributed to "a lack of transparency and accountability around the payments that companies are making to governments, and the revenues that governments are receiving from those companies" (EITI 2005, p. 2). It is rooted in the belief that good governance is necessary in order to ensure that royalties from oil, gas, and mining projects are used to foster economic growth and poverty reduction. However, as has already been noted, the majority of Africa's EITI signatories are countries with long histories of corruption, civil violence, and/or dictatorships, including Chad, Equatorial Guinea, and Sierra Leone.

In reflecting upon several policy solutions implemented to date to prevent the onset of a resource curse, Weinthal and Luong (2006, p. 41) observe that most "have largely failed because making the state a better 'manager' of its mineral wealth requires institutions that promote transparency, accountability, and oversight—that is, institutions that are widely absent in [these] developing countries." Thus, in many cases, to "improve transparency . . . whereby citizens can hold their governments to account for the use of those revenues" (Weinthal and Luong 2006, p. 41), as the EITI intends to do, will require more wholesale policy and administrative changes to take place, which those driving the initiative are incapable of facilitating on their own. Many regimes in sub-Saharan Africa have retained control over populations by systematically depriving them of wealth, and hoarding oil and mineral proceeds: achieving the objectives of the EITI would require these very governments to change their attitudes towards their citizens.

Proponents of the EITI argue that its implementation "is a step towards better governance—often the first step—and can support wider improvements in transparency and accountability within an implementing country" (EITI 2005, p.25). This, however, assumes that host governments are interested in tackling these problems in the first place, and that citizens are capable of mobilizing changes in government policy should it be discovered that mineral and oil revenues are being embezzled. Advocates boldly assert that "political instability caused by opaque governance is a clear threat to investments" (EITI 2005, p. 5). But this was far from being the case in the 1990s in sub-Saharan Africa. At the turn of the century, 19 of the region's 48 countries were ruled by the same individuals who had held power in the single-party days before 1990 (van de Walle 2002), a list which includes eight EITI potential signatories (Cameroon, Congo-Brazzaville, Equatorial Guinea, Gabon, Ghana, Guinea, Mauritania, and Chad). In fact, among Africa's 16 EITI potential signatory countries, in the 1990s, democratic elections led to a change in government in only Mali and São Tomé. The majority of the remaining signatories were either politically unstable and/or fraught with corruption.

If the authors of the EITI are correct in assuming that political instability and corruption threaten investment in the extractive industries, then the undemocratic regimes that proliferated in sub-Saharan Africa in the 1990s—many of which still exist today—would have been hard pressed to lure private investment and support. But this is far from being the case. For example, during the 1990s, there was much controversy over financing by the IFC of the Delaware-based oil company, Pecten, in its exploitations in the Lokélé oilfields in Cameroon, a country that is widely regarded to be the most corrupt in the world (Assiga-Ateba 2001). Likewise, in 2000, the IFC also approved financing for the controversial US$3.7 billion

Chad–Cameroon Pipeline, the largest private sector investment in sub-Saharan Africa to date, in two countries that are plagued by human rights abuses. However, perhaps the most telling example of how political instability failed to discourage investment in extractive industry projects in 1990s sub-Saharan Africa is Nigeria, where widespread corruption and community backlash—including the controversial execution of activist Ken Saro-Wiwa, whose death has become iconic in the Ogoni peoples' epic struggle against Shell in the Niger Delta (Pegg 2000; Frynas 2001)— has not discouraged oil investment in the country. As Frynas (1998, p. 457) stated at the time, "Nigeria experiences serious political instability, yet Shell is expanding its investment in the country", furthermore explaining that "political instability has not deterred Shell from investing in the country and may have been beneficial to the company [helping it] . . . maintain its market position in Nigeria" (Frynas 1998, p. 458). The ability of the region's undemocratic regimes to attract, unimpeded, outside capital for the development of extractive industry projects raises an important question: what do proponents of the EITI mean by good governance?

At the beginning of the 1990s, the development agenda began espousing the good governance message for the first time. With the Cold War having ended, there was no longer a need to keep dictators such as Mobutu (then President of Zaïre) in power. The donor community, therefore, began reinventing the notion of good governance, a move described by Rhodes (1996, p. 656) as "the latest flavour of the month at the World Bank, shaping its lending policy towards Third World countries", and deemed necessary to facilitate the launch of a new generation of "political conditionalities" (Doornbos 2001). The World Bank would thus be instrumental in ushering in an era of good governance and political conditionality in which donors, responding to increased media attention toward dictatorial regimes squandering foreign aid, began "offering support for the removal of authoritarian governments and the spread of democracy in the Third World, largely through the form of electoral assistance, but also judicial reform, training of the media, and support for civil society, among other things" (Marquette 2001, p. 395). The good governance agenda includes promoting transparency, accountability, fairness, and ownership, values which the Bank and major bilateral donors believed at the time would translate into a broader objective to improve the rule of law, political accountability, and flows of information between governments and citizens (Woods 2000).

The Bank's own charter, however, requires it to remain apolitical—that is, to only factor economic issues into decision-making, and not to interfere in the political affairs of any member country (Marquette 2001). Thus, its approach hitherto to governance reform has emphasized such issues as efficiency in public administration, rule of law, transparency, and accountability to ensure economic growth and development, while at the same time devoting little attention to evaluating the legitimacy of governments and power structures, how policy decisions are arrived at, and how equitable economic systems are in developing countries (Nanda 2006). The shortcomings of this approach soon became apparent: realization that "the idea of posing political conditionalities was easier in theory than in practice" (Doornbos 2001, p. 101) and that "sociocultural and political contexts in the recipient countries and not the Western donors' preferences [should] primarily shape the

[aid] agenda" (Nanda 2006, p. 274). Because the transparency of political processes and the "idea of level playing fields did not easily match with prevailing political cultures and configurations of power [in developing countries]" (Doornbos 2001, p. 101), in positioning themselves to secure aid, recipient governments in sub-Saharan Africa, "paid lip service to conditionalities for promoting transparency and political reform" (Nanda 2006, p. 274). In response to calls for increased transparency and democracy, "many authoritarian regimes skillfully transformed themselves into dominant parties within façade-type multi-party systems" (Doornbos 2001, p. 101).

In implementing policy mechanisms such as the EITI, the lesson learned from the Bank's experience in the 1990s is that promoting good governance requires the regimes entrenched within host countries to be unequivocally committed to embracing it—that is, committed to: efficient public service, an independent judicial system, and legal framework to enforce contracts; accountable administration of public funds; respect for the law and human rights at all levels of government; and a pluralistic institutional structure (Rhodes 1996). The era of good governance ushered in by the World Bank, however, appears to have induced little change in sub-Saharan Africa. As Chabal (2002, p. 457) explains, "[the] multiparty politics ... that have taken place have not resulted in the widespread systemic political change that has been widely anticipated", with pluralist elections held solely to satisfy donors threatening to withhold aid for failure to embrace democracy. Past experience suggests that there must be a foundation of good governance in place in order for policy interventions such as EITI to be effective: in other words, that contrary to the views of its proponents, good governance, transparency, and improved government–citizen relations are keys to implementing a successful EITI, and not the opposite. Officials at the UK's Department for International Development (DfID), the World Bank, and other Western donors, however, insist that the EITI is the key to ameliorating the resource curse epidemic plaguing resource-rich sub-Saharan Africa today.

For ordinary Africans, the call for host governments to take the lead on implementing the EITI certainly sends mixed signals. On the one hand, it signifies that oil companies and donors are aware that corruption is a problem that plagues mineral- and petroleum-rich countries of the region, and are demanding that host governments become more transparent and take greater responsibility for their actions. On the other hand, with no penalties in place for inappropriate behaviour, those driving the EITI indicate a willingness to work with corrupt leaders who have long siphoned mineral and/or petroleum revenues for personal gain. Why should they change their attitudes with the onset of a voluntary initiative if there is no evidence that corruption discourages foreign investment in the extractive industries?

The position here is that the EITI alone is incapable of facilitating improved governance in resource-rich sub-Saharan Africa. More sweeping changes are needed, including more progressive regimes, in order to ensure that oil and mineral revenues are managed more prudently.

3 Can the EITI Facilitate Improved Economic Performance in Sub-Saharan Africa?

The effectiveness of a voluntary pact, such as the EITI, in sub-Saharan Africa, where extractive industries have long been operating in corrupt environments, is open to considerable debate. There are additional hurdles that must be overcome, which vary according to the different resource economies in the region: petro, lootable, and conventional. This section of the chapter explores some of these challenges at greater length.

3.1 Petro-Economies

Ross (2001, p. 356) makes a convincing case that the "oil-impedes-democracy claim is both valid and statistically robust"—that "in other words, oil *does* hurt democracy", doing greater damage in poor states than in rich ones. The views of Ross and supporters, however, run counter to the rhetoric espoused by proponents of the EITI: that certain petro-economies (i.e., those rich in oil and/or natural gas) in sub-Saharan Africa are actively embracing transparency and good governance. While pledging to be more democratic, what cannot be overlooked is that this list of supposed trailblazers includes Chad, Equatorial Guinea, and Gabon—countries whose ruling elites have long embezzled funds from petroleum projects at the expense of state and community development.

Despite the optimism resonating in EITI circles, there are few signs to suggest that this group of countries would diligently follow through with their pledges to improve transparency and good governance. This would require these regimes to cease their dictatorial ruling, abandon their corrupt practices, and stop siphoning revenues from oil projects for personal enrichment; in many respects, this corruption appears to be endemic and may therefore be challenging to correct. As Shaxson (2007) suggests, it is undoubtedly those countries where the problem of corruption is the most entrenched that will be most resistant to changing the status quo. This dilemma is compounded by the voluntary approach to disclosure espoused by the initiative. Shaxson notes that "Under EITI, corrupt governments can choose whether or not to publish data: some have, some haven't, and some have just pretended to" (Shaxson 2007, p. 218).

The case of Nigeria reveals how superficially proponents of the EITI have analyzed the corruption now rampant in petroleum-rich sub-Saharan Africa. Former President Olusegun Obasanjo's alleged move to integrate Nigeria into the EITI machinery stemmed from pressures exerted by the donor and NGO communities to address the corruption associated with payments and revenues generated from oil production: an estimated 50% of oil revenues were being squandered, stolen, or siphoned away by corrupt officials (FOI 2005). There are few signs, however, that this corruption has been alleviated. Since a return to civilian rule in 1999, the country's 36 state governments have received US$36 billion in federal allocations, with the 774 Local Government Councils receiving an additional US$23.4 billion (HRW 2007); in 2004, US$6 billion were transferred to state authorities, a third of which went to the major oil producing Delta, Rivers, Bayelsa, and Akwa Ibom states (ICG 2006). Yet, 31 of the country's 36 governors, including all those presiding in the state governments of these four oil producing states, face possible charges of corruption after leaving office (HRW 2007).

The paradox in Nigeria—it being the world's eighth largest producer of oil, yet being a net importer of fuel—has given rise to an extensive debate about corruption in the country. At the centre of many discussions is the Nigerian National Petroleum Corporation (NNPC), whose executives were involved in a US$1.5 million conspiracy with the Texas-based construction and petroleum engineering contractor Willbros in February 2005,[5] and, more recently, were the subject of a US$7 billion controversy.[6] The NNPC, which was established in 1977 as a vehicle for facilitating partnerships with foreign multinationals, receives some 57% of crude oil produced, most of which it exports. The proceeds are deposited into the Central Bank of Nigeria and then shared by the three levels of government (Gary and Karl 2003). President Umaru Yar'Adua's newly-elected government has pledged to dismantle the NNPC on the grounds that it both "produces crude oil in partnership with foreign oil companies, but also imports fuel and acts as a regulator and administrator of the oil sector",[7] which are judged to be conflicting responsibilities. The relevance of this, however, is unclear: while apparently committed to eradicating corruption, President Yar'Adua's ascension to power was itself dubious, reportedly fuelled by election rigging and intimidation (Rawlence and Albin-Lackey 2007). Significantly, Yar'Adua, who is of Obasanjo's People's Democratic Party (PDP), faces the challenge of eradicating the deeply-rooted corruption in the country's petroleum economy, and carrying out the commitment made by his predecessor regarding Nigeria becoming signatory to the EITI.

In February 2004, Obasanjo made Nigeria the first country to voluntarily subscribe to the principles of the EITI, a move that made him extremely popular in donor circles, and which has helped to improve the country's international image. But oil revenues continue to be embezzled in Nigeria, with sparingly few funds reaching the poverty-stricken inhabitants of the Niger Delta where extraction takes place. Continued tensions between government and oil company officials on the one hand, and the impoverished Ogoni people on the other hand, are testament to the latter's frustration with the former over siphoned revenues earmarked for poverty alleviation in the Delta (e.g., Boele et al. 2001a, b; Ikelegbe 2001; Omeje 2005).

---

[5]"Nigeria: $1.5m NNPC Bribe Scandal—Willbros Official Pleads Guilty" http://allafrica. com/stories/200609190387.html (accessed 11 November 2007)."Nigeria: $1.5m NNPC Bribe Scandal—Willbros Official Pleads Guilty" http://allafrica.com/stories/200609190387.html (accessed 11 November 2007).

[6]"NNPC GMD In $7 Billion Contract Controversy" http://www.ocnus.net/cgi-bin/exec/view. cgi?archive=109&num=28228 (accessed 14 November 2007).

[7]"Nigeria scraps state-oil company" http://news.bbc.co.uk/2/hi/africa/6970395.stm (accessed 19 November 2007).

As challenging as addressing the corruption in Nigeria may seem, the situation in Equatorial Guinea and Chad—two other petro-economies in sub-Saharan Africa that have also committed to the EITI—is comparatively bleaker. Moreover, the oil-producing states of Cameroon and Gabon face similar rampant levels of corruption and a lack of transparency. Point source resources, such as oil and natural gas, have high barriers to entry in production, and therefore tend to be inaccessible to rebel groups and are thus more likely to enrich governments (Duruigbo 2005; Di John 2007). Scholars reason that because supplies of oil and natural gas cannot readily be seized, it becomes second nature for the governments of petro-economies to engage in abnormal rent-seeking behaviour (Soderling 2006), driven by the belief that "political power can only be sustained as long as oil revenues flow" (Gary and Karl 2003, p. 24). Excessive rents can give rise to a situation where a government no longer has a need to tax citizens, who in turn lose the incentive to demand accountability of those who spend tax revenues (Duruigbo 2005); this, by extension, breeds corrupt practices in government circles.

A second area in question in the region's petro-economies is the commitment of multinational oil companies. As Shaxson (2007, p. 218) notes, "the (oil) companies love EITI—it takes the pressure off them and puts it onto African governments to disclose"—in other words, that its voluntary approach to disclosure in effect lets these multinationals off the hook. To date, only 25% of the world's top 50 oil and gas companies have signed up to the EITI (Doane and Holder 2007), which can be interpreted as the industry having a complete disregard for the environments where it operates. In fact, there is ample evidence that suggests that multinationals are willing to forge questionable deals with the corrupt bureaucrats in petroleum-rich sub-Saharan Africa. Frynas (2004), for example, reports that in Equatorial Guinea, World Bank audits revealed discrepancies between what companies were supposed to pay to government and how much was actually paid. In Chad, following the decision of the World Bank to dispense US$100 million in royalties, the government forced two oil companies to pay alleged tax arrears, while continuing to pay normal business taxes—these two things together may increase the country's revenue by more than 50% (van Dijk 2007). These cases shed light on the lengths to which foreign multinationals will go to extract oil in the region: a willingness to engage in questionable negotiations with corrupt governments, and to appease these governments with bribes and finances in order to ensure continued operation. In these, and similar, situations, impoverished communities are being deprived of badly-needed finances.

3.2 Lootable Economies

Four EITI signatories in sub-Saharan Africa—Sierra Leone, Liberia, Madagascar, and the DRC—are considered to be lootable economies: countries containing pockets of lucrative, easily-accessible mineral wealth, such as gemstones and diamonds. Several scholars (e.g., Reno 2002; Ross 2004b) suggest that there is a strong correlation between such resources and political disorder; and, by extension, that lootable wealth has the propensity to fuel greed-based insurgency in collapsed states (see Collier and Hoeffler 2004; Berdal 2005; Korf 2005; Regan and Norton 2005). These scholars have explored the links between lootable resources and civil violence, drawing different conclusions from compiled datasets on wars and intermittent conflicts. Humphreys (2005), for example, noted that diamonds tend to *shorten* civil wars by facilitating military victories, not negotiated settlements. Ross (2004a, b), on the other hand, observed that lootable resources could make conflict so profitable that one or more combatants lose their incentive to reach a peace settlement, views which are reinforced by Le Billon (2001, 2006).

While an informative debate has indeed coalesced around the issue of civil violence and resource lootability, comparatively little attention has been paid to identifying ways in which to facilitate good governance and transparency in lootable economies. The civil wars that have surfaced in recent years in many areas of lootable sub-Saharan Africa are manifestations of what Allen (1999, p. 377) terms spoils politics, which occurs when the primary goal of parties competing for power is self-enrichment. The factors that Snyder and Bhavnani (2005, p. 565) claim determine "the ability of rulers to get the revenue with which to govern and, hence, maintain political order" in lootable settings are also key determinants of a nation's susceptibility to spoils politics. These factors include: its resource profile, especially whether non-lootable resources—goods with high economic barriers to entry—are also available to rulers as a source of revenue; the robustness of economic institutions, and whether lootable resources such as diamonds are extracted by difficult-to-tax artisans or by large, taxable companies; and the ways in which rulers spend the revenue accrued, and in particular whether it is mismanaged or spent prudently.

Past evidence suggests that a commitment to good governance and increased transparency may, in fact, have little bearing on changing the ways in which resources are managed and exploited in lootable economies. In particular, and as explained previously, the relative ease of extraction and transport associated with lootable resources, and their high weight-to-value ratio, make them an attractive focus for powerful and often well-armed interest groups and their networks. As has been apparent not only in Sierra Leone and the DRC, but also other lootable economies such as Angola and Liberia, many groups benefiting from the extraction of lootable minerals are firmly entrenched, their positions strengthened and maintained by the resources derived from the exploitation of "exclusionary spatial enclaves" (Ferguson 2006). It is therefore unlikely that the disclosure of payments would rapidly dissolve these powerful networks or facilitate the derivation of greater benefit from resource extraction by citizens.

While it is both easy and economically feasible for mining companies to control deep pit mining (as is the case of South Africa and Botswana), in lootable environments, such as Sierra Leone's alluvial diamond fields, where there are relatively few diamonds per hectare, people tend to reside where the diamonds are found, and labour intensive extraction methods—often involving nothing beyond picks, shovels, buckets, and sieves—are common. Ferguson (2006) argues that, whereas transnational capital "hops" over "unusable" Africa to find its way to mineral-extraction enclave spaces, alluvial diamonds are more difficult to exploit using enclave methods because they are diffuse resources and less spatially concentrated. But this has not discouraged powerful actors with vested interests from exploiting, for example, Sierra Leone's diamond wealth. Critics have noted that the lootability of the country's diamonds not only induced decades of diamond smuggling in the country but has also been responsible for rampant corruption in the government, and has funded the decade-long civil war, deprived the country of millions of dollars in development funds, exacerbated instability in the West African sub-region overall, and distorted Sierra Leoneans' basic sense of governance.

The concern, however, is that the corruption in Africa's lootable economies is so deeply-rooted that it may be beyond the capability of governments to correct. In the case of Sierra Leone, since the discovery of diamonds in the 1930s, political elites have found ways to capitalize on the diamond economy, often strengthening their own political positions through patrimonial networks and clientelism, a process that Bayart (1993) refers to as "the politics of the belly". When former president Siaka Stevens's All Peoples Congress party came to power in 1968, diamond wealth was used to reward his political supporters, which reduced the industry to a parastatal rife with corruption and smuggling. Under Stevens, official diamond exports declined from 1.7 million carats in the 1960s, to a mere 50,000 carats by 1985 (Temple 2006). As the country became deprived of economic resources and the bureaucratic state became hollowed out, patrimonial networks became even more firmly entrenched. Toward the end of Stevens's tenure, when economic and social development had collapsed completely, and the majority of the population had become completely cut off from the benefits of the diamond industry, it became increasingly necessary for him to secure his own political position by rewarding a small group of elites and a growing personal security force with diamond wealth.

Corruption and the "criminalization of the state" (Bayart et al. 1999) eventually set the stage for the brutal civil war during the 1990s, with diamonds playing a key role in fuelling and prolonging the conflict, and various parties financing their efforts through mining activities. But while large-scale civil violence has since ceased in Sierra Leone, diamonds continue to be looted: the present post-war mining situation is particularly chaotic, with numerous reports of illegal mining and significant smuggling of diamonds across international borders. The EITI, proponents argue, could help correct situations like this, provided that it fits "comfortably within the legal framework of a well-functioning revenue administration alongside fiscal control mechanisms" (EITI 2005, p. 28). It is claimed, against this background, that the "EITI should not involve extraordinary demands on the government" (EITI 2005, p. 28), the assumption being that in the case of Sierra Leone, the Kimberley Process Certification Scheme (KPCS), which itself is a voluntary pact, *has* helped to strengthen policies and facilitate improved transparency with respect to the extraction and marketing of alluvial diamonds.

There is little denying that the KPCS has led to some improvement, but smuggling continues to be rampant in the country, threatening to undermine the entire exercise. For example, in 2002 it was estimated that up to 90% of the country's diamond production was exported illegally, amounting to over US$350–$400 million in lost funds (MSI 2004). Some industry observers suggest that despite the KPCS, up to 50% of Sierra Leone's diamonds will continue to leave the country illegally. In a recent speech in parliament, former President Kabbah acknowledged that corruption still plagues Sierra Leone, explaining that "the entire country has crumbled beneath the burden of warfare, economic ruin, rampant corruption, and autocracy."[8] But to assume, as Dr. Shekou M. Sesay, Minister of Presidential Affairs in Sierra Leone, put it at the country's EITI Keynote Address, that "... the EITI is a laudable idea, which if implemented to the spirit and letter will ensure prudent management of the dividends from mining and go a long way to alleviate extreme poverty among [our] people",[9] would be premature. There are several benefactors of the ongoing embezzlement in the country, and policy flaws which the KPCS has failed to correct. For the EITI to be effective in Sierra Leone and other lootable economies in sub-Saharan Africa, the dynamics of the production chain must be unearthed, and sources of corruption identified. Only then will the government be in a position "to enshrine EITI legally; create new transparency, revenue and industry policy legislation; or make changes to existing EITI related policies and legislation" (EITI 2005, p. 18).

This raises a second concern: can the Sierra Leonean government realistically overcome the obstacles—which proponents of the EITI have too often cursorily overlooked—that prevent improved governance and transparency in lootable settings? The magnitude of this challenge was broached in the same keynote speech delivered by Dr. Shekou M. Sesay, who, in response to the country's commitment to implementing the EITI, explained that:

> In pledging our commitment to the EITI, we remain conscious of the fact that its implementation comes with its own challenges—it will require consideration of some complicated issues, for example contract, confidentiality, the need for new regulation framework data, and legislation; established modalities for gathering the required data; and building capacity in government and civil society to be able to competently engage in the process and use the information produced.[8]

It is unlikely, however, that, despite its commitment and enthusiasm, the government will be capable of addressing these issues on its own, in large part because of the complex network of actors that have long sustained the corruption and lack of transparency that characterize local diamond extraction. In 2006, a World Bank delegation reviewed the challenges of dissolving these complex networks, and concluded that significant capacity-building is needed if the EITI is to be moved forward in Sierra Leone, the implication being that a commitment to transparency will

---

[8] Kabbah, A.A.T. "Defining Moments of my Presidential Journey", Address delivered on the occasion of the Special Session of the Second Parliament of the Republic of Sierra Leone, 19 June 2007. http://statehouse-sl.org/speeches/last-parl-june2007.html (accessed 19 September 2007).

[9] http://www.eitransparency.org/UserFiles/File/sierraleone/sierraleonne_keynoteaddress_shekou_sesay.pdf

be insufficient to break the negative linkages between natural resource wealth, poor governance, unequal development, and conflict.

Finally, presuming that increased transparency is the solution to empowering the citizenry of lootable economies such as Sierra Leone and the DRC, and is a key to attracting investment in diamond extraction, how do proponents of the EITI propose to go about achieving this? The Kimberley Process provides little foundation for the initiative, and if anything, magnifies the weaknesses of voluntary schemes as regulatory mechanisms in lootable settings: there is no mandatory impartial monitoring mechanism associated with the scheme, which, in effect, allows the industry to monitor itself. Perhaps the most significant shortcoming of the KCPS is its failure to implement and monitor diamond control systems. Certification must be supported by robust control systems in order to ensure that diamonds are not used for the purposes of spoils politics. Reports from the DRC and Côte d'Ivoire—two countries that have recently been engulfed in conflict, and are still considered to be fragile states—suggest that despite an embargo, diamonds continue to flow from both countries and may be being exported by other KPCS participants.

In summary, the position here is that for the EITI to be effective in facilitating improved revenue management and transparency in lootable economies, a sound policy framework must be in place. While the challenges of implementing the EITI in Sierra Leone are significant, they could be even greater in the DRC, where tribal divisions appear to be more pronounced and, as Fairhead (2004, p. 298) explains, there is "the most documented case of corporate culpability in transforming governance and fuelling conflict." The conclusion drawn from the EITI Sourcebook (2005) is that proponents of the EITI have not yet come to grips with how firmly entrenched production and trading networks are in African lootable economies, and the level of institutional commitment needed to facilitate improvement. It stipulates that "however incomplete, publication of available revenue—preferably from the extractive sector—would serve as a baseline for judging progress in improving data quality" (EITI 2005, p. 18). On the contrary, the key to improving revenue management in countries such Sierra Leone and the DRC is not to invest in partial and inaccurate reporting as a solution, and subsequently superimpose the EITI atop problematic schemes such as the KPCS in order to articulate "a baseline for judging progress", but rather to strengthen these schemes before making such declarations.

3.3 Conventional Mineral Producers

The conventional mineral producers—defined here as developing countries endowed with mainly unlootable, non-fuel minerals—are being increasingly overlooked in studies of the resource curse in sub-Saharan Africa. A possible explanation for this is that while countries such as Zambia, Ghana, and Tanzania may be illustrative examples of Dutch Disease, at the same time, they do not conform to many of the more recent ideas presented on warlordism, civil violence, and greed in resource-rich economies. The dismissal of the relevance of conventional mineral producers in the resource curse literature, now replete with discussions of petroleum-rich and lootable economies, is reflected in Jensen and Wantchekon (2004, p. 818), who, in explaining "why an abundance of natural resources increases competition for control of the state", refer to Mali, now Africa's third largest gold producer, as a resource poor country. The performance of Africa's conventional mineral producers is being rapidly overshadowed by issues such as civil violence in the DRC, oil in the "New Gulf", the reconstruction of Sierra Leone, and diamonds in Angola, which are perhaps more exciting and topical areas for researchers.

But while the importance of conventional mineral producers may be losing attention in the literature, they remain a key component of the EITI agenda. Of the 16 African potential signatories, four fit into the category of conventional mineral producers: Ghana (gold), Niger (gold, uranium, and industrial minerals), Guinea (bauxite), and Mali (gold). Donor countries, the World Bank, and other drivers of the EITI see the continued transparency and disclosure of mineral payments in these countries—particularly Ghana—as a key to bringing other conventional mineral producers such as Tanzania and Zambia on board. Ghana recently disclosed details about its mineral revenues (Boas & Associates 2007), covering the period January–June 2004, a move which proponents of the EITI have lauded as significant. What is less clear, however, is how this transparency will affect Ghana's people, or, more specifically, improve their lives.

The position being taken by supporters of the EITI appears clear: that mining projects are generating sufficient revenue capable of bolstering the country's economy, but that the people of Obuasi, Tarkwa, and other mining towns are deriving little benefit from company–government transactions, potentially because of corruption. While an inquiry (Murphy 2007) into the report of Boas & Associates (2007) has revealed several omissions and problem areas requiring clarification during the period of study, what additional disclosure of past transactions should reveal is that mismanagement of funds accrued from mining projects has never been a major problem in Ghana. What subsequent reports *are* likely to expose, however, is not that a corrupt government is embezzling finances, but rather that the country is deriving minimal benefit from its large-scale mining projects because of the generous tax breaks it provides to incoming companies, and the low royalty payments it receives as compensation for production. This could put the burgeoning body of literature on Africa's mineral codes, which continues to be overlooked in debates on the performance of resource-rich developing world economies, into the spotlight.

The current mining boom in countries such as Ghana, Tanzania, and Mali is largely owed to the overhaul of legislation for the benefit of investors—namely, the redesign of mining policies to provide generous incentives and favourable terms concerning security of tenure, ownership/marketing of mineral, surface/land access, and import/export policy (Filho and Vilhena 2002). But while these reforms have succeeded in attracting requisite levels of foreign investment, which have facilitated marked increases in mineral production, host governments are netting only small shares of the resulting profits. What these cases illustrate is that the poor economic performance of Africa's conventional mineral producers is more a result of inequitable mining codes than poor governance—that even in situations where revenue mismanagement may be taking place, the quantities of money available to embezzle are low.

All signs point to the idea that the same multinational mining corporations that have championed the EITI process have exploited an advantageous position: armed with state-of-the-art technologies and the support of the World Bank, companies have negotiated favourable terms over the past two decades for mining gold, bauxite, and copper in countries such as Ghana, Guinea, Tanzania, Zambia, and Mali, in each case, taking advantage of a country's financial crisis and inability to harvest its mineral riches with obsolete equipment. The case of Ghana is striking because "it illustrates [how] a mining boom may be accompanied by . . . a much lower contribution to GDP than might have been expected, 2–3%, while mining represents approximately 40% of total merchandise exports earnings since 1992" (Campbell 2003b, p. 7). As Akabzaa and Darimani (2001) and Campbell (2006) explain, this situation has arisen in large part because of the Minerals and Mining Law, enacted in 1986. Although this legislation has facilitated an 800-fold increase in national gold production in the past 10–15 years, the country has failed to benefit from the growing multinational presence in the mining sector and accompanying influxes of foreign investment, largely because of the tax breaks provided to incoming operators. The law not only scaled down corporate income tax liability but also: reduced corporate income tax from 50–55% to 45% in 1986, and even further to 35% in 1994; permitted investors to recoup 75% of initial capital allowance in the first year of operation; decreased the royalty rate from 6% of the value of minerals won to 3%; and abolished the mineral duty (5%), import duty (5–35%), and foreign exchange tax (33–75%).

Although Ghana's large-scale mining companies produced not less that US$5.2 billion in gold between 1990 and 2002 (calculated from Yakubu 2003), as reported by the Bank of Ghana, the government received only US$68.6 million in royalty payments and US$18.7 million in corporate income taxes from these companies during this period (Bank of Ghana 2003). The government retains and reinvests 80% of mining revenues, and allocates only 9% of royalties to community development (Hilson and Nyame 2006). At the country's most recent EITI meeting in Accra, on 15 January 2007, there was extensive lobbying among company officials and officers from the Chamber of Mines, the representative body of the country's mining sector, for the government to increase the amount of money it dispenses for community development purposes from 9 to 30% of royalties. While this could stimulate some improvements in mining communities, it would, at the same time, deprive the government of badly-needed finances. These discussions have also been instrumental in drawing attention to the government as the reason behind the underdeveloped state of the country's mining communities, while at the same time shifting focus away from the underlying cause of the problem: that the sector's overall economic contribution is small.

Many other conventional mineral producers which the EITI is attempting to get on board are experiencing similar problems. For example, in Zambia, the royalty tax on copper is a paltry 0.6%, reduced from 2% in 2002 in order to encourage more mining activity in the Copperbelt. Moreover, as Calì and te Velde (2007) explain, since the implementation of the 1995 Mines and Minerals Act, companies engaged in copper and cobalt extraction are only required to pay a corporate tax of 25%, compared to 35% previously. Similarly, in Tanzania the explanation behind why the current gold boom has failed to generate significant economic growth is more strongly associated with mineral policies than corruption. In addition to having in place a 3% royalty, the government has waived import duties on equipment, charges inexpensive rates for ground rents (US$20/km^2), permits a 1-year grace period for payment of value-added tax, and perhaps most significantly, allows 100% foreign ownership and unrestricted repatriation of profits (Forster and Bills 2002). Not surprisingly, with these agreements in place, increased gold production has had minimal impact on GDP: despite experiencing a marked increase in gold exports from US$3.34 million to US$120.53 million between 1998 and 2000, the contribution of the mining sector to national GDP still hovered at just over 2% at the beginning of 2001.[10] This figure has since increased modestly to 3.5%[11] in large part because US$700 million in gold continues to be exported annually.[12]

In summary, the picture painted by DfID, the World Bank, and other supporters of the EITI is that the poor economic performance of Africa's conventional mineral producers, much like its lootable and petroleum-rich counterparts, is a result of mismanagement and embezzlement of mineral revenues. But, as argued in this section of the chapter, in potential EITI signatory countries such as Ghana, as well as other conventional mineral producers such as Tanzania and Zambia, poor economic performance could be linked to policies which, despite bolstering (mineral) production, fail to provide significant economic returns for host governments. If, indeed, this is the case, should the World Bank, bilateral donors, and multinational mining companies be held accountable for taking advantage of the vulnerability of these poor countries, and pressuring their governments to draft these inequitable mineral policies? Further disclosure of payments in Ghana and other conventional mineral producers may reveal that it is not a case of their governments siphoning earnings generated from extractive industry projects, but rather that there are few revenues to seize. The key to these countries improving their economic performance, therefore, lies in an overhaul of mineral taxation agreements.

4 Conclusions

The objective of this chapter has been to draw attention to the challenges associated with implementing the EITI in sub-Saharan Africa. The EITI attempts to facilitate good governance and improve transparency in mineral-rich countries that have underperformed economically. Proponents maintain that if the blueprints developed

---

[10]http://www.nationsencyclopedia.com/Africa/Tanzania-MINING.html (accessed 13 November 2007).

[11]http://www.ippmedia.com/ipp/observer/2007/03/11/86083.html (accessed 12 November 2007).

[12]http://metalsplace.com/news/?a=3396 (accessed 12 October 2007).

are followed, signatory countries will achieve marked economic improvements, and their citizens will also derive numerous benefits.

This discussion, however, has argued that those driving the EITI have diagnosed the challenge far too superficially in sub-Saharan Africa. Supporters of the EITI, including Western governments and the World Bank, assert that its implementation will enable the citizens of developing countries to hold their governments account-able in the event that petroleum and/or mineral revenues are embezzled. The pact, however, is merely voluntary, which makes it unclear as to how governments *can* be held accountable. Moreover, supporters assume that governments are willing to be held accountable for mismanagement of petroleum and/or mineral revenues in the first place. Some of the African signatories to the EITI, including Equatorial Guinea and Chad, are among the most oppressive regimes in the world, and have long marginalized their citizens. These countries are unlikely to embrace any doc-trine of good governance.

The EITI is still in its embryonic stages. It would therefore be premature at this point to declare that it is a complete failure, incapable of facilitating economic improvements and greater accountability in resource-rich sub-Saharan Africa. This chapter, however, argues that fundamental changes must take place in the region in order for the aims of the EITI to be fulfilled, foremost being a commitment to good governance in host countries. The analysis has also drawn attention to the challenges of implementing the EITI in different landscapes: petro-economies, lootable settings, and conventional environments. Detailed analysis of each category of mineral economy in sub-Saharan Africa reveals that a unique set of challenges must be overcome in order to implement the EITI, and that, while host govern-ments are indeed contributing to the problem, other parties are also responsible for the region's poor economic performance. It is concluded that the EITI is a pol-icy mechanism that could prove effective with accompanying institutional changes in host economies, but which will unlikely facilitate reduced corruption, improve governance, and increase transparency on its own in resource-rich sub-Saharan Africa.

References

Akabzaa, T., and Darimani, A. 2001. Impacts of Mining Sector Investment in Ghana: A Study of the Tarkwa Mining Region. Accra: SAPRI, Third World Network.

Allen, C. 1999. Warfare, endemic collapse and state violence in Africa. Review of African Political Economy, 81, pp. 367–384.

Assiga-Ateba, E.M. 2001. Economic analysis of corruption in Cameroon, Canadian. Journal of Development Studies, 22(3), pp. 721–745.

Atkinson, G., and Hamilton, K. 2003. Savings, growth and the resource curse hypothesis. World Development, 31(11), pp. 1793–1807.

Auty, R. 1994. Industrial-policy reform in 6 large newly industrializing countries – The resource curse thesis. World Development, 22(1), pp. 11–26.

Auty, R. 1995. Industrial-policy, sectoral maturation, and postwar economic-growth in Brazil – The resource curse thesis. Economic Geography, 71(3), pp. 257–272.

Bank of Ghana. 2003. Report on the Mining Sector. Bank of Ghana, Report No. 1(3), Accra: Bank of Ghana.

Bayart, J. 1993. The State in Africa: The Politics of the Belly. London: Longman Group.

Bayart, J., Ellis, S., and Hibou, B. 1999. The Criminalization of the State in Africa, Oxford: James Currey.

Berdal, M. 2005. Beyond greed and grievance—and not too soon… A review essay. Review of International Studies, 31(4), pp. 687–698.

Boas & Associates. 2007. The Aggregation of Payments and Receipts of Mining Benefits in Ghana. Prepared by Boas & Associates. Accra: Ministry of Finance and Economic Planning.

Boele, R., Fabig, H., and Wheeler, D. 2001a. Shell, Nigeria and the Ogoni. A study in unsustainable development I. The story of Shell, Nigeria and the Ogoni—Environment, economy, relationships: Conflicts and prospects for resolution. Sustainable Development, 9(2), pp. 74–86.

Boele, R., Fabig, H., and Wheeler, D. 2001b. Shell, Nigeria and the Ogoni. A study in unsustainable development II. Corporate social responsibility and 'stakeholder management' versus a rights-based approach to sustainable development. Sustainable Development, 9(3), pp. 121–135.

Calì, M., and te Velde, D.W. 2007. Is Zambia contracting Dutch Disease? Working Paper 279. London: Overseas Development Institute.

Campbell, B. 2003a. African mining codes questioned. Mining Journal, February 14th, pp. 106–109.

Campbell, B. 2003b. Factoring in governance is not enough. Mining codes in Africa, policy reform and corporate responsibility. Minerals and Energy, 18(3), pp. 2–13.

Campbell B. 2006. Good governance, security and mining in Africa. Minerals and Energy – Raw Materials Report, 21, pp. 31–44.

Caruso, E., Colchester, M., MacKay, F., Hildyard, N., and Nettleton, G. 2003. Extracting Promises: Indigenous Peoples, Extractive Industries and the World Bank. Washington, DC: Extractive Industries Review Report http://www.worldbank.org/ogmc/files/eirreport/volume6indigenous.pdf

Chabal, P. 2002. The quest for good government and development in Africa: Is NEPAD the answer? International Affairs, 78(3), pp. 447–462.

Collier, P., and Hoeffler, A. 1998. On economic causes of civil war. Oxford Economic Papers, 50(4), pp. 563–573.

Collier, P., and Hoeffler, A. 2004. Aid, policy and growth in post-conflict societies. European Economic Review, 48(5), pp. 1125–1145.

Collier, P., and Hoeffler, A. 2005. Resource rents, governance, and conflict. Journal of Conflict Resolution, 49(4), pp. 625–633.

Van de Walle, N. 2002. Africa's range of regimes. Journal of Democracy, 13(2), pp. 66–80.

Di John, J. 2007. Oil abundance and violent political conflict: A critical assessment. Journal of Development Studies, 43(6), pp. 961–986.

Doane, D., and Holder, A. 2007. Why Corporate Social Responsibility is Failing Children. London: Save the Children UK.

Doornbos, M. 2001. 'Good Governance': The rise and fall of a policy metaphor? Journal of Development Studies, 37(6), pp. 93–108.

Duruigbo, E. 2005. The World Bank, multinational oil corporations and the resource curse in Africa, University of Pennsylvania. Journal of International Economic Law, 26(1), pp. 1–67.

Elbadawi, I., and Sambanis, N. 2002. How much war will we see? Explaining the prevalence of civil war. Journal of Conflict Resolution, 46(3), pp. 307–334.

Extractive Industries Transparency Initiative (EITI). 2005. Extractive Industries Transparency Initiative Sourcebook. UK: Extractive Industries Transparency Initiative (EITI).

Extractive Industries Transparency Initiative (EITI). 2006. Final Report International Advisory Group. UK: Extractive Industries Transparency Initiative (EITI).

Fairhead, J. 2004. Achieving sustainability in Africa, pp. 292–306, in Targeting Development: Critical Perspectives on the Millennium Development Goals (Ed. R. Black and H. White). London: Routledge.

Ferguson, J. 2006. Global Shadows: Africa in the Neoliberal World Order. Durham: Duke University Press.

Filho, C., and Vilhena, C. 2002. Main considerations in the formulation of mining policies to attract foreign investment. Applied Earth Science: IMM Transactions Section B, 111(3), pp. 177–182.

Forster, J.J., and Bills, J.H. 2002. Comparison of the impact of the fiscal regime on gold projects in Tanzania and Burkina Faso. Transactions of the Institution of Mining and Metallurgy, 111, pp. 195–199.

Friends of the Earth (FOI). 2005. Flaring in Nigeria: A Human Rights, Environmental and Economic Monstrosity. Amsterdam: Friends of the Earth.

Frynas, J.G. 1998. Political instability and business: Focus on Shell in Nigeria. Third World Quarterly, 19(3), pp. 457–478.

Frynas, J.G. 2001. Corporate and state responses to anti-oil protests in the Niger Delta. African Affairs, 100, pp. 27–54.

Frynas, J.G. 2004. The oil boom in equatorial Guinea. African Affairs, 103(413), pp. 527–546.

Gary, I., and Karl, T.L. 2003. Bottom of the Barrel: Africa's Oil Boom and the Poor. London: Catholic Relief Services.

Gelb & Associates. 1988. Oil Windfalls: Blessing or Curse? London: Oxford University Press.

Heller, T.C. 2006. African transitions and the resource curse: An alternative perspective. Economic Affairs, 26(4), pp. 24–33.

Hilson, G., and Nyame, F. 2006. Gold mining in Ghana's forest reserves: A report on the current debate. Area, 38(2), pp. 175–185.

Human Rights Watch (HRW). 2007. Chop Fine: The Human Rights Impact of Local Government Corruption and Mismanagement in Rivers State, Nigeria. London: Human Rights Watch.

Humphreys, M. 2005. Natural resources, conflict, and conflict resolution—Uncovering the mechanisms. Journal of Conflict Resolution, 49(4), pp. 508–537.

International Crisis Group (ICG). 2006. Nigeria: Want in the Midst of Poverty. Brussels: International Crisis Group.

Ikelegbe, A. 2001. Civil society, oil and conflict in the Niger Delta region of Nigeria: ramifications of civil society for a regional resource struggle. Journal of Modern African Studies, 39(3), pp. 437–469.

Jensen, N., and Wantchekon, L. 2004. Resource wealth and political regimes in Africa. Comparative Political Studies, 37(7), pp. 816–841.

Korf, B. 2005. Rethinking the greed-grievance nexus: Property rights and the political economy of war in Sri Lanka. Journal of Peace Research, 42(2), pp. 201–217.

Le Billon, P. 2001. The political ecology of war: Natural resources and armed conflicts. Political Geography, 20(5), pp. 561–584.

Le Billon, P. 2005. Aid in the midst of plenty: Oil wealth, misery and advocacy in Angola. Disasters, 29(1), pp. 1–25.

Le Billon, P. 2006. Securing transparency—Armed conflicts and the management of natural resource revenues. International Journal, 62(1), pp. 93–107.

Management Systems International (MSI). 2004. Integrated Diamond Management in Sierra Leone: A Two-Year Pilot Project. Report prepared for assistance of the United States Agency for International Development. Washington, DC: USAID.

Marquette, H. 2001. Corruption, democracy and the World Bank. Crime, Law and Social Change, 36, pp. 395–407.

Mikesell, R. 1997. Explaining the resource curse, with special reference to mineral-exporting countries. Resources Policy, 23(4), pp. 191–199.

Moody-Stuart, M. 2004. The curse of oil? Proceedings of the Geologists Association, 115, pp. 1–5.

Murphy, R. 2007. Ghana's EITI—Delivering on the Promise? A Review of the First Report on the Aggregation/Reconciliation of Mining Benefits in Ghana Prepared for UK Department for International Development. London: DfID.

Nanda, V. 2006. The 'good governance' concept revisited. Annals of the American Academy of Political and Social Science, 603, pp. 269–283.

Omeje, K. 2005. Oil conflict in Nigeria: Contending issues and perspectives of the local Niger Delta people. New Political Economy, 10(3), pp. 321–334.

Pegg, S. 2000. Ken Saro-Wiwa: Assessing the multiple legacies of a literary interventionist. Third World Quarterly, 21(4), pp. 701–708.

Pegg, S. 2003. Poverty Reduction or Poverty Exacerbation? World Bank Group Support for Extractive Industries in Africa. Washington, DC: Oxfam America.

Pegg, S. 2006a. Mining and poverty reduction: Transforming rhetoric into reality. Journal of Cleaner Production, 14(3–4), pp. 376–387.

Pegg, S. 2006b. Can policy intervention beat the resource curse? Evidence from the Chad-Cameroon pipeline. African Affairs, 105(418), pp. 1–25.

Poverty Reduction and Environmental Management (PREM). 2007. The Resource Curse in Mineral-Based Economies: The Case of the Copperbelt in Zambia. The Netherlands: PREM.

Rawlence, B., and Albin-Lackey, C. 2007. Nigeria's 2007 general elections: Democracy in retreat. African Affairs, 106(424), pp. 497–506.

Regan, P.M., and Norton, D. 2005. Greed, grievance, and mobilization in civil wars. Journal of Conflict Resolution, 49(3), pp. 319–336.

Reno, W. 2002. The politics of insurgency in collapsing states. Development and Change, 33(5), pp. 837–858.

Rhodes, R.A.W. 1996. The new governance: Governing without government. Political Studies, 44(4), pp. 652–667.

Ross, M. 2001. Does oil hinder democracy? World Politics, 53, pp. 325–361.

Ross, M.L. 2004a. What do we know about natural resources and civil war? Journal of Peace Research, 41(3), pp. 337–356.

Ross, M.L. 2004b. How do natural resources influence civil war? Evidence from thirteen cases. International Organization, 58 (1), pp. 35–67.

Ross, M. 2006. A closer look at oil, diamonds and civil war. Annual Review of Political Science, 9, pp. 265–300.

Sachs, J.D., and Warner, A. 1995. Natural Resource Abundance and Economic Growth. National Bureau for Economic Research (NBER). Working Paper No. 5398. Cambridge: National Bureau for Economic Research.

Sachs, J. D., and Warner, A. 1997. Natural Resource Abundance and Economic Growth. Center for International Development (HIID). Cambridge: Harvard University.

Sachs J.D., and Warner, A. 2001. The curse of natural resources. European Economic Review, 45, pp. 827–838.

Shaxson, N. 2005. New approaches to volatility: Dealing with the 'resource curse' in sub-Saharan Africa. International Affairs, 81(2), pp. 311–324.

Shaxson, N. 2007. Poisoned Wells: The Dirty Politics of African Oil. New York: Palgrave Macmillan.

Soderling, L. 2006. After the oil: Challenges ahead in Gabon. Journal of African Economies, 15(1), pp. 117–148.

Snyder, R., and Bhavnani R. 2005. Diamonds, blood, and taxes – A revenue-centered framework for explaining political order. Journal of Conflict Resolution, 49(4), pp. 563–597.

Temple, H. 2006. Livelihoods Report. Report prepared for DfID, Freetown. London: DfID.

van Dijk, H. 2007. Political deadlock in Chad. African Affairs, 106, pp. 697–703.

Weber-Fahr, M. 2002. Treasure or Trouble? Mining in Developing Countries. Mining and Development Series. Washington, DC: World Bank/International Finance Corporation.

Weinthal, E., and Luong, P.J. 2006. Combating the resource curse: An alternative solution to managing mineral wealth. Perspectives on Politics, 4, pp. 35–53.

Wheeler, D. 1984. Sources of stagnation in sub-Saharan Africa. World Development, 12(1), pp. 1–23.

Woods, N. 2000. The challenge of good governance for the IMF and the World Bank themselves. World Development, 28(5), pp. 23–41.

Yakubu, B.R. 2003. Regularisation of small-scale mining in Ghana: Technical approach and its shortcomings. Communities and Small Scale Mining (CASM) 3rd AGM and Learning Event. Ghana: CASM.

Index

J.P. Richards (ed.), *Mining, Society, and a Sustainable World*,   493
DOI 10.1007/978-3-642-01103-0_BM2, © Springer-Verlag Berlin Heidelberg 2009